ELECTROCHEMISTRY FOR CHEMISTS

ELECTROCHEMISTRY FOR CHEMISTS

Second Edition

DONALD T. SAWYER
Texas A&M University

ANDRZEJ SOBKOWIAK
Rzeszow University of Technology

JULIAN L. ROBERTS, Jr.
University of Redlands

A Wiley-Interscience Publication
JOHN WILEY & SONS, INC.
New York / Chichester / Brisbane / Toronto / Singapore

This text is printed on acid-free paper.

Library of Congress Cataloging in Publication Data:

Sawyer, Donald T.
Electrochemistry for chemists / Donald T. Sawyer, Andrzej
Sobkowiak, Julian L. Roberts, Jr.—2nd ed.
p. cm.
Rev. ed. of: Experimental electrochemistry for chemists. 1974
Includes bibliographical references.
ISBN 0-471-59468-7 (cloth)
1. Electrochemistry. I. Sobkowiak, Andrzej. II. Roberts, Julian
L. III. Sawyer, Donald T. Experimental electrochemistry for
chemists. IV. Title.
QD553.S32 1995
541.3'7—dc20 95-2738

Printed in the United States of America

10 9 8 7 6 5 4 3 2

PREFACE

Since the previous edition of this monograph (*Experimental Electrochemistry for Chemists*, 1974), the utilization of electrochemistry for chemical characterization has undergone a revolution. Chemists, biologists, and materials scientists make regular use of potentiometry and cyclic voltammetry to detect and evaluate the redox (reduction–oxidation) thermodynamics, the electron-transfer stoichiometry and kinetics, and the valence states of redox-active molecules. Although the basic principles and methodologies for useful electrochemical measurements remain unchanged, the currently available commercial instrumentation, electrodes, solvents, and reagents make possible routine application of electrochemical measurements for the characterization of chemical systems and materials that were outside the realm of consideration 20 years ago. Because of this, many new applications of electrochemical measurements have been perfected in recent years. These substantive changes and advances have made much of our earlier volume obsolete and seriously deficient with respect to instrumentation, solvents, reagents, and applications.

The focus of this edition remains the same as that for the first, namely, to make electrochemistry an attractive, useful characterization methodology for chemists [comparable to infrared (IR), nuclear magnetic resonance (NMR) spectroscopy, and mass spectrometry (MS)]. The goal is to outline the basic principles and modern methodology of electrochemistry in such a way that the uninitiated may gain sufficient background to use electrochemical methods for the study of chemical systems. Thus chemical problems that are amenable to an electrochemical approach are introduced as representative examples.

This monograph has been developed for the nonelectrochemist who is interested in the advantages and use of electrochemical methods for research and laboratory measurements. With such a pragmatic goal, our purpose has not been to develop the theory in detail. Therefore, we have limited our presentation to those relations that allow the effective treatment of experimental data. Sufficient references are given for those who wish to pursue the theory.

Twenty years ago the main applications of electrochemistry were trace-metal analysis (polarography and anodic stripping voltammetry) and selective-ion assay (pH, pNa, p*K* via potentiometry). A secondary focus was the use of voltammetry to characterize transition-metal coordination complexes (metal–ligand stoichiometry, stability constants, and oxidation–reduction thermodynamics). With the commercial development of (1) low-cost, reliable potentiostats; (2) pure, inert glassy-carbon electrodes; and (3) ultrapure, dry aprotic solvents, molecular characterization via electrochemical methodologies has become accessible to nonspecialists (analogous to carbon-13 NMR and GC/MS).

The goal of this volume is to provide (1) an introduction to the basic principles of electrochemistry (Chapter 1), potentiometry (Chapter 2), voltammetry (Chapter 3), and electrochemical titrations (Chapter 4); (2) a practical, up-to-date summary of indicator electrodes (Chapter 5), electrochemical cells and instrumentation (Chapter 6), and solvents and electrolytes (Chapter 7); and (3) illustrative examples of molecular characterization (via electrochemical measurements) of hydronium ion, Brønsted acids, and H_2 (Chapter 8); dioxygen species (O_2, $O_2^{\cdot-}$/HOO·, HOOH) and H_2O/HO^- (Chapter 9); metals, metal compounds, and metal complexes (Chapter 10); nonmetals (Chapter 11); carbon compounds (Chapter 12); and organometallic compounds and metalloporphyrins (Chapter 13). The later chapters contain specific characterizations of representative molecules within a class, which we hope will reduce the "barriers of unfamiliarity" and encourage the reader to make use of electrochemistry for related chemical systems.

This enterprise would not have been possible without the assistance, support, and encouragement of colleagues, current and former students and postdoctoral associates, and the staff of our respective universities. In particular, the dedicated efforts of Monica Gonzales and Lori Locke to produce the final manuscript are gratefully acknowledged. Drs. Xiu Liu and John Hage provided assistance with the graphics and equations, and Debbie Bush produced preliminary drafts for several chapters. Closure of what appeared to be an interminable project was made possible by a grant-in-aid from the Alcoa Foundation.

A special acknowledgment to Dr. Ted Hoffman [Chemistry Editor of Wiley-Interscience for 25 years until retirement; November 1, 1994] for the encouragement and support that made possible the first edition, and for the faith and patience that allowed the present volume to be realized. We are especially grateful for his (1) understanding of the academics' propensity to overcommit and (2) ability to give a positive dimension to missed deadlines. The editorial staff of Wiley-Interscience have been helpful and understanding in their assistance to the authors.

Finally, a sincere and personal acknowledgment to our wives, Shirley, Krystyna, and Jane, for their indulgence, patience, and support in what appeared to be a never-ending project.

DONALD T. SAWYER
ANDRZEJ SOBKOWIAK
JULIAN L. ROBERTS, JR.

College Station, Texas
Rzeszow, Poland
Redlands, California
November 1994

CONTENTS

CHAPTER 1

INTRODUCTION AND FUNDAMENTALS

1.1 INTRODUCTION

The technology for the interconversion of chemical energy and electrical energy has been utilized since the midnineteenth century. This conversion is accomplished by ionic current flow in an electrolyte solution between two electrodes connected to each other via an external circuit with an electrical load or current source. Batteries, fuel cells, and corrosion processes convert the energy of chemical reactions into electrical energy. Electrolysis, electroplating, and some forms of electroanalysis reverse the direction of conversion, using electrical energy to produce a net chemical change. The basic principles and quantitative relationships (voltage, current, charge conductance, capacitance, and concentration) for electrochemical phenomena were empirically elucidated by Michael Faraday and other European scientists before the discovery of the electron (J. J. Thompson, 1893) and the development of chemical thermodynamics (G. N. Lewis, 1923). Building on this foundation, the utilization of electrochemical phenomena for thermodynamic characterization and analysis of molecules and ions (electroanalytical chemistry) began at the beginning of this century [potentiometry (1920) and polarography (1930)]. Relationships that describe the techniques of potentiometry and polarography derive directly from solution thermodynamics. In the case of polarography, there is a further dependence on the diffusion of ionic species in solution. The latter is the basis of conductivity measurements, another area that traces its origin to the nineteenth century. These quantitative relationships make it possible to apply electrochemistry to the detailed characterization of chemical species and processes in the solution phase.

During the past four decades the dynamics and mechanisms of electron-transfer processes have been studied via the application of transition-state theory to the kinetics for electrochemical processes. As a result, both the kinetics of the electron-transfer processes (from solid electrode to the solution species) as well as of pre- and post-electron-transfer homogeneous processes can be characterized quantitatively.

By the use of various transient methods, electrochemistry has found extensive new applications for the study of chemical reactions and adsorption phenomena. Thus a combination of thermodynamic and kinetic measurements can be utilized to characterize the chemistry of heterogeneous electron-transfer reactions. Furthermore, heterogeneous adsorption processes (liquid–solid) have been the subject of intense investigations. The mechanisms of metal ion complexation reactions also have been ascertained through the use of various electrochemical impulse techniques.

The so-called renaissance of electrochemistry has come about through a combination of modern electronic instrumentation and the development of a more pragmatic theory implemented with the data-processing and computational power of computers. Within the area of physical chemistry, numerous thermodynamic studies of unstable reaction intermediates have made use of modern electrochemistry. In addition, extensive studies of the kinetics of electron-transfer processes in aqueous and nonaqueous media have been accomplished. The electrochemical characterization of adsorption phenomena has been of immense benefit to the understanding of catalytic processes.

Some of the most exciting applications of electrochemistry have occurred in the areas of organic and inorganic chemistry, and biochemistry. The applications have ranged from mechanistic studies to the synthesis of unstable or exotic species. The control of an oxidation or reduction process through electrochemistry is much more precise than is possible with chemical reactants. Within the area of inorganic chemistry, electrochemistry has been especially useful for the determination of formulas of coordination complexes and the electron-transfer stoichiometry of new organometallic compounds. Electrochemical synthesis is increasingly important to the field of organometallic chemistry.

During the past 40 years there have been numerous exciting extensions of electrochemistry to the field of analytical chemistry. A series of selective-ion potentiometric electrodes have been developed, such that most of the common ionic species can be quantitatively monitored in aqueous solution. A highly effective electrolytic moisture analyzer provides continuous online assays for water in gases. Another practical development has been the voltammetric membrane electrode for dioxygen (O_2), which responds linearly to the partial pressure of O_2, either in the gas phase or in solution. The use of an immobilized enzyme (glucose oxidase) on an electrode sensor to assay glucose in blood is another extension of electrochemistry to practical analysis.

1.2 NOMENCLATURE AND CLASSES OF ELECTROCHEMICAL METHODOLOGY

Modern electrochemistry has evolved to the extent that it has a diverse set of specialized terms and symbols. The latter are defined in Table 1.1 as used in most contemporary electrochemical literature and in this book. Because of the rapid expansion in specialized electrochemical methodology and its application to chemical problems, a nomenclature has evolved for their categorization. This is outlined in Table 1.2 and provides an overview of the complete realm of electrochemical methodology. Within this table, key references to the major monographs for each specialized type of electrochemistry are included.[1-36] These references provide the theory and details of applications to complement the introductory and practical presentation of this book.

TABLE 1.1 Definitions of Symbols Used in Electrochemistry

A	amperes
A	area
a	activity
a	$nF\nu/RT$
AC	alternating current
C	concentration
C^0	concentration at the electrode surface
C^b	bulk concentration
C	capacitance
D	diffusion coefficient
DC	direct current
DME	dropping-mercury electrode
d	density
E	potential
E°	standard potential
$E^{\circ\prime}$	formal potential
$E_{1/2}$	half-wave potential (polarography)
E_{p}	peak potential
e^-	electron
exp	exponential
erf	error function
F	faraday
F	farad
f	frequency
G	Gibbs free energy
H	enthalpy
h	height
i	current
i_{d}	diffusion current
i_{lim}	limiting current

TABLE 1.1 (*Continued*)

I	diffusion current constant (polarography)
j	current density
K	equilibrium constant
k_f	forward homogeneous rate constant
k_b	backward homogeneous rate constant
k_c	heterogeneous electrochemical rate constant for the cathodic process
k_a	heterogeneous electrochemical rate constant for the anodic process
k_s	heterogeneous electrochemical rate constant at $E^{\circ\prime}$
m	mass flow rate of DME
NHE	normal-hydrogen electrode
n	number of electrons
n_a	number of electrons in rate-determining step
ox	oxidized species
pH	negative logarithm of hydronium ion (H_3O^+) activity
Q	coulombs of charge
R	gas constant
R	resistance
r	radius
red	reduced species
S	entropy
SCE	saturated-calomel electrode
T	temperature [kelvins (K)]
t	time
t_d	drop time (polarography)
$t_+ t_-$	transference numbers
$u_+ u_-$	ion mobilities
V	voltage (volts)
V	volume
v	velocity
z	charge of an ion
α	transfer coefficient (symmetry parameter)
γ	activity coefficient
$\gamma\pm$	mean activity coefficient
δ	mobility of an ion
Δ	thickness of a diffusion layer
Δ	differential
ΔG_{BF}	the free energy of homolytic bond formation
ΔH_{DBE}	the enthalpy of homolytic bond dissociation (dissociative bond energy)
ϵ	dielectric constant
Λ	equivalent conductance
λ	conductance of an ion
η	overpotential
ν	scan rate; kinematic viscosity
τ	transition time (chronopotentiometry)
ω	angular velocity
Ω	ohms

TABLE 1.2 Outline and Nomenclature for the Methodology of Electrochemistry

Methods	Controlled Variable	Measured Variable	Refs.
Potentiometry	$i = 0$	E	1–3
Controlled potential			
Voltammetry	E	i	4–7
Polarography (DME)	E	i	7–11
Single-sweep	$E(t)$	i	5,7,11–14
Cyclic-sweep	$E(t)$	i	5, 7, 11–13, 15–17
Rotated-disk and ring–disk electrodes	E, ω	i	6,7,13,18–20
Pulse	$\Delta E(t)$	Δi	11, 21–23
Chronoamperometry	E-step	$i(t)$	4, 5, 7, 11, 24
Chronocoulometry	E-step	$\int idt$	5,11,24,25
Controlled-potential electrolysis	E	$\int idt$	26,27
Electrolysis	V	Weight	27,28
Controlled current			
Chronopotentiometry	i	$E(t)$	5,7,17,29–31
Galvanostatic	i	$E(t = 0)$	32
Coulometric titrations	i	$(it)/F$	27,33
Conductivity	V(AC)	i(AC)	34–36

1.3 SIGN AND GRAPHICAL CONVENTIONS

Nothing has caused more difficulty than the sign convention of electrochemistry. The problem probably began with Benjamin Franklin's experiments and has evolved through many historical and regional stages. The result is that the electrochemical literature requires exceptional expertise or a "guidebook" to avoid confusion. Although the approach followed in this book is outlined in subsequent chapters, the basic rationale can be stated here.

All cells are considered as the combination of two half-cells, with each of the latter represented by a half-reaction written as a reduction:

$$ox + ne^- \longrightarrow red \tag{1.1}$$

By use of an electromotive series ($E°$ values) for standard half-reactions written as reductions (see Chapter 2) the potential of each half-cell can be calculated by means of the Nernst equation:

$$E = E° + \frac{RT}{nF} \ln \frac{[ox]}{[red]} \tag{1.2}$$

On this basis the half-cell with the more positive value will be the positive electrode (and the other the negative electrode); the algebraic difference of the two half-cell potentials equals the cell potential.

Another area of confusion in electrochemical literature is the way current–voltage curves ("voltammograms") are displayed. Historically, they were often oriented with increasingly negative potentials directed to the right of the origin, and with cathodic (net reduction) currents directed upward from the origin, as shown in Figure 1.1. This convention is widely used in monographs and textbooks of electrochemistry and is the convention in this book for the display of current–voltage curves. However, the opposite convention is arguably more consistent with the usual format of graphs and is beginning to be more widely used, particularly in the international literature. In this "rational" convention, positive potentials are directed to the right of the origin and anodic (net oxidation) currents are taken as positive (directed upward from the origin). Whatever convention is used, the axes should be clearly labeled so that there can be no doubt in the reader's mind about the convention employed.

1.4 UTILIZATION OF ELECTROCHEMISTRY FOR CHEMICAL CHARACTERIZATION

An increasing number of chemists use electrochemistry as a characterization technique in a fashion analogous to their use of infrared, ultraviolet (UV)–visible, NMR, and electron spin resonance (ESR) spectroscopy. One of the main purposes of this book is to encourage this trend, and to provide practical

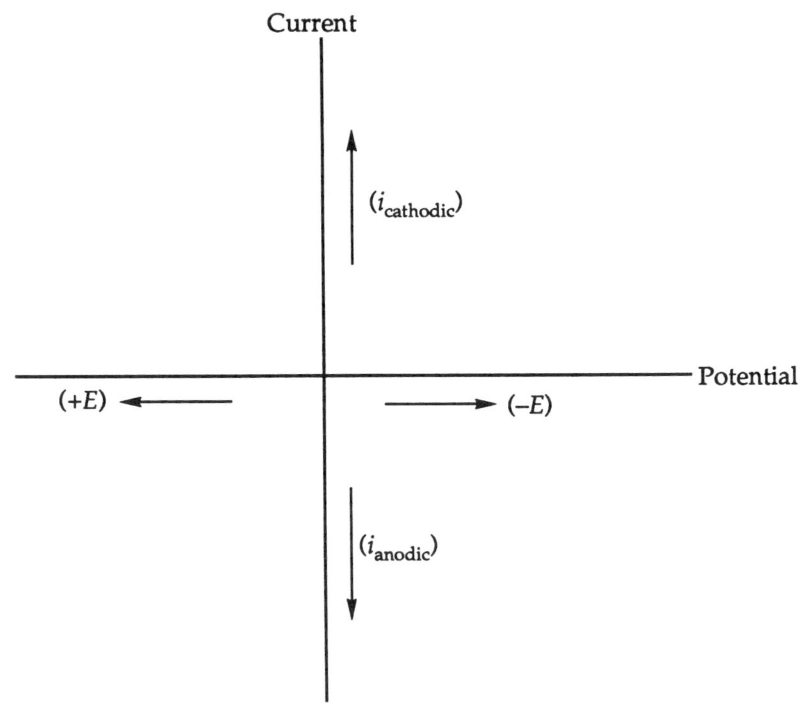

Figure 1.1 Graphical system for electrochemical data presentation in this monograph.

insights to electrochemical methodology. At this point a brief outline of the approach for solving chemical questions by means of electrochemical measurements may be helpful and illustrative.

Some of the chemical questions that are amenable to treatment by electrochemistry include (1) the standard potentials ($E°$) of the compound's oxidation–reduction reactions, (2) evaluation of the solution thermodynamics of the compound, (3) determination of the electron stoichiometry of the compound's oxidation–reduction reactions, (4) evaluation of the heterogeneous electron-transfer kinetics and mechanisms of the compound, (5) determination of the effect of solvent and electrode material and its preconditioning on the electron-transfer kinetics, (6) study of reaction and product adsorption processes in relation to heterogeneous catalysis, (7) study of pre- and postchemical reactions (thermodynamics and kinetics) that are associated with the electron-transfer reaction of a compound, (8) preparation and study of unstable intermediates, (9) evaluation of the valence of the metal in new compounds, (10) determination of the formulas and stability constants of metal complexes, (11) evaluation of $M{-}X$, $H{-}X$, and $O{-}Y$ covalent-bond-formation energies ($-\Delta G_{BF}$), and (12) studies of the effects of solvent, supporting electrolyte, and solution acidity on oxidation–reduction reactions. To answer those questions, the investigator must select the proper electrochemical method and sample conditions. Frequently the solubility and stability of the sample dictate the solvent system that must be used. After selection of an electrochemically compatible solvent, a noninterfering soluble supporting electrolyte must be added (Chapter 6 discusses solvents and supporting electrolytes that are commonly utilized in electrochemistry).

As the later chapters indicate, a given question concerning a chemical system usually can be answered by any one of several electrochemical techniques. However, experience has demonstrated that there is a most convenient or reliable method for a specific kind of data. For example, polarography with a static or dropping-mercury electrode remains the most reliable electrochemical method for the quantitative determination of trace-metal ion concentrations. This is true for two reasons: (1) the reproducibility of the dropping-mercury electrode is unsurpassed and (2) the reference literature for analysis by polarography surpasses that for any other electrochemical method by at least an order of magnitude.

Often the first step in the electrochemical characterization of a compound is to ascertain its oxidation–reduction reversibility. In our opinion, cyclic voltammetry is the most convenient and reliable technique for this and related qualitative characterizations of a new system, although newer forms of pulse polarography may prove more suitable for quantitative determination of the electrochemical parameters. The discussion in Chapter 3 outlines the specific procedures and relationships. The next step in the characterization usually is the determination of the electron stoichiometry of the oxidation–reduction steps of the compound. Controlled-potential coulometry (discussed in Chapter 3) provides a rigorously quantitative means for such evaluations.

With respect to chemical steps prior to the electron-transfer step, chronopotentiometry offers a convenient technique. The methods of measurement and the quantitative relationships are outlined in Chapter 4. Post-electron-transfer reactions to the electron-transfer step are most conveniently characterized by cyclic voltammetry (see Chapter 3). Although the techniques of cyclic voltammetry and chronopotentiometry both provide a means for the qualitative detection of adsorption processes at an electrode, the coulostatic method and chronocoulometry are the methods of choice for quantitative measurements of adsorption.

Because electrochemistry provides a unique controlled means of adding or subtracting electrons to or from a compound, it can be used to produce transiently stable species for study by other physical methods such as optical and ESR spectroscopy and mass spectroscopy. Conversely, electrochemistry is an especially sensitive means for the detection of reaction products from photolysis and pyrolysis reactions.

Many other applications of electrochemical methods for chemical characterization are presented in the following chapters. The state of utilization is such that for many research groups in the fields of organic chemistry and inorganic chemistry, electrochemistry has become a characterization tool as essential as infrared and NMR spectroscopy. This is quickly becoming true for several areas of biochemistry, especially enzymology.

Utility of Electrochemical Methodologies

- Electrochemical methods are well established and use relatively inexpensive equipment to produce unique characterization information for molecules and chemical systems: qualitative (speciation) and quantitative analytical data, thermodynamic data (equilibrium constants), and kinetic data (heterogeneous and homogeneous reaction rates).
- Electrochemical methods are sensitive; they are able to detect submicromolar concentrations and subpicomole amounts of electroactive material.
- Electrochemical methods are selective; they are able to control the potential of an electrode, which makes it possible to determine the electrochemical "spectrum" of electroactive species in solution, analogous to probing the energy states of a molecule with light via spectroscopy.

Limitations of Electrochemical Methodologies

- Electrochemical reactions take place at the interface between two phases; this interface region is about 1 nm thick and is called the *double layer* because it is a region in which layers of charge of opposite sign exist. Most of the potential (voltage) difference between the electrodes of an electrochemical cell exists at the electrode–solution interfaces (double layers). The double layer behaves in many ways like a capacitor. When the potential of an electrode is suddenly changed, some of the current that flows charges the double-layer capacitance. The use of ultramicroelectrodes essentially eliminates this problem.
- A new kind of dynamic process is added to the picture compared to reactions in homogeneous solution. The electrochemical experiment allows one to control the rate of electron transfer across the interface via adjustment of the potential across the interface. This potential difference creates an intense electric field whose magnitude is of the order of 1 V nm^{-1} or about 10^9 V m^{-1}.
- Because the electron transfer takes place at the interface rather than in the bulk of the solution, every electrochemical procedure must take account of the mass transport of the electroactive species to the interface by diffusion, convection (stirring), and migration (the movement of ions in an electric field). Migration is usually minimized by adding a dissolved salt, called the *supporting electrolyte*, to make the solution a good ionic conductor.

1.5 THE FUNDAMENTALS

Electrochemistry is the science of electron transfer across a solution–electrode interface. At the cathode ions or molecules are transformed within the interface via reaction with electrons (from the electrode) to produce reduced molecules or ions [e.g., $H_3O^+ + e^- \longrightarrow H\cdot + H_2O$; $H_2O + e^- \longrightarrow H\cdot + HO^-$, and in aprotic media $O_2 + e^- \longrightarrow O_2^{\overline{\cdot}}$; $Cu^{II}(bpy)_2^{2+} + e^- \longrightarrow Cu^{I}(bpy)_2^{+}$; $Fe^{III}Cl_3 + e^- \longrightarrow Fe^{II}Cl_3^-$]. At the anode molecules or ions (from the solution) are transformed within the interface to produce electrons (at the electrode surface) and oxidized ions and molecules (e.g., $2\,H_2O \longrightarrow H_3O^+ + HO\cdot + e^-$; $Fe^{II}Cl_3^- \longrightarrow Fe^{III}Cl_3 + e^-$). The resultant electrons move from the anode through the wires of the external circuit to the cathode as electronic current (amperes; coulombs per second). Within the solution phase the current is carried by the ions of the supporting electrolyte (positive ions toward the cathode and negative ions toward the anode). The limitation of ionic current in the solution phase (between the anode and the cathode), which is the defining difference for electrochemistry and electronics, is due to the incompatibility of electrons and electrolyte solutions.

The Hydrated Electron. In aqueous media electrons become hydrated in 10^{-12} s to become the ultimate base, nucleophile, and reductant:[37]

$$e^- + n\,H_2O \longrightarrow e_{aq}^- \qquad E°, -3.0 \text{ V vs. NHE} \tag{1.3}$$

The conjugate acid of the hydrated electron is the hydrogen atom (H•), which is the effective reductant under acidic conditions:

$$e_{aq}^- + H_3O^+ \longrightarrow H\cdot + H_2O \qquad k, 2.3 \times 10^{10}\ M^{-1}\ s^{-1} \tag{1.4}$$

$$e_{aq}^- + H_2O \longrightarrow H\cdot + HO^- \qquad k, 1.9 \times 10^{1}\ M^{-1}\ s^{-1} \tag{1.5}$$

From electrochemistry the respective standard reduction potentials in water are[38]

$$e^- + H_3O^+ \longrightarrow H\cdot + H_2O \qquad E°, -2.10 \text{ V vs. NHE} \tag{1.6}$$

$$e^- + H_2O \longrightarrow H\cdot + HO^- \qquad E°, -2.93 \text{ V vs. NHE} \tag{1.7}$$

and in acetonitrile[39]

$$e^- + H_3O^+ \longrightarrow H\cdot + H_2O \qquad E°', -1.58 \text{ V vs. NHE} \tag{1.8}$$

$$e^- + H_2O \longrightarrow H\cdot + HO^- \qquad E°', -3.90 \text{ V vs. NHE} \tag{1.9}$$

Hydrated electrons also are highly reactive with many components of aqueous biological matrices:[37]

$$e_{aq}^- + O_2 \longrightarrow O_2^{\overline{\cdot}} \qquad k, 1.9 \times 10^{10}\ M^{-1}\ s^{-1} \tag{1.10}$$

$$e_{aq}^- + CO_2 \longrightarrow \cdot CO_2^- \qquad k, 7.7 \times 10^{9}\ M^{-1}\ s^{-1} \tag{1.11}$$

$$e_{aq}^- + HOOH \longrightarrow H\cdot + HOO^- \qquad k, 1.1 \times 10^{10}\ M^{-1}\ s^{-1} \tag{1.12}$$

$$e_{aq}^- + RSH \longrightarrow HS^- + R\cdot \qquad k, 1.1 \times 10^{10}\ M^{-1}\ s^{-1} \tag{1.13}$$

Whereas for HO—H the free energy of bond formation ($-\Delta G_{BF}$) is 464 kJ mol^{-1}[,40], within $(H_3O)^+$ the O—H bond energy is about 301 kJ mol^{-1}. The difference accounts for the dramatic increase in the reactivity of e^-_{aq} with the hydronium ion [Eqs. (1.4) and (1.5)]. On the basis of such considerations the reaction rates of solvated electrons with phenols (*R*PhO—H; $-\Delta G_{BF}$, 339–351 kJ mol^{-1}) and carbonic acid [(HO)C(O)(O—H); $-\Delta G_{BF}$, 410 kJ mol^{-1}] should be large:

$$e^-_{aq} + R\text{PhOH} \longrightarrow \text{H}\cdot + R\text{PhO}^- \qquad k \sim 10^{10}\ \text{M}^{-1}\ \text{s}^{-1} \tag{1.14}$$

$$e^-_{aq} + (\text{HO})_2\text{C(O)} \longrightarrow \text{H}\cdot + \text{HOC(O)O}^- \qquad k \sim 10^{7}\ \text{M}^{-1}\ \text{s}^{-1} \tag{1.15}$$

Hence, tyrosine residues (*R*PhOH) in proteins and biological fluids with 3–5 mM $(HO)_2C(O)$ will be effective traps for hydrated electrons in biological systems. When O_2 reacts with hydrated electrons in aqueous media, the product ($O_2^{\cdot-}$) reacts with water at diffusion controlled rates to produce HOO·

$$e^-_{aq} + \text{O}_2 \longrightarrow \text{O}_2^{\cdot-} \underset{pK_{\text{HA}}\ 4.9}{\overset{\text{H}_2\text{O}}{\rightleftharpoons}} \text{HOO}\cdot + \text{HO}^- \qquad k \sim 10^{10}\ \text{M}^{-1}\ \text{s}^{-1} \tag{1.16}$$

This indicates that any hydrated electrons in aerobic biology that escape trapping by the CO_2, carbonic acid, and tyrosine and cysteine residues will produce finite fluxes of the soft radical, HOO·.

In aqueous solutions the hydrated electron interacts with water to form a hydrogen atom and a hydroxide ion (the conjugate base of H_2O):

$$e^-_{aq} + \text{H}_2\text{O} \rightleftharpoons \text{H}\cdot + \text{HO}^- \qquad K,\ 2 \times 10^5 \tag{1.17}$$

which means that H· is the conjugate acid of e^-_{aq} with a pK_{HA} value of 9.3.[37]

Electron Transfer in Electrochemistry. In electrochemical cells electron transfer occurs within the electrode–solution interface, with electron removal (oxidation) at the anode, and with electron introduction (reduction) at the cathode. The current through the solution is carried by the ions of the electrolyte, and the voltage limits are those for electron removal *from* and electron insertion *into* the solvent–electrolyte [e.g., $H_2O/(H_3O^+)(ClO_4^-)$; $(Na^+_{aq})(^-OH)$; $(Na^+_{aq})(Cl^-)$]:

$$2\ \text{H}_2\text{O} \overset{-e^-}{\rightleftharpoons} [\text{H}_2\text{O}(\text{H}_2\text{O}\cdot)^+] \rightleftharpoons$$

$$\text{H}_3\text{O}^+ + \text{HO}\cdot \quad (E^\circ)_{\text{pH}0},\ +2.72\ \text{V vs. NHE} \tag{1.18}$$

$$(E^\circ)_{\text{pH}7},\ +2.31\ \text{V}$$

$$\text{HO}^- \overset{-e^-}{\rightleftharpoons} \text{HO}\cdot \quad (E^\circ)_{\text{pH}14},\ +1.89\ \text{V} \tag{1.19}$$

$$\text{Cl}^- \overset{-e^-}{\rightleftharpoons} \text{Cl}\cdot \quad E^\circ,\ +2.41\ \text{V} \tag{1.20}$$

In the gas phase electron removal from atoms is limited by their ionization potential (e.g., H·, 13.6 eV; K·, 4.3 eV; Na·, 5.1 eV; Cu·, 7.7, 20.3 eV; Ag, 7.6 eV; Fe, 7.9, 16.2, 30.7 eV).[41] However, in the solution phase electron removal (oxidation) from the solvent may be facilitated by the presence of substrate atoms (rather than be from them).

For example, with pH 0 water the process of Eq. (1.18) is shifted −4.82 V when hydrogen atoms are present

$$H\cdot + 2\ H_2O \overset{-e^-}{\rightleftharpoons} [(H_2O)H_3O^+] \rightleftharpoons H_3O^+ + H_2O \qquad E°,\ -2.10\ V \tag{1.21}$$

and −1.92 V with a silver electrode

$$Ag_{(s)} + 2\ H_2O \overset{-e^-}{\rightleftharpoons} [(H_2O)Ag(OH_2)^+] \rightleftharpoons (H_2O)Ag^I{-}OH_2^+ \qquad E°,\ +0.80\ V \tag{1.22}$$

Water oxidation at pH 7 with a silver electrode is shifted −2.04 V:

$$Ag_{(s)} + 2\ H_2O \overset{-e^-,\ +0.27\,V}{\rightleftharpoons} [(H_2O)Ag(OH_2)^+] \rightleftharpoons Ag^I{-}OH_{(s)} + H_3O^+ \tag{1.23}$$

which indicates that the HO• product of Eq. (1.18) is stabilized by formation of a 197-kJ-mol^{-1} covalent bond with a silver atom [($-\Delta G_{BF}$), (2.04 × 96.5 kJ/eV) = 197] [the $(H_2O)Ag^I{-}OH_2^+$ bond of Eq. (1.22) has a $-\Delta G_{BF}$ value of 184 kJ mol^{-1}]. Likewise, the oxidation of Cl^- [Eq. (1.20)] is facilitated at a silver electrode via formation of a $Ag^I{-}Cl$ covalent bond [$-\Delta G_{BF}$, (2.19 × 96.5) = 213 kJ mol^{-1}]:

$$Ag_{(s)} + Cl^- \overset{-e^-}{\rightleftharpoons} Ag^I{-}Cl_{(s)} \qquad E°,\ +0.22\ V \tag{1.24}$$

Hence, the precipitation of $AgCl_{(s)}$ is favored on the basis of the differential bond energetics [Eq. (1.24) − Eq. (1.22)]:

$$Ag^I(OH_2)_2^+ + Cl^- \longrightarrow AgCl_{(s)} + 2\ H_2O \tag{1.25}$$

Reductive electron transfer in an electrochemical cell occurs by insertion of an electron from the electrode (cathode) into the solution matrix within the double layer of the electrode–solution interface [e.g., $H_2O/(H_3O^+)$-(ClO_4^-); $(Na_{aq}^+)(ClO_4^-)$]:

$$H_3O^+ \overset{e^-}{\rightleftarrows} H\cdot + H_2O \qquad (E^\circ)_{pH0},\ -2.10\ \text{V vs. NHE} \qquad (1.26a)$$

$$(E^\circ)_{pH7},\ -2.51\ \text{V} \qquad (1.26b)$$

The reductive processes of Eq. (1.26) may be facilitated by the presence of substrates to stabilize the H-atom product. For example, in pH 0 water the process of Eq. (1.26a) is shifted by +4.82 V when hydroxyl radicals ($HO\cdot$) are present

$$H_3O^+ + HO\bullet \overset{e^-}{\rightleftarrows} \underset{(-\Delta G_{BF},\ 464\ \text{kJ mol}^{-1})}{HO{-}H + H_2O} \quad E^\circ,\ +2.72\ \text{V} \qquad (1.27)$$

and by +2.10 V at a platinum electrode:

$$H_3O^+ + Pt \overset{e^-}{\rightleftarrows} \underset{(-\Delta G_{BF},\ 201\ \text{kJ mol}^{-1})}{Pt{-}H + H_2O} \qquad E^\circ,\ 0.00\ \text{V} \qquad (1.28)$$

In the presence of benzoquinone (Q) the shift is +2.80 V:

$$2\,H_3O^+ + Q \overset{2e^-}{\rightleftarrows} \underset{(-\Delta G_{BF},\ 272\ \text{kJ mol}^{-1})}{HO{-}C_6H_4{-}OH} \qquad (E^\circ)_{pH\,0},\ +0.70\ \text{V} \qquad (1.29)$$

The free electron interacts with all atoms and molecules that have finite electron affinities to produce anions, and thus is unstable in all except the most inert liquids. Electrochemistry attests to this general axiom and provides a convenient means for evaluation of the energetics for the addition of an electron to solvent molecules and to species at the electrode–solution interface, for example:

$$e^- + H_2O_{(l)} \longrightarrow H\cdot + HO^- \qquad E^\circ,\ -2.93\ \text{V vs. NHE}$$

$$\xrightarrow{e^-,\ H_2O} H{-}H + HO^- \qquad (1.30)$$

$$e^- + CO_2(MeCN) \longrightarrow \left[\cdot C(=O)O^- \longleftrightarrow \cdot C(O^-)=O\right] \qquad E^\circ,\ -1.86\ \text{V vs. NHE} \qquad (1.31)$$

$$e^- + Me_2\ddot{S}(O) \longrightarrow H_2C{=}\ddot{S}(O^-)Me + H\cdot \qquad (1.32)$$

$$e^- + Me_2\ddot{N}CH(O) \longrightarrow [Me_2\ddot{N}\dot{C}H(O^-)] \longrightarrow Me_2N{\equiv}CO^- + H\cdot \quad (1.33)$$

$$e^- + MeCN \longrightarrow H_2C{=}C{=}\ddot{N}{:}^- + H\cdot \quad (1.34)$$

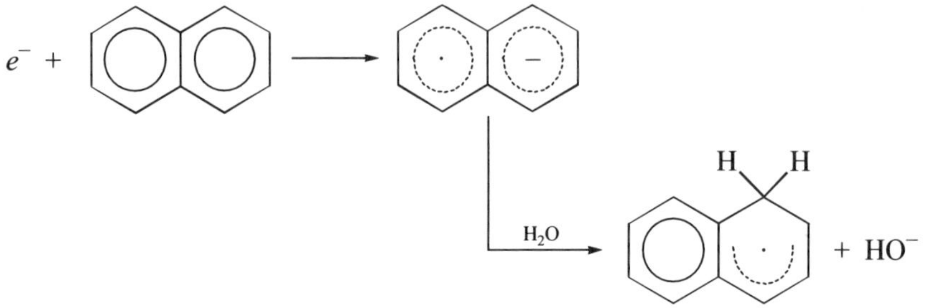

$$e^- + \text{naphthalene} \longrightarrow \text{[naphthalene radical anion]} \xrightarrow{H_2O} \text{[dihydronaphthyl radical, H H]} + HO^- \quad (1.35)$$

Hence, reductive electrochemistry converts electrons (e^-) via the solution matrix at the interface to atoms and anions. The solution outside the inner double layer never is exposed to an electron. Some examples of such *inner-double-layer electron transfer* include

$$H_2O + e^- \longrightarrow H\cdot + HO^- \qquad E°, -2.93 \text{ V vs. NHE} \quad (1.36)$$

$$H_3O^+ + e^- \longrightarrow H\cdot + H_2O \qquad E°, -2.10 \text{ V} \quad (1.37)$$

$$(H_2O)_5{}^{2+}Fe^{III}(OH) + H_3O^+ + e^- \xrightarrow[H_2O]{pH\,1} Fe^{II}(OH_2)_6^{2+} + H_2O \qquad E°', +0.71 \text{ V} \quad (1.38)$$

$$HO\cdot + H_3O^+ + e^- \xrightarrow[H_2O]{pH\,1} 2\,H_2O \qquad E°', +2.66 \text{ V} \quad (1.39)$$

$$-\Delta G_{BF}\,[(H_2O)_5^{2+}Fe^{III}{-}OH] = [E^{\circ\prime}_{39} - E^{\circ\prime}_{38}] \times 96.5 = 188 \text{ kJ mol}^{-1} \quad (1.40)$$

$$\underset{-\Delta G_{BF} = 117\,\text{kJ mol}^{-1}}{O_3Mn^{VII}{-}OH} + H_3O^+ + e^- \xrightarrow{pH\,1} Mn^{VI}O_3 + 2\,H_2O \qquad E°', +1.45 \text{ V} \quad (1.41)$$

An important consideration in the reactions of Eqs. (1.39) and (1.41) is that the oxidant in each is the hydronium ion (H_3O^+) and that the reduction potential is determined by the H—OH bond energy ($-\Delta G_{BF}$) of the product H_2O, minus the Mn—OH bond energy [Eq. (1.41)].

1.6 NUCLEOPHILE–ELECTROPHILE ELECTRON TRANSFER

Inner-sphere electron transfer always involves nucleophilic addition or substitution with an electrophile via a single-electron-transfer (SET) step:[42]

$$HO^- + PhC(O)Cl \longrightarrow [Ph(HO)C(O^-)Cl] \longrightarrow PhC(O)OH + Cl^- \quad (1.42)$$

$$HO^- + n\text{-BuBr} \xrightarrow{S_N2} HOBu\text{-}n + Br^- \quad (1.43)$$

With saturated electrophiles [n-BuBr, Eq. (1.43)] the SET is to a less nucleophilic leaving group [$N{:}^- + R{:}\ddot{X}{:}(E) \rightarrow N{:}R + {:}\ddot{X}{:}^-$]. The driving force for nucleophile–electrophile electron-transfer reactions is the redox potential for the nucleophile in the solution matrix ($HO^-/HO\cdot$, +0.92 V vs. NHE in MeCN)[42] *plus* the nucleophile–substrate bond-formation free energy ($-\Delta G_{BF}$, n-Bu—OH; ~349 kJ mol^{-1}):

$$\begin{aligned} \text{Nucleophilicity} &= (E_{1/2})_{ox} - \frac{-\Delta G_{BF(N-E)}}{96.5} \\ &= (+0.92\ \text{V}) - [349\ \text{kJ mol}^{-1}/96.5\ \text{kJ mol}^{-1}(\text{eV})^{-1}] \\ &= (+0.92 - 3.62)\ \text{V} = -2.70\ \text{V vs. NHE} \end{aligned} \quad (1.44)$$

In contrast, the measure of the electrophilicity for a molecule or ion is its one-electron reduction potential in the solution matrix:

$$\text{Electrophilicity} = (E_{1/2})_{red} \quad (1.45)$$

$$\text{e.g., } n\text{-BuBr} + e^- \longrightarrow n\text{-Bu}\cdot + Br^- \qquad E_{1/2},\ -2.14\ \text{V vs. NHE}$$

$$n\text{-BuI} + e^- \longrightarrow n\text{-Bu}\cdot + I^- \qquad E_{1/2},\ -1.78\ \text{V vs. NHE}$$

Hence, nucleophile–electrophile redox reactions for HO^- with n-BuBr and n-BuI are both exoergonic:

$$n\text{-BuBr} + HO^- \longrightarrow n\text{-BuOH} + Br^- \quad (1.46)$$

$$\begin{aligned} -\Delta G_{reac} &= [(\text{electrophilicity})_{n\text{-BuBr}} - (\text{nucleophilicity})_{HO^-}] \\ &\quad \times 96.5\ \text{kJ mol}^{-1}(\text{eV})^{-1} \\ &= [(-2.14\ \text{V}) - (-2.70\ \text{V})] \times 96.5 \\ &= 54\ \text{kJ mol}^{-1} \end{aligned}$$

$$n\text{-BuI} + HO^- \longrightarrow n\text{-BuOH} + I^- \quad (1.47)$$

$$-\Delta G_{reac} = [(-1.78\ \text{V}) - (-2.70\ \text{V})] \times 96.5 = 88\ \text{kJ mol}^{-1}$$

TABLE 1.3 Nucleophilicity and Electrophilicity of Molecules and Ions[a]

Nucleophile	$(E_{1/2})_{ox}$, V vs. NHE (MeCN)	$(E_{1/2})_{ox}$, V vs. NHE (H_2O)	Electrophile	$(E_{1/2})_{red}$, V vs. NHE (MeCN)	$(E_{1/2})_{red}$, V vs. NHE (H_2O)
e^-_{aq}	−3.9	−2.9	$H_2O^+\cdot$	+3.2	+2.7
$K\cdot$	−2.9	−2.9	$Ph^+\cdot CH_2OH$	+2.2	
$Na\cdot$	−2.1	−2.7	Au^+	+1.6	+1.8
$Li\cdot$	−2.0		$(Cl_8TPP^+\cdot)Fe^{IV}{=}O$	+1.5	
$(TPP^-\cdot)Co^-$	−1.7		(compound I)		
$H\cdot$	−1.6	−2.1	$Fe^{III}(bpy)_3^{3+}$	+1.3	+1.1
$(TPP^-\cdot)Fe^-$	−1.4		$HO\cdot$	+0.9	+1.9
$(TPP)Fe^-$	−0.8		$Fe^{III}(PA)_3$	+0.4	
$(TPP)Co^-$	−0.6		$(TPP)Fe^{III}(py)_2^+$	+0.4	
$O_2^-\cdot$	−0.7	−0.2	$(TPP)Fe^{III}Cl$	+0.2	
$PhCH_2S^-$	0.0		MV^{2+}	−0.2	
HOO^-	0.0	+0.7	$(Cl_8TPP)Fe^{IV}{=}O$	−0.3	
PhO^-	+0.3		(compound II)		
Me_3N	+0.7		AQ (anthraquinone)	−0.6	
HO^-	+0.9	+1.9	O_2	−0.7	−0.2
$MeC(O)O^-$	+1.5		CCl_4	−0.9	
PhOH	+1.7		$PhCH_2Br$	−1.4	
pyridine	+2.0		$PhCl_6$	−1.4	
Cl^-	+2.2	+2.4	H_3O^+	−1.6	−2.1
HOOH	+2.3	+1.0	*t*-BuI	−1.5	
H_2O	+3.0	+2.3	$PhCH_2Cl$	−1.7	
			n-BuI	−1.9	
			c-$C_6H_{11}Br$	−1.9	
			t-BuBr	−2.0	
			n-BuBr	−2.2	
			n-BuCl	−2.5	
			PhCl	−2.6	
			H_2O	−3.9	−2.9

[a]Strongest or most reactive at top of lising.

Table 1.3 summarizes the oxidation potentials for *nucleophiles* and the reduction potentials for *electrophiles* in acetonitrile (MeCN) and water (H_2O). The relative positions for each listing will be the same in other solvents, but nucleophiles that are stronger than HO^- will be "leveled" to its nucleophilicity in aqueous media, and electrophiles in water-containing matrices that are stronger than $[(H_2O)(H_2O\cdot)^+ \rightleftharpoons H_3O^+ + HO\cdot]$ will be "leveled" to its reactivity (e.g., $PhH^{+}\cdot + 2\,H_2O \longrightarrow HPh\cdot OH + H_3O^+$).

1.7 THE DYNAMICS OF ELECTRON TRANSFER (KINETICS AND THERMODYNAMICS)

The rate of electron transfer at an electrode–solution interface for the direct reduction of an oxidized species (*ox*) to its reduced state (*red*) [e.g., $Cu^{II}(bpy)_2^{2+} \xrightarrow{e^-} Cu^{I}(bpy)_2^{+}$]

$$ox + n\,e^- \underset{k_a}{\overset{k_c}{\rightleftharpoons}} red \qquad (1.48)$$

is a function of the concentration of the oxidized species and its heterogeneous electron-transfer rate constant (k_c; cathodic process).[4,43] The latter is a function of the potential difference across the electrode–solution interface (ΔE), which is directly proportional to the activation energy for reduction ($-\Delta G_c^{\ddagger}$). Only a fraction of ΔE is effective for accelerating the rate of reduction, which is represented by a symmetry parameter, α [transfer coefficient, $0.0 < \alpha < 1.0$ (usually about 0.5)]:[44]

$$(\Delta G_c^{\ddagger})_{total} = (\Delta G_c^{\ddagger})_{\Delta E=0} + (\Delta G_c^{\ddagger})_{\Delta E} = (\Delta G_c^{\ddagger})_{\Delta E=0} + \alpha n_a \Delta E F \qquad (1.49)$$

(where n_a = number of electrons in the rate-limiting step; usually one).

The flux of *ox* that is reduced at the electrode (v_c) has the dimensions of moles per unit area (cm^2) per second, and

$$v_c = \frac{d(C_{ox})_0}{A\,dt} = k_c(C_{ox})_0 = \frac{j_c}{nF} \qquad (1.50)$$

where $(C_{ox})_0$ is concentration (mol cm^{-3}) at the electrode surface, A is area of the electrode (cm^2), and j_c is net cathodic (reductive) current density (amperes/cm^2).

At equilibrium ($\Delta E = 0$)

$$(v_c)_{\Delta E=0} = (C_{ox})_0 \frac{\kappa k T}{h} \exp \frac{-(\Delta G_c^{\ddagger})_{\Delta E=0}}{RT} \qquad (1.51)$$

(where κ is the transmission coefficient within the activated complex theory, k is the Boltzmann constant, h is the Planck constant), which, in combination with Eq. (1.50), gives an expression for the heterogeneous electron-transfer rate constant when ΔE is zero:[45]

$$(k_c)_{\Delta E=0} = \frac{\kappa k T}{h} \exp \frac{-(\Delta G_c^{\ddagger})_{\Delta E=0}}{RT} \tag{1.52}$$

In turn, combination of these equations gives expressions for the flux of *ox* reduction (v_c) and cathodic current density (j_c) for a potential difference ΔE:

$$\begin{aligned} v_c &= (C_{ox})_0 \frac{\kappa k T}{h} \exp \frac{-(\Delta G_c^{\ddagger})_{\Delta E=0}}{RT} \exp \frac{-\alpha n_a \Delta E F}{RT} \\ &= (C_{ox})_0 (k_c)_{\Delta E=0} \exp \frac{-\alpha n_a \Delta E F}{RT} \end{aligned} \tag{1.53}$$

$$j_c = (C_{ox})_0 n F (k_c)_{\Delta E=0} \exp \frac{-\alpha n_a \Delta E F}{RT} \tag{1.54}$$

The expressions for the reverse anodic process (oxidation) of Eq. (1.48) (*red* $\xrightarrow{k_a}$ *ox* + ne^-) follow from similar arguments:

$$v_a = (C_{red})_0 (k_a)_{\Delta E=0} \exp \frac{(1-\alpha) n_a \Delta E F}{RT} \tag{1.55}$$

$$j_a = (C_{red})_0 n F (k_a)_{\Delta E=0} \exp \frac{(1-\alpha) n_a \Delta E F}{RT} \tag{1.56}$$

When the cathodic current density (j_c) is equal to the anodic current density (j_a), the net current flow across the electrode–solution interface is zero and the net flux of *ox* and *red* is zero. For this unique condition (zero net current) the current densities represent the equilibrium exchange current density (j_0):

$$j_c = j_a = j_0 \tag{1.57}$$

which is associated with the equilibrium potential difference, ΔE_e. Thus

$$j_0 = j_c = (C_{ox})_0 n F (k_c)_{\Delta E=0} \exp \frac{-\alpha \Delta E_e F}{RT} \tag{1.58}$$

$$= j_a = (C_{red})_0 n F (k_a)_{\Delta E=0} \exp \frac{(1-\alpha) \Delta E_e F}{RT} \tag{1.59}$$

The difference between ΔE and ΔE_e is the activation overpotential (η):

$$\eta = \Delta E - \Delta E_e \tag{1.60}$$

These relationships can be combined to give an expression for the net current density (j), which by definition is equal to $j_a - j_c$ in terms of the activation overpotential (substitute $\eta + \Delta E_e = \Delta E$), which is referred to as the *Butler–Volmer equation*:

$$j = j_a - j_c = j_0 \left[\exp \frac{(1-\alpha)n_a F \eta}{RT} - \exp \frac{-\alpha n_a F \eta}{RT} \right] \tag{1.61}$$

Figure 1.2 illustrates these relationships for a typical *ox/red* couple. Exchange current densities (j_0) for redox couples at room temperature range from 10 A cm^{-2} for well-behaved systems down to 10^{-6} μA cm^{-2} for H_3O^+/H_2 at a mercury electrode.

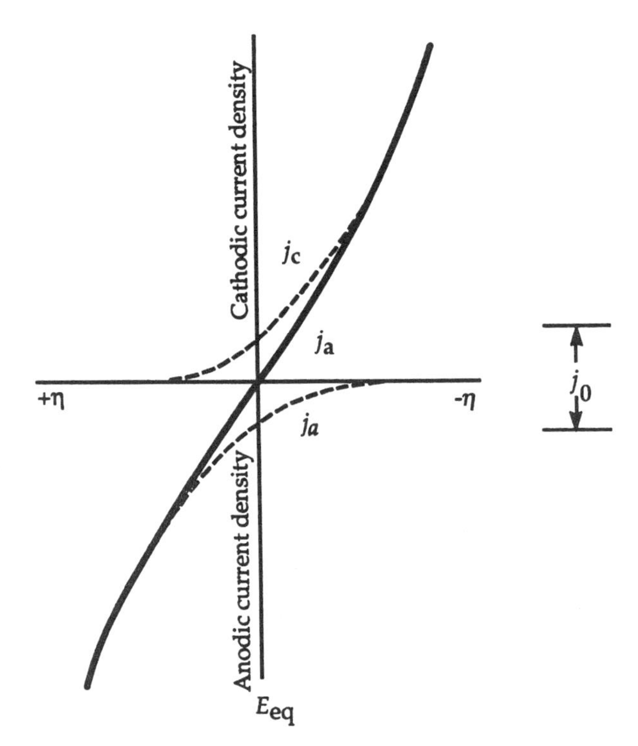

Figure 1.2 Cathodic and anodic components of current as a function of electrode polarization η. Conditions: $C_{ox} = C_{red} = 10^{-6}$ mol cm^{-3}, $\alpha = 0.5$, $T = 298$ K, and $j = 10^{-6}$ A cm^{-2}.

Both k_c and k_a vary exponentially with potential difference:

$$k_c = (k_c)_{\Delta E=0} \exp \frac{-\alpha n_a \Delta EF}{RT} \tag{1.62}$$

$$k_a = (k_a)_{\Delta E=0} \exp \frac{(1-\alpha)n_a \Delta EF}{RT} \tag{1.63}$$

At equilibrium ($j = 0$) and for the special case when $(C_{ox})_0 = (C_{red})_0$

$$(k_c)_{\Delta E=0} = (k_a)_{\Delta E=0} = k_s \tag{1.64}$$

This simple rate constant (k_s) is the defined value of k_c and k_a at the formal potential, $\Delta E^{\circ\prime}$. Then

$$(k_c)_{\Delta E=0} = k_s \exp\left[\frac{-\alpha n_a F(\Delta E - \Delta E^{\circ\prime})}{RT}\right] \tag{1.65}$$

$$(k_a)_{\Delta E=0} = k_s \exp\left[\frac{(1-\alpha) n_a F(\Delta E - \Delta E^{\circ\prime})}{RT}\right] \tag{1.66}$$

$$j = nFk_s \left\{ (C_{red})_0 \exp\left[\frac{(1-\alpha) n_a F(\Delta E - \Delta E^{\circ\prime})}{RT}\right] - (C_{ox})^0 \exp\left[\frac{-\alpha n_a F(\Delta E - \Delta E^{\circ\prime})}{RT}\right]\right\} \tag{1.67}$$

Redox couples in aqueous solutions at room temperature have k_s values that range from 0.1 cm s^{-1} for $Fe^{III}(CN)_6^{3-}/Fe^{II}(CN)_6^{4-}$ to 10^{-5} cm s^{-1} for $Cr^{III}(OH_2)_6^{3+}/Cr^{II}(OH_2)_6^{2+}$.

At the equilibrium potential, $\Delta E = \Delta E_e$, $j = 0$, $C^0_{ox} = C^b_{ox}$, and $C^0_{red} = C^b_{red}$, where C^0 denotes the concentration of a species at the electrode surface, whereas C^b its bulk concentration. Then

$$\frac{C^b_{ox}}{C^b_{red}} = \exp\left[\frac{-nF(\Delta E - \Delta E^{\circ\prime})}{RT}\right] \tag{1.68}$$

and

$$\Delta E_e = \Delta E^{\circ\prime} + \frac{RT}{nF} \ln \frac{C^b_{ox}}{C^b_{red}}$$

$$= \Delta E^{\circ\prime} + \frac{0.05915}{n} \log \frac{[ox]}{[red]} \quad \text{(Nernst equation for a half-reaction)}$$

$$\tag{1.69}$$

Also, at the equilibrium potential (ΔE_e)

$$\begin{aligned} j_0 &= nF\, k_s (C^b_{ox})^{(1-\alpha)} (C^b_{red})^{\alpha} \\ &= nF\, k_s C \quad (\text{when } C_{ox} = C_{red}) \end{aligned} \tag{1.70}$$

REFERENCES

1. Kolthoff, I. M.; Furman, N. H.; *Potentiometric Titrations*, 2nd ed., Wiley, New York, 1931.
2. Furman, N. H., ''Potentiometry,'' in *Treatise on Analytical Chemistry*, Vol. 4, Kolthoff, I. M.; Elving, P. J., eds., Interscience: New York, 1963, Chap. 45.
3. Kissel, T. R., ''Potentiometry: Oxidation–Reduction, pH Measurements, and Ion-Selective Electrodes,'' in *Physical Methods of Chemistry*, 2nd ed., Vol. II, *Electrochemical Methods*, Rossiter, B. W.; Hamilton, J. F., eds., Wiley, New York, 1986, pp. 53–189.
4. Bard, A. J.; Faulkner, L. R., *Electrochemical Methods*, Wiley: New York, 1980.
5. Macdonald, D. D., *Transient Techniques in Electrochemistry*, Plenum Press, New York, 1977.
6. Galus, Z., ''Voltammetry with Stationary and Rotated Electrodes,'' in *Physical Methods of Chemistry*, 2nd ed., Vol. II, *Electrochemical Methods*, Rossiter, B. W.; Hamilton, J. F., eds., Wiley, New York, 1986, pp. 191–272.
7. Galus, Z., *Fundamentals of Electrochemical Analysis*, 2nd ed., Ellis Horwood Ltd., Chichester (UK), 1994.
8. Kolthoff, I. M.; Lingane, J. J., *Polarography*, 2nd ed., Interscience, New York, 1952.
9. Heyrovsky, J.; Kuta, J., *Principles of Polarography*, Academic Press, New York, 1966.
10. Meites, L., *Polarographic Techniques*, 2nd ed., Interscience, New York, 1965.
11. Heineman, W. P.; Kissinger, P. T., ''Large-Amplitude Controlled-Potential Techniques,'' in *Laboratory Techniques in Electroanalytical Chemistry*, Kissinger, P. T.; Heineman, W. R., eds., Marcel Dekker, New York, 1984, pp. 51–127.
12. Adams, R. N., *Electrochemistry at Solid Electrodes*, Marcel Dekker, New York, 1969.
13. Greef, R.; Peat, R.; Peter, L. M.; Pletcher, D.; Robinson, J., *Instrumental Methods in Electrochemistry*, Ellis Horwood Ltd., Chichester, 1985.
14. Vlcek, A. A.; Volke, J.; Pospisil, L.; Kalvoda, R., ''Polarography,'' in *Physical Methods of Chemistry*, 2nd ed., Vol. II. *Electrochemical Methods*, Rossiter, B. W.; Hamilton, J. F., eds., Wiley, New York, 1986, pp. 797–886.
15. Nicholson, R. S.; Shain, I., *Anal. Chem.* **1964,** *36,* 706.
16. Nicholson, R. S., *Anal. Chem.* **1965,** *37,* 1351.
17. Brown, E. R.; Sandifer, J. R., ''Cyclic Voltammetry, AC Polarography, and Related Techniques,'' in *Physical Methods of Chemistry*, 2nd ed., Vol. II, *Electrochemical Methods*, Rossiter, B. W.; Hamilton, J. F., eds., Wiley, New York, 1986, pp. 273–432.

18. Levich, V. G., *Physiochemical Hydrodynamics*, Prentice-Hall, Englewood Cliffs, N.J., 1963.
19. Albery, W. J.; Bruckenstein, S.; Napp, D. T., *Trans. Faraday Soc.* **1966,** *62,* 1932.
20. Albery, W. J.; Hitchman M. L., *Ring-disc Electrodes*, Clarendon Press, Oxford, 1977.
21. (a) Burge, D. E., *J. Chem. Ed.* **1990,** *47,* A81; (b) Parry, E. P.; Osteryoung, R. A., *Anal. Chem.*, **1965,** *37,* 1634.
22. Dean, W. O.; Osteryoung, R. A., *Anal. Chem.* **1971,** *43,* 1879.
23. Kissinger, P. T., "Small-Amplitude and Related Controlled-Potential Techniques," in *Laboratory Techniques in Electroanalytical Chemistry*, Kissinger, P. T.; Heineman, W. R., eds., Marcel Dekker, New York, 1984, pp. 143–161.
24. Murray, R. W., "Chronoamperometry, Chronocoulometry, and Chronopotentiometry," in *Physical Methods of Chemistry*, 2nd ed., Vol. II, *Electrochemical Methods*, Rossiter, B. W.; Hamilton, J. F., eds., Wiley, New York, 1986, pp. 525–589.
25. Osteryoung, J.; Osteryoung, R. A., *Electrochim. Acta* **1971,** *16,* 525.
26. Lingane, J. J., *Electroanalytical Chemistry*, 2nd ed., Interscience, New York, 1958, pp. 351–391, 450–483.
27. Meites, L., "Controlled-Potential Electrolysis and Coulometry," in *Physical Methods of Chemistry*, 2nd ed., Vol. II, *Electrochemical Methods*, Rossiter, B. W.; Hamilton, J. F., eds., Wiley, New York, 1986, pp. 433–523.
28. Lingane, J. J., *Electroanalytical Chemistry*, 2nd ed., Interscience, New York, 1958, pp. 196–233, 392–415.
29. Lingane, J. J., *Electroanalytical Chemistry*, 2nd ed., Interscience, New York, 1958, pp. 617–638.
30. Delahay, P., "Chronoamperometry and Chronopotentiometry," in *Treatise on Analytical Chemistry*, Vol. 4, Kolthoff, I. M.; Elving, P. J., eds., Interscience, New York, 1963, Chap. 44.
31. Heineman, W. P.; Kissinger, P. T., "Large-Amplitude Controlled Current Techniques," in *Laboratory Techniques in Electroanalytical Chemistry*, Kissinger, P. T.; Heineman, W. R., eds., Marcel Dekker, New York, 1984, pp. 129–142.
32. Delahay, P., *New Instrumental Methods in Electrochemistry*, Interscience, New York, 1954, pp. 186–189.
33. Lingane, J. J., *Electroanalytical Chemistry*, 2nd ed., Interscience, New York, 1958, pp. 484–616.
34. Lingane, J. J., *Electroanalytical Chemistry*, 2nd ed., Interscience, New York, 1958, pp. 168–195.
35. Shedlovsky, T.; Shedlovsky, L., "Conductometry," in *Physical Methods of Chemistry*, Vol. 1, Part IIA, Weissberger, A.; Rossiter, B. W., eds., Interscience, New York, 1971, Chap. III.
36. Spiro, M., "Conductance and Transference Determinations," in *Physical Methods of Chemistry*, 2nd ed., Vol. II, *Electrochemical Methods*, Rossiter, B. W.; Hamilton, J. F., eds., Wiley, New York, 1986, pp. 663–796.
37. Buxton, G. V.; Greenstock, C. L.; Helman, W. P.; Ross, A. B., *J. Phys. Chem. Ref. Data.* **1988,** *17,* 513–886.

38. Parsons, R., *Handbook of Electrochemical Constants*, Butterworths, London, 1959.
39. Sawyer, D. T., *Oxygen Chemistry*, Oxford University Press, New York, 1991, Chap. 3.
40. Sawyer, D. T., *J. Phys. Chem.* **1989,** *93,* 7977.
41. Lide, D. R., ed., *CRC Handbook of Chemistry and Physics*, 71st ed., CRC, Boca Raton, Fla., 1990, pp. **10**-180, 181, **10**-210, 211.
42. Sawyer, D. T.; Roberts, J. L., Jr., *Acc. Chem. Res.*, **1988,** *21,* 469.
43. Delahay, P., *Double Layer and Electrode Kinetics*, Interscience, New York, 1965.
44. Bockris, J. O'M.; Nagg, Z., *J. Chem. Educ.* **1973,** *50,* 839.
45. Laidler, K. J., *J. Chem. Educ.* **1970,** *47,* 600.

CHAPTER 2

POTENTIOMETRIC MEASUREMENTS

2.1 INTRODUCTION

Use of the potential of a galvanic cell to measure the concentration of an electroactive species developed later than a number of other electrochemical methods. In part this was because a rational relation between the electrode potential and the concentration of an electroactive species required the development of thermodynamics, and in particular its application to electrochemical phenomena. The work of J. Willard Gibbs[1] in the 1870s provided the foundation for the Nernst equation.[2] The latter provides a quantitative relationship between potential and the ratio of concentrations for a redox couple ([*ox*]/[*red*]), and is the basis for potentiometry and potentiometric titrations.[3] The utility of potentiometric measurements for the characterization of ionic solutions was established with the invention of the glass electrode in 1909 for a selective potentiometric response to hydronium ion concentrations.[4] Another milestone in the development of potentiometric measurements was the introduction of the hydrogen electrode for the measurement of hydronium ion concentrations;[5] one of many important contributions by Professor Joel Hildebrand. Subsequent development of special glass formulations has made possible electrodes that are selective to different monovalent cations.[6–8] The idea is so attractive that intense effort has led to the development of electrodes that are selective for many cations and anions, as well as several gas- and bioselective electrodes.[9] The use of these electrodes and the potentiometric measurement of pH continue to be among the most important applications of electrochemistry.

At an early date Nernst also introduced the concept of potentiometry with polarized electrodes,[10] which, together with the many other specialized forms of potentiometric measurements for a wide range of chemical systems, has been thoroughly discussed and reviewed in the definitive monograph by Kolthoff and Furman.[11] Unfortunately, this valuable work is out of print; it is still surprisingly current, although published 63 years ago.

2.2 PRINCIPLES AND FUNDAMENTAL RELATIONS

Potentiometric measurements are based on thermodynamic relationships and more particularly the Nernst equation, which relates potential to the concentration of electroactive species. For our purposes it is most convenient to consider the redox process that occurs at a single electrode, although two electrodes are always essential for an electrochemical cell. However, by considering each electrode individually, the two electrode processes are easily combined to obtain the entire cell process. Furthermore, confusion can be minimized if the half-reactions for electrode processes are written in a consistent manner. Here, these are always reduction processes with the oxidized species reduced by n electrons to give a reduced species:

$$ox + ne^- \longrightarrow red \tag{2.1}$$

For such a half-reaction the free energy is given by the relation

$$\Delta G = \Delta G^\circ + RT \ln \frac{[red]}{[ox]} \tag{2.2}$$

where $-\Delta G$ indicates the tendency for the reaction to go to the right; R is the gas constant and in the units appropriate for electrochemistry has a value of 8.317 J mol^{-1} K^{-1}; T is the temperature of the system in kelvins (K); and the logarithmic terms in the bracketed expression represent the activities (effective concentrations) of the electroactive pair at the electrode surface. The free energy of this half-reaction is related to the electrode potential E by the expression

$$-\Delta G = nFE; \qquad -\Delta G^\circ = nFE^\circ \tag{2.3}$$

The quantity ΔG° is the free energy of the half-reaction when the activities of the reactant and product have values of unity and is directly proportional to the standard half-cell potential for the reaction as written. It also is a measure of the equilibrium constant for the half-reaction assuming the activity of electrons is unity:

$$-\Delta G^\circ = RT \ln K \tag{2.4}$$

An extensive summary of E° values is presented in the classic monograph by Latimer,[12] followed by other compilations;[13–16] that by Bard et al.[15] is the most complete.[17] Table 2.1 lists the most important couples. Standard potentials are

TABLE 2.1 Electromotive Series for Aqueous Solutions at 25°C and 1 atm

Reaction	$E°$, V versus NHE
$Li^{I}(OH_2)_4^+ + e^- \rightleftarrows Li(s) + 4\ H_2O$	−3.04
$K^{I}(OH_2)_4^+ + e^- \rightleftarrows K(s) + 4\ H_2O$	−2.92
$Rb^{I}(OH_2)_6^+ + e^- \rightleftarrows Rb(s) + 6\ H_2O$	−2.92
$Cs^{I}(OH_2)_6^+ + e^- \rightleftarrows Cs(s) + 6\ H_2O$	−2.92
$Ba^{II}(OH_2)_6^{2+} + 2\ e^- \rightleftarrows Ba(s) + 6\ H_2O$	−2.92
$Sr^{II}(OH_2)_6^{2+} + 2\ e^- \rightleftarrows Sr(s) + 6\ H_2O$	−2.89
$Ca^{II}(OH_2)_6^{2+} + 2\ e^- \rightleftarrows Ca(s) + 6\ H_2O$	−2.84
$Na^{I}(OH_2)_4^+ + e^- \rightleftarrows Na(s) + 4\ H_2O$	−2.71
$Mg^{II}(OH_2)_6^{2+} + 2\ e^- \rightleftarrows Mg(s) + 6\ H_2O$	−2.36
$Be^{II}(OH_2)_4^{2+} + 2\ e^- \rightleftarrows Be(s) + 4\ H_2O$	−1.97
$Al^{III}(OH_2)_6^{3+} + 3\ e^- \rightleftarrows Al(s) + 6\ H_2O$	−1.67
$Ti^{II}(OH_2)_6^{2+} + 2\ e^- \rightleftarrows Ti(s) + 6\ H_2O$	−1.63
$Mn^{II}(OH_2)_6^{2+} + 2\ e^- \rightleftarrows Mn(s) + 6\ H_2O$	−1.18
$V^{II}(OH_2)_6^{2+} + 2\ e^- \rightleftarrows V(s) + 6\ H_2O$	−1.13
$Zn^{II}(OH_2)_6^{2+} + 2\ e^- \rightleftarrows Zn(s) + 6\ H_2O$	−0.76
$Se(s) + 2\ e^- \rightleftarrows Se^{2-}$	−0.67
$Pb^{II}O(s) + H_2O + 2\ e^- \rightleftarrows Pb(s) + 2\ HO^-$	−0.58
$S(s) + 2\ e^- \rightleftarrows S^{2-}$	−0.45
$Fe^{II}(OH_2)_6^{2+} + 2\ e^- \rightleftarrows Fe(s) + 6\ H_2O$	−0.44
$In^{III}(OH_2)_6^{3+} + 2\ e^- \rightleftarrows In^{I}(OH_2)_6^+$	−0.44
$Cr^{III}(OH_2)_6^{3+} + e^- \rightleftarrows Cr^{II}(OH_2)_6^{2+}$	−0.42
$Cd^{II}(OH_2)_6^{2+} + 2\ e^- \rightleftarrows Cd(s) + 6\ H_2O$	−0.40
$Ti^{III}(OH_2)_6^{3+} + e^- \rightleftarrows Ti^{II}(OH_2)_6^{2+}$	−0.37
$Pb^{II}I_2(s) + 2\ e^- \rightleftarrows Pb(s) + 2\ I^-$	−0.36
$Pb^{II}(SO_4)(s) + 2\ e^- \rightleftarrows Pb(s) + SO_4^{2-}$	−0.35
$In^{III}(OH_2)_6^{3+} + 3\ e^- \rightleftarrows In(s) + 6\ H_2O$	−0.34
$Tl^{I}(OH_2)_6^+ + e^- \rightleftarrows Tl(s) + 6\ H_2O$	−0.34
$O_2(g) + e^- \rightleftarrows O_2^{-}\cdot$	−0.33
$Co^{II}(OH_2)_6^{2+} + 2\ e^- \rightleftarrows Co(s) + 6\ H_2O$	−0.28
$Pb^{II}Br_2(s) + 2\ e^- \rightleftarrows Pb(s) + 2\ Br^-$	−0.28
$Pb^{II}Cl_2(s) + 2\ e^- \rightleftarrows Pb(s) + 2\ Cl^-$	−0.27
$Ni^{II}(OH_2)_6^{2+}\ 2\ e^- \rightleftarrows Ni(s) + 6\ H_2O$	−0.26
$V^{III}(OH_2)_6^{3+} + e^- \rightleftarrows V^{II}(OH_2)_6^{2+}$	−0.25
$Ag^{I}I(s) + e^- \rightleftarrows Ag(s) + I^-$	−0.15
$Sn^{II}(OH_2)_4^{2+} + 2\ e^- \rightleftarrows Sn(s) + 4\ H_2O$	−0.14
$Pb^{II}(OH_2)_4^{2+} + 2e^- \rightleftarrows Pb(s) + 4\ H_2O$	−0.13
$In^{I}(OH_2)_6^+ + e^- \rightleftarrows In(s) + 6\ H_2O$	−0.13
$Se(s) + 2\ H_3O^+ + 2\ e^- \rightleftarrows H_2Se + 2\ H_2O$	−0.11
$Hg_2^{I}I_2(s) + 2\ e^- \rightleftarrows 2\ Hg(l) + 2\ I^-$	−0.04
$2\ H_3O^+(aq) + 2\ e^- \rightleftarrows H_2(g) + 2\ H_2O$	0.00
$Cu^{I}Br(s) + e^- \rightleftarrows Cu(s) + Br^-$	+0.03

TABLE 2.1 (*Continued*)

Reaction	$E°$, V versus NHE
$Ag^{I}Br(s) + e^- \rightleftarrows Ag(s) + Br^-$	+0.07
$Ag^{I}SCN(s) + e^- \rightleftarrows Ag(s) + SCN^-$	+0.09
$Hg^{II}O(s) + H_2O + 2\,e^- \rightleftarrows Hg(l) + 2\,HO^-$	+0.10
$Cu^{I}Cl(s) + e^- \rightleftarrows Cu(s) + Cl^-$	+0.12
$Hg_2^{I}Br_2(s) + 2\,e^- \rightleftarrows 2\,Hg(l) + 2\,Br^-$	+0.14
$S(s) + 2\,H_3O^+ + 2\,e^- \rightleftarrows H_2S + 2\,H_2O$	+0.14
$Sn^{IV}(OH_2)_4^{4+} + 2\,e^- \rightleftarrows Sn^{II}(OH_2)_4^{2+}$	+0.15
$Cu^{II}(OH_2)_4^{2+} + e^- \rightleftarrows Cu^{I}(OH_2)_4^{+}$	+0.16
$Ag^{I}Cl(s) + e^- \rightleftarrows Ag(s) + Cl^-$	+0.22
$Hg_2^{I}Cl_2(s) + 2\,e^- \rightleftarrows 2\,Hg(l) + 2\,Cl^-$	+0.27
$Ag^{I}N_3(s) + e^- \rightleftarrows Ag(s) + N_3^-$	+0.29
$V^{IV}(O)(OH_2)_4^{2+} + 2\,H_3O^+ + e^- \rightleftarrows V^{III}(OH_2)_6^{3+} + H_2O$	+0.34
$Cu^{II}(OH_2)_4^{2+} + 2\,e^- \rightleftarrows Cu(s) + 4\,H_2O$	+0.34
$Fe^{III}(CN)_6^{3-} + e^- \rightleftarrows Fe^{II}(CN)_6^{4-}$	+0.36
$O_2(g) + 2\,H_2O + 4\,e^- \rightleftarrows 4\,HO^-$	+0.40
$Ag_2^{I}CrO_4(s) + 2\,e^- \rightleftarrows 2\,Ag(s) + CrO_4^{2-}$	+0.45
$H_2SO_3 + 4\,H_3O^+ + 4\,e^- \rightleftarrows S(s) + 7\,H_2O$	+0.50
$4\,H_2SO_3 + 4\,H_3O^+ + 6\,e^- \rightleftarrows S_4O_6^{2-} + 10\,H_2O$	+0.51
$Cu^{I}(OH_2)_4^{+} + e^- \rightleftarrows Cu(s) + 4\,H_2O$	+0.52
$I_2 + 2\,e^- \rightleftarrows 2\,I^-$	+0.54
$I_3^- + 2\,e^- \rightleftarrows 3\,I^-$	+0.54
$H_3As^{V}O_4 + 2\,H_3O^+ + 2\,e^- \rightleftarrows H_3As^{III}O_3 + 3\,H_2O$	+0.56
$Mn^{VII}O_4^- + e^- \rightleftarrows Mn^{VI}O_4^{2-}$	+0.56
$Cu^{II}(OH_2)_6^{2+} + Cl^- + e^- \rightleftarrows CuCl(s) + 6\,H_2O$	+0.56
$Mn^{VII}O_4^- + 2\,H_2O + 3\,e^- \rightleftarrows Mn^{IV}O_2(s) + 4\,HO^-$	+0.60
$Hg_2^{I}SO_4(s) + 2\,e^- \rightleftarrows 2\,Hg(l) + SO_4^{2-}$	+0.61
$Cu^{II}(OH_2)_6^{2+} + Br^- + e^- \rightleftarrows CuBr(s) + 6\,H_2O$	+0.65
$O_2(g) + 2\,H_3O^+ + 2\,e^- \rightleftarrows HOOH + 2\,H_2O$	+0.69
$Tl^{III}(OH_2)_6^{3+} + 3\,e^- \rightleftarrows Tl(s) + 6\,H_2O$	+0.72
$Fe^{III}(OH_2)_6^{3+} + e^- \rightleftarrows Fe^{II}(OH_2)_6^{2+}$	+0.77
$Hg_2^{I}(OH_2)_2^{2+} + 2\,e^- \rightleftarrows 2\,Hg(l) + 2\,H_2O$	+0.80
$Ag^{I}(OH_2)_2^{+} + e^- \rightleftarrows Ag(s) + 2\,H_2O$	+0.80
$Hg^{II}(OH_2)_6^{2+} + 2e^- \rightleftarrows Hg(l) + 6\,H_2O$	+0.85
$Cu^{II}(OH_2)_6^{2+} + I^- + e^- \rightleftarrows Cu^{I}I(s) + 6\,H_2O$	+0.86
$2\,Hg^{II}(OH_2)_6^{2+} + 2\,e^- \rightleftarrows Hg_2^{I}(OH_2)_2^{2+} + 10\,H_2O$	+0.91
$Pd^{II}(OH_2)_4^{2+} + 2\,e^- \rightleftarrows Pd(s) + 4\,H_2O$	+0.91
$Hg^{II}O(s) + 2\,H_3O^+ + 2\,e^- \rightleftarrows Hg(l) + 3\,H_2O$	+0.93
$V^{V}(O)_2(OH_2)_4^{+} + 2\,H_3O^+ + e^- \rightleftarrows V^{IV}(O)(OH_2)_4^{2+} + 3\,H_2O$	+1.00
$Br_2(l) + 2\,e^- \rightleftarrows 2\,Br^-$	+1.06
$2\,IO_3^- + 12\,H_3O^+ + 10\,e^- \rightleftarrows I_2(s) + 18\,H_2O$	+1.19
$Mn^{IV}O_2(s) + 4\,H_3O^+ + 2\,e^- \rightleftarrows Mn^{II}(OH_2)_6^{2+}$	+1.23

TABLE 2.1 ***(Continued)***

Reaction	$E°$, V versus NHE
$O_2(g) + 4\ H_3O^+ + 4\ e^- \rightleftarrows 6\ H_2O$	+1.23
$Tl^{III}(OH_2)_6^{3+} + 2\ e^- \rightleftarrows Tl^{I}(OH_2)_6^+$	+1.25
$Au^{III}(OH_2)_6^{3+} + 2\ e^- \rightleftarrows Au^{I}(OH_2)_2^+ + 4\ H_2O$	+1.36
$Cl_2(g) + 2\ e^- \rightleftarrows 2\ Cl^-$	+1.36
$Cr_2^{VI}O_7^{2-} + 14\ H_3O^+ + 6\ e^- \rightleftarrows 2\ Cr^{III}(OH_2)_6^{3+} + 9\ H_2O$	+1.36
$2\ ClO_3^- + 12\ H_3O^+ + 10\ e^- \rightleftarrows Cl_2(g) + 18\ H_2O$	+1.47
$Pb^{IV}O_2(s) + 4\ H_3O^+ + 2\ e^- \rightleftarrows Pb^{II}(OH_2)_6^{2+}$	+1.47
$2\ BrO_3^- + 12\ H_3O^+ + 10\ e^- \rightleftarrows Br_2(l) + 18\ H_2O$	+1.48
$Mn^{III}(OH_2)_6^{3+} + e^- \rightleftarrows Mn^{II}(OH_2)_6^{2+}$	+1.50
$Mn^{VII}O_4^- + 8\ H_3O^+ + 5\ e^- \rightleftarrows Mn^{II}(OH_2)_6^{2+} + 6\ H_2O$	+1.51
$Au^{III}(OH_2)_6^{3+} + 3\ e^- \rightleftarrows Au(s) + 6\ H_2O$	+1.52
$Pb^{IV}O_2(s) + SO_4^{2-} + 4\ H_3O^+ + 2\ e^- \rightleftarrows Pb^{II}(SO_4)(s) + 6\ H_2O$	+1.70
$Ce^{IV}(OH_2)_6^{4+} + e^- \rightleftarrows Ce^{III}(OH_2)_6^{3+}$	+1.72
$Co^{III}(OH_2)_6^{3+} + e^- \rightleftarrows Co^{II}(OH_2)_6^{2+}$	+1.92
$F_2(g) + 2\ e^- \rightleftarrows 2\ F^-$	+2.87

thermodynamic quantities that usually are evaluated via caloriometry and the relationship of Eq. (2.3).

When Eqs. (2.2) and (2.3) are combined, the resulting Nernst expression relates the half-cell potential to the effective concentrations of the redox couple:

$$E = E° - \frac{RT}{nF} \ln \frac{[red]}{[ox]} = E° + \frac{RT}{nF} \ln \frac{[ox]}{[red]} \tag{2.5}$$

The activity of a species is indicated as the symbol of the species enclosed in brackets. This quantity is equal to the concentration of the species times a mean activity coefficient:

$$[M^{a+}] = a_{M^{a+}} = \gamma_{\pm}\, C_{M^{a+}} \tag{2.6}$$

Although there is no straightforward and convenient method for evaluating activity coefficients for individual ions, the Debye–Hückel relationship permits an evaluation of the mean activity coefficient ($\gamma_{\pm}$), for ions at low concentrations (usually <0.01 M)

$$\log \gamma_{\pm} = -0.509\, z^2 \frac{\sqrt{\mu}}{1 + \sqrt{\mu}} \tag{2.7}$$

where z is the charge on the ion and μ is the ionic strength

$$\mu = \tfrac{1}{2} \sum C_i z_i^2 \tag{2.8}$$

More rigorous methods for the calculation of aqueous activity coefficients are available.[18]

The reaction of an electrochemical cell always involves a combination of two redox half-reactions such that one species oxidizes a second species to give the respective redox products. Thus the overall cell reaction can be expressed by a balanced chemical equation

$$a\ ox_1 + b\ red_2 \longrightarrow c\ red_1 + d\ ox_2, \qquad K_{\text{equil}} \tag{2.9}$$

However, electrochemical cells are most conveniently considered as two individual half-reactions, whereby each is written as a reduction in the form indicated by Eqs. (2.1)–(2.5). When this is done and values of the appropriate quantities are inserted, a potential can be calculated for each half-cell electrode system. Then that half-cell reaction with the more positive potential will be the positive terminal in a galvanic cell and the electromotive force of that cell will be represented by the algebraic difference between the potential of the more positive half-cell and the potential of the less positive half-cell:

$$E_{\text{cell}} = E_{(\text{more positive})} - E_{(\text{less positive})} = E_1 - E_2 \tag{2.10}$$

Insertion of the appropriate forms of Eq. (2.5) into Eq. (2.10) gives an overall expression for the cell potential

$$E_{\text{cell}} = E_1^\circ - E_2^\circ + \frac{RT}{nF} \ln \frac{[ox_1]^a\ [red_2]^b}{[ox_2]^d\ [red_1]^c} \tag{2.11}$$

The equilibrium constant for the chemical reaction expressed by Eq. (2.9) is related to the difference of the standard half-cell potentials by the relation

$$\ln K_{\text{equil}} = \frac{nF}{RT}(E_1^\circ - E_2^\circ) \tag{2.12}$$

To apply potentiometric measurements to the determination of the concentration of electroactive species, a number of conditions have to be met. The basic measurement system must include an indicator electrode that is capable of monitoring the activity of the species of interest, and a reference electrode that gives a constant, known half-cell potential to which the indicator-electrode potential can be referred. The voltage resulting from the combination of these two electrodes must be measured in a manner that minimizes the amount of current drawn by the measuring system. For low-impedance electrode systems a conventional potentiometer is satisfactory. However, electrochemical measurements with high-impedance electrode systems, and in particular the glass-membrane electrode, require the use of an exceedingly high-input-impedance measuring instrument (usually an electrometer amplifier with a current drain of less than 10^{-12} A). Because of the logarithmic nature of the Nernst equation, the measuring instrumentation must have considerable sensitivity. For example, a one-electron half-reaction at 25°C gives a voltage change of 59.1 mV for a 10-fold change in the concentration of the electroactive species. Another important point is that the potential response is directly dependent on the temperature of the measuring system. Thus, if the correct temperature is not used in the Nernst expression, large absolute errors can be introduced in the measurement of the activity for an electroactive species.

2.3 ELECTRODE SYSTEMS

The indicator electrodes for potentiometric measurements traditionally have been categorized into three separate classes. "First-class" electrodes consist of a metal immersed in a solution that contains the metal ion. These electrode systems provide a direct response to the ion or species to be measured:

$$E = E° + \frac{RT}{nF} \ln [M^{n+}]; \qquad M^{n+} + n\,e^- \rightleftharpoons M(s) \tag{2.13}$$

Therefore, the primary electrode reaction includes the sensed species. Such electrodes give a direct response according to the Nernst equation for the logarithm of the activity of the species. The details of electrode fabrication and their characteristics are discussed in Chapter 5. An extensive authoritative treatment is provided by Ives and Janz.[19]

Electrodes classified as "second-class" electrode systems are those in which the electrode is in direct contact with a slightly soluble salt of the electroactive species such that the potentiometric response is indicative of the concentration of the inactive anion species. Thus the silver/silver-chloride electrode system, which is representative of this class of electrodes, gives a potential response that is directly related to the logarithm of the chloride ion activity

$$E = E° - \frac{RT}{nF} \ln [Cl^-]; \qquad AgCl(s) + e^- \rightleftharpoons Ag(s) + Cl^- \tag{2.14}$$

even though it is not the electroactive species. This is true because the chloride ion concentration, through the solubility product, controls the activity of the silver ion, which is measured directly by the potentiometric silver-electrode system.

"Third-class" electrodes are really a specialized case of second-class electrodes. They consist of the metal being in direct contact with a slightly soluble salt of the metal, which is then used to monitor the activity of an electroinactive metal ion in equilibrium with a more soluble salt that includes the same anion as the electrode–salt system. For example, the concentration of calcium ions in equilibrium with solid calcium oxalate may be monitored using a silver/silver oxalate electrode system. The concentration of calcium ion affects the concentration of oxalate ion, which in turn controls the concentration of silver ion; the latter is monitored by the potentiometric silver-electrode system. A second and extremely useful example of a third-class electrode is the mercury/mercury(II)–EDTA (ethylenediaminetetraacetic acid) electrode system,[20,21] which is used as a sensing system for the potentiometric titration of electroinactive metal ions with EDTA. Because the mercury(II)–EDTA complex is one of the most stable of those encompassing the common divalent metal ions it is the least dissociated system. Hence, when calcium ion is titrated with EDTA, the concentration of calcium ion controls the equilibrium concentration of the EDTA anion in solution, which in turn directly controls the free concentration of mercuric ion. The latter is monitored by a mercury-electrode system to give a direct measure of the calcium ion concentration. This type of system can be applied to most of the divalent ions that form moderately strong complexes with EDTA. It is extremely useful in the development of potentiometric titration systems for mixtures of ions.

Because any potentiometric electrode system ultimately must have a redox couple (or an ion-exchange process in the case of membrane electrodes) for a meaningful response, the most common form of potentiometric electrode systems involves oxidation–reduction processes. Hence, to monitor the activity of ferric ion [iron(III)], an excess of ferrous iron [iron(II)] is added such that the concentration of this species remains constant to give a direct Nernstian response for the activity of iron(III). For such redox couples the most common electrode system has been the platinum electrode. This tradition has come about primarily because of the historic belief that the platinum electrode is totally inert and involves only the pure metal as a surface. However, during the past decade it has become evident that platinum electrodes are not as inert as long believed and that their potentiometric response is frequently dependent on the history of the surface and the extent of its activation. The evidence is convincing that platinum electrodes, and in all probability all metal electrodes, are covered with an oxide film that changes its characteristics with time. Nonetheless, the platinum electrode continues to enjoy wide popularity as an "inert" indicator of redox reactions and of the activities of the ions involved in such reactions.

For many systems the gold electrode is as satisfactory as the platinum electrode. Both rhodium and palladium as well as carbon have been used for specialized systems as "inert" potentiometric electrodes.

For those redox couples that involve a metal ion plus the metal, the logical electrode system is the metal itself. In other words, if the measured quantity is to be cupric ion [copper(II)], a practical indicator electrode is a piece of copper metal. All second-class electrodes involve an active metal in combination with an insoluble compound or salt. Thus, the silver/silver chloride electrode actually is a silver/silver ion electrode system that incorporates the means to control the silver ion concentration through the chloride ion concentration [Eq. (2.14)]. A related form of this is the antimony electrode, which involves antimony and its oxide (an adherent film on the surface of the antimony–metal electrode) such that the activity of antimony ion is controlled by the pH of the solution. This is an illustration of a second-class electrode that is used to monitor hydronium-ion (H_3O^+) concentration [or more properly hydroxide ion (HO^-) concentration].

On occasion tungsten-wire electrodes are used to monitor a redox reaction. These usually are oxide-covered systems that respond to the activity of the species of a redox couple; however, the response seldom obeys the Nernst equation.

A number of the most common potentiometric electrode systems and their applications are summarized in Table 2.2. Additional information is available in the biennial reviews of *Analytical Chemistry* within the section on potentiometric measurements.

One of the most important and extensively used indicator-electrode systems is the glass-membrane electrode that is used to monitor hydronium ion activity. Although developed in 1909, it did not become popular until reliable electrometer amplifiers were developed in the 1930s. Figure 2.1 gives a schematic representation of this electrode and indicates that the primary electrode system is a silver/silver chloride (or mercury/mercurous chloride) electrode in contact with a known and fixed concentration of hydrochloric acid (usually about 0.1 M). When the outside surface of the glass membrane is exposed to an ionic solution, a response for the hydronium ion activity $[H_3O^+]$ is obtained that follows the Nernst expression. Although there has been considerable debate about the mechanism of response for the glass electrode, the current thinking invokes an ion-exchange process that involves the hydroxyl groups on the surface of the glass. Thus, the population of protons on the outside surface on the membrane affects the population on the inside of the membrane, which generates a membrane potential that is indicated by the silver/silver chloride. At one time there was a belief that hydronium ions actually penetrated the glass membrane from the outside to the interior. However, experiments with labeled systems establish that this is not true. Further support for the ion-exchange mechanism is provided by the realization that glass electrodes are not specific for hydronium ions, but give only a selective response. In other words, other

TABLE 2.2 Potentiometric Electrode Systems

Electrode	Couple	Application
$Pt/H_2, H_3O^+$	$2\ H_3O^+ + 2\ e^- \rightleftarrows H_2 + H_2O$	pH; P_{H_2}
$Pt/M^{m+}, M^{(m-n)+}$	$M^{m+} + n\ e^- \rightleftarrows M^{(m-n)+}$	$[M^{m+}]$; $[M^{(m-n)+}]$
M/M^{n+}	$M^{n+} + n\ e^- \rightleftarrows M(s)$	$[M^{m+}]$
$Ag/AgX/X^-$	$AgX(s) + e^- \rightleftarrows Ag(s) + X^-$	$[X^-]$
$Hg/Hg_2Cl_2/Cl^-$	$Hg_2Cl_2(s) + 2\ e^- \rightleftarrows 2\ Hg(s) + 2\ Cl^-$	$[Cl^-]$
$Ag/Ag_2C_2O_4/CaC_2O_4/C_2O_4^{2-}$	$Ag_2C_2O_4(s) + 2\ e^- \rightleftarrows 2\ Ag(s) + C_2O_4^{2-}$	$[C_2O_4^{2-}]$; $[Ca^{2+}]$
	$Ca^{2+} + C_2O_4^{2-} \rightleftarrows CaC_2O_4(s)$	
$Hg/HgY^{2-}, M^{n+}$	$HgY^{2-} + 2\ e^- \rightleftarrows Hg(s) + Y^{4-}$	$[Y^{4-}]$; $[M^{n+}]$
	$M^{n+} + Y^{4-} \rightleftarrows MY^{(n-4)}$	Y = EDTA

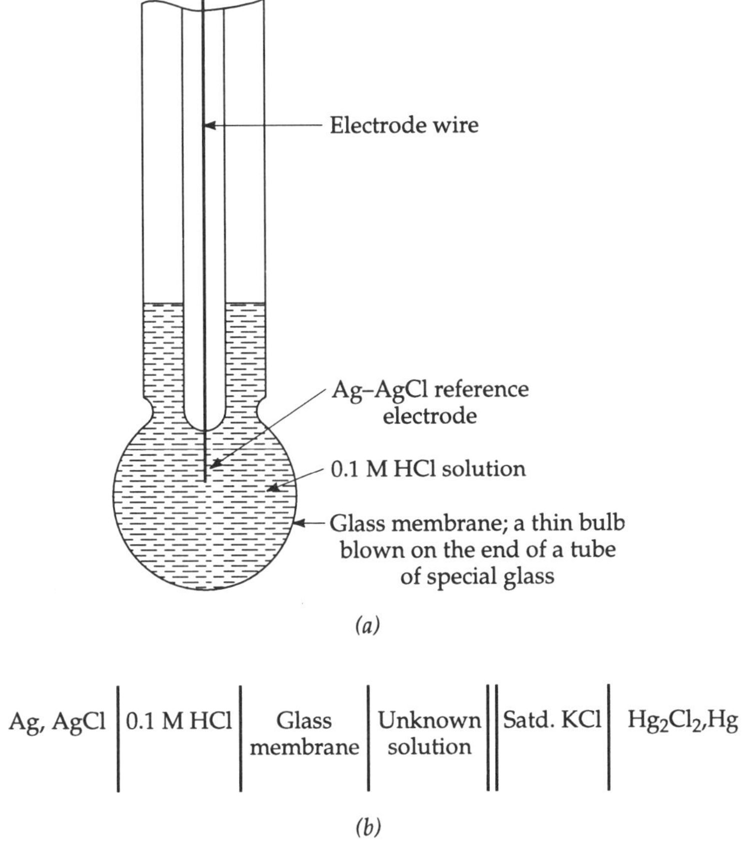

Figure 2.1 Glass electrode and its cell schematic in association with a reference electrode.

metal ions, in particular those of sodium and lithium, cause a response from glass electrodes through an equivalent ion-exchange process.

The unwanted response of glass electrodes to metal ions, particularly alkali metal ions, has prompted the development of specialized glass membranes that have an enhanced selective response for monovalent cations. A parallel approach has been in the development of ion-exchange membranes (prepared from organic, polymer-based ion-exchange resins) that give a selective response to specific cations and anions. The main advantages of the membranes are resistance to high acid and alkali concentrations and high conductivity. The membranes, however, show little selectivity between ions. A related form of the membrane electrode is the inorganic-crystal electrode. An example is the lanthanum fluoride (LaF_3) electrode that senses fluoride ion concentrations, which has become the standard sensor for fluoride determinations in water analysis. It represents a combination of a membrane electrode with ion-exchange characteristics and an incomplete form of a second-class electrode in which lanthanum fluoride is the insoluble material that responds to the free fluoride ion concentration in the test sample. Other examples of this form of electrode include AgX/Ag_2S and MS/Ag_2S electrodes, which give response to monovalent anions and divalent cations, respectively.

In contrast to solid-membrane electrodes, liquid-membrane electrodes can extract counterions from the solution phase into the membrane phase. Selectivity is provided by the charged nature of the membrane carriers and arises from the competitive degree of extractability of various counterions. Totally liquid systems can be employed but are impractical. Instead, a porous support or an inert polymer support is used in most commercial electrodes.

These and other ion-selective electrodes are discussed in Chapter 5. A large selection of such electrodes is commercially available, and the biennial reviews of *Analytical Chemistry* include a section that describes new types of electrodes and their applications.

Gas-selective electrodes are a particularly important application of the glass electrode. For example, the carbon dioxide electrode is a self-contained system with a glass electrode and a concentric silver–silver chloride electrode enclosed by a CO_2 permeable membrane. The latter holds a thin film of bicarbonate solution in contact with the glass membrane, which provides a junction to the silver/silver chloride reference electrode. The electrode, which is illustrated schematically by Figure 2.2, has found extensive application in monitoring CO_2 levels in blood and probably will find increasing application in other systems that require continuous measurement of CO_2 partial pressures. The electrode response is based on the reaction

$$CO_2 + 2\,H_2O \rightleftharpoons H_3O^+ + HOC(O)O^- \tag{2.15}$$

such that changes in the partial pressure of carbon dioxide cause an attendant change in the concentration of hydronium ion. Thus, with a fixed concentration of bicarbonate, the electrode provides a direct potentiometric response to the partial pressure of carbon dioxide. Other gas-monitoring electrode systems based on similar processes should be possible. For example, an ammonia (NH_3) electrode might well be developed with the converse of the reactions indicated for the CO_2 electrode. Thus an ammonium ion (NH_4^+) electrolyte would be used such that changes in pH would be proportional to changes in the partial pressure of NH_3. These electrodes as well as SO_x-, NO_x-, HF-, H_2S-, and HCN-sensing electrodes are commercially available.

The gas-sensing electrodes also are used for the potentiometric measurement of biologically important species. An enzyme is immobilized at or near the gas probe. The gas sensor measures the amount of characteristic gas produced by the reaction of the analyzed substance with the enzyme. For example, an enzyme electrode for urea [$NH_2C(O)NH_2$] determination is constructed by the immobilization of *urease* onto the surface of an ammonia-selective electrode. When the electrode is inserted into a solution that contains urea, the enzyme catalyzes its conversion to ammonia:

$$NH_2CONH_2 + H_2O \xrightarrow{\text{urease}} 2\,NH_3 + CO_2 \tag{2.16}$$

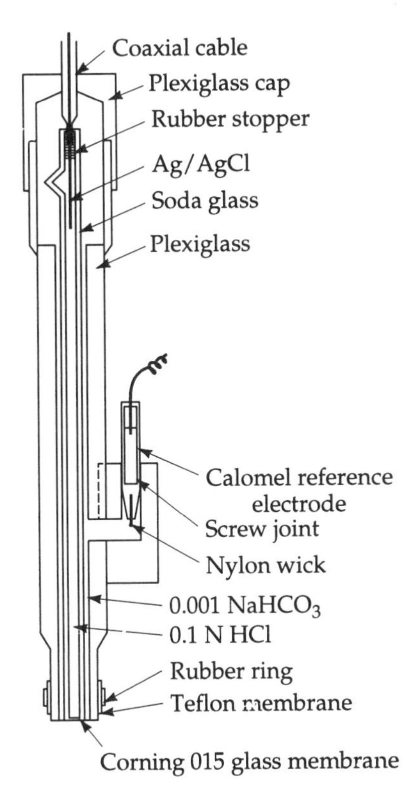

Figure 2.2 Potentiometric CO_2 electrode. Response proportional to that of p_{CO_2} $(-\log[CO_2])$.

The generated ammonia diffuses through the gas-permeable membrane and is detected by the probe of the electrode. A steady-state signal is reached when the rate of NH_3 diffusion from the electrode equals the rate of its generation. Another example of an enzyme electrode is one sensitive to glucose; its surface contains immobilized *glucose oxidase*. The latter catalyzes the oxidation of glucose by dioxygen to produce gluconic acid and hydrogen peroxide. The choice of membrane system depends on whether dioxygen or hydrogen peroxide is to be determined. Probes of this type have been developed to measure the concentrations of creatinine, glutamine, histidine, phenylalanine, amygdalin, adenosine, adenosine monophosphate, tyrosine, acetylcholine, and other biological molecules. Several reviews of enzyme electrodes are available.[22–24]

In general a necessary part of a potentiometric measurement is the coupling of a reference electrode to the indicating electrode. The ideal reference electrode has a number of important characteristics: (1) a reproducible potential, (2) a low-temperature coefficient, (3) the capacity to remain unpolarized when small currents are drawn, and (4) inertness to the sample solution. If the reference electrode must be prepared in the laboratory, a convenient and reproducible system is desirable.

Although the standard half-cell reactions are all referenced to the standard hydrogen electrode (NHE or SHE), this is an exceedingly awkward reference electrode. It has been selected because its potential falls in the middle of those for the most common half-reactions in water and because highly reproducible potentials can be duplicated with rigorous care by equally careful workers in other laboratories. Furthermore, because it consists of a platinized platinum electrode over which hydrogen gas is passed in combination with a known activity of hydronium ion, it can be combined in a number of cases with other half-cells without a liquid junction. These three factors undoubtedly justify its selection as the ultimate reference electrode for fundamental measurements. Figure 5.8 illustrates a common form of the hydrogen electrode. The electrode reaction is a typical redox half-reaction that includes the oxidized and reduced forms of hydrogen. By controlling the hydrogen partial pressure at a fixed level, this becomes an indicating electrode for hydronium ion activity. Conversely, by controlling the activity of hydronium ion in the sample solution, one might suppose that this electrode would indicate partial pressures of dissolved hydrogen. Although this is true, in principle, the platinized platinum surface gives a very slow response to changes in hydrogen partial pressure; this is a major reason for using a platinized electrode when its main function is to monitor hydronium ion activity.

For most potentiometric measurements either the saturated calomel reference electrode or the silver/silver chloride reference electrode are used. These electrodes can be made compact, are easily produced, and provide reference potentials that do not vary more than a few millivolts. The discussion in Chapter 5 outlines their characteristics, preparation, and temperature coefficients. The silver/silver chloride electrode also finds application in nonaqueous titrations, although some solvents cause the silver chloride film to become soluble. Some have utilized reference electrodes in nonaqueous solvents that are based on zinc or silver couples. From our own experience, aqueous reference electrodes are as convenient for nonaqueous systems as are any of the prototypes that have been developed to date. When there is a need to rigorously exclude water, double-salt bridges (aqueous/nonaqueous) are a convenient solution. This is true even though they involve a liquid junction between the aqueous electrolyte system and the nonaqueous solvent system of the sample solution. The use of conventional reference electrodes does cause some difficulties if the electrolyte of the reference electrode is insoluble in the sample solution. Hence the use of a calomel electrode saturated with potassium chloride in conjunction with a sample solution that contains perchlorate ion can cause erratic measurements due to the precipitation of potassium perchlorate at the junction. Such difficulties normally can be eliminated by using a double junction that inserts another inert electrolyte solution between the reference electrode and the sample solution (e.g., a sodium chloride solution).

For measurement of redox couples, a frequently overlooked but convenient reference electrode is a conventional glass pH electrode (assuming that the sample solution system contains a constant level of acidity). Such an electrode provides an extremely inert and stable reference potential that is completely indifferent to most redox species. However, the glass electrode requires the use of an electrometer amplifier such as that contained in pH meters.

Where potentiometric measurements are used to monitor a titration rather than as an absolute measure of a constituent's activity, a number of other monitoring systems that are based on potentiometric principles are possible. For example, if a platinum electrode is coupled with a tungsten wire, sharp potential breaks often occur at the equivalence point for a redox titration. This undoubtedly comes about because the platinum electrode behaves in a reversible Nernstian fashion, while the tungsten electrode tends to be irreversible and unresponsive to changes in the activity of redox ions. Many other bimetallic pairs have been used, but the platinum–tungsten combination is the most popular and provides a highly inert electrode pair. Another approach for monitoring potentiometric titrations is to obtain a differential-type measurement by using two identical electrodes. For example, a pair of platinum electrodes (with one of them shielded from the sample solution when an increment of titrant is added) give a differential potential step per increment, which becomes a maximum at the equivalence point. This can be accomplished by inserting one of the platinum electrodes in a shielded compartment such that mixing with the sample solution is slow. An alternative form is to immerse the tip of the burette in the sample solution and have one of the electrodes inserted in the bore of the tip itself. An elegant version of the differential titration system is the Pinkhoff–Treadwell method, which uses as a "reference electrode" an indicator electrode in a solution with the exact composition that will exist at the equivalence point of the titration. This is connected by a salt bridge to the sample solution where an identical indicator electrode is used to monitor the sample. The measuring circuit is simply a sensitive galvanometer or voltmeter that monitors the signal and indicates when it becomes zero at the equivalence point. If current is being measured, the direction of current flow will change at the equivalence point. Extremely precise results are possible under ideal conditions with this simple measuring system. Figure 2.3 illustrates a differential titration system that utilizes a medicine dropper.

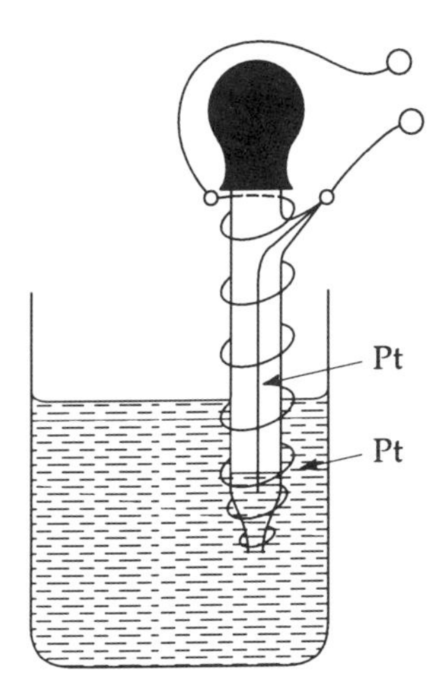

Figure 2.3 Differential potentiometric electrode system for titrations. After each addition of titrant and reading, the solution in the dropper is exchanged with the bulk solution.

Another approach to differential measurements is the use of two identical electrodes in the sample solution with a small polarizing current passed between them, usually 5 or 10 μA. Most modern pH meters have provision for a polarizing current built into them via appropriate connections at the back of the meter. With dual-polarized electrodes (using a constant current), there generally is a sharp change in potential at the equivalence point. This results because either a reversible couple is destroyed during the course of the titration such that a much higher potential is generated at one of the electrodes, or there is an absence of a reversible couple for one of the electrodes until the equivalence point is reached. A common application of the dual-polarized potentiometric indicating system is in the Karl Fischer titration (for the assay of H_2O). Up to the equivalence point, there is a large potential on the indicating meter that decreases sharply at the equivalence point because an excess of iodine is introduced into the sample solution. The I_2/I^- couple is reversible, and the anode clearly has plenty of iodide available to be oxidized to iodine prior to the equivalence point. However, the cathode reaction cannot be the reduction of iodine to iodide because none is available prior to the equivalence point. Hence some other species must be reduced to pass the small current; in general, this is the solvent that requires a high potential. When an excess of iodine is introduced into the sample solution, the potential drops to almost zero to provide a dramatic indication of the equivalence point. Undoubtedly dual-polarized potentiometric endpoint detection systems should be used more widely than they have been. Unfortunately, predictability is limited, and one must try each system to establish whether the method is applicable. Other examples of their application are discussed in Chapter 4.

2.4 APPLICATIONS OF POTENTIOMETRY

Although potentiometric measurements frequently are applied as a means of endpoint detection in potentiometric titrations, the purpose of this volume is not to review the many titration procedures that utilize potentiometric measurements. However, a brief summary of the species that can be monitored through the use of potentiometric measurements will illustrate the potential applications.

Although all potentiometric measurements (except those involving membrane electrodes) ultimately are based on a redox couple, the method can be applied to oxidation–reduction processes, acid–base processes, precipitation processes, and metal ion complexation processes. Measurements that involve a component of a redox couple require that either the oxidized or reduced conjugate of the species to be measured be maintained at a constant and known activity at the electrode. If the goal is to measure the activity of silver ion in a solution, then a silver wire coupled to the appropriate reference electrodes makes an ideal potentiometric system. Likewise, if the goal is to monitor iron(III) concentrations with a platinum electrode, a known concentration of iron(II) must be present in the sample solution such that potential varies only with the iron(II) concentration.

Table 2.1 summarizes a number of redox couples that are well behaved in aqueous solutions and provide a means for monitoring the indicated species by potentiometric measurements. This can be either in the form of monitoring a titration or as a direct absolute measurement of activity. Although the tabulations of standard potentials[12–17] imply that the listing should be much more comprehensive, most of the couples tabulated are not well behaved in an electrochemical sense and do not provide a Nernstian response under normal laboratory conditions. The vast majority of the data tabulated is based on other than electrochemical measurements.

The major application of the potentiometric method is for acid–base measurements in both aqueous and nonaqueous solvent systems. Although the glass electrode is universally the most common indicating electrode system for such measurements, many other electrodes have been developed. However, except in extremely specialized circumstances, none of these provides the reliability and precision that is afforded by the glass electrode. In the absence of interfering substances the quinhydrone electrode (an equimolar combination of quinone and hydroquinone with a gold-foil electrode) provides a simple monitoring system for measurements up to pH 8. However, the presence of oxidizing or strongly reducing ions in the sample system will interfere, as is true for the hydrogen gas electrode and most other systems that are an alternative to the glass electrode. Table 2.1 indicates that a number of redox couples involve hydronium ion. Any of these can be used if the oxidized and reduced species are introduced into the sample solution in controlled amounts. For such conditions the response of the electrode will be dependent on the hydronium ion activity. The iodate–iodide couple is an example of such a system.

For adverse conditions (in terms of either temperature or vibration) the antimony electrode has proved useful, particularly for industrial processes with extreme environmental problems. The electrode is not particularly reliable for precise measurements, but its simple form (consisting of antimony metal embedded in an insulating material) allows pH measurements under such adverse conditions. The principle of the electrode is based on a half-reaction whereby the metal and its metal oxide are both insoluble and the electrode's response is dependent on hydronium ion activity:

$$Sb_2O_3 + 6\ H_3O^+ + 6\ e^- \longrightarrow 2\ Sb + 9\ H_2O \tag{2.17}$$

Potentiometric redox measurements are often performed in nonaqueous or mixed-solvent media. For such solvents various potentiometric sensors have been developed, which, under rigorously controlled conditions, give a Nernstian response over a wide ranges of activities, particularly in buffered solutions. There are some experimental limitations, such as with solvent purification and handling or use of a reference electrode without salt bridges, but there also are important advantages. Solutes may be more soluble in such media, and redox properties of the species may be altered in comparison with aqueous solutions. pH measurements in nonaqueous solvents almost without exception use the glass electrode in combination with an appropriate reference electrode, frequently the silver/silver chloride electrode. In general, the response of the glass electrode follows the Nernst expression in nonaqueous solvents and is an accurate representation of the changes in activity of hydronium ion. Unfortunately, few, if any, standard buffers are available to calibrate pH meters for nonaqueous measurements. Thus nonaqueous pH measurements are meaningful only for monitoring the course of an acid–base titration or relative to some reference measurement made within the individual laboratory. Little, if any, confidence can be attached to absolute pH measurements in nonaqueous systems. The application of ion-selective electrodes in nonaqueous media has been limited, but the response for several cations and anions is usable, especially when 10–20% of water is added to pure nonaqueous solvent. Limited use of liquid-membrane electrodes in such media arises from the solubility of electrode-system components in organic media. A recent review[25] describes the application of potentiometric sensors to the characterization of nonaqueous solvents.

Second-class electrodes—that is, those whose response is dependent on the change in concentration of an anion that gives an insoluble salt with the metal ion of the indicator electrode—provide a general means for monitoring the concentrations of anions. Some of those half-reactions, which are well-behaved electrochemically and provide means for the potentiometric monitoring of anion species, are summarized in Table 2.1. This table also includes a tabulation of redox reactions that are useful in monitoring the concentration of ligands that can complex metal ions. Consideration of these indicates one of the difficulties with absolute potentiometric measurements as a measure of metal ion activity. If one is concerned purely with the actual activity of free metal ion, these measurements are meaningful. However, if the measurement is taken to represent the total metal ion content of the solution, both as a free solvated ion and as its various complexes, then highly erroneous conclusions can be made. Thus the ability to monitor the concentration of ligands should be recognized as a pitfall if one does not take account of this in the use of potential measurements to monitor metal ion concentrations.

A recent and rapidly developing extension of potentiometry is in the area of membrane-type indicator electrodes. These include (1) specialized glass electrodes that respond to ions other than the hydronium ion, (2) ion-exchange membranes, and (3) single-inorganic-crystal membranes. Each year the selectivity and reliability are improved for this important class of electrode. In particular, the development of ion-exchange membranes that provide selective response for a number of anions has made new areas of analysis amenable to potentiometric measurements. This has been particularly important for the biomedical field where nondestructive, highly specific potentiometric measurements are desirable. Furthermore, the potentiometric method, because of its continuous nature, is particularly attractive to those concerned with in vivo monitoring of biological substances.

Table 5.10 summarizes the presently available electrodes categorized as glass, ion-exchange membrane, crystal membrane, and liquid membrane. These electrodes can be used either for direct potentiometric measurements of ionic activity after calibration of the Nernst expression for the particular electrode or to monitor a potentiometric titration when a selected reaction that involves the monitored ion is available. Table 5.10 also indicates the common interfering ions. Several instrument companies are endeavoring to develop potentiometric-membrane electrodes to monitor directly ions in body fluids.

Because potentiometry (through the Nernst equation) gives a response that is proportional to the logarithm of the activity of the electroactive ion, the accuracy and precision are more limited than for many methods that give a direct proportional response. Thus, for a one-electron redox process an order-of-magnitude change in activity gives a potential change of 59.1 mV (at room temperature), a 10% change in activity gives a change of 2.5 mV, and a 1% change in activity gives a change of only 0.25 mV. This has prompted many efforts to improve the accuracy of potentiometry and has led to the development of differential methods of various types. One of the most effective of these is the method of Malmstadt, which utilizes two identical indicator electrodes: one immersed in the sample solution and the other immersed in a reference solution containing pure solvent and an inert electrolyte. To make the measurement a sensitive voltage-measuring device is attached to the two indicator electrodes and a standardized solution of the constituent species is added to the reference half-cell with a precision micrometer syringe until the indicated voltage reaches a null point. By knowing the concentration of the added constituent and the volume, one can compute the concentration of the constituent ion in the unknown solution. This approach allows dilute sample systems to be analyzed, and the level of precision frequently is higher than by direct potentiometric measurements.

Potentiometry has found extensive application over the past half-century as a means to evaluate various thermodynamic parameters. Although this is not the major application of the technique today, it still provides one of the most convenient and reliable approaches to the evaluation of thermodynamic quantities. In particular, the activity coefficients of electroactive species can be evaluated directly through the use of the Nernst equation (for species that give a reversible electrochemical response). Thus, if an electrochemical system is used without a junction potential and with a reference electrode that has a well-established potential, then potentiometric measurement of the constituent species at a known concentration provides a direct measure of its activity. This provides a direct means for evaluation of the activity coefficient (assuming that the standard potential is known accurately for the constituent half-reaction). If the standard half-reaction potential is not available, it must be evaluated under conditions where the activity coefficient can be determined by the Debye–Hückel equation.

Another important application is the use of potentiometric measurements for the evaluation of thermodynamic equilibrium constants. In particular, the dissociation constants for weak acids and weak bases in a variety of solvents are

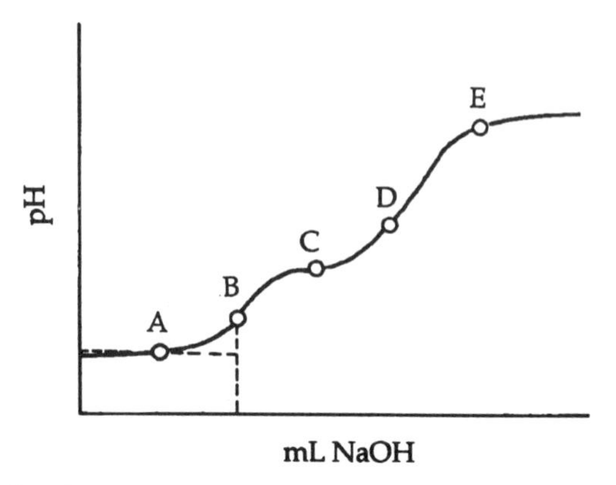

Figure 2.4 Titration curve for a polyprotic acid titrated with NaOH.

evaluated conveniently with a pH–electrode measuring system. The most precise approach is to perform an acid–base titration such that the titration curve can be recorded. Figure 2.4 illustrates the titration curve for a polyprotic acid that has two breaks; actually it is for orthophosphoric acid and illustrates the principles for the evaluation of such constants.

Point *A* on the titration curve is half-way to the first equivalence point with 0.5 mol of base added per mole of phosphoric acid. If the pH at this point were between pH > 5 and pH < 9, the concentration of H_3PO_4 and of $H_2PO_4^-$ ion could be assumed to be equal. However, because the pH is approximately pH 3, the concentration of H_3PO_4 at this point is less than that of the $H_2PO_4^-$ ion by an amount equivalent to about twice the hydronium ion concentration. Thus, the correct expressions at the half-equivalence point for the equilibrium concentrations are

$$[H_2PO_4^-] = \frac{M_{HO^-} V_{HO^-}}{v + V_{HO^-}} + [H_3O^+] \tag{2.18}$$

$$[H_3PO_4] = \frac{M_{HO^-} V_{HO^-}}{v + V_{HO^-}} - [H_3O^+] \tag{2.19}$$

$$K_1 = \frac{[H_3O^+]\,[H_2PO_4^-]}{[H_3PO_4]}$$

$$= [H_3O^+]\,\frac{[(M_{HO^-} V_{HO^-})/(v + V_{HO^-}] + [H_3O^+]}{[(M_{HO^-} V_{HO^-})/(v + V_{HO^-})] - [H_3O^+]} \tag{2.20}$$

where M_{HO^-} is the molarity of the base, V_{HO^-} its volume, and v the initial sample volume.

At point *C* on the titration curve, 1.5 mol of base have been added per mole of phosphoric acid and the pH indicates that the concentration of $H_2PO_4^-$ ion equals the concentration of HPO_4^{2-} ion. Thus the correct expression for the equilibrium constant at this point is

$$K_2 = \frac{[H_3O^+]\,[HPO_4^{2-}]}{[H_2PO_4^-]} = [H_3O^+] \tag{2.21}$$

At point E on the titration curve, 2.5 of base per mole of phosphoric acid have been added. Again, it is erroneous to assume that the PO_4^{3-} ion concentration equals the HPO_4^{2-} ion concentration at this half-equivalence point because a significant fraction of the base added beyond the second equivalence point (D in Figure 2.4) has not reacted with the HPO_4^{2-} ion. Thus, the proper expression for the third dissociation constant for phosphoric acid is given by the relations

$$[PO_4^{3-}] = \frac{M_{HO^-}\,[V_{HO^-} - (V_{HO^-})_D]}{V_{HO^-} + v} - [HO^-] \tag{2.22}$$

$$[HPO_4^{2-}] = \frac{M_{HO^-}\,[V_{HO^-} - (V_{HO^-})_D]}{V_{HO^-} + v} + [HO^-] \tag{2.23}$$

$$K_3 = \frac{[H_3O^+]\,[PO_4^{3-}]}{[HPO_4^{2-}]}$$

$$= [H_3O^+]\,\frac{\{M_{HO^-}\,[V_{HO^-} - (V_{HO^-})_D]/(V_{HO^-} + v)\} - [HO^-]}{(\{M_{HO^-}\,[V_{HO^-} - (V_{HO^-})_D]\}/(V_{HO^-} + v)) + [HO^-]} \tag{2.24}$$

Similar expressions can be developed for other weak acids and weak bases to permit the evaluation of their dissociation constants. The constants normally are evaluated at points on the titration curve where the pH is changing slowly relative to the added titrant; equivalence points are to be avoided because of the significant experimental errors that are possible at this point on the titration curve.

Obviously one could measure the pH of a known concentration of a weak acid and obtain a value of its hydronium ion activity, which would permit a direct evaluation of its dissociation constant. However, this would be a one-point evaluation and subject to greater errors than by titrating the acid halfway to the equivalence point. The latter approach uses a well-buffered region where the pH measurement represents the average of a large number of data points. Similar arguments can be made for the evaluation of solubility products and stability constants of complex ions. The appropriate expression for the evaluation of solubility products again is based on the half-equivalence point of the titration curve for the particular precipitation reaction [$Ag^I(OH_2)_2^+$ represents the titrant]:

$$Ag^I(OH_2)_2^+ + Cl^- \rightleftharpoons AgCl(s) + 2\ H_2O \tag{2.25}$$

$$[Cl^-] = \frac{M_{Cl^-}\ V_{Cl^-}}{V_{Cl^-} + v} + [Ag^I(OH_2)_2^+] \tag{2.26}$$

$$K_{sp} = [Ag^I(OH_2)_2^+]\ [Cl^-] = [Ag^I(OH_2)_2^+] \left[\frac{M_{Cl^-}\ V_{Cl^-}}{V_{Cl^-} + v} + [Ag^I(OH_2)_2^+] \right] \tag{2.27}$$

In the use of potentiometry for the evaluation of stability constants for complex ions, the expressions can become extremely complicated if multiequilibria are present. For a simple one-to-one complex a direct potentiometric titration curve again provides the most satisfactory route to an accurate evaluation of the constant. The curve looks similar to that for an acid–base titration, and the appropriate point to pick is the half-equivalence point. If the complex is extremely stable, then the amount of free metal ion at this point on the titration curve (ligand titrated with metal ion) is sufficiently low that it can be disregarded. If not, it must be handled in a way similar to the first point on the titration curve for phosphoric acid. Assuming that it is a stable complex, at the first half-equivalence point the concentration of complexed metal ion will be equivalent to that of the free ligand. The potential will give a direct measure of the free metal ion and allow the stability constant for the complex to be evaluated at the half-equivalence point:

$$[Ag^I(en)^+] = \frac{M_{Ag^+}\ V_{Ag^+}}{V_{Ag^+} + v} - [Ag^I(OH_2)_2^+] \tag{2.28}$$

$$[en] = \frac{M_{Ag^+}\ V_{Ag^+}}{V_{Ag^+} + v} + [Ag^I(OH_2)_2^+] \tag{2.29}$$

and

$$K_f = \frac{[Ag^I(en)^+]}{[Ag^I(OH_2)_2^+]\ [en]} \tag{2.30}$$

For multistep complexation reactions and for ligands that are themselves weak acids, extremely involved calculations are necessary for the evaluation of the equilibrium expression from the individual species involved in the competing equilibria. These normally have to be solved by a graphical method or by computer techniques.[26,27] Discussion of these calculations at this point is beyond the scope of this book. However, those who are interested will find adequate discussions in the many books on coordination chemistry, chelate chemistry, and the study and evaluation of the stability constants of complex ions.[20,21,28-30] The general approach is the same as outlined here; namely, that a titration curve is performed in which the concentration or activity of the substituent species is monitored by potentiometric measurement.

TABLE 2.3 Formal Potentials

Reaction	$E^{\circ\prime}$ V versus NHE	Solution Composition
$Ag^{I}(OH)_2^+ + e^- \rightleftarrows Ag(s) + 2\ H_2O$	+0.77	1 M H_2SO_4
	+0.79	1 M $HClO_4$
$Ag^{II}(OH_2)_4^{2+} + e^- \rightleftarrows Ag^{I}(OH_2)_2^+ + 2\ H_2O$	+1.93	4 M HNO_3
	+2.00	4 M $HClO_4$
$Ag^{I}I(s) + e^- \rightleftarrows Ag(s) + I^-$	−0.14	1 M KI
$H_3As^{V}O_4 + 2\ H_3O^+ + 2\ e^- \rightleftarrows HAs^{III}O_2 + 4\ H_2O$	+0.58	1 M HCl or 1 M $HClO_4$
$Au^{I}Cl_2^- + e^- \rightleftarrows Au(s) + 2\ Cl^-$	+1.11	1 M Cl^-
$Au^{III}Cl_4^- + 2\ e^- \rightleftarrows Au^{I}Cl_2^- + 2\ Cl^-$	+0.93	1 M HCl
$Ce(IV) + e^- \rightleftarrows Ce(III)$	+0.06	2.5 M K_2CO_3
	+1.28	1 M HCl
	+1.44	1 M H_2SO_4
	+1.60	1 M HNO_3
	+1.70	1 M $HClO_4$
$Co^{III}(OH_2)_6^{3+} + e^- \rightleftarrows Co^{II}(OH_2)_6^{2+}$	+1.82	8 M H_2SO_4
	+1.85	4 M HNO_3
$Co^{III}(NH_3)_6^{3+} + e^- \rightleftarrows Co^{II}(NH_3)_6^{2+}$	+0.06	7 mM NH_3 + 1 M NH_4Cl
$Co^{III}(en)_3^{3+} + e^- \rightleftarrows Co^{II}(en)_3^{2+}$	−0.20	0.1 M en + 0.1 M KNO_3
$Cr(III) + e^- \rightleftarrows Cr(II)$	−0.26	Satd. (saturated) $CaCl_2$
	−0.37	0.1–0.5 M H_2SO_4
	−0.40	5 M HCl
$Cr^{III}(CN)_6^{3-} + e^- \rightleftarrows Cr^{II}(CN)_6^{4-}$	−1.13	1 M KCN
$Cr^{VI}O_4^{2-} + 2\ H_2O + 3\ e^- \rightleftarrows Cr^{III}O_2^- + 4\ HO^-$	−0.12	1 M NaOH

TABLE 2.3 (*Continued*)

Reaction	$E^{\circ\prime}$ V versus NHE	Solution Composition
$Cr_2^{VI}O_7^{2-} + 14\ H_3O^+ + 6\ e^- \rightleftarrows 2\ Cr^{III}(OH_2)_6^{3+} + 9\ H_2O$	+0.84	0.1 M $HClO_4$
	+1.02	1 M $HClO_4$
	+0.93	0.1 HCl
	+1.00	1 M HCl
	+1.08	3 M HCl
	+0.92	0.1 M H_2SO_4
	+1.15	4 M H_2SO_4
$Cu^I(CN)_3^{2-} + e^- \rightleftarrows Cu(s) + 3\ CN^-$	−1.00	7 M KCN
$Cu^ICl_2^- + e^- \rightleftarrows Cu(s) + 2\ Cl^-$	+0.18	1 M HCl
$Cu(II) + e^- \rightleftarrows Cu(I)$	+0.01	1 M NH_3 + 1 M NH_4^+
	+0.30	0.1 M py + 0.1 M pyH^+
	+0.52	1 M KBr
$Cu^{II}(C_2O_4)_2^{2-} + 2\ e^- \rightleftarrows Cu(s) + 2\ C_2O_4^{2-}$	+0.06	1 M $K_2C_2O_2$
$Cu^{II}(EDTA)^{2-} + 2\ H_3O^+ + 2\ e^- \rightleftarrows Cu(s) + H_2(EDTA)^{2-} + 2\ H_2O$	+0.13	0.1 M EDTA; pH 4–5
$Eu(III) + e^- \rightleftarrows Eu(II)$	−0.43	0.1 M HCOOH
	−0.92	0.1 M EDTA; pH 6–8
$Fe(III) + e^- \rightleftarrows Fe(II)$	+0.46	2 M H_3PO_4
	+0.68	1 M H_2SO_4
	+0.71	0.5 M HCl
	+0.64	5 M HCl
	+0.53	10 M HCl
	+0.73	1 M $HClO_4$
	+0.01	1 M $K_2C_2O_4$, pH 5
	+0.07	0.5 M Na_2tart, pH 5–6
	−0.68	10 M NaOH
$Fe^{III}(CN)_6^{3-} + e^- \rightleftarrows Fe^{II}(CN)_6^{4-}$	+0.56	0.1 M HCl
	+0.71	1 M HCl
	+0.72	1 M $HClO_4$

$Fe^{VI}O_4^{2-} + H_2O + 3\,e^- \rightleftarrows Fe^{III}O_2^- + 4\,HO^-$	+0.55	10 M NaOH
$Fe^{III}(EDTA)^+ + e^- \rightleftarrows Fe^{II}(EDTA)$	+0.12	0.1 M EDTA; pH 4–6
$2\,H_3O^+ + 2\,e^- \rightleftarrows H_2(g) + 2\,H_2O$	+0.005	1 M HCl or 1 M $HClO_4$
$Hg_2^I(OH_2)_2^{2+} + 2\,e^- \rightleftarrows 2\,Hg(l) + 2\,H_2O$	+0.78	1 M $HClO_4$
$I_3^- + 2\,e^- \rightleftarrows 3\,I^-$	+0.54	0.5 M H_2SO_4
$2\,ICl_2^- + 2\,e^- \rightleftarrows I_2 + 4\,Cl^-$	+1.06	1 M HCl
$Ir^{IV}Cl_6^{2-} + e^- \rightleftarrows Ir^{III}Cl_6^{3-}$	+1.02	1 M HCl
$Mn^{II}(CN)_6^{4-} + e^- \rightleftarrows Mn^{I}(CN)_6^{5-}$	−1.08	1 M NaCN
$Mn^{III}(OH_2)_6^{3+} + e^- \rightleftarrows Mn^{II}(OH_2)_6^{2+}$	+1.50	7.5 M H_2SO_4
	−0.24	1.5 M NaCN
$Mo(IV) + e^- \rightleftarrows Mo(III)$	+0.10	4.5 M H_2SO_4
$Mo^{V}(CN)_8^{3-} + e^- \rightleftarrows Mo^{IV}(CN)_8^{4-}$	+0.80	0.25 M KBr, KCl, or KNO_3
$Mo(V) + 2\,e^- \rightleftarrows$ Mo(III)(green)	−0.25	2 M HCl
(red)	+0.11	2 M HCl
$Mo(VI) + e^- \rightleftarrows Mo(V)$	+0.53	2 M HCl
	+0.70	8 M HCl
	+0.50	2 M KSCN + 1 M HCl
$NO_3^- + 3\,H_3O^+ + 2\,e^- \rightleftarrows HNO_2 + 4\,H_2O$	+0.92	1 M HNO_3
$Nb^{V}O^{3+} + 2\,H_3O^+ + 6\,H_2O + 2\,e^- \rightleftarrows Nb^{III}(OH_2)_6^{3+} + 3\,H_2O$	−0.34	2–6 M HCl or 1.5–3 M H_2SO_4
$Nb(V) + e^- \rightleftarrows Nb(IV)$	−0.21	12 M HCl
$Ni^{II}(CN)_4^{2-} + e^- \rightleftarrows Ni^{I}(CN)_4^{3-}$	−0.82	1 M KCN
$Np(IV) + e^- \rightleftarrows Np(III)$	+0.14	1 M HCl
	+0.15	1 M $HClO_4$
$Np(V) + e^- \rightleftarrows Np(IV)$	+0.74	1 M $HClO_4$
$Np^{VI}O_2^{2+} + e^- \rightleftarrows Np^{V}O_2^+$	+1.14	1 M HCl
	+1.14	1 M $HClO_4$
$Os(IV) + e^- \rightleftarrows Os(III)$	+0.35	2 M HBr
$Os(VI) + 2\,e^- \rightleftarrows Os(IV)$	+0.66	0.5 M HCl
	+0.84	5 M HCl
	+0.97	9 M HCl

TABLE 2.3 (*Continued*)

Reaction	$E^{\circ\prime}$ V versus NHE	Solution Composition
$Os(VIII) + 4\,e^- \rightleftarrows Os(IV)$	+0.79	5 M HCl
$Pb^{II}(OH_2)_4^{2+} + 2\,e^- \rightleftarrows Pb(s) + 4\,H_2O$	−0.32	1 M NaOAc
$Pb^{IV}O_3^{2-} + H_2O + 2\,e^- \rightleftarrows Pb^{II}O_2^{2-} + 2\,HO^-$	+0.21	8 M KOH
$Pb^{II}SO_4(s) + 2\,e^- \rightleftarrows Pb(s) + SO_4^{2-}$	−0.29	1 M H_2SO_4
$Pd^{II}(OH_2)_4^{2+} + 2\,e^- \rightleftarrows Pd(s) + 4\,H_2O$	+0.99	4 M $HClO_4$
$Pd^{IV}Br_6^{2-} + 2\,e^- \rightleftarrows Pd^{II}Br_4^{2-} + 2\,Br^-$	+0.99	1 M KBr
$Pd^{IV}I_6^{2-} + 2\,e^- \rightleftarrows Pd^{II}I_4^{2-} + 2\,I^-$	+0.48	1 M KI
$Po(IV) + 4\,e^- \rightleftarrows Po(s)$	+0.60	1 M HCl
	+0.80	1 M HNO_3
$Po(IV) + 2\,e^- \rightleftarrows Po(II)$	+0.70	1 M HCl
$Pt^{IV}Br_6^{2-} + 2\,e^- \rightleftarrows Pt^{II}Br_4^{2-} + 2\,Br^-$	+0.64	1 M NaBr
$Pt^{IV}Cl_6^{2-} + 2\,e^- \rightleftarrows Pt^{II}Cl_4^{2-} + 2\,Cl^-$	+0.72	1 M NaCl
$Pt^{IV}I_6^{2-} + 2\,e^- \rightleftarrows Pt^{II}I_4^{2-} + 2\,I^-$	+0.39	1 M NaI
$Pt^{IV}(SCN)_6^{2-} + 2\,e^- \rightleftarrows Pt^{II}(SCN)_4^{2-} + 2\,SCN^-$	+0.47	1 M NaSCN
$Pu(IV) + e^- \rightleftarrows Pu(III)$	+0.40	1 M HOAc + 1 M NaOAc
	+0.50	1 M HF
	+0.59	0.6 M H_3PO_4 + 1 M HCl
	+0.75	1 M H_2SO_4
	+0.92	1 M HNO_3
	+0.97	1 M HCl
$Pu^{VI}O_2^{2+} + 4\,H_3O^+ + 2\,e^- \rightleftarrows Pu^{IV}(OH_2)_6^{4+}$	+1.05	1 M HCl
	+1.04	1 M $HClO_4$
$Pu^{VI}O_2^{2+} + e^- \rightleftarrows Pu^{V}O_2^{+}$	+0.92	1 M $HClO_4$
p-Benzoquinone + $2\,H_3O^+ + 2\,e^- \rightleftarrows$ hydroquinone + $2\,H_2O$	+0.70	1 M HCl or 1 M $HClO_4$
$Re^+ + 2\,e^- \rightleftarrows Re^-$	−0.23	0.4–2 M H_2SO_4

$Re^{IV}Cl_6^{2-} + e^- \rightleftarrows Re^{III}Cl_4^- + 2\ Cl^-$	−0.25	1 M HCl
$Re(V) + 2\ e^- \rightleftarrows Re(III)$	+0.14	2 M NaCN
$Rh(IV) + e^- \rightleftarrows Rh(III)$	+1.43	1 M H_2SO_4
$Rh(VI) + 2\ e^- \rightleftarrows Rh(IV)$	+1.5	0.1 M H_2SO_4
$Ru^{III}(CN)_6^{3-} + e^- \rightleftarrows Ru^{II}(CN)_6^{4-}$	+0.80	0.5 M H_2SO_4
$Ru^{III}(NH_3)_6^{3+} + e^- \rightleftarrows Ru^{II}(NH_3)_6^{3+}$	+0.07	CF_3SO_3Na
	+0.05	0.1 M $NaBF_4$
$Ru(IV) + e^- \rightleftarrows Ru(III)$	+0.91	0.5 M HCl
	+0.86	2 M HCl
$SO_4^{2-} + 4\ H_3O^+ + 2\ e^- \rightleftarrows SO_2 + 6\ H_2O$	+0.07	1 M H_2SO_4
$Sb^{III}O_2^- + 2\ H_2O + 3\ e^- \rightleftarrows Sb(s) + 4\ HO^-$	+0.67	10 M KOH
$Sb(V) + 2\ e^- \rightleftarrows Sb(III)$	+0.75	3.5 M HCl
	+0.82	6 M HCl
$Sb^{V}O_3^- + H_2O + 2\ e^- \rightleftarrows Sb^{III}O_2^- + 2\ HO^-$	−0.59	10 M NaOH
$Sn^{II}Cl_4^{2-} + 2\ e^- \rightleftarrows Sn(s) + 4\ Cl^-$	−0.19	1 M HCl
$Sn^{IV}(OH_2)_6^{4+} + 2\ e^- \rightleftarrows Sn^{II}(OH_2)_6^{2+}$	+0.14	1 M HCl
	+0.13	2 M HCl
$Te(IV) + 4\ e^- \rightleftarrows Te(s)$	+0.56	2 M HCl
$Ti^{IV}OCl^+ + 2\ H_3O^+ + 3\ Cl^- + e^- \rightleftarrows Ti^{III}Cl_4^- + 3\ H_2O$	−0.09	1 M HCl
	+0.24	6 M HCl
$Ti(IV) + e^- \rightleftarrows Ti(III)$	−0.01	0.2 M H_2SO_4
	+0.12	2 M H_2SO_4
	+0.2	4 M H_2SO_4
	−0.05	1 M H_3PO_4
	−0.15	5 M H_3PO_4
	−0.24	0.1 M KSCN
$Tl^{III}(OH_2)_6^{3+} + 2\ e^- \rightleftarrows Tl^{I}(OH_2)_6^{+}$	+0.89	0.1 M HCl + 0.9 M $HClO_4$
	+0.78	1 M HCl

TABLE 2.3 (*Continued*)

Reaction	$E^{\circ\prime}$ V versus NHE	Solution Composition
$U(IV) + e^- \rightleftarrows U(III)$	−0.64	1 M HCl
	−0.63	1 M $HClO_4$
$U^{VI}O_2^{2+} + 4\ H_3O^+ + 2\ e^- \rightleftarrows U(IV) + 2\ H_2O$	+0.41	0.5 M H_2SO_4
$U^{VI}O_2^{2+} + e^- \rightleftarrows U^{V}O_2^+$	+0.06	0.1 M Cl^-
$V^{III}(EDTA)^- + e^- \rightleftarrows V^{II}(EDTA)^{2-}$	−1.02	0.001–0.02 M EDTA
$V(III) + e^- \rightleftarrows V(II)$	−0.27	1 M HCl
	−0.29	12 M HCl
	−0.22	0.1–1 M NH_4SCN
	−0.33	Satd. H_2Tart
	−0.27	0.5 M H_2SO_4
	−0.24	1 M $HClO_4$
	−0.49	7.3 M H_3PO_4
	−0.78	0.001–0.2 M EDTA, pH 5–8.3
	−0.21	Satd. $H_2NNH_2 \cdot 2\ HCl$
	−0.89	1 M $K_2C_2O_4$, pH 3.5–6.5
	−0.62	Satd. KHPhthal
	−0.22	0.1–1 M NH_4SCN
$V^{IV}(O)^{2+} + 2\ H_3O^+ + 6\ H_2O + e^- \rightleftarrows V^{III}(OH_2)_6^{3+} + 3\ H_2O$	+0.36	1 M H_2SO_4
$V(V) + e^- \rightleftarrows V(IV)$	+1.02	1 M HCl
	+1.02	1 M $HClO_4$
	+1.00	1 M H_2SO_4
$W(V) + 2\ e^- \rightleftarrows W(III)$(green)	+0.10	12 M HCl
(red)	−0.20	12 M HCl
$W(V) + e^- \rightleftarrows W(IV)$	−0.30	12 M HCl
$W(VI) + e^- \rightleftarrows W(V)$	+0.26	Concd. (concentrated) HCl
$Yb(OH_2)_n^{3+} + e^- \rightleftarrows Yb(OH_2)_n^{2+}$	−1.15	0.1 M NH_4Cl

Potentiometry also is a direct means to evaluate the standard potential for half-reactions ($E°$) and has been applied for appropriate reversible systems. Such measurements require corrections for activity coefficients or extrapolation of the data to infinite dilution. Again, direct measurements in which equal molar concentrations of the oxidized or reduced species are introduced into the system provide a simple approach to such evaluations and are as precise as those obtained by less direct methods. However, $E°$ values also can be extracted from potentiometric titration data. For example, in the titration of $Fe(OH_2)_6^{2+}$ ion with $Ce(OH_2)_n^{4+}$ ion, the $Fe(OH_2)_6^{3+}$ ion concentration equals the $Fe(OH_2)_6^{2+}$ concentration at the half-equivalence point; the half-reaction potential, assuming activities to be equal to concentrations, is given directly by the potential of the indicator electrode relative to the reference electrode. If the latter is a standard hydrogen electrode, the measured potential is equal to the $E°$ for the $Fe(OH_2)_6^{3+}/Fe(OH_2)_6^{2+}$ couple. As indicated at the beginning of this chapter, the evaluation of the $E°$ for a half-reaction provides a direct measure of the free energy for the half-reaction relative to the free energy for the reduction of hydronium ion to hydrogen gas. Likewise, a combination of any pair of $E°$ values or of the free-energy values permits the evaluation of the equilibrium constant and the standard free energy for a redox reaction.

From a practical standpoint it is often useful to have the observed potential in the medium of measurement for the condition of equal concentrations of the oxidized and reduced species of a half reaction. Such potentials are known as formal potentials, $E°'$, rather than standard potentials, and are not purely thermodynamic quantities. The term *formal potential* comes from the tradition of having the supporting electrolyte at a one formal concentration. However, other stated solution conditions are also included in many listings. Thus the indicated potential is what one would expect at the half-equivalence point under actual titration conditions. In other words, activity corrections have not been made. Table 2.3 summarizes a number of formal potentials for commonly encountered half-reactions.

REFERENCES

1. *Collected Works of J. Willard Gibbs*, Yale University Press, New Haven, Conn., 1948.
2. Nernst, W., *Z. Phys. Chem.* **1889,** *4,* 129.
3. Behrend, R., *Z. Phys. Chem.* **1893,** *11,* 466.
4. Haber, F.; Klemensiewicz, Z., *Z. Phys. Chem.* **1919,** *67,* 385.
5. Hildebrand, J. H., *J. Am. Chem. Soc.* **1913,** *35,* 847.
6. Eisenman, G.; Rudin, D. O.; Casby, I. U., *Science* **1957,** *126,* 831.
7. Schwabe, K.; Suschke, H. D., *Angew. Chem. Int. Ed.* **1964,** *3,* 36.
8. Eisenman, G., ed., *Glass Electrodes for Hydrogen and Other Cations*, Marcel Dekker, New York, 1967.

9. Pranitis, D. M.; Telting-Diaz, M.; Meyerhoff, M. E., *Crit. Rev. Anal. Chem.* **1992,** *23,* 163.
10. Nernst, W.; Merriam, E. S., *Z. Phys. Chem.* **1905,** *52,* 235.
11. Kolthoff, I. M.; Furman, N. H., *Potentiometric Titrations*, 2nd ed., Wiley, New York, 1931.
12. Latimer, W. M., *The Oxidation States of the Elements and Their Potentials in Aqueous Solutions*, 2nd ed., Prentice-Hall, New York, 1952.
13. Meites, L., "Potentiometry," in *Handbook of Analytical Chemistry*, Meites, L., ed., McGraw-Hill, New York, 1963, pp. 5-6.
14. Charlot, G.; Collumeau, A.; Marchon, M. J. C., *Selected Constants. Oxidation-reduction Potentials of Inorganic Substances in Aqueous Solution*, Butterworths, London, 1971.
15. Dobos, D., *Electrochemical Data*, Akademiai Kiado, Budapest, 1975.
16. Bratsch, S. G., *J. Phys. Chem. Ref. Data* **1989,** *18,* 1.
17. Bard, A. J.; Parsons, R.; Jordan, J., *Standard Potentials in Aqueous Solution*, Marcel Dekker, New York, 1985.
18. Kissel, T. R., "Potentiometry: Oxidation-Reduction, pH Measurements, and Ion-Selective Electrodes," in *Physical Methods of Chemistry*, 2nd ed., Vol. II, *Electrochemical Methods*, Rossiter, B. W.; Hamilton, J. F., eds., Wiley, New York, 1986, pp. 53–189.
19. Ives, D. J. G.; Janz, G. J., *Reference Electrodes*, Academic Press, New York, 1961.
20. Reilley, C. N.; Schmid, R. W., *Anal. Chem.* **1958,** *30,* 947.
21. Reilley, C. N.; Schmid, R. W.; Lamsen, D. W., *Anal. Chem.* **1958,** *30,* 953.
22. Kobos, R. K., "Potentiometric Enzyme Methods," in *Selective Electrodes in Analytical Chemistry*, Vol. 2, Freiser, H., ed., Plenum Press, New York, 1980, pp. 1–84.
23. Guilbault, G. G., *Ion-Select. Elect. Rev.* **1982,** *4,* 187.
24. Rechnitz, G. A., *Anal. Chim. Acta* **1986,** *180,* 28.
25. Coetzee, J. F.; Deshmukh, B. K.; Liao, C. C., *Chem. Rev.* **1990,** *90,* 827.
26. Sabatini, A.; Vacca, A.; Gans, P., *Coord. Chem. Rev.* **1992,** *120,* 389.
27. Martell, A. E.; Motekaitis, R. J., *The Determination and Use of Stability Constants*, VCH, Weinheim, 1988.
28. Rossotti, F. J. C.; Rossotti, H., *The Determination of Stability Constants*, McGraw-Hill, New York, 1961, pp. 127–170.
29. Meloun, M.; Havel, J.; Hoegfeldt, E., *Computation of Solution Equilibria: A Guide to Methods in Potentiometry, Extraction, and Spectrophotometry*, Ellis Horward Ltd., Chichester, 1988.
30. Martell, A. E.; Motekaitis, R. J., *Coord. Chem. Rev.* **1990,** *100,* 323.

CHAPTER 3

CONTROLLED-POTENTIAL METHODS

3.1 INTRODUCTION

For chemists, the second important application of electrochemistry (beyond potentiometry) is the measurement of species-specific [e.g., iron(III) and iron(II)] concentrations. This is accomplished by an experiment in which the electrolysis current for a specific species is independent of applied potential (within narrow limits) and controlled by mass transfer across a concentration gradient, such that it is directly proportional to concentration ($i = kC$). Although the contemporary methodology of choice is cyclic voltammetry, the foundation for all voltammetric techniques is polarography (discovered in 1922 by Professor Jaroslov Heyrovsky; awarded the Nobel Prize for Chemistry in 1959). Hence, we have adopted a historical approach with a recognition that cyclic voltammetry will be the primary methodology for most chemists.

Control of Potential and Measurement of Current. With the formulation of the laws of electrolysis by Michael Faraday in 1834, the basis for relating electrolysis currents to chemical quantities was established. Although the concept of electrolysis was known prior to then, its utility in terms of chemical analysis depended on a quantitative relationship between current and equivalents of substance. Because an electrolysis current always necessitates mass transfer to or away from the electrode, the formulation of equations for diffusion by Fick was an important event in developing quantitative relationships.[1] With the laws of electrolysis and diffusion established, Heyrovsky combined these in a preferred form to provide a practical analytical method, namely, polarography.[2] His real contribution beyond combining the important concepts of Faraday and Fick was to realize that a reproducible and continuously renewed electrode surface was essential for electrochemistry to be a reliable analytical tool. Another important factor was the realization that a diffusion-controlled current could best be established by using a combination of a large nonpolarized electrode with a small working electrode. Hence, the dropping mercury electrode became an important part of the effectiveness of polarography. Ilkovic developed the equation that relates diffusion currents to the parameters involved in the dropping mercury electrode,[3] and thereby provided insight to the variables that affect the current response for polarographic analysis.

Because polarography rapidly provides information about the best conditions to perform an electrolysis with well-defined products, electrolysis conditions frequently are adjusted to take advantage of polarographic data. Thus, the technique of controlled-potential electrolysis often makes use of a mercury pool as a working electrode. By stirring this pool vigorously and reproducing the conditions of the polarographic data, the control potential and supporting-electrolyte conditions can be idealized to obtain the maximum yield. Therefore, based on the vast amount of electrochemical data from polarography as well as solid-electrode voltammetry, numerous practical electrochemical syntheses of inorganic and organic compounds have been developed.

Although solid-electrode voltammetry had serious limitations because of the changing character of solid-electrode surfaces, it provides the ability to go to much more positive potentials than are possible with the mercury electrode. Contemporary solid electrodes, particularly glassy carbon, allow one to obtain reproducible results (see Chapter 5).

Another limitation of solid electrodes has been their complex diffusion-current response relative to time with slow-sweep voltammetry. The development of a capillary hanging-mercury-drop electrode (HMDE) by Kemula and Kublik,[4,5] together with modern electronic instrumentation, allowed the principles of voltage-sweep voltammetry and cyclic voltammetry to be established. The success has been such that this has become one of the most important research tools for electrochemists concerned with the kinetics and mechanisms of electrochemical processes. These important contributions by Nicholson and Shain[6,7] rely, as have all electrochemical kinetic developments, on the pioneering work by Eyring et al.[8]

A number of specialized electrode systems have been developed in the last decade that provide practical extensions of the principles of voltammetry. These include the rotated disk electrode as well as the ring–disk configuration. The latter is particularly effective for studying pre- and post-electron-transfer processes. Because the limiting current is proportional to the square root of the rotation rate, a much broader kinetic window is available relative to that with a stationary electrode. Another useful electrode system uses close-spaced electrodes (of the order of a few micrometers apart in the most refined design) such that the electrolysis products of one electrode rapidly diffuse to the second electrode, where they can be reelectrolyzed. This configuration also provides the possibility of almost instantaneously electrolyzing the contents in the narrow space between the electrodes. Traditional voltammetric electrodes (macroelectrodes) like the dropping mercury electrode and disks of platinum, gold, or glassy carbon embedded in an insulator have areas of 0.25 cm^2 (0.4-cm diameter) or greater. The miniaturization of electrodes to diameters of 1–10 μm (microelectrodes) and even several nanometers (ultramicroelectrodes) permits (1) much more rapid electrochemical measurements (minimization of capacitive currents and R–C (resistance–capacitance) time constants), (2) their use as sensors for *in vivo* measurements, and (3) studies of electrode kinetics under steady-state conditions.

3.2 PRINCIPLES AND FUNDAMENTAL RELATIONS

Diffusion to a Planar Electrode. The basic approach in controlled-potential methods of electrochemistry is to control in some manner the potential of the working electrode while measuring the resultant current, usually as a function of time. When a potential sufficient to electrolyze the electroactive species completely is applied to the electrode at ($t = 0$), the concentration at the electrode surface is reduced to zero and an electrode process occurs, for example

$$ox + n\,e^- \rightleftharpoons red \tag{3.1}$$

where *ox* and *red* respectively represent oxidized and reduced forms of an electroactive species. Passage of current requires material to be transported to the electrode surface as well as away from it. Thus relationships must be developed that involve the flux and diffusion of materials; this is appropriately accomplished by starting with Fick's second law of diffusion

$$\frac{\partial C_{(x,t)}}{\partial t} = \frac{D \partial C_{(x,t)}}{\partial x^2} \tag{3.2}$$

where D represents the diffusion coefficient, C the concentration of the electroactive species at a distance x from the electrode surface, and t the amount of time that the concentration gradient has existed. Through the use of Laplace transforms with initial and boundary conditions

$$\text{For } t = 0 \text{ and } x \geq 0, \quad C = C^{\mathrm{b}}$$
$$\text{For } t \geq 0 \text{ and } x \to 0, \quad C \to C^{\mathrm{b}}$$
$$\text{For } t > 0 \text{ and } x = 0, \quad C = 0$$

Equation (3.2) can be solved to give a relationship for concentration in terms of parameters x and t:

$$C_{(x,t)} = C^b \operatorname{erf} \frac{x}{2D^{1/2}t^{1/2}} \tag{3.3}$$

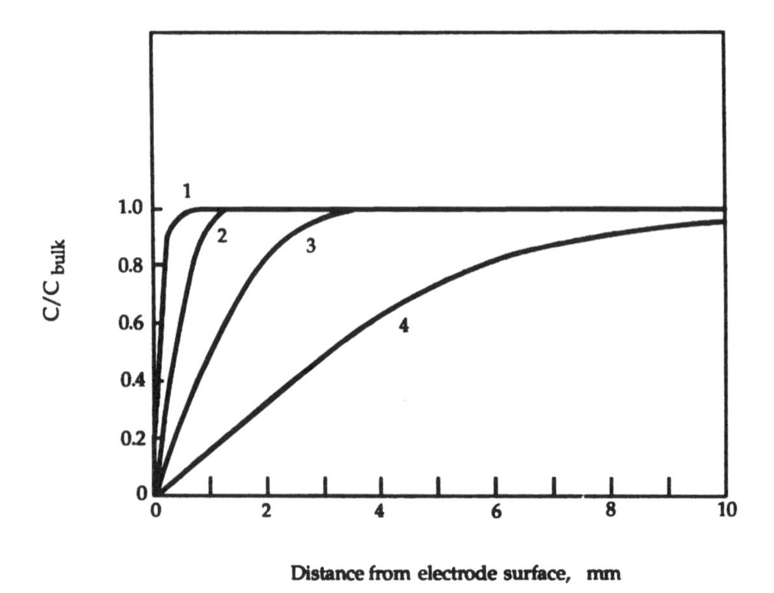

Figure 3.1 Concentration–distance curves for different periods of linear diffusion to a planar-electrode surface. Diffusion times: (1) 10 s; (2) 100 s; (3) 1000 s; (4) 10,000 s. Data for a diffusion coefficient D of 1×10^{-5} cm^2 s^{-1}.

where C^{b} is the bulk concentration of the electroactive species. Figure 3.1 illustrates concentration gradients at a planar-electrode surface (exposed to a solution with an electroactive species at a bulk concentration of C^{b}) as a function of time [from Eq. (3.3)].

By taking the derivative of Eq. (3.3) for the proper boundary condition, namely, at the electrode surface ($x = 0$), the diffusion gradient at the electrode surface is expressed by the relation

$$\left(\frac{\partial C}{\partial x}\right)_{(0,t)} = \frac{C^{\mathrm{b}}}{\pi^{1/2}D^{1/2}t^{1/2}} \tag{3.4}$$

This flux of material crossing the electrode boundary can be converted to current by the expression

$$i = nFAD\,\frac{\partial C_{(0,t)}}{\partial x} \tag{3.5}$$

where n is the number of electrons involved in the electrode reaction, F the faraday constant, and A the area of the electrode. When Eq. (3.4) is substituted into this relation, a complete expression for the current that results from semi-infinite linear diffusion is obtained (the Cottrell equation for a planar electrode)

$$i = \frac{nFAC^{b}D^{1/2}}{\pi^{1/2}t^{1/2}} \tag{3.6}$$

This relationship holds for any electrochemical process that involves semiinfinite linear diffusion and is the basis for a variety of electrochemical methods (e.g., polarography, voltammetry, and controlled-potential electrolysis). Equation (3.6) is the basic relationship used for solid-electrode voltammetry with a preset initial potential on a plateau region of the current–voltage curve. Its application requires that the electrode configuration be such that semiinfinite linear diffusion is the controlling condition for the mass-transfer process.

3.3 POLAROGRAPHY

The most important and extensively studied form of voltammetry has been polarography. The potential–time dependence that is used for polarographic measurements is presented in Figure 3.2 (solid line). The potential is scanned from E_1 to E_2 to obtain a current response that qualitatively and quantitatively

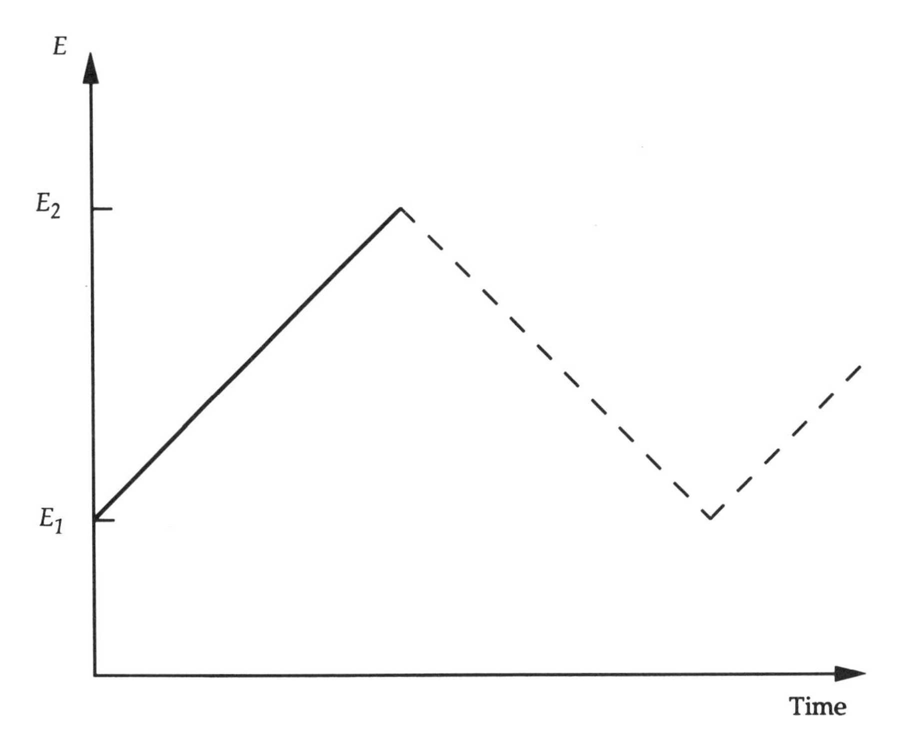

Figure 3.2 Potential–time profile used for polarography and linear-sweep voltammetry (solid line) and cyclic voltammetry (both solid and dashed lines).

characterizes the electroactive species present. The vast body of data from polarographic measurements can be adopted by other electroanalytical methods. Moreover, pulse polarographic methods and anodic stripping analysis, which are still used for determination of trace amounts of metal ions, are closely related to polarography. The theory of polarography has been described in several classical[9–11] and recent monographs.[12–14] Its unique characteristic is that it uses a dropping-mercury electrode, such that the electrode surface is continuously renewed in a well-defined and regulated manner to give reproducible effective electrode areas as a function of time. The diffusion-current equation [Eq. (3.6)] can be extended to include a dropping-mercury electrode by appropriate substitution for the area of the electrode. Thus the volume of the drop for a dropping-mercury electrode is given by the relationship

$$V = \frac{4}{3}\pi r^3 = \frac{mt}{d} \tag{3.7}$$

where r = radius of the drop of mercury
m = mass flow rate of mercury from orifice of capillary
t = life of the drop
d = density of mercury under the experimental conditions

When this equation is solved for r and the latter is substituted into the equation for the area of a sphere, an expression for the area of the dropping-mercury electrode drop as a function of the experimental parameters is obtained

$$A = (4\pi)^{1/3}3^{2/3}m^{2/3}t^{2/3}d^{-2/3} \tag{3.8}$$

This then can be substituted into Eq. (3.6) to give a calculated diffusion current for the dropping-mercury electrode:

$$(i_t)_{\text{calc}} = 462\ nC^{\text{b}}D^{1/2}m^{2/3}t^{1/6} \tag{3.9}$$

Actually, experimental results indicate that the constant in Eq. (3.9) is too small by a factor of $\sqrt{7/3}$. We now realize that this $\sqrt{7/3}$ quantity is not empirical but is the appropriate contribution due to the growth of the mercury drop into the solution away from the capillary orifice. Thus, the correct diffusion current expression for a dropping-mercury electrode is

$$i = 706nC^{\text{b}}D^{1/2}m^{2/3}t^{1/6} \tag{3.10}$$

which gives the current at any time up to the lifetime of the drop. If the drop time t_{d} is substituted, the well-known Ilkovic equation results

$$i_{\text{d}} = 706\ nC^{\text{b}}D^{1/2}m^{2/3}t_{\text{d}}^{1/6} \tag{3.11}$$

where i_d is in A (amperes) if D is in $cm^2\ s^{-1}$, C^b is in mol cm^{-3}, m is in mg/s, and t is in s. Alternatively, i_d is in μA when C^b is in mM.

In the original work on polarography well-damped ballistic galvanometers were used to record the current because potentiometric recorders were not available. The characteristics of a ballistic galvanometer are such that it gives the average current for a dropping-mercury electrode as used in polarography (see Figure 3.3). Integrating Eq. (3.10) over the life of a drop permits the average current to be evaluated:

$$\overline{i_d} = \frac{6}{7} i_d = 607\ nC^b D^{1/2} m^{2/3} t_d^{1/6} \tag{3.12}$$

It is noteworthy that much of the polarographic literature is tabulated on the basis of Eq. (3.12). In spite of this almost all modern recording polarographs use potentiometric stripchart recorders, which have time constants such that the maximum of the polarographic oscillations is the instantaneous current. Furthermore, the mode of response of a potentiometric recorder is such that

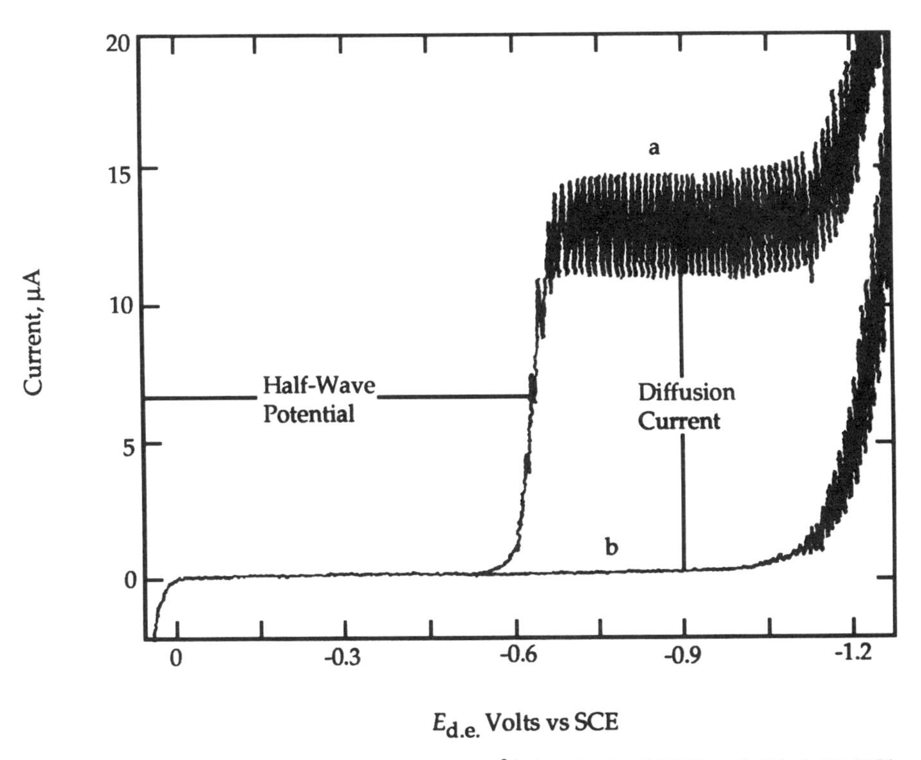

Figure 3.3 Polarograms for (*a*) 0.5 mM Cd^{2+} ion in 1 M HCl and (*b*) 1 M HCl alone.

the mean of the oscillation is not equal to the average current. This is true because ballistic galvanometers obey Hooke's law (in terms of rate of movement), while strip-chart recorders use constant-speed drive motors for the pen. Unless a ballistic galvanometer is used for recording current, Eq. (3.12) and the envelope of the maxima of the polarographic current oscillations are the appropriate means for polarographic measurements.

Although the experimental conditions for diffusion-controlled current may be in effect for a polarographic measurement, the resultant current may not be controlled purely by diffusional processes. A convenient way to test whether this is true is to vary the height of the mercury column. The fluid flow characteristics of a capillary with a hydrostatic head are such that the diffusion current is directly proportional to the square root of the height of the column (small corrections for the surface tension of mercury on glass and for the hydrostatic backpressure of the water-immersed portion of the capillary are necessary for the most precise measurements)

$$i_d = \text{constant} \times h^{1/2} \tag{3.13}$$

Thus, if three or four different heights of a mercury column are used for the dropping-mercury electrode, the resulting current can be tested by Eq. (3.13). If kinetic or catalytic complications are present, the current will not adhere to this relationship.

For systems where kinetic, adsorptive, or catalytic effects complicate a simple diffusion-controlled process, polarography provides a means of verifying the existence of these complications. Hence, if there is a prechemical process that is kinetically limiting, the observed current will be less than that anticipated from Eq. (3.11). Furthermore, tests based on Eq. (3.13) will indicate that the current is more or less independent of the height of the mercury column. With adsorptive effects the diffusion current tends to be larger than would be anticipated from Eq. (3.11), unless the rate of adsorption becomes a limiting part of the process. Again, such currents will not obey Eq. (3.13). Currents that result from catalytic processes will take many different forms and cannot be predicted a priori. However, these systems will not in general obey Eqs. (3.11) and (3.13).

The other factor that can show the influence of kinetic, catalytic, and adsorption effects on a diffusion-controlled process is the temperature coefficient.[10] The effect of temperature on a diffusion current can be described by differentiating the Ilkovic equation [Eq. (3.11)] with respect to temperature. The resulting coefficient is described as $[\ln (i_{d,2}/i_{d,1})/(T_2 - T_1)]$, which has a value of +0.013 deg^{-1}. Thus, the diffusion current increases about 1.3% for a one-degree rise in temperature. Values that range from 1.1 to 1.6% $°\text{C}^{-1}$, have been observed experimentally. If the current is controlled by a chemical reaction the values of the temperature coefficient can be much higher (the Arrhenius equation predicts a two- to threefold increase in the reaction rate for a 10-degree rise in temperature). If the temperature coefficient is much larger than 2% $°\text{C}^{-1}$, the current is probably limited by kinetic or catalytic processes.

The temperature coefficients for currents that are limited by adsorption vary and may have negative values.

The polarographic current–potential wave illustrated by Figure 3.3 conforms to the Nernst equation for reversible electrochemical processes. However, it is more convenient to express the concentrations at the electrode surface in terms of the current i and the diffusion current i_d. Because i_d is directly proportional to the concentration of the electroactive species in the bulk and i at any point on the curve is proportional to the amount of material produced by the electrolysis reaction, these quantities can be directly related to the concentration of the species at the electrode surface. For a generic reduction process [Eq. (3.1)] the potential of the electrode is given by the Nernst equation:

$$E = E^\circ + \frac{RT}{nF} \ln \frac{C_{ox(0,t)}}{C_{red(0,t)}} \tag{3.14}$$

Initially the solution contains only *ox* (concentration C_{ox}^{b}). When a potential is applied and reductive current flows, the oxidized form of the electroactive species diffuses toward the electrode. From Eq. (3.10) it follows that

$$i = 706n[C_{ox}^{b} - C_{ox(0,t)}]\, D_{ox}^{1/2} m^{2/3} t_d^{1/6} = i_d - 706nC_{ox(0,t)} D_{ox}^{1/2} m^{2/3} t_d^{1/6} \tag{3.15}$$

Hence

$$C_{ox(0,t)} = \frac{i_d - i}{706nD_{ox}^{1/2} m^{2/3} t_d^{1/6}} \tag{3.16}$$

After reduction of *ox*, its reduced form (*red*) diffuses into either the bulk of solution or into the mercury to from an amalgam. In either case

$$i = 706n[C_{red(0,t)} - C_{red}^{b}]D_{red}^{1/2} m^{2/3} t_d^{1/6} \tag{3.17}$$

But the reduced form of the electroactive species was not present in the solution before electrolysis; therefore, $C_{red}^{b} = 0$ and

$$C_{red(0,t)} = \frac{i}{706nD_{red}^{1/2} m^{2/3} t_d^{1/6}} \tag{3.18}$$

Substitution of Eqs. (3.16) and (3.18) into the Nernst equation [Eq. (3.14)] gives

$$E = E^\circ + \frac{RT}{nF} \ln \left(\frac{D_{red}}{D_{ox}}\right)^{1/2} + \frac{RT}{nF} \ln \frac{i_d - i}{i} \tag{3.19}$$

At the half-height of a polarographic wave ($i = i_d/2$) the corresponding potential is defined as the half-wave potential ($E_{1/2}$). Therefore, Eq. (3.19) takes the form

$$E = E_{1/2} + \frac{RT}{nF} \ln \frac{i_d - i}{i} \tag{3.20}$$

For the reduction of a simple solvated metal ion to its amalgam, $E_{1/2}$ is given by

$$E_{1/2(s)} = E_s^\circ + \frac{RT}{nF} \ln \frac{\gamma_{ion} D_a^{1/2}}{\gamma_a D_{ion}^{1/2}} \tag{3.21}$$

where γ_{ion} is the activity coefficient for the ion and γ_a is the activity coefficient for the amalgamated species. The diffusion coefficients for the amalgam and ionic species also are a part of this expression. The standard reduction potential is for reduction of the ion to the amalgamated species. These expressions also hold for the reduction of an ion to a lower oxidation state, but require that the appropriate value be used. For well-behaved polarograms, the number of electrons in the reduction process can be determined by taking the difference between the $\frac{3}{4}$-wave potential and the $\frac{1}{4}$-wave potential

$$E_{3/4} - E_{1/4} = \frac{-0.056}{n} \quad \text{at } 25°\text{C} \tag{3.22}$$

Equation (3.20) also can be extended to conditions where the electroactive species is a complexed metal ion. For such conditions the half-wave potential for the complexed species, $E_{1/2(c)}$, is related to that for the uncomplexed species, $E_{1/2(s)}$, by the relationship

$$E_{1/2(c)} = E_{1/2(s)} + \frac{RT}{nF} \ln K_{diss} - p \frac{RT}{nF} \ln (\gamma_x C_x) \tag{3.23}$$

where K_{diss} = dissociation constant for the complex
p = number of ligands per metal ion
γ_x = activity coefficient of the ligand
C_x = concentration of the ligand

A similar relationship holds for the reduction of a complexed species to another ionic complexed species:

$$E_{1/2(c)} = E_{1/2(s)} + \frac{RT}{nF} \ln \frac{K_{diss}^{ox}}{K_{diss}^{red}} - (p - q) \frac{RT}{nF} \ln (\gamma_x C_x) \tag{3.24}$$

where q represents the number of ligands per reduced ion. This expression indicates that only the ratio of dissociation constants can be determined; some independent means must be used to evaluate one of them.

Unfortunately, a large number of substances reduced at the dropping-mercury electrode do not behave reversibly; in other words, they do not behave according to equilibrium thermodynamics. For a totally irreversible process (one in which the kinetics for the backreaction are essentially equal to zero), the current–potential relationship takes the form

$$E = E_{1/2} + \frac{0.916\ RT}{\alpha n_a F} \ln \frac{i_d}{i} \tag{3.25}$$

where α (with values between 0 and 1) is the transfer coefficient (indicates the symmetry of the potential-energy function for the transition state) and n_a is the number of electrons involved in the rate-determining step of the reduction process. For such a system the half-wave potential does not represent the standard potential, but is related to the kinetic parameters for the reduction process by the expression

$$E_{1/2} = \frac{RT}{\alpha n_a F} \ln \frac{1.35 k_c t_d^{1/2}}{D^{1/2}} \tag{3.26}$$

where k_c is the heterogeneous rate constant for the reduction process at a potential of 0.00 V versus NHE and t_d is the drop time for the dropping-mercury electrode. A simplified form of Eq. (3.20) can be used to evaluate the quantity αn_a:

$$E_{3/4} - E_{1/4} = -\frac{0.052}{\alpha n_a} \quad \text{at } 25°\text{C} \tag{3.27}$$

whereby the potential at the $\frac{3}{4}$ wave height minus that at the $\frac{1}{4}$-wave height is measured.

The theory of polarography has been adopted for reversible, quasireversible, and irreversible processes, and also for those complicated by kinetic, catalytic, and adsorption processes. The detailed description of these processes can be found in the monographs on polarography.[10–14]

Current-Sampled Polarography. As follows from the Ilkovic equation [Eq. (3.10)], the current at the electrode that results from the electrolysis of an electroactive species (Faradaic current) increases proportionally to $t^{1/6}$ during the life of the mercury drop. However, the electrical double layer is a capacitor that must be charged as potential is applied via a charging current (i_{cc})

$$i_{cc} = 0.00567\ C_i(E_z - E)\ m^{2/3} t^{-1/3} \tag{3.28}$$

where E_z is the potential of zero charge and C_i is the capacitance of the double layer. Thus, i_{cc} contributes to the total current registered at the electrode. Figure 3.4 illustrates the Faradaic, charging, and total current at a dropping mercury

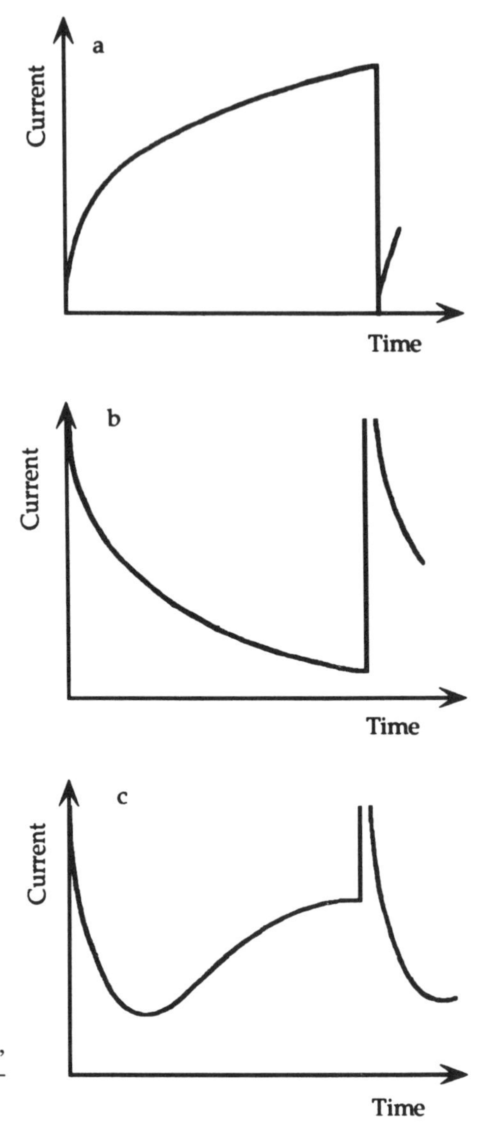

Figure 3.4 Faradaic (*a*), capacitive (*b*), and total (*c*) current at the dropping-mercury electrode.

electrode. Hence, by recording only the current near the end of the drop life instead of the entire current–time curve, one can minimize the contribution of the charging current to the total current. The method is called *current-sampled polarography* (sometimes tast polarography).[12,13] The approach enhances sensitivity and eliminates the serration caused by the continuous mercury drop fall and drop growth. Polarograms obtained are simpler and can be evaluated more exactly then in normal polarography. In the method the drop time is enforced at some fixed value by mechanical dislodging each drop just after the current sample is taken.

AC Polarography. The imposition of an AC voltage across an electrochemical cell results in an AC current whose magnitude is proportional to the ionic concentration of the solution and to the ionic mobility of the ions. This serves as the basis for the techniques of conductometry and conductometric titrations. Another dimension of AC electrochemistry is small-amplitude AC polarography. This consists of the imposition of a sinusoidal AC voltage of about 10 mV (frequency 100–1000 Hz) on the linear voltage ramp of classical polarography. The resulting AC current is measured as the fundamental harmonic, or it can be measured as the second harmonic. Figure 3.5 illustrates the different response curves for an electrochemically reversible system. These curves are the envelopes of the dropping-mercury electrode current oscillations. Effective use of AC polarography requires a degree of electrochemical expertise beyond the scope of this monograph. However, the method is becoming more popular

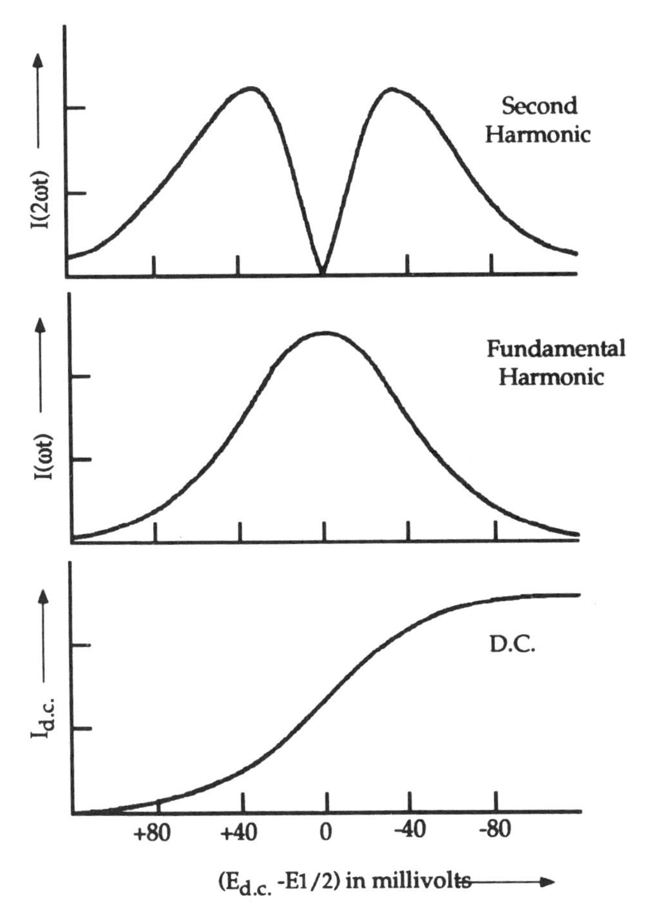

Figure 3.5 Conventional (bottom), fundamental-harmonic AC, and second-harmonic AC polarograms for a reversible couple at a dropping-mercury electrode. Curves represent the envelope of the upper limits of the current oscillations.

for special analytical problems and for the study of the fundamentals of electrode processes; reviews and discussions are available.[12,13,15-17]

Square-Wave Polarography. The highest degree of sensitivity (and instrumental complexity) is achieved with square-wave polarography.[12,17] The method is similar to AC polarography with a linearly increasing DC potential applied to a dropping mercury electrode, but in this method a square-wave voltage is superimposed (frequency ~225 Hz; amplitude ~30 mV). Scan rates up to 5 Vs^{-1} allow a complete voltammogram to be determined on a single drop. Metal ions at concentrations at low as 4×10^{-5} mM can be detected and assayed by this method. Its sensitivity for irreversible processes is an order of magnitude smaller than that for reversible processes.

Pulse Polarography. Another derivative of AC measurements is the method of pulse polarography. This takes two forms: (1) *normal*-pulse polarography and (2) *differential*-pulse polarography. Form 1 involves the imposition of square voltage pulses, each of increasing magnitude, on a static DC voltage (see Figure 3.6*a*). The resulting current response is illustrated by Figure 3.7*a*. For differential-pulse polarography, uniform square voltage pulses are imposed on the linear voltage ramp of classical polarography (see Figure 3.6*b*) to yield the current response that is illustrated in Figure 3.7*b*. The purpose of the pulse technique is to minimize the amount of capacitative current (from the charging of the electrochemical double layer) in the current measurement. Figure 3.6*c* indicates the current–time response relative to the potential–time poise for the differential-pulse method. By sampling of the current near the end of the pulse, most of the capacitative charging current has decayed.

Pulse polarography has enjoyed increasing popularity because of the availability of commercial instruments. Furthermore, there is general agreement that the differential form of the method provides the greatest analytical sensitivity in the electrochemical field. Review articles provide more detail of the method and its applications.[12,13,17-20]

Stripping Analysis. The analysis of extremely dilute concentrations requires some type of preconcentration step prior to actual determination. For metal ions this can be accomplished by the electrolytic deposition of the metals on an electrode. Analytical information can be obtained via a redissolution (stripping) process. The route of the stripping analysis is shown in Figure 3.8. In the first preconcentration step the electrolysis, at a sufficiently negative potential, is carried out for a precisely known time with efficient stirring, followed by a short time without stirring to allow the solution to become quiet. The potential is then scanned in a positive direction to give anodic current. Usually a hanging-mercury-drop electrode (HMDE) is used for the measurements. The resulting current–time curve is an anodic voltammogram with well-developed peaks that correspond to the successive reoxidation of the analyzed elements

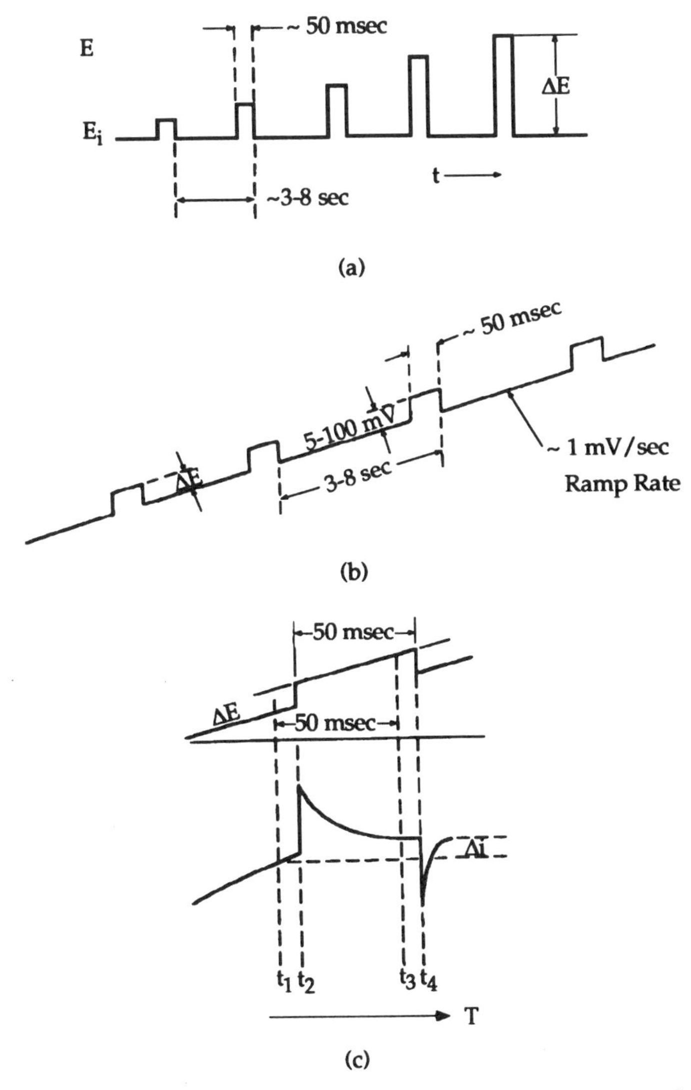

Figure 3.6 Potential–time sequence for (*a*) normal-pulse polarogram and (*b*) differential-pulse polarogram. The current–time response for the latter is given by (*c*), with t_1 and t_3 the times at which current is measured, t_2 the time at which pulse is applied, and t_4 the time at which pulse is removed.

from the amalgam. Details of the technique and reference data on analyzed elements and compounds are available.[12,21,22]

Polarographic AC and pulse polarographic techniques as well as stripping analysis are effective tools for the determination of trace levels of metal ions. Table 3.1 provides a comparison of the sensitivity and usefulness of the various methods.[12,21–23] For metal ions, stripping voltammetry usually is the method

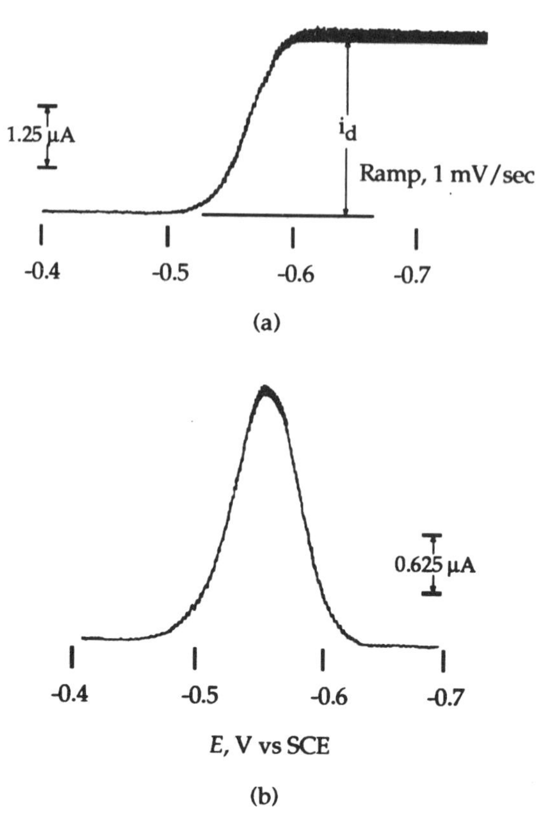

Figure 3.7 Pulse polarograms for 10^{-4} M Cd^{II} ion: (*a*) normal and (*b*) differential modes.

of choice, but requires more time for the preelectrolysis. Differential-pulse polarography and square-wave polarography are the most sensitive techniques with detection limits of 1×10^{-7} M. Recent work[24] has shown that this concentration limit can be lowered by two or three orders of magnitude by interfacing a minicomputer to the differential-pulse polarograph. The general characteristics and controlling parameters for pulse polarography have been discussed in an earlier article.[25]

3.4 LINEAR SWEEP AND CYCLIC VOLTAMMETRY

The potential–time relation for voltammetric measurements is presented in Figure 3.2. With linear-sweep voltammetry, the potential is linearly increased between potentials E_1 and E_2. Cyclic voltammetry is an extension of linear-sweep voltammetry with the voltage scan reversed after the current maximum (peak) of the reduction process has been passed. The voltage is scanned negatively beyond the peak and then reversed in a linear positive sweep. Such a

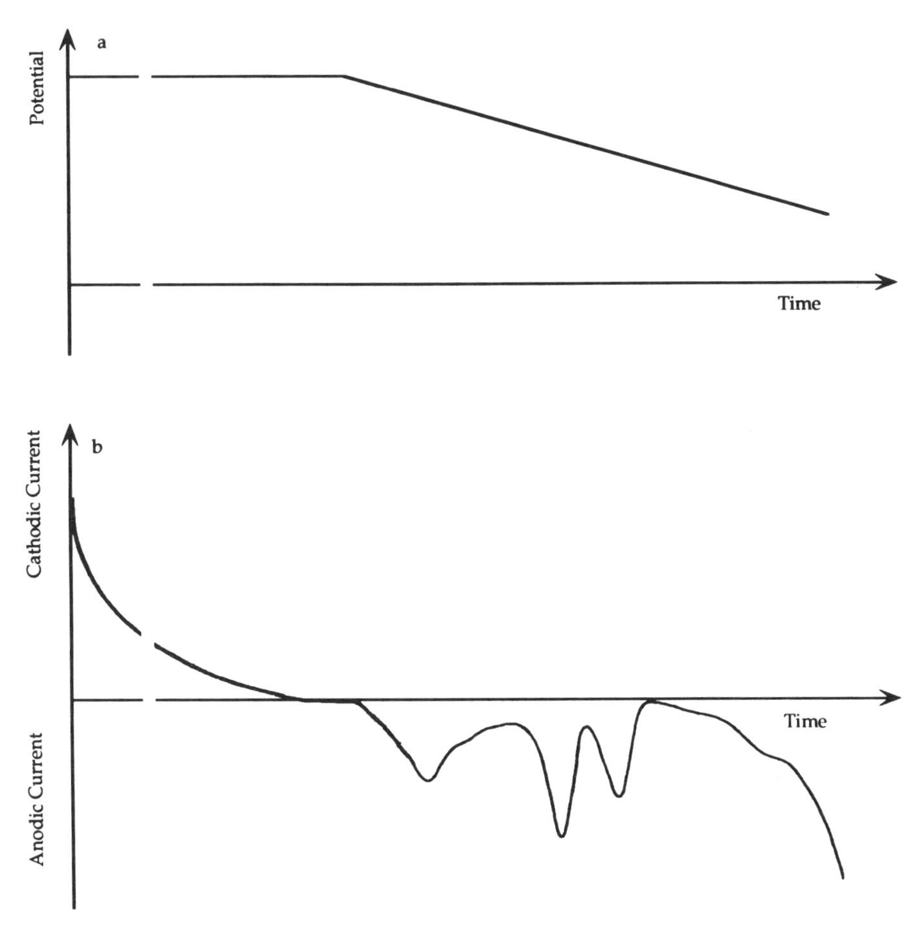

Figure 3.8 The principle of stripping voltammetry: (*a*) applied potential; (*b*) current response.

technique provides even more information about the properties and characteristics of the electrochemical process and also gives insight into any complicating side processes such as pre- and post-electron-transfer reactions as well as kinetic considerations. Whereas in classical polarography the voltage scan rate is about 1 V min^{-1}, linear-sweep voltammetry uses scan rates of ≤ 100 V s^{-1} for conventional microelectrodes (and $\leq 10{,}000$ V s^{-1} for ultramicroelectrodes; 10^{-6}-m diameter). For scan rates of up to about 1 V s^{-1} an X–Y recorder can be used, but above this a fast transient recorder, an oscilloscope, or a computer is required.

Figure 3.9 illustrates the shape of a cyclic voltammogram with an electrode of fixed area. The voltammogram is characterized by a peak potential E_p, at which the current reaches a maximum value and by the value of the current i_p. When the reduction process is reversible, the peak current is given by the

TABLE 3.1 Comparison of Polarographic Methods for Determination of Metal Ions

	Limit of Detection (M)	Concentration Limit Recommended for Quantitative Analysis (M)	Comments
Conventional DC	2×10^{-6}	6×10^{-6}	—
Phase-sensitive AC with natural time	5×10^{-7}	1×10^{-6}	$\Delta E = 10$ mV; frequency ≈ 100 Hz recommended; far superior to nonphase-sensitive AC polarography
Square-wave	5×10^{-9}	5×10^{-8}	Sensitive method, but requires complicated differential pulse
Pulse	5×10^{-7}	1×10^{-7}	Superior to DC methods $< 10^{-5}$ M; long drop times most favorable
Derivative pulse	$\approx 5 \times 10^{-7}$	—	Poor reproducibility at low concentrations; not recommended for trace analysis
Differential pulse	1×10^{-7}	—	Sensitive technique but waves broad under some conditions; long drop times most favorable
Stripping voltammetry	10^{-7}–10^{-10}	—	The most sensitive technique but requires preelectrolysis

relation

$$i_p = 0.4463\, nFA(Da)^{1/2}C^b \tag{3.29}$$

with

$$a = \frac{nF\nu}{RT} = \frac{n\nu}{0.026} \quad \text{at } 25°\text{C} \tag{3.30}$$

where ν is the scan rate in volts per second. This relation results from the set of differential equations for Fick's second law of diffusion (with the appropriate initial and boundary conditions for *ox* and *red*). Thus, in terms of the adjustable parameters, the peak current is given by the Randles–Sevcik equation:

$$i_p = 2.69 \times 10^5 n^{3/2} A D^{1/2} C^b \nu^{1/2} \quad \text{at } 25°\text{C} \tag{3.31}$$

where i_p is in A, A in cm^2, D in cm^2 s^{-1}, C^b in mol cm^{-3}, and ν in V s^{-1}.

The derivation of the expression for the voltammetric current–potential curve is even more mathematically challenging, but several solutions of the appro-

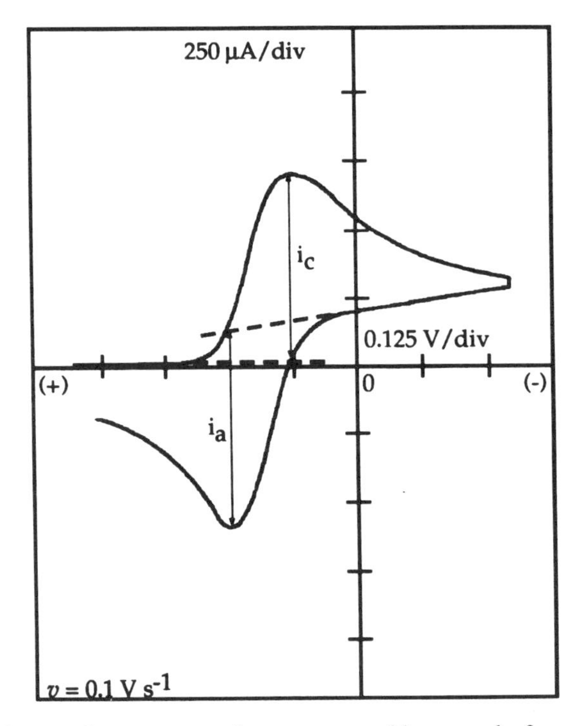

Figure 3.9 Linear voltage-sweep voltammogram with reversal of sweep direction to give a cyclic voltammogram. Initial sweep direction to more negative potential.

priate differential equations have appeared during the past three decades. Nicholson and Shain[6] revolutionized the voltammetric experiment with their elegant development and demonstration of linear-sweep and cyclic voltammetry. In their approach the current–potential curve is presented as

$$i = nFAC^{\text{b}}(\pi Da)^{1/2}\chi(at) \tag{3.32}$$

The relation between $\pi^{1/2}\chi(at)$ and $n(E - E_{1/2})$ is given in a tabulated form in Table 3.2, a is defined by Eq. (3.30), and $E_{1/2}$ is the half-wave potential of the electrochemical process. For simplicity the term $\pi^{1/2}\chi(at)$ can be presented as χ_{rev},[26] and Eq. (3.32) takes the form

$$i = nFAC^{\text{b}}(Da)^{1/2}\chi_{\text{rev}} \tag{3.33}$$

For a given potential (E) the value of χ_{rev} can be obtained from Table 3.2. A similar approach was used to obtain quantitative relations of irreversible processes as well as for those complicated by chemical reactions.[6,7]

For a reversible process the peak potential can be related to the polarographic half-wave potential $E_{1/2}$ by the expression

TABLE 3.2 Current Function Values for Cyclic Voltammetry of Reversible Processes (α = 0.5)[6]

$n(E - E_{1/2})$, mV	$\chi_{rev} = \pi^{1/2}\chi(at)$	$n(E - E_{1/2})$, mV	$\chi_{rev} = \pi^{1/2}\chi(at)$
120	0.009	−5	0.400
100	0.020	−10	0.418
80	0.042	−15	0.432
60	0.084	−20	0.441
50	0.117	−25	0.445
45	0.138	−28.5	0.4463
40	0.160	−30	0.446
35	0.185	−35	0.443
30	0.211	−40	0.438
25	0.240	−50	0.421
20	0.269	−60	0.399
15	0.298	−80	0.353
10	0.328	−100	0.312
5	0.335	−120	0.280
0	0.380	−150	0.245

$$E_p = E_{1/2} - 1.11\frac{RT}{nF} = E_{1/2} - \frac{0.0285}{n} \quad \text{at } 25°\text{C} \tag{3.34}$$

Another useful parameter of the voltammetric curves is the half-peak potential $E_{p/2}$, which is the potential at which the registered current reaches half its maximum value. For a reversible process $E_{1/2}$ is located half-way between E_p and $E_{p/2}$.

The ratio of the peak current for the cathodic process relative to the peak current for the anodic process is equal to unity ($i_{p,c}/i_{p,a} = 1$) for a reversible electrode process. For measurement of the peak current for the anodic process, the extrapolated baseline going from the foot of the cathodic wave to the extension of this cathodic current beyond the peak must be used as a reference, as illustrated by Figure 3.9. Another approach to measuring peak-current ratios is illustrated by Figure 3.10.

For the condition

$$|E_\lambda - E_{p/2}| \geq \frac{0.141}{n} \tag{3.35}$$

where E_λ is the extent of the voltage sweep. The difference in the peak potentials between the anodic and cathodic processes of the reversible reaction is given by the relationship

$$|\Delta E_p| = |E_{p,a} - E_{p,c}| = \frac{0.059}{n} \quad \text{at } 25°\text{C} \tag{3.36}$$

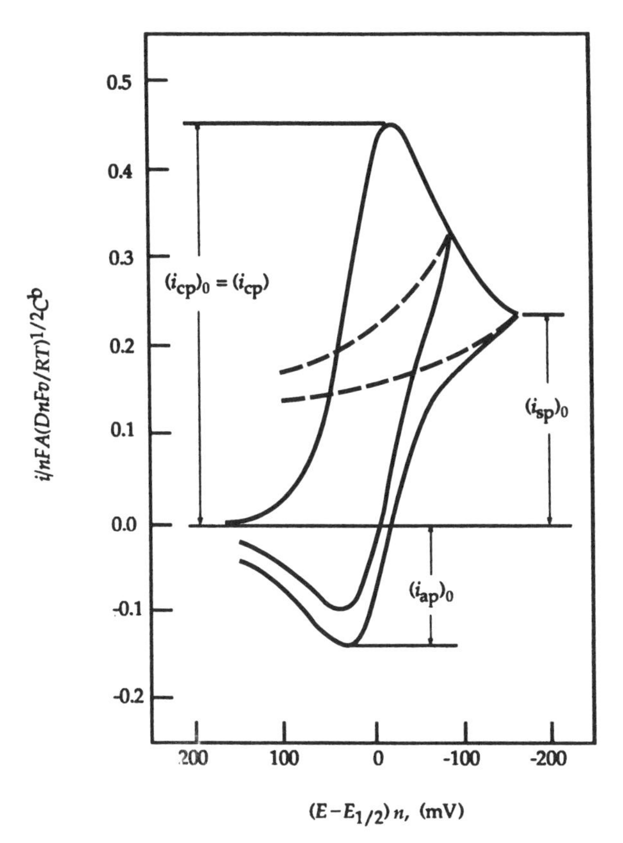

Figure 3.10 Method for measurement of peak currents and peak-current ratios of cyclic voltammograms.

which provides a rapid and convenient means for determining the number of electrons involved in the electrochemical reaction. For a reversible system i_p is a linear function of $\sqrt{\nu}$, and also E_p is independent of ν.

The peak current for an irreversible reduction may be expressed in terms of a heterogeneous rate constant k_s and the peak potential by the relation

$$i_p = 0.227\ nFAC^b k_s \exp\left[\frac{-\alpha n_a F}{RT}(E_p - E')\right] \tag{3.37}$$

where k_s represents the heterogeneous rate constant of the electrode process for the condition that the electrode has a potential equal to the formal potential for the electrode process E'. A somewhat simpler form of this expression is possible if the instantaneous current of the wave is measured at a point that is less than 10% that of the peak current. For this condition the measured current is related to the kinetic parameters by the expression

$$i = nFAC^{b}k_s \exp \frac{-\alpha n_a F}{RT} (E_p - E_i) \tag{3.38}$$

where E is the potential at the measured current and E_i is the potential at which the scan was initiated.

Cyclic voltammetry is of particular value for the study of electrochemical processes that are limited by finite rates of electron transfer. The quantitative relationships derived by Nicholson and Shain[7] allow the evaluation of kinetic parameters for such rate-limited processes via cyclic voltammetry. A particularly useful function for such measurements is given by the relation

$$\psi = \frac{\gamma_D^{\alpha} k_s}{\sqrt{\pi a D_{ox}}} = f(n\Delta E_p) \tag{3.39}$$

where

$$\gamma_D = \frac{D_{ox}}{D_{red}} \tag{3.40}$$

and the quantity a is that given by Eq. (3.30). Values of the function in Eq. (3.39) are given in Table 3.3. Thus, by adjustment of the cyclic scan rate such that the peak potential difference times the number of electrons involved in the electrochemical process approaches 100 mV, a sensitive measure of the kinetic parameters is possible. This clearly is one of the most convenient methods to evaluate rapid heterogeneous electron-transfer rate constants.

Because of the dynamic nature of voltage-sweep voltammetry, irreversible

TABLE 3.3 Analytical Function Values for Cyclic Voltammetry of Electron-Transfer-Limited Processes (α = 0.5).[6]

($n\Delta E_p$) (mV)	Ψ
61	20.00
63	7.00
64	6.00
65	5.00
66	4.00
68	3.00
72	2.00
84	1.00
92	0.75
105	0.50
121	0.35
141	0.25
212	0.10

processes give an expression for the peak current distinctly different from those for reversible systems:

$$i_p = 2.99 \times 10^5 n\ (\alpha n_a)^{1/2} A D^{1/2} C^b \nu^{1/2} \quad \text{at } 25^\circ\text{C} \tag{3.41}$$

where n_a represents the number of electrons in the rate-controlling step and α is the transfer coefficient (normally with a value between 0.3 and 0.7). The latter two quantities can be evaluated by taking the difference between the peak potential and the half-peak potential:

$$E_p - E_{p/2} = -1.857 \frac{RT}{\alpha n_a F} = -\frac{0.048}{\alpha n_a} \quad \text{at } 25^\circ\text{C} \tag{3.42}$$

An alternative approach is to scan the voltammogram at two different rates. Under these conditions α and n_a may be evaluated by the expression

$$(E_p)_2 - (E_p)_1 = \frac{RT}{\alpha n_a F} \ln \sqrt{\frac{\nu_1}{\nu_2}} \tag{3.43}$$

For irreversible systems the peak potential of a reduction process is shifted toward more negative potentials by about 0.030 V for a decade increase in the scan rate [Eq. (3.43)]. By analogy, a peak of an anodic process is shifted toward more positive potentials. The most characteristic feature of a cyclic voltammogram of a totally irreversible system is the absence of a reverse peak. However, it does not necessarily imply an irreversible electron transfer but could be due to a fast following chemical reaction.

Cyclic voltammetry is one of the most reliable electrochemical approaches to elucidate the nature of electrochemical processes, and to provide insights into the nature of processes beyond the electron-transfer reaction. Several investigations[27–29] have extended this method to the study of the chemical kinetics for chemical processes that precede or follow the electron-transfer process, as well as for the study of various adsorption effects that occur at the electrode surface. However, these are sufficiently complicated that those interested should consult the original papers or recent reviews.[13,14,30–38] Some simple, general cases are discussed in this chapter, and other examples are included in later chapters.

For a process in which a chemical reaction precedes a reversible electron transfer

$$\text{Y} \underset{k_b}{\overset{k_f}{\rightleftharpoons}} ox + n\, e^- \rightleftharpoons red \tag{3.44}$$

[CE (chemical–electron-transfer) mechanism] the shape of the resulting peak depends on the chemical reaction rate. If the chemical reaction is very slow, the current is kinetically controlled, and therefore a wave instead of a peak is observed. If however, the chemical reaction is very fast, the voltammogram is that for a diffusion-controlled electron transfer. For an intermediate condition, the concentration of the electroactive species (and the current) is partially controlled by the chemical reaction and its maximum value is lower than that for the diffusion-controlled process. Hence, for a reduction process an increase of ν causes $i_{p,c}/\nu^{1/2}$ to decrease.

An electrochemical process that is followed by a chemical reaction represents an EC (electron-transfer–chemical) mechanism

$$ox + ne^- \rightleftharpoons red \underset{k_b}{\overset{k_f}{\rightleftharpoons}} Y \tag{3.45}$$

The simplest situation is when the electron transfer is totally irreversible or when the rate of the electrochemical step is much larger than the rate of chemical reaction. For such situations a reverse peak is not observed. If a post-electron-transfer process destroys the product before the reverse scan occurs, the ratio of the cathodic peak current to the anodic peak current will be greater than unity. At low scan rates an anodic peak may not be observed, but becomes detectable after an increase in the scan rate. Relations have been developed to evaluate the rate constants for post-electron-transfer reversible chemical reactions [Eq. (3.36)]:

$$E_p = E_{1/2} - \frac{RT}{nF}\left[0.780 + \ln K\sqrt{\frac{nF\nu}{RT(k_f + k_b)}} - \ln(1 + K)\right] \tag{3.46}$$

and irreversible chemical reactions:

$$E_p = E_{1/2} - \frac{RT}{nF}\left(0.780 - \ln\sqrt{\frac{k_f RT}{nF\nu}}\right) \tag{3.47}$$

where $K = k_f/k_b$.

With electrocatalytic processes (EC′) the following chemical reaction regenerates the electroactive species:

$$ox + n\,e^- \rightleftharpoons red \underset{Z}{\overset{k}{\rightarrow}} ox + W \tag{3.48}$$

(This assumes that Z and W are not electroactive in the potential range under investigation.) Under pseudo-first-order-conditions ([Z] >> [ox]) and when k is small or ν is large, the chemical process has no effect and the reversible couple of peaks is observed. For larger values of k or as ν is decreased, the height of the forward peak increases and the height of the reverse peak decreases. In the limit the forward peak is replaced by a plateau with a height that is independent of the concentration of Z and the scan rate.

The most interesting for chemists probably is the ECE mechanism, which involves a chemical step between two electron-transfer reactions:

$$ox + n_1\, e^- \rightleftharpoons red \xrightarrow{k} ox_1 \overset{n_2 e^-}{\rightleftharpoons} red_1 \tag{3.49}$$

This type of reaction sequence is common for most multielectron processes [e.g., Cu(II) → Cu(s); O_2 → HOOH; *R*CH(O) → *R*CH_2OH]. If both electron transfers are reversible and the chemical reaction is irreversible, the first peak should indicate an n_1-electron process for small k values and an $(n_1 + n_2)$-electron process for large k values. Therefore, an increase of the scan rate should decrease the apparent number of electrons involved in the overall process. Often the second reduction step occurs at a less negative potential than the first, which means that only a single irreversible peak is observed. Potential-scan reversal can provide anodic peaks and data for *red* and red_1.

Adsorption. In the examples discussed above it was assumed that the reactants and products are soluble in the solution and that surface processes (adsorption of the reactants or products, and phase formation and removal) can be neglected. However, if the shape of the peak is unusual (sharp), the electrochemical reaction probably is complicated by a surface process. Cyclic voltammetry is especially sensitive to such phenomena and is a useful characterization tool.[13,14] Adsorption of an electroactive species usually favors the electrode reaction that takes place at a lower potential.

If only adsorbed forms of *ox* and *red* are electroactive, then the voltammetric peaks are sharp and symmetrical (the current rises from zero to a maximum value and then falls again to zero, and there is little or no peak separation).[39–41] The symmetrical, sharp peak results from the fixed amount of the reactant that is adsorbed on the electrode. The values of i_p, E_p, and the peak width depend on the type of adsorption isotherm involved and relative strength of the adsorption of oxidized and reduced species on the electrode surface. If the adsorption is described by a Langmuir isotherm, $E_{p,c} = E_{p,a}$ and the peak current is described by

$$i_{p,c} = \frac{n^2 F^2 A \Gamma_{ox}}{4\, RT}\, \nu \tag{3.50}$$

where Γ_{ox} is the surface excess of *ox* before the potential scan is initiated. *Note:* The peak current is proportional to ν and not to $\sqrt{\nu}$. The area under the peak corresponds to the charge associated with the reduction of the adsorbed layer of *ox*, which allows determination of the surface excess of *ox*:

$$\Gamma_{ox} = \frac{Q}{nF} \tag{3.51}$$

The more common case is that both adsorbed and solution species are electroactive. The voltammograms usually show symmetrically shaped peaks that

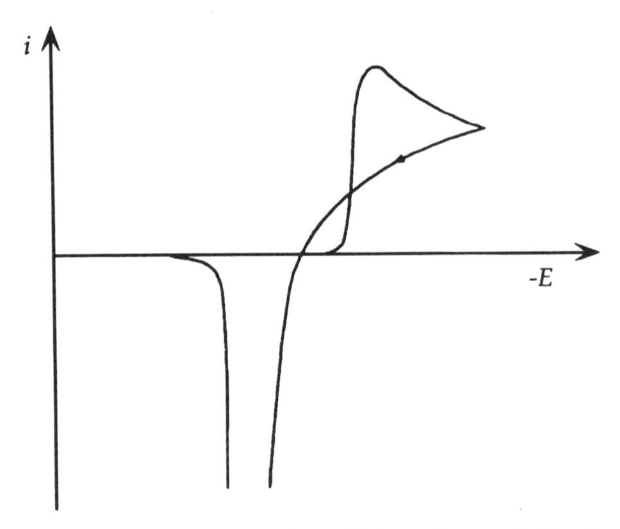

Figure 3.11 Cyclic voltammogram of the oxidation of a metal film deposited on the electrode.

correspond to the electrode reaction for the adsorbed species in combination with the normal peak for soluble species.[42–44]

The deposition of metals or other phases from dilute solutions onto the surface of an electrode is a special case of the adsorptive process. The deposition peak is similar to that for any soluble electroactive species. However, the reverse peak is large, sharp, and narrow (usually off the chart; see Figure 3.11). This occurs because all the reacting substance is on the electrode surface; the diffusion process is eliminated and Eq. (3.50) applies.

3.5 MICROELECTRODE VOLTAMMETRY

Disk-shaped voltammetric electrodes with a radius of less than 10 μm (called *microelectrodes* or *ultramicroelectrodes*) were developed in the late 1970s.[45] However, electrodes of any geometry give the same response if they are sufficiently small. These electrodes are usually prepared by sealing a carbon fiber in a glass capillary with epoxy resin. However, microelectrodes fabricated from platinum, gold, and tungsten wires also have been used. Because of their very small size, microelectrodes exhibit properties different from those of electrodes for conventional voltammetry. Because of their minute dimensions, these electrodes have extremely small currents (on the order of nano- rather than microamperes), which allows electrochemistry to be done under unusual conditions. Thus, the microelectrodes can be used in solutions of high resistance, or even without supporting electrolyte. The latter condition extends the available potential window such that normally inert molecules can be electrolyzed, such as in oxidation of alkanes.[46] Furthermore, the small electrode area reduces proportionately the double-layer capacitance and RC time constant. For these reasons microelectrodes are particularly useful for fast voltammetric measurements.[47] Such electrodes make possible in vivo measurements in clinical analysis. Because they draw extremely small currents, they are practically nondestructive toward the species that is electrolyzed. Moreover, the concentration of an electroactive species is perturbed by the electrolysis to an extremely small distance from the microelectrode. Hence, the current at microelectrodes is not affected by movement in solution, which allows their use in flowing streams. The small size of microelectrodes also enhances the signal-to-noise ratio and thereby extends detection limits.

Microelectrodes also are used for the determination of diffusion coefficients. In these experiments a pulsed waveform usually is used with the current measured at a single potential (at which the process is controlled only by diffusion). Under these conditions, the current obeys the relation

$$i = 4\pi rnFC^{b}D\left[r\left(\frac{1}{\pi Dt}\right)^{1/2} + 1\right] \tag{3.52}$$

where r is the radius of the electrode and other symbols have their usual meaning. This equation indicates that at sufficiently long times the current is independent of time, and that the time for a steady-state current depends on the radius of the microelectrode. At steady-state conditions the diffusion coefficient is directly proportional to the current [Eq. (3.52)].

The main limitation to the use of microelectrode voltammetry is the need to measure small DC currents without noise interference. This usually requires that measurements be done in a Faraday cage (a shield against electronic noise). Recent reviews give good insight into the theory and application of microelectrode voltammetry.[48–51]

3.6 RING–DISK VOLTAMMETRY

The advantages of a solid electrode of fixed area that functions in the voltammetric experiment with a constant diffusion-layer thickness have led to the development of the rotated-disk and ring–disk electrodes.[52–54] By rotation of a disk, the electrode diffusion layer becomes fixed such that the current is constant as a function of time and does not decay [in contrast to conventional voltammetry; Eq. (3.6)]. Voltammetry with such an electrode system gives a current–potential wave that is analogous to a polarogram and follows the relationship

$$E = E_{1/2} + \frac{RT}{nF}\ln\frac{i_{\text{lim}} - i}{i} \tag{3.53}$$

for reversible processes, where i_{lim} is the limiting current on the plateau. The magnitude of the latter is given by the relationship

$$i_{\lim} = 0.62\ nFAD^{2/3}C^{b}\omega^{1/2}\nu^{-1/6} \tag{3.54}$$

where ω is the angular velocity of the disk, s^{-1} [$\omega = 2\pi$ rps (revolutions per second)]; ν is the kinematic viscosity, $cm^2\ s^{-1}$; and the other quantities have their usual meanings. The current increases with the rate of rotation as long as the process is diffusion-controlled. Thus, increasing the rate of rotation until the limiting current ceases to obey Eq. (3.54) provides a measure of the electron-transfer kinetics.

With a concentric pair of electrodes the product from the disk electrode, which is produced at a given potential, is conveyed centrifically to the ring electrode (Figure 3.12). The latter usually is controlled at a different potential such that the product can be monitored. Because the relationship between the ring current and the disk current has been quantitatively established, the ring–disk electrode provides a means of measuring the kinetics of post-electron-transfer reactions of electrode products.

The relationship between the disk current (i_D) and the ring current (i_R) depends on the rate of movement (velocity) of the product species from the disk to the ring electrode. However, only a fraction of the disk-electrode products will reach the ring-electrode surface. Thus, each ring–disk electrode must be calibrated with a well-behaved reversible couple to determine its collection efficiency (N):

$$N = \frac{i_R}{i_D} \tag{3.55}$$

Thus, N represents the fraction of the disk-electrode-formed product (i_R). Couples that are used for the determination of the collection efficiency include $Fe^{II}(CN)_6^{4-}/Fe^{III}(CN)_6^{3-}$, Br^-/Br_3^-, hydroquinone/quinone, and $Fe^{II}(Cp)_2/Fe^{III}(Cp)_2^+$, where HCp represents cyclopentadiene. The collection efficiency N depends only on the electrode geometry (the disk radius and the inner and outer ring radii).

The utility of ring–disk voltammetry for the study of post-electron-transfer chemical reactions is illustrated with an evaluation of a rapid disproportionation rate constant ($HOO\cdot + HOO\cdot \rightarrow HOOH + O_2$; $HOO\cdot$ generated by oxidation of HOOH at a glassy-carbon disk electrode).[55] Figure 3.13*a* illustrates cyclic voltammograms for anhydrous HOOH in dry MeCN. The anodic peak current increases linearly with the HOOH concentration and is proportional to the square root of the scan rate for a given HOOH concentration. After the peak potential is passed during the voltage scan, there is no smooth decay of the current; a current plateau occurs, which indicates the presence of secondary electrode processes before the onset of the solvent-oxidation front.

The voltammetric oxidation of HOOH at a rotated-disk electrode yields a peaked anodic wave with a half-wave potential ($E_{1/2}$) of + 2.1 V versus SCE. The maximum current ($i_{\lim}$) for HOOH oxidation at the rotated-disk electrode is directly proportional to the concentration of HOOH, and is consistent with a first-order diffusion-controlled process.

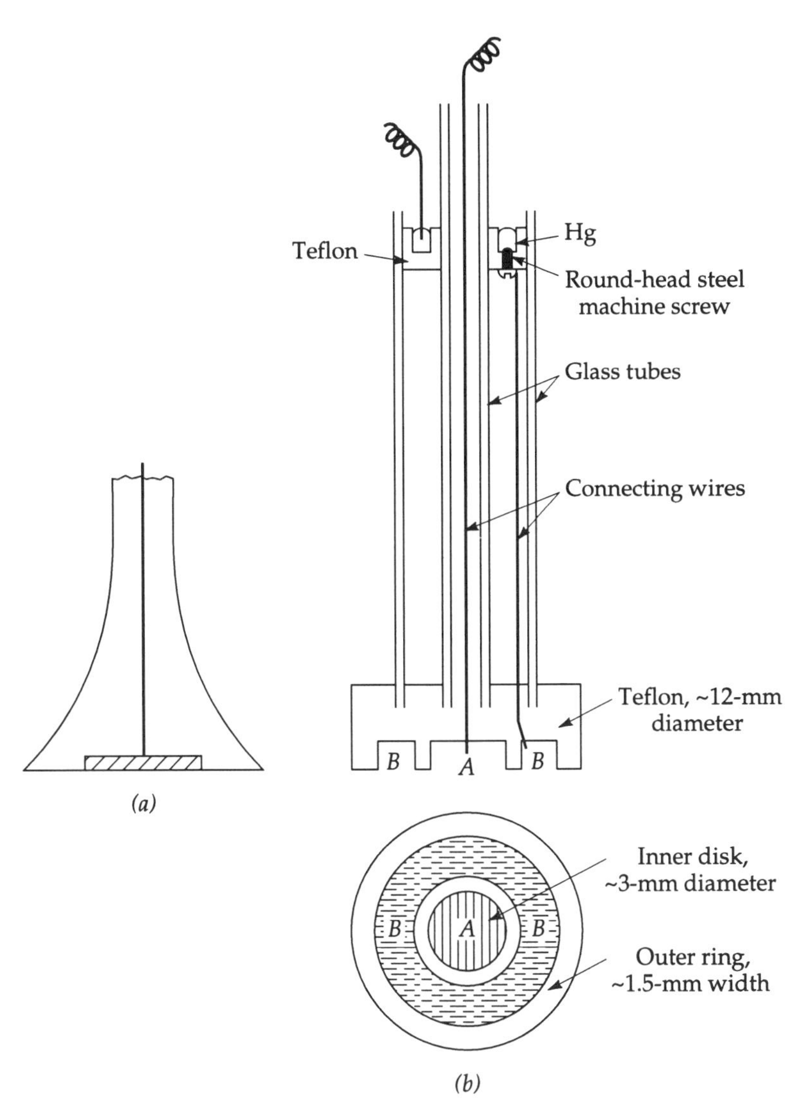

Figure 3.12 Construction of rotated-disk and ring–disk electrodes: (*a*) hydrodynamic disk electrode for high rotation speeds; (*b*) side and bottom views of a ring–disk electrode with wells for carbon paste or press-fit noble metals.

In a rotated ring–disk electrode experiment where the disk is controlled at +2.6 V versus SCE for the oxidation of HOOH and the ring electrode is controlled at −1.4 V versus SCE for the reduction of the oxidation products from HOOH, the observed collection efficiency ($N = I_R/I_D$) is 0.384. This is slightly less than the theoretical value of 0.418 for the electrode. The products from HOOH oxidation at the glassy-carbon disk electrode (E_D, + 2.6 V vs. SCE) can be characterized at the ring by scanning its potential from +1.0 to

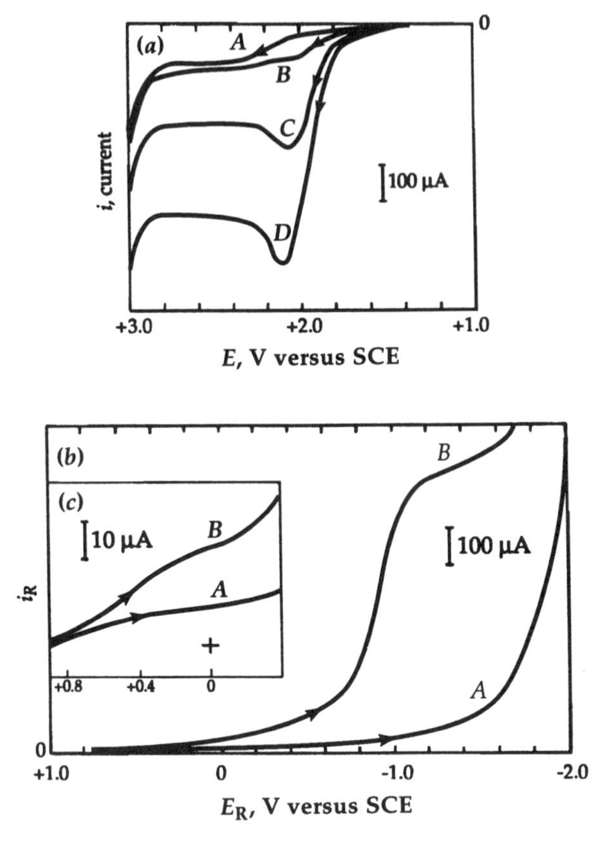

Figure 3.13 Electrochemical oxidation of HOOH and reduction of its products at GC electrodes in MeCN (0.1 M TEAP): (*a*) linear-sweep anodic voltammograms for (*A*) 0, (*B*) 0.3, (*C*) 1.7, and (*D*) 3.3 mM HOOH (scan rate 2 V min^{-1}; electrode area, 0.46 cm^2); (*b*) rotated-ring electrode cathodic voltammogram (scan rate 10 mV s^{-1}) of the product from the oxidation of 4 mM HOOH at the rotated-disk electrode (rotation rate 1600 rpm) for (*A*) E_D disconnected, and (*B*) E_D = +2.6 V versus SCE; (*c*) rotated-ring electrode cathodic voltammogram (scan rate 10 mV s^{-1}) of the products from the oxidation of 1 mM HOOH at the rotated-disk electrode (rotation rate 4900 rpm) for (*A*) disconnected and (*B*) E_D = +2.6 V versus SCE.

−2.0 V versus SCE (Figure 3.13*b*). The first and second of the three reduction waves ($E_{1/2}$, −0.4 V vs. SCE) are better resolved at a lower concentration of HOOH and a higher rotation rate (Figure 3.13*c*). The second ($E_{1/2}$, −0.2 V vs. SCE) and third ($E_{1/2}$ = −0.95 V vs. SCE) reductions are characteristic of O_2 reduction in the presence of protons for this solvent and electrode material. Introduction of O_2 into the cell results in a proportionate increase for the currents of these waves.

The potential for the first reduction wave ($E_{1/2}$, +0.4 V vs. SCE) is consistent with that for HOO·. If the collection efficiencies (N_{exptl}) are determined with the ring-electrode potential at +0.1 V versus SCE, a smooth decay in the N_{exptl} values is obtained as the bulk HOOH concentration is increased. A plot of $1/N_{\text{exptl}}$ against HOOH concentration yields a straight line as shown in Figure 3.14. The HOO·/disproportionation reaction-rate constant (k) can be calculated from these data if (1) the oxidation of HOOH is a diffusion-controlled process (indicated by the scan-rate dependence of voltammetric peak current) and (2) the only decay path for HOO· is its disproportionation in solution while traversing from the disk electrode to the ring electrode. The disproportionation

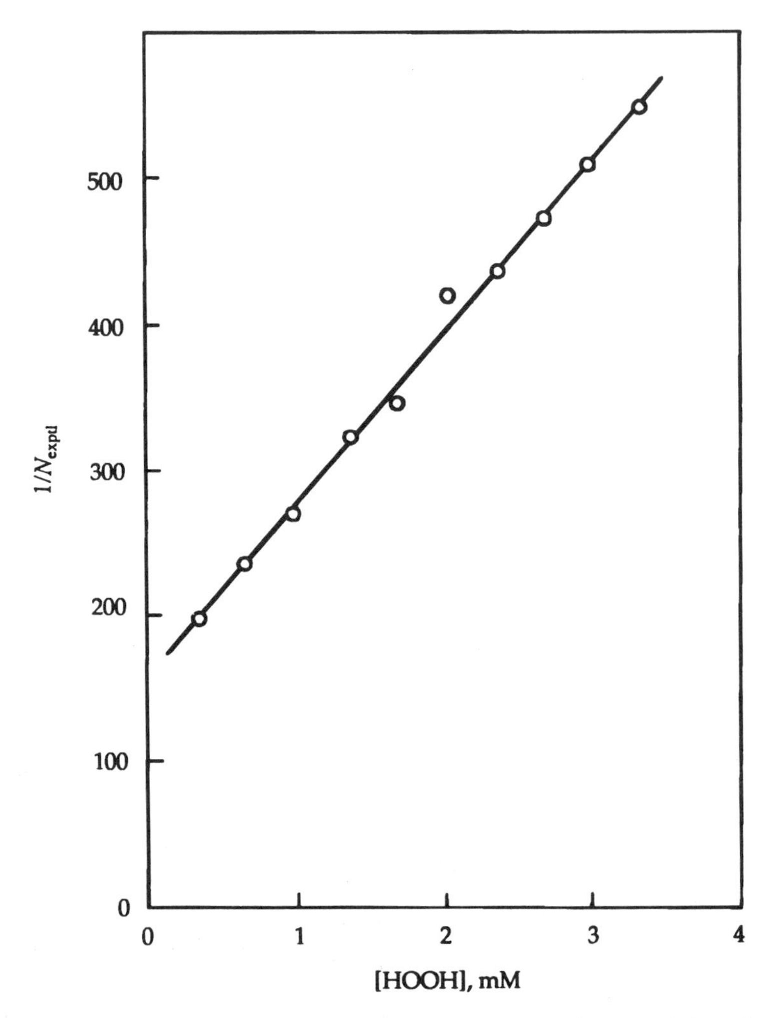

Figure 3.14 Experimental collection efficiencies (N_{exptl}) at the rotated-ring electrode as a function of HOOH concentration for the product from the oxidation of HOOH at the rotated-disk electrode. Control conditions for GC ring–disk electrode: rotation rate, 4900 rpm; E_D, +2.6 V versus SCE; E_R, +0.1 V versus SCE.

of HOO· follows a simple second-order rate law

$$\frac{1}{[\mathrm{HOO}\cdot]_t} - \frac{1}{[\mathrm{HOO}\cdot]_0} = kt \tag{3.56}$$

where $[\mathrm{HOO}\cdot]_t$ is the concentration at the inner edge of the ring electrode after a reaction time t and $[\mathrm{HOO}\cdot]_0$ is the surface concentration at the disk (or zero-time concentration). For a diffusion-controlled process, the latter (at the limiting current condition for the oxidation of HOOH) is equal to the bulk HOOH concentration.

The experimental collection efficiencies N_{exptl}, when compared to the theoretical N, make it possible to relate the two HOO· concentrations ($[\mathrm{HOO}\cdot]_t$ and $[\mathrm{HOO}\cdot]_0$):

$$[\mathrm{HOO}\cdot]_t = \frac{N_{\mathrm{exptl}}}{N}[\mathrm{HOO}\cdot]_0 \tag{3.57}$$

Substitution in Eq. (3.56) gives

$$\frac{1}{N_{\mathrm{exptl}}} = \frac{1}{N} + \frac{kt}{N}[\mathrm{HOOH}] \tag{3.58}$$

From the slope of a plot of $1/N_{\mathrm{exptl}}$ versus [HOOH] the value for the second-order rate constant (k) has an approximate value of $(1.0 \pm 0.5) \times 10^7\ \mathrm{M}^{-1}\ \mathrm{s}^{-1}$. This result is based on a value of 0.418 for N and an estimate of 27 ms for the time to traverse the gap between the ring electrode and disk electrode.[56]

3.7 CHRONOAMPEROMETRY AND CHRONOCOULOMETRY

In voltammetric methods the potential is scanned between selected potentials (E_1 and E_2) with the responding current recorded. In contrast, with potential-step methods, the potential of the working electrode is changed instantaneously between potentials E_1 and E_2 (Figure 3.15*a*), and either the current–time (chronoamperometry) or charge–time (chronocoulometry) curve is recorded. Typical responses are presented in Figure 3.15*b*,*c*. Chronocoulometry is equivalent to the integral of the chronoamperometric signal. As such, the chronocoulometric curve contains no more information than that from chronoamperometry, but its interpretation is simpler. Chronoamperometry has no unique analytical utility, but it is useful for the evaluation of diffusion coefficients, rates of electrode processes, adsorption parameters, and rates of coupled chemical reactions. Measurements should be made over as long a time period as possible to ensure reliable results. During the first 100–300 μs much of the current is due to charging the double layer. At longer times natural convection disturbs the current response. Therefore, the responses over the range from 1 ms to 10 s

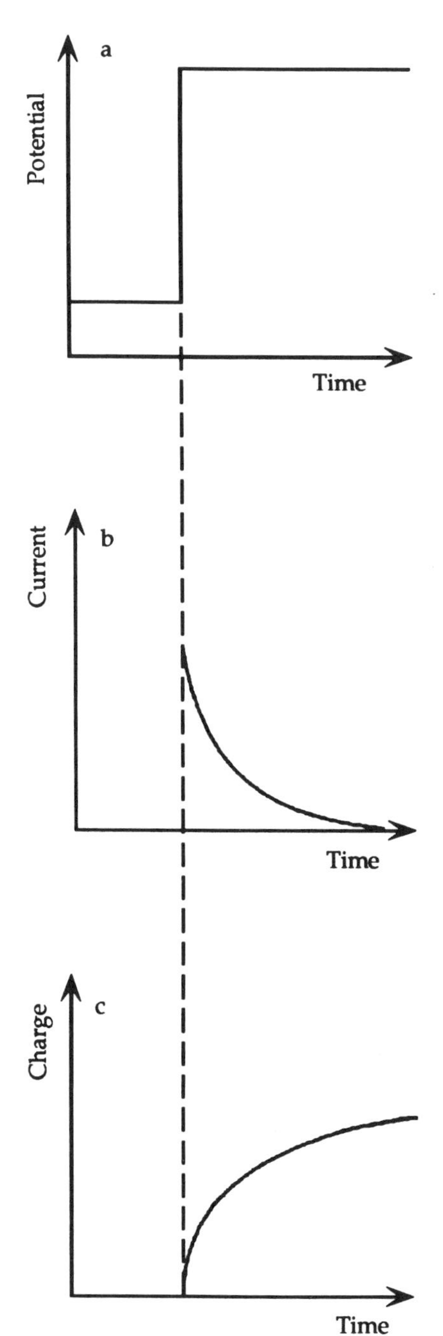

Figure 3.15 Chronoamperometry and chronocoulometry: (*a*) excitation potential step; (*b*) chronoamperometric response; (*c*) chronocoulometric response.

commonly are recorded. Such short times usually require the use of an oscilloscope or microcomputer.

The potentials E_1 and E_2 should be chosen in such way that at E_1 no electrode process occurs and at E_2 the electrode reaction of an electroactive species takes place. If the rate of the electrode process is controlled only by diffusion, the Cottrell equation [Eq. (3.6)] can be applied. Therefore, the observed current should be a linear function of $t^{-1/2}$ with the intercept at the origin (a test for diffusion control). The diffusion coefficient of the electroactive species is directly proportional to the slope of the curve. The heterogeneous rate constant of a kinetically limited electrode reaction (k_c or k_a) also can be evaluated.

The chronocoulometry and chronoamperometry methods are most useful for the study of adsorption phenomena associated with electroactive species. Although less popular than cyclic voltammetry for the study of chemical reactions that are coupled with electrode reactions, these "chrono-" methods have merit for some situations. In all cases each step (diffusion, electron transfer, and chemical reactions) must be considered. For the simplification of the data analysis, conditions are chosen such that the electron-transfer process is controlled by the diffusion of an electroactive species. However, to obtain the kinetic parameters of chemical reactions, a reasonable mechanism must be available (often ascertained from cyclic voltammetry). A series of recent monographs provides details of useful applications for these methods.[13,37,57]

3.8 CONTROLLED-POTENTIAL BULK ELECTROLYSIS

Because of the extensive amount of data available from the polarographic and voltammetric literature, the optimum conditions for macroscopic electrolyses often are established. In particular, controlled-potential electrolysis at a mercury pool can be approached with predictable success on the basis of available polarographic information for the system of interest. An electrolysis can be accelerated by maximizing the electrode surface area and minimizing the thickness of the diffusion layer. However, the same electrode material must be used as in polarography. Thus, a conventional approach in controlled-potential electrolysis is the use of a mercury pool stirred as vigorously as possible with a magnetic stirring bar to minimize the concentration gradient. Under such conditions the decay of the current as well as the decay of the concentration of the electroactive species is given by the relation

$$\frac{i_t}{i_{t=0}} = \frac{C_t}{C_{t=0}} = \exp\left[\left(\frac{-DA}{V\Delta x}\right)t\right] \tag{3.59}$$

where V is the volume of the solution to be electrolyzed and Δx is the thickness of the concentration gradient. Thus, the current and concentration decay exponentially. Under idealized conditions 90% of the electroactive species will be electrolyzed in approximately 20 min. Increases in the temperature as well as in the electrode area relative to the solution volume will accelerate the rate of electrolysis. The fundamentals of controlled-potential bulk electrolysis are given by Lingane,[58] and the method is discussed in recent monographs.[13,59]

3.9 METHODOLOGY

With the exception of the indicating-electrode systems the instrumentation for polarography and voltammetry is basically the same. The unique feature for polarography is the dropping-mercury electrode, which has a number of advantages but also introduces certain difficulties that must be overcome if reliable data are to be obtained. The advantages of this electrode are that it provides a continually renewed surface of liquid mercury that eliminates electrode contamination. Furthermore, the high hydrogen overvoltage on mercury permits a number of ions to be reduced before solvent or hydronium ions are reduced. Thus, it extends the range of cathodic processes that may be studied. Because of the repetitive formation of the drop of mercury at the end of a capillary tube, sufficient stirring is provided such that the maxima of the current oscillations remain constant at a given potential and do not decay with time. This is in contrast to a static microelectrode whose current is inversely proportional to the square root of time, due to semiinfinite linear diffusion.

If a solid microelectrode is used in a static position, the voltammetric current–voltage curve will give a peak just past the half-wave potential after which the current will decay as a function of the square root of time. In the preceding section relations are given that relate the peak current to the concentration as well as to the rate of voltage scan. By agitation or rotation of the solid electrode, conditions can be obtained to yield a current plateau, similar to the polarographic wave obtained with a dropping-mercury electrode, that will be directly proportional to the concentration of the active species (with the current constant and independent of time). In the early days of voltammetry this form of electrode had the wire mounted such that its axis was perpendicular to the axis of rotation; commercial equipment used a 600-rpm (revolutions per minute) motor to rotate this electrode with contact being made through a mercury pool at the center of the axis of rotation. Since then rotated-disk and rotated ring–disk voltammetry has utilized a disk electrode made out of platinum, other metal, or glassy carbon that is sealed in the end of glass or plastic, and of a larger area; frequently as large as a 0.5 in. in diameter (Figure 3.12). With careful control of the geometry, conditions can be developed such that the limiting current on the voltammetric plateau is directly proportional to both the concentration and the square root of the rate of rotation. Under these conditions the limiting current follows the relationship of Eq. (3.54).

To summarize, voltammetric electrodes have taken many special forms; the necessary conditions are that they be of a geometry and be used in such a way that the current that flows is diffusion-controlled. Thus, the electrodes can be in the form of disks, cylinders, spheres, or any other shape that fulfills the necessary experimental conditions. The disk electrode provides the means to give a linear diffusion-controlled response over a wide dynamic range; in contrast, cylindrical and spherical electrodes have only a limited set of conditions that yield such a response. The use of closely spaced parallel pairs of electrodes is a specialized case and is more related to controlled-potential electrolysis than to voltammetry. However, conditions can be adjusted such that the principles of voltammetry apply (the current is diffusion-controlled).

In voltammetry the second necessary part of the electrode system is an adequate reference electrode. Such an electrode must have a well-established and invariant potential that is not affected by the passage of significant quantities of current. During the first decade of polarography the major application was the measurement of diffusion currents rather than the evaluation of kinetic and thermodynamic quantities. Thus, a mercury pool at the bottom of the electrolysis cell frequently served as a reference electrode. This system was subject to a number of problems; these include a variable potential and the introduction of electrolysis products that might ultimately interfere with the sample system under study at the indicator electrode. Today most voltammetric measurements utilize a saturated calomel electrode (SCE) connected to the sample solution by a large salt bridge. Convenient electrode and cell systems for polarography and voltammetry are discussed in Chapter 5.

When voltammetry measurements are made in nonaqueous solvents, the problems of an adequate reference electrode are compounded. Until the 1960s the most common reference electrode was the mercury pool, because of its convenience rather than because of its reliability. With the advent of sophisticated electronic voltammetric instrumentation, more reliable reference electrodes have been possible, especially if a three-electrode system is used. Thus, variation of the potential of the counter electrode is not a problem if a second non-current-carrying reference electrode is used to monitor the potential of the sensing electrode. If three-electrode instrumentation is used, any of the conventional reference electrodes common to potentiometry may be used satisfactorily. Our own preference is a silver chloride electrode connected to the sample solution by an appropriate noninterfering salt bridge. The one problem with this system is that it introduces a junction potential between the two solvent systems that may be quite large. However, such a reference system is reproducible and should ensure that two groups of workers can obtain the same results.

Because polarographic and voltammetric measurements assume that the mass transfer occurs only by diffusion, any electrostatic field effects that will attract or repulse the electroactive species must be avoided. The significant electrostatic fields that exist between the indicator electrode and the reference electrode must be dissipated by introducing an inert supporting electrolyte. Normally, a 100-fold excess of supporting electrolyte relative to the sample species is a minimum requirement if migration currents are to be avoided. Almost any strong electrolyte can serve as the supporting electrolyte in voltammetry, but over the past decade approximately a dozen such systems have evolved to meet

TABLE 3.4 Examples of Polarographic Supporting Electrolytes

Electrolyte	Effective Voltage Range, V versus SCE
1 M KCl	+0.1 to −2.1
1 M HCl	+0.1 to −1.4
12 M HCl	−0.4 to −1.0
1 M NH_3 and 1 M NH_4Cl	−0.1 to −1.8
1 M NaOH	−0.1 to −2.0
1 M NaF	+0.2 to −2.0
1 M KCN	−0.6 to −1.9
2.5 M H_3PO_4	+0.1 to −1.3
1 M KSCN	−0.3 to −1.6
0.3 M Na_3(citrate) and 0.1 M NaOH	−0.2 to −1.8
0.25 M Na_2(tartrate), pH 9	0.0 to −1.8
2 M HOAc and 2 M NH_4OAc, 0.01% gelatin	+0.3 to −1.3
0.3 M $N(C_2H_4OH)_3$ and 0.1 M KOH	−0.3 to −1.3
0.1 M pyridine and 0.1 M pyridinium chloride	+0.2 to −1.2

the needs of practicing polarographers. These are summarized in Table 3.4 and indicate the range of media that are available for voltammetric studies.

With the exception of perchlorate media almost all supporting electrolyte systems have some tendency to complex metal ions. Thus, the half-wave potential of such ions will be affected significantly by the composition of the supporting electrolyte. Furthermore, if the electroactive species has any acidic or basic properties, the acidity of the supporting electrolyte will also have a direct effect on the potential at which many species will be reduced. Clearly the effect on the half-wave potential of the supporting electrolyte can be used to advantage for the identification and selective detection of species in mixtures. A number of tabulations have been published that summarize the voltammetric behavior of the most common substances in the presence of a variety of supporting electrolytes.[10,11] Extensive summaries of polarographic and voltammetric data for inorganic[60] and organic[61,62] compounds have been compiled. Extensive electrochemical data are also included in the *Encyclopedia of Electrochemistry of the Elements.*[63]

Because O_2 is electroactive at almost all electrodes, it is a serious interference in voltammetric and electrochemical measurements. For this reason means must be provided to deaerate the sample solution before meaningful voltammetric measurements can be made. Although for many years further purification of tank nitrogen was necessary to remove traces of oxygen, today prepurified nitrogen is sufficiently inexpensive and available to eliminate this problem. For alkaline solutions sodium sulfite provides a convenient means for directly eliminating any interfering O_2. A final precaution is that the deaeration of organic solvents is far more difficult than that of aqueous solutions; this is primarily because of the much greater solubility of O_2 in such solvents relative to water.

During the past 20 years increasing interest in the use of electrochemical methods for organic and biochemical systems has promoted the application of nonaqueous solvents for voltammetric and polarographic measurements. A major problem is to find a system such that an adequate amount of supporting electrolyte may be introduced and dissociated to prevent migration currents. For some time the most popular solvent systems were mixtures of alcohols and water, as well as mixtures of dioxane and water. However, with improved instrumentation and the use of three-electrode systems pure nonaqueous solvents have become popular. Thus, polar materials such as acetonitrile, dimethyl sulfoxide, and dimethylformamide have sufficiently high dielectric constants to make them useful solvents for electrochemical measurements, including voltammetric studies. A number of electrolytes are soluble in these solvents, and are sufficiently dissociated to make effective supporting electrolytes. Each of them must be purified to avoid interferences (see Chapter 6). With such solvents the most popular supporting electrolytes are lithium chloride, lithium perchlorate, and various tetraalkylammonium salts, including tetraethylammonium perchlorate, tetramethylammonium chloride, and tetrabutylammonium iodide.

If the half-wave potentials or the peak potentials in polarography and voltammetry are to be used for identification purposes as well as for the evaluation of thermodynamic quantities, they must represent the actual electrode potential relative to the reference electrode. Unfortunately, with the passage of current and the existence of a finite resistance in all electrochemical cells, the recorded current–potential curve includes the iR drop across the cell if a two-electrode form of instrumentation is used. For most electrochemical measurements the upper limit for polarographic currents is approximately 20 μA. Assuming a cell resistance of 200 Ω, the recorded half-wave potential would be in error by -4 mV as a result of iR losses. Clearly this approaches the limits of accuracy of most instrumentation and is negligible. On the other hand, use of a nonaqueous solvent may result in a cell resistance of 20,000 Ω, which for a 20-μA current would give an iR drop of -0.4 V, and thereby preclude meaningful identification as well as thermodynamic data without making precise corrections. Because corrections of this order of magnitude are almost impossible with any degree of precision, the practical alternative for high-resistance systems is employment of a three-electrode system, in which the indicator-electrode potential is referred to a non-current-carrying reference electrode. Through the use of operational amplifiers for the construction of electronic polarographs and potentiostats such three-electrode instrumentation is commonplace and commercially available. However, complete compensation demands judicious placement of the nonworking reference electrode because the iR losses represent a field gradient between the indicator electrode and the current-carrying counter electrode. Thus, placement of the reference electrode must be as close as physically possible to the surface of the indicator electrode. To do this directly is awkward and unnecessary. A more convenient arrangement is the use of a Luggin capillary, which is normally constructed out of glass or plastic tubing. Figure 3.16 illustrates such a capillary as well as its appropriate placement in an electrochemical cell. The principle of this device is that the electric-field

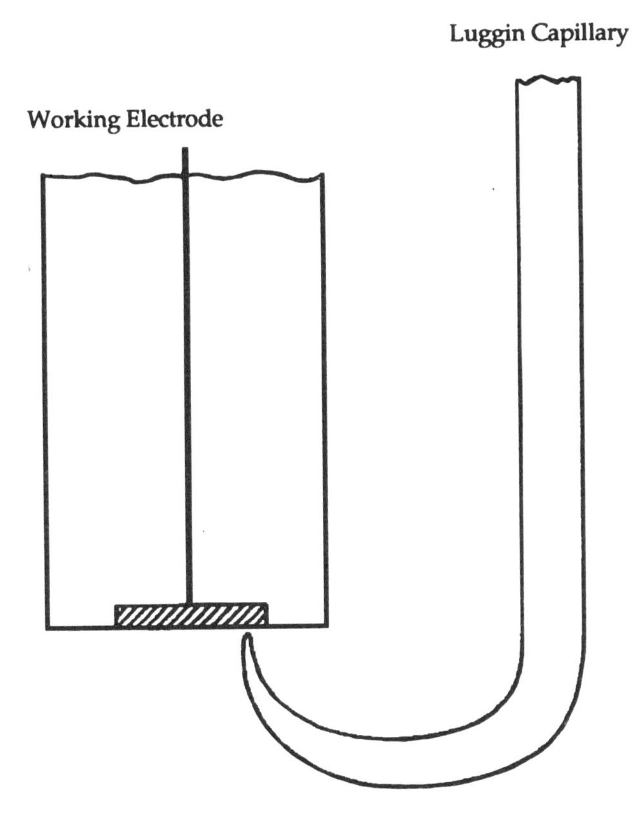

Figure 3.16 Luggin capillary and its placement relative to the working electrode.

gradient represented by the *iR* loss is not transmitted through the glass or plastic tubing because it is an insulator. Hence, if the tip of the Luggin capillary is within a millimeter or so of the surface of the indicator electrode, the electrolyte within the Luggin capillary does not contribute any *iR* loss because current is not flowing and the potential at the electrode surface is effectively referenced to that of the reference electrode within the Luggin capillary.

The instrumentation for voltammetry in the early days of its use normally provided a means to scan the applied potential or the potential of the indicator electrode with a motor-driven slidewire or by electronic means. Prior to the advent of recording instrumentation, manual instruments used a hand-adjusted potentiometer to obtain a point-by-point scan, with the current monitored by a ballistic galvanometer. Today commercial instrumentation makes use of operational amplifiers or digital electronics to provide linear potential scans (ranging from 0.001 to 10,000 V s^{-1}). Traditionally, the scan rates in polarography and voltammetry were about 0.1 V min^{-1}. This evolved in part because of the limitation in recording devices during the early years of polarography. However, there is a tendency to increase scan rates, and many modern instruments record 1-V polarograms in 3–5 min. With the advent of voltage-sweep and cyclic voltammetry this acceleration of scan rate has progressed to the point where it is not unreasonable to scan with rates as high as 100 V s^{-1} with fast recorders.

Linear-sweep and cyclic voltammetry requires instrumentation that allows adjustment of the initial potential and the terminal potential for a voltage sweep. In the case of cyclic voltammetry the instrumentation must allow reversal of the sweep at variable points along the potential scan. For moderate scan rates, this can be accomplished manually; at more rapid scan rates, microswitches are useful; and at the most rapid rates, transistor switching circuits and computer control are necessary. In general, voltage-sweep and cyclic voltammograms are most conveniently recorded with an *X–Y* recorder that has a slewing speed of at least 25 cm s^{-1}. For the fastest scan rates a memory oscilloscope or microcomputer is necessary to obtain an accurate record of the current–voltage curve.

Another specialized form of voltammetry involves the use of either a rotated-disk or a ring–disk indicating electrode. With this type of electrode the current is directly proportional to the square root of the rate of rotation if it is a diffusion-controlled process. To obtain complete adherence to the square-root relationship, a hydrodynamically sound design for the electrode is essential.[43] Figure 3.12 illustrates the geometric features that have been found to give reliable performance for rotation rates as high as 10,000 rpm.

The ring–disk electrode represents a specialized variant of the rotated-disk electrode and involves a concentric ring electrically insulated from the disk electrode. The advantage of this form of the electrode is that products produced by the disk electrode may be monitored at the ring electrode. Thus the potential of the ring can be controlled at a level different from that of the disk electrode to give a specific response for a transient or unstable product. Although this requires a potentiostat separate from that controlling the potential of the disk electrode, the ring–disk electrode has found extensive application for the study for the kinetics and mechanisms of post-electron-transfer reactions. Figure 3.12 indicates the details of construction.

Another recent specialized form of electrode configuration is known as the *thin-layer electrolysis system.*[64] With this arrangement two electrodes are placed sufficiently close to each other that the product at one diffuses rapidly to the second electrode, where it may be monitored. Figure 3.17 illustrates one form of this system in which a precision micrometer is used as a support of one electrode and provides the means for adjusting the gap between the electrodes. In its simplest application this system provides the means of evaluating the electron stoichiometry of an electrode reaction. It also, like the ring–disk electrode system, provides a convenient means to monitor the products of the opposite electrode reaction and to evaluate the kinetics of post-electron-transfer reactions. With adequate electronic instrumentation the potential of both electrodes can be controlled independently such that one or the other will be controlling the magnitude of the electrolysis current. General control of the poten-

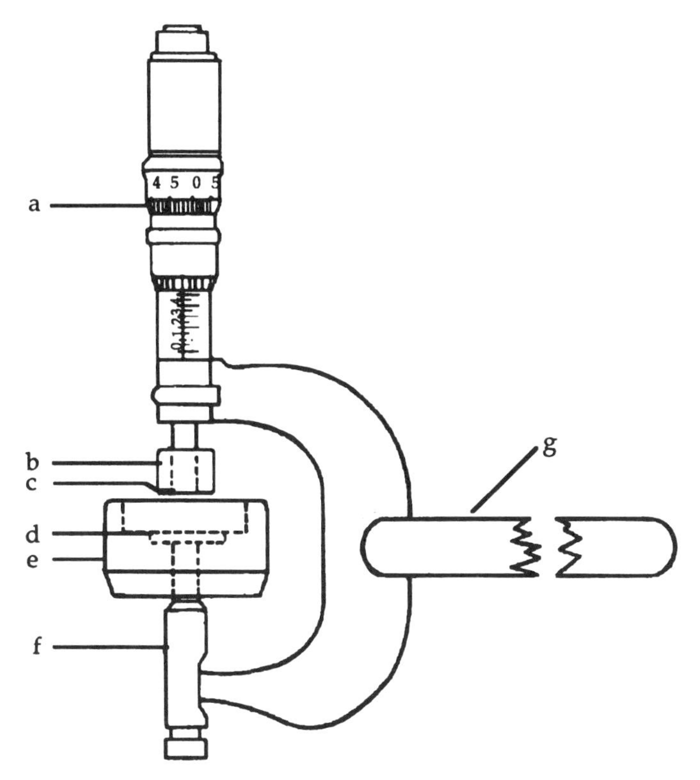

Figure 3.17 Micrometer thin-layer electrode system: (*a*) outer thimble for readings to 1×10^{-4} in; (*b*) press-fit Teflon collar; (*c*) platinum face of micrometer spindle; (*d*) flat, glass disk in Teflon cup against the face of the detachable anvil; (*e*) machined Teflon cup; (*f*) Starrett No. 212 detachable anvil; (*g*) rod for mounting cell assembly.

tial is accomplished with conventional potentiostats. A crucial feature in parallel-electrode electrolysis is the maintenance of a precisely parallel configuration. Obviously this becomes an increasingly demanding specification as the spacing is made smaller and smaller. To date systems have been developed with gaps as small as 20 μm.

One of the oldest forms of electrochemistry involves the analysis of materials by electrodeposition. The principles are extremely simple but still provide some of the highest levels of precision possible in analysis. Usually the electrolysis potential is provided by either a DC-motor generator or a battery with a crude potentiometer to control the applied voltage. Provisions are not made to compensate the *iR* drop across the cell, and selectivity usually is obtained by control of the acidity and of the supporting electrolyte of the electrolysis solution. A further degree of selectivity is provided by the working electrode and its overpotential for certain reactions. Thus, mercury, because of its high overpotential for the reduction of hydronium ion, provides a wide range of potentials for the reduction of most metal ions before the reduction of hydronium ion interferes. In contrast, if platinum is used only electropositive ions such as copper(II) will be reduced without hydronium ion interfering.

As an example, a sample that contains a mixture of copper(II) and nickel(II) salts can be analyzed by first electrolyzing the sample solution under acidic conditions with platinum electrodes such that the copper is plated onto a platinum gauze electrode. Because the solution is acidic, hydronium ion is reduced before nickel ion and there is no interference. After the electrolysis for copper is completed, the electrolysis solution can be neutralized and made basic with ammonia. Having determined the copper and removed it from the platinum electrode, one can electrolyze the remaining basic electrolysis solution to plate nickel on the platinum electrode.

Classical electrodeposition has limited application today because of the development of more sophisticated methods. It is a slow and tedious process, and there is a need for large samples. However, it still represents one of the most precise quantitative techniques for the determination of copper, cadmium, silver, and nickel. The best conditions for the determination of these and other ions by electrodeposition are summarized in Table 3.5.

The mercury-pool electrode is especially attractive because of its renewable surface and because it allows the use of a stirring bar to vigorously stir the electrode–solution interface. Its major limitation relative to a platinum electrode is its high mass and the awkwardness of rinsing and of weighing a liquid electrode relative to a solid electrode.

The rate of electrodeposition is dependent on a number of factors, and these are predictable to only a limited degree. However, the thickness of the diffusion layer must be minimized to obtain a rapid electrolysis. This is accomplished by vigorously stirring and by the use of electrodes with large surface areas. An increase in temperature enhances the rate of electrolysis because it increases the mobility of the electroactive species. The use of high ionic concentrations minimizes the iR drop between electrodes and also improves the electrolysis rate. The orientation and geometry of the electrodes is important to insure a uniform and adherent plate. Depolarizers frequently are introduced to prevent formation of interfering products from the counter electrode (see Table 3.5).

With the development of more sophisticated electronic instrumentation, a higher degree of control has been made possible through the development of the potentiostat. The basic operating principle of a potentiostat involves a three-electrode assembly such that the working electrode's potential is monitored relative to a closely spaced reference electrode. In the simplest form this potential is controlled by manually adjusting the applied voltage across the working electrodes while monitoring the potential of the working electrode relative to the reference electrode. However, this form of controlled-potential electrolysis is not only extremely tedious, but it is inadequate in terms of precise potential control. Even with the most "religious" attention the dexterity of an individual is such as to give an effective response time of at least 0.1 s.

The next stage in the development of the potentiostat was to replace the

TABLE 3.5 Electrolysis Conditions for the Determination of Metal Ions

Metal Ion	Solution Conditions	Electrolysis Conditions	Interferences
Ag(I)	200 mL of 1 M HNO_3, 0.5 M $NaNO_2$	2 A, 1 h	As, Hg, Se, Te
	10% KCN	2 A, 20 min	Au, Bi, Cd, Co, Cu, Hg, Ni, Zn
Au(III)	160 mL of 0.6 M HCl, 0.15 M $H_2OH \cdot HCl$	2–3 A, 20 min	Bi, Cd, Cu, Hg, Pb, Pt, Sb, Sn, NO_3^-
	100 mL of 0.2 M KCN	0.3–0.5 A, 30 min 30–50°C	Cu, Pd, Pt, Ag, Bi, Cd, Co, Hg, Ni, Zn
Bi(III)	200 mL of 0.5 M $HClO_4$, 5 mL satd. $N_2H_4 \cdot H_2SO_4$	1 A, 1 h	Ag, As, Cd, Cu, Hg, Pb, Sb, Sn
Cd(II)	200 mL of 0.1 M H_2SO_4, 10 mL 0.1% gelatin (Cu/Pt cathode)	3 A, 40 min	Cu, Ag, Au, Bi, Hg, Pb, Pt, Sb, Sn
Co(II)	100 mL of 2 M NH_3, 0.6 M NH_4Cl, 0.015 M $NH_2OH \cdot HCl$	2–5 A, 45 min	Cu, Ni, Zn, Pd, Tl
Cu(II)	200 mL of 0.5 M H_2SO_4, 0.25 M HNO_3	2 A, 1 h	Ag, As, Au, Bi, Hg, Mo, Pt, Pd, Sb, Se, Te, Sn, *N*-oxides, Cl^-
Fe(II)	100 mL of 0.2 M H_2SO_4, 0.4 M $(NH_4)_2C_2O_4$	6 A, 30 min	Co, Cu, Mn, Ni
Fe(III)	100 mL of 1 M HCl, 1 M H_3PO_4, NH_3 to neutralize plus 15 mL, NH_3 excess, diluted to 200 mL	2 A, 45 min	As, Co, Cu, Mo, Ni, Sb, W, Zn
Hg(II)	100 mL of 1.5 M HNO_3 (Au cathode)	1 A, 45 min	Ag, Au, Bi, Cd, Cu, Pb, Pt, Sb, Sn
Ni(II)	100 mL of 6 M NH_3, 0.1 M $N_2H_4 \cdot 2\ HCl$ (Cu/Pt cathode)	0.5 A, 3 h	Co, Cu, Fe, In, Pd, Tl, Zn, NO_3^-
Pb(II)	100 mL of 1 M H_3PO_4, 10 mL 0.1% gelatin (Cu/Pt cathode)	0.4 A	Bi, Cu, Sb
Pb(II)	100 mL of 1 M HNO_3	2.4 A, 1.5 h, 70–80°C (on anode as PbO_2)	Fe, Mn, Tl
Pd(II)	150 mL of 2.5 M NH_3 (Ag/Pt cathode)	0.05 A, 12 h (unstirred)	Co, Cu, Fe, Hg, In, Ni, Tl, Zn

TABLE 3.5 (*Continued*)

Metal Ion	Solution Conditions	Electrolysis Conditions	Interferences
Pt(IV)	100 mL of 0.05 M H_2SO_4 (Cu/Pt cathode)	0.01–0.03 A, 5 h 70°C (unstirred)	In, Ni, Tl, Zn
Rh(III)	100 mL of 0.05 M H_2SO_4 (Ag/Pt cathode)	8 A, 15 min	Ag, An, Bi, Cd, Cu, Hg, Pb, Pt, Sb, Sn, Tl
Sn(II)	200 mL of 1 M H_2SO_4, 0.25 M $HClO_4$, 0.3 M HCl, 0.35 M $NH_2OH \cdot HCl$ (Cu/Pt cathode)	2 A	Ag, As, Au, Bi, Cd, Cu, Hg, Pb, Pt, Sb, Tl
Tl(I)	180 mL of 0.25 M HNO_3, 0.05 M benzoic acid (Hg/Pt cathode)	5 A, 15 min, 45°C	Ag, As, Au, Bi, Cd, Hg, Pb, Pt, Sb, Sn
U(VI)	100 mL of 0.5 M NH_4OAc, pH 6–7	0.2 A, 90°C	Mo
Zn(II)	125 mL of 0.12 M KCN, 2.5 M NH_3 (Cu/Pt cathode)	3 A, 30 min	Ag, Au, Bi, Cd, Co, Cu, Hg, Ni
Zn(II)	100 mL of 0.2 M KNa(tartrate) plus sufficient KOH to dissolve ppt (Cu/Pt cathode)	0.3 A, 45 min	Bi, Cu, Fe, Pb, Sn

human operator by a servomechanical system, whereby microswitches are placed on a recorder to monitor the working-electrode potential such that when preset limits are approached these activate a motor-driven rheostat that controls the applied voltage. Again the mechanical nature and slow response-time limit the applicability of this type of potentiostat. During the past decade electronic potentiostats, based to a large extent on operational amplifiers, have provided sophisticated instrumentation with response times of less than 1 ms, applied potential capabilities of up to 100 V, and currents of up to several amperes. Several commercial instruments based on these modern approaches are available. An ideal potentiostat allows one to set the electrode potential to a precision of ± 1 mV over a potential range from $+3$ to -3 V, to have electronics able to apply up to 100 V for high-resistance solutions and to deliver currents of at least 1 A, and to have a response time close to 1 ms.

The majority of controlled-potential electrochemistry has been carried out at mercury-pool electrodes. This is because of the vast amount of reference data available from polarography. Furthermore, the uniform and reproducible surface, and the high voltage for solvent reduction make the mercury pool particularly attractive relative to solid electrodes. As with electrodeposition, controlled-potential electrolysis rates are dependent on electrode area, stirring rates, solution volume, solution temperature, and supporting electrolyte. If the diffusion layer is uniform and the applied potential is such that one is on the diffusion plateau, the electrolysis obeys the relation

$$i_t = i_{t=0} \exp\left[\left(\frac{-DA}{V\Delta x}\right)t\right] \tag{3.60}$$

where Δx is the thickness of the diffusion layer. Under the best obtainable conditions 99.9% of the material in a sample can be electrolyzed in approximately 29 min. Because of the nonlinear nature of the electrolysis current, its integration is not simple.

To decrease the size of the samples and to avoid the necessity of weighing the electrode to determine the amount of a substituent, a number of devices to integrate the electrolysis current have been developed. An example of a simple, but sensitive and accurate integrator is illustrated in Figure 3.18. This is based on the O_2/H_2 coulometer and relies on the stoichiometric electrolysis of acidified water to give a reproducible volume of gas under standard conditions per coulomb of electricity.[58] Although cumbersome, it is capable of good precision when corrected for barometric pressure and the vapor pressure of water. Obviously, if controlled-potential electrolysis is used for the conversion of a species from one ionic state to another, it is essential to have some current-integrating system because there is no means for weighing the amount of material electrolyzed. Through the use of standard resistors to convert the electrolysis current to a potential, any integrating device capable of integrating potential–time curves can be used for current integration. The most popular integrating device for electrochemistry consists of operational amplifiers in an

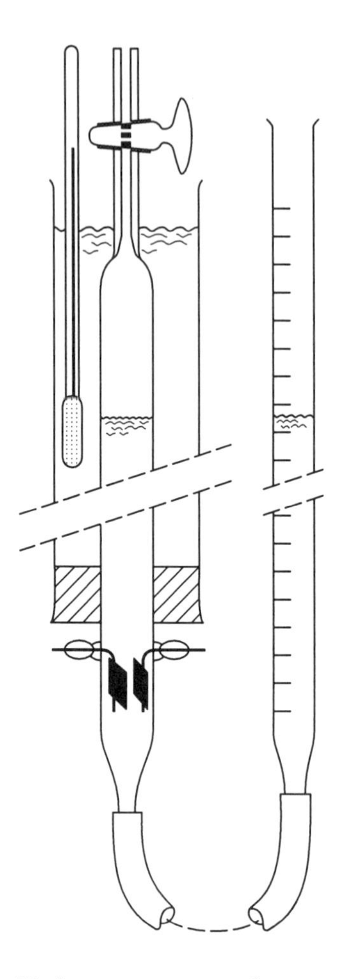

Figure 3.18 The cell contains 1 M H_2SO_4 and platinized platinum electrodes. A thermometer monitors the temperature of the water jacket of the cell, and a conventional burette measures the volume of displaced electrolyte solution.

Hydrogen–oxygen coulometer

R–C circuit. Assuming the use of operational amplifier instrumentation, this is a simple and inexpensive adjunct to provide for effective integration of potential–time curves.

3.10 APPLICATION OF CONTROLLED-POTENTIAL METHODS

To date the most extensive application of electrochemical methods with controlled potential has been in the area of qualitative and quantitative analysis. Because a number of monographs have more than adequately reviewed the literature and outlined the conditions for specific applications, this material is not covered here. In particular, inorganic applications of polarography and voltammetry have been discussed in great detail in the classic monograph by Kolthoff and Lingane.[9] Excellent compilations of polarographic half-wave potentials and diffusion current constants are presented in the monograph by Meites.[10] The two-volume work by Kolthoff and Lingane also provides a good, but somewhat dated, review of organic and biological applications of polarographic methods. A more recent treatment is offered in the two-volume work by Kolthoff and Zuman.[65] A brief summary of the best conditions for the polarographic determination of a number of inorganic and organic substances is presented in Table 3.6. This represents an extremely limited summary; there are many other conditions that are satisfactory and that may provide the degree of selectivity needed for a specific analysis. These interested should refer to the recent compilations.[60–63]

An important specialized type of voltammetric system is a self-contained cell for the determination of O_2 in the gas or solution phases. This is the so-called Clark electrode,[66,67] which consists of a platinum or gold electrode in the end of a support rod that is covered by an O_2 permeable membrane (polyethylene or Teflon) such that a thin film of electrolyte is contained between the electrode surface and the membrane. A concentric tube provides the support for the membrane and the means to contain an electrolyte solution in contact with a silver–silver chloride reference electrode. The Clark device has found extensive application to monitor O_2 partial pressure in blood, the atmosphere, and in sewage plants. By appropriate adjustment of the applied potential it gives a voltammetric current plateau that is directly proportional to the O_2 partial pressure. The membrane material prevents interference from electroactive ions as well as from surface-contaminating biological materials. Figure 3.19 illustrates one configuration for this important device.

Controlled-potential electrolysis provides the features of electrodeposition plus the ability to carry out analyses more rapidly while using smaller samples. Furthermore, by integrating the current–time curve the necessity to plate a weighable amount of substituent is eliminated. One of the most important applications of controlled-potential electrolysis is the evaluation of the number of electrons involved in the electrode reaction (n).

In addition to the analytical applications discussed above, controlled-potential methods are used for the evaluation of thermodynamic data and diffusion coefficients in both aqueous and nonaqueous solvents. Polarographic and voltammetric methods provide a convenient and straightforward means for evaluation of the diffusion coefficients in a variety of media. The requirements are that the current be diffusion-controlled, the number of electrons in the electrode reaction be known, and the concentration of the electroactive species and the area of electrodes be known. With these conditions satisfied, diffusion coefficients can be evaluated rapidly over a range of temperatures and solution conditions.

Voltammetric methods also provide a convenient approach for establishing the thermodynamic reversibility of an electrode reaction and for the evaluation of the electron stoichiometry for the electrode reaction. As outlined in earlier

TABLE 3.6 Optimum Conditions for Polarographic Determination of Inorganic and Organic Substrates[a]

A. Inorganic Ions

Ion	Supporting Electrolyte	$E_{1/2}$ versus SCE	I
Ag(I)	Dilute HNO_3, $NaNO_3$, $HClO_4$, or $NaClO_4$	>0	—
	1 M NaF, 0.01% gelatin	>0	2.35
	12 M HCl	>0	—
		−0.65	—
	0.1 M KNO_3	>0	2.50
As(III)	0.1 M NH_3, 0.1 M NH_4Cl	−1.71	—
	1 M Na_3cit, 0.1 M NaOH	(−0.31)	—
	1 M HCl, 0.0001% methylene blue	−0.43	6.04
		−0.67	12.00
	12 M HCl	>0	3.94
		−0.55	—
	0.5 M KOH, 0.025% gelatin	(−0.26)	−3.82
	1 M HNO_3, 0.01% gelatin	−0.70	—
		−1.00	8.80
	7.3 M H_3PO_4	−0.46	—
		−0.71	—
	0.5 M H_2SO_4, 0.01% gelatin	−0.70	—
		−1.00	8.40
	1 M H_2tart, 1 M HCl	−0.40	4.32
		−0.67	—
	1 M Na_2tart, 0.8 M NaOH	(−0.31)	−2.87
	0.1 M EDTA, pH 6–8	1.60	—
Au(I)	0.1 M KCN	−1.46	—
	0.1 M KOH	−1.16	—

Au(III)	2 M HOAc, 2 M NH_4OAc, 0.01% gelatin	>0	—
	0.1 M KCN	>0	—
		−1.40	—
	1 M NaF, 0.01% gelatin	>0	—
Ba(II)	0.1 M LiCl or Et_4NI	−1.92	3.58
Bi(III)	2 M HOAc,2 M NH_4OAc, 0.01% gelatin	−0.25	3.50
	0.25 M $(NH_4)_3$cit, pH 5	−0.52	—
	1 M NaF, pH 0.7–2.1, 0.01% gelatin	−0.07	4.88
	1 M HCl, 0.01% gelatin	−0.09	5.23
	12 M HCl	−0.45	—
	1 M HNO_3	−0.01	4.59
	0.7 M $HClO_4$	+0.02	—
	0.1 M KHphthalate	−0.23	—
	0.5 M H_2SO_4	−0.03	4.31
	0.5 M Na_2tart, pH 4.5, 0.01% gelatin	−0.29	3.12
	0.3 M triethanolamine, 0.1 M KOH	−0.74	—
	0.1 M Nagluconate, 1 M NaOH	−0.80	3.65
	0.1 M EDTA, 1 M K_2CO_3	−0.78	—
	0.1 M EDTA, 2 M NaOAc	−0.70	—
Br^-	0.1 M KNO_3	(+0.12)	—
BrO_3^-	0.1 M KCl	−1.78	—
	0.1 M $BaCl_2$ or $CaCl_2$	−1.53	—
	0.1 M KCl, 0.001–0.1 M $LaCl_3$	−0.82	—
	0.1 M H_2SO_4, 0.2 M KNO_3	−0.41	—
Cd(II)	2 M HOAc, 2 M NH_4OAc, 0.01% gelatin	−0.65	2.30
	1 M NH_3, 0.1 M KNO_3	−0.78	3.85
	1 M NH_3, 1 M NH_4Cl	−0.81	3.68
	5 M $CaCl_2$	−0.80	—
	0.1 M KCl or HCl	−0.60	3.51
	1 M KCl or HCl	−0.64	3.58

TABLE 3.6 (*Continued*)

A. Inorganic Ions (*Continued*)			
Ion	Supporting Electrolyte	$E_{1/2}$ versus SCE	I
	0.5 M K_3cit, pH 7	−0.71	—
	1 M Na_3cit, 0.1 M NaOH	−1.46	—
	1 M KCN	−1.18	—
	1 M ethylenediamine, 0.1 M KNO_3, 0.01% gelatin	−0.93	3.13
	1 M NaF, 0.01% gelatin	−0.63	2.99
	1 M KI	−0.74	—
	0.1 M KNO_3	−0.58	3.53
	1 M KNO_3 or HNO_3	−0.59	—
	7.3 M H_3PO_4	−0.71	—
	0.1 M pyridine, 0.1 M KNO_3, 0.01% gelatin	−0.59	—
	0.5 M Na_2tart, pH 4.5, 0.01% gelatin	−0.64	2.34
	0.5 M Na_2tart, pH 8.8, 0.01% gelatin	−0.64	2.34
	2 M KSCN	−0.66	—
	0.3 M triethanolamine, 0.1 M KOH	−0.82	—
Ce(III)	0.1 M LiCl or $(CH_3)_4NBr$	−2.00	—
Ce(IV)	12 M HCl	>0	—
		−0.68	—
	7.3 M H_3PO_4	>0	—
	0.1 M solutions of strong acids	>0	—
Cl^-	0.1 M KNO_3	(+0.25)	—
ClO^-	Neutral 0.5 M K_2SO_4	+0.08	—
ClO_2^-	1 M KCl, 0.05 M $LaCl_3$	−1.02	—
	1 M NaOH	−1.00	—
CN^-	0.1 M KOH	(−0.36)	−3.00

$(CN)_2$	0.1 M NaOAc	−1.20	—
Co(II)	0.1 M NaOAc	−1.19	—
	1 M NH_3, 1 M NH_4Cl, 0.004% gelatin	−1.29	—
	5 M $CaCl_2$	−0.82	—
	0.1–1 M KCl or NaCl, 0.01% gelatin	−1.20	—
	1 M Na_3cit, 0.1 M NaOH	−1.45	—
	1 M KCN	−1.45	—
	0.1 M ethylenediamine, 0.1 M KNO_3	(−0.46)	—
	1 M NaF, pH 3–6, 0.01% gelatin	−1.38	2.75
	1 M KOH	−1.43	—
	10 N NaOH	−1.54	—
	0.1 M KHphthalate	−1.24	—
	0.1 M pyridine, 0.1 M pyridinium chloride	−1.06	—
	0.1 M K_2SO_4	−1.21	—
	1 M K_2SO_4	−1.43	—
	1 M neutral or weakly acidic tartrate	−1.60	—
	1 M KSCN	−1.04	—
Co(III)	1 M $K_2C_2O_4$, 0.2 M NH_4OAc, 0.5 M HOAc, 0.02% gelatin	>0	1.38
	0.3 M EDTA, 0.3 M pyridine, 0.3 M pyridinium chloride	>0	—
$Co(NH_3)_6^{3+}$	2 M NH_3, 2 M NH_4Cl	−0.28	—
		−1.30	—
	0.1 M KCl	−0.26	1.78
		−1.20	5.38
	5 M $CaCl_2$	−0.26	—
		−0.88	—
	0.5 M K_3cit	−0.36	1.23
		−1.52	3.66
	10 M NaOH	−0.35	—
		−1.54	—

TABLE 3.6 (*Continued*)

Ion	Supporting Electrolyte	$E_{1/2}$ versus SCE	I
A. Inorganic Ions (*Continued*)			
	0.1 M KNO_3	−0.24	1.74
		−1.21	5.36
	0.1 M K_2SO_4	−0.46	1.60
		−1.23	4.77
	0.1 M K_2tart	−0.31	1.50
		−1.32	4.35
Cr(II)	Saturated $CaCl_2$	(−0.51)	−0.47
	1 M KCl, 0.005% gelatin	(−0.40)	−1.54
	1 M KCN, 0.005% gelatin	(−1.38)	—
	0.5 M HC(O)OH, pH 1.8, 0.005% gelatin	(−0.30)	−1.54
	0.1 M Nasalicylate, 0.1 M NaOH, 0.005% gelatin	(−1.23)	−1.17
	0.01 M $(CH_3)_4NBr$ or Et_4NBr,	(−0.43)	−1.54
	0.005% gelatin	−1.58	—
	1 M KSCN, 0.005% gelatin	(−0.80)	−1.64
Cr(III)	1 M NH_3, 1 M NH_4Cl,	−1.43	—
	0.004% gelatin	−1.71	—
	Sautrated $CaCl_2$, 0.005% gelatin	−0.51	—
	1 M Na_3cit, 0.1 M NaOH	−1.45	—
	1 M KCN	−1.38	1.55
	7.3 M H_3PO_4	−1.02	—
	0.1 M pyridine, 0.1 M pyridinium perchlorate, pH 5.4, 0.02% gelatin	−0.95	—
	0.05 M Na_2SO_4	−1.01	—
	0.001 M Me_4NBr or	−1.14	—
	Et_4NIBr, 0.005% gelatin	−1.58	—
	1 M KSCN, 0.005% gelatin	−0.99	—

Cr(VI)	1 M NH_3, 1 M NH_4Cl, 0.1 M KCl	−0.20	—
		−1.60	—
	0.1 M KCl, 0.0003% sodium methyl red	−0.30	—
		−1.00	5.95
		−1.80	12.00
	1 M Na_3cit, 0.1 M NaOH	−0.83	—
		−1.49	—
	1 M NaF, pH 9.3, 0.01% gelatin	−0.26	2.98
		−1.10	—
	12 M HCl	>0	—
		−0.61	—
	0.2 M NaOH	−0.97	—
	1 M NaOH	−0.85	5.72
	1.5 M NaOH, 3% mannitol	−0.83	—
Cs(I)	0.1 M Me_4NCl or Et_4NOH	−2.09	—
Cu(I)	1 M NH_3, 1 M NH_4Cl	(−0.22)	—
		−0.50	—
	0.1 M KCl, 50% pyridine	−0.52	—
Cu(II)	2 M HOAc, 2 M NH_4OAc, 0.01% gelatin	−0.07	3.10
	1 M NH_3, 1 M NH_4Cl, 0.004% gelatin	−0.24	—
		−0.51	3.75
	1 M K_2CO_3, pH 9.5–11	−0.20	—
	5 M $CaCl_2$	−0.33	—
	0.1 M KCl or HCl, 0.01% gelatin	+0.04	3.23
	1 M KCl or HCl, 0.01% gelatin	+0.04	—
		+0.22	3.39
	0.5 M K_3cit, pH 9–11	−0.37	—
	1 M Na_3cit, 0.1 M NaOH	−0.50	—
	0.1 M ethylenediamine, 0.1 M KNO_3, 0.01% gelatin	−0.51	3.56
	0.1 M diethylenetriamine, 0.1 M KNO_3, 0.01% gelatin	−0.54	3.15
	1 M NaF, pH 5, 0.01% gelatin	0.00	3.48

TABLE 3.6 (*Continued*)

A. Inorganic Ions (*Continued*)			
Ion	Supporting Electrolyte	$E_{1/2}$ versus SCE	I
	12 M HCl	>0	—
		−0.71	—
	1 M KOH or NaOH	−0.41	2.91
	1.5 M NaOH, 3% mannitol	−0.60	—
	0.1 M KNO_3	+0.02	3.41
	1 M KNO_3 or HNO_3, 0.01% gelatin	−0.01	3.24
	0.25 M $(NH_4)_2C_2O_4$, pH 10	−0.19	—
		−0.42	—
	1 M $K_2C_2O_4$, pH 5–10	−0.27	—
	0.1 M $NaClO_4$ or $HClO_4$	+0.01	—
	7.3 M H_3PO_4	−0.09	—
	0.1 M KHphthalate	−0.10	—
	0.1 M pyridine, 0.1 M pyridinium perchlorate, pH 5.4, 0.02% gelatin	+0.05	—
		−0.25	—
	0.1 M $Na_4P_2O_7$, pH 4.8, 0.0006 tropeoline 00	−0.11	2.51
	0.05 M $Na_4P_2O_7$, 0.1 M NaOH	−0.30	—
	0.5 M H_2SO_4, 0.01% gelatin	0.00	2.12
	0.25 M $(NH_4)_2$tart, pH 10	−0.22	—
		−0.43	—
	0.5 M Na_2tart, pH 4	−0.06	2.37
	0.5 M Na_2tart, pH 10	−0.30	—
	0.5 M KNatart, 1 M KOH	−0.52	—

	0.1 M KSCN	−0.02	—
		−0.39	—
	1 M KSCN	−0.62	—
	0.3 M triethanolamine, 0.1 M KOH	−0.53	—
	0.1 M EDTA, 1 M K_2CO_3, pH 9.5	−0.50	—
	0.1 M EDTA, 2 M NaOAc	−0.39	—
	0.25 M EDTA, pH 7	−0.41	2.83
	0.25 M EDTA, pH 9–11.3	−0.47	—
Eu(III)	0.1 M NH_4Cl	−0.67	1.47
	0.1 M EDTA, pH 6–8	−1.17	1.30
	0.1 M EDTA, 0.2 M NaOH	−1.22	1.50
Fe(II)	1 M NH_3, 1 M NH_4Cl	(−0.34)	—
		−1.49	—
	0.05 M $BaCl_2$ or 0.1 M KCl	−1.30	—
	1 M Na_3cit, pH 7	(−0.34)	−0.78
	0.1 M KHF_2	+0.11	—
	1 M KOH, 8% mannitol	(−1.09)	—
		−1.55	—
	0.2–1 M $K_2C_2O_4$, pH 5	(−0.24)	−137
	1 M NH_4ClO_4	−1.46	
	0.5 M $(NH_4)_2$tart, 1 M NH_3, 0.005% gelatin	(−0.62)	—
		−1.53	—
	0.5 M Na_2tart, pH 5.6, 0.005% gelatin	(−0.17)	—
		−1.50	—
	1 M KSCN	−1.40	—
	0.04–0.4 M EDTA, pH 4	(−0.12)	−1.45
Fe(III)	5 M $CaCl_2$, pH 3.5	>0	—
		−1.20	—
	0.15 M Na_3cit, pH 6	−0.18	—
	0.5 M Na_3cit, pH 10	−0.85	—

TABLE 3.6 (*Continued*)

Ion	Supporting Electrolyte	$E_{1/2}$ versus SCE	I
	A. Inorganic Ions (*Continued*)		
	0.1 M KHF_2	−0.54	—
	1 M KF, pH 7	−0.74	—
		−1.44	—
	12 M HCl	>0	—
		−0.65	—
	1 M K_2malonate, pH 8.2	−0.30	1.25
	3 M KOH, 3% mannitol	−1.12	—
		−1.74	—
	0.2 M $K_2C_2O_4$ or $Na_2C_2O_4$, pH 4, 0.005% gelatin	−0.24	1.50
	1 M H_3PO_4, 1 M NaH_2PO_4, 0.01% gelatin	−0.03	—
	7.3 M H_3PO_4	+0.06	—
	0.1 M $Na_4P_2O_7$, pH 10	−0.82	1.02
	0.1 M Nasulfosalicylate, 0.5 M $NaBO_2$, 0.4 M $NaClO_4$, pH 9	−1.50	—
	0.5 M $(NH_4)_2$tart, 1 M NH_3, pH 9.7, 0.005% gelatin	−0.98	—
		−1.53	—
	0.1 M H_2tart, pH 2.0	+0.12	1.55
	0.5 M Na_2tart, pH 6, 0.005% gelatin	−0.19	1.11
		−1.52	—
	0.5 M Na_2tart, pH 9, 0.005% gelatin	−1.20	—
		−1.73	—
	0.7 M KSCN, 0.02 M H_2SO_4, 0.004% gelatin	>0	1.48
	0.3 M triethanolamine, 1 M NH_3, 0.9 M NH_4Cl	−0.50	—
		−1.60	—

	0.3 M triethanolamine, 0.1 M KOH	−1.01	—
	0.1 M EDTA, 2 M NaOAc	−0.12	—
	0.04 M EDTA, pH 4.5–6.5	−0.14	1.46
	Dilute sol. of KCl, HCl, $HClO_4$, etc.	>0	—
$Fe(CN)_6^{3-}$	0.1 M KCl	>0	1.79
Ga(III)	1 M NH_3, 1 M NH_4Cl	−1.60	—
	0.1 M KCl	−1.10	—
	1 M KCN	−1.29	—
	0.001 M HCl	−1.20	—
Gd(III)	0.1 M KCl or LiCl, 0.01% gelatin	−1.75	3.70
Ge(II)	6 M HCl	−0.45	—
	0.1 M NH_3, 0.1 M NH_4Cl	−1.72	—
		−1.90	—
	0.5 M NH_3, 1 M NH_4Cl	−1.45	—
		−1.70	—
	0.1 M EDTA, pH 6–8, 10^{-4} M fuchsin	−1.30	—
H(I)	0.1–0.5 M KCl, LiCl, or $NaClO_4$	−1.58	5.60
Hg(I)	0.1 M HNO_3	>0	3.68
	1 M HCl, KCl, $HClO_4$, $NaClO_4$, or NaF	>0	—
Hg(II)	0.1 M HNO_3	>0	3.48
	2 M HOAc, 2 M NH_4OAc, 0.01% gelatin; 1 M NaF, 0.01% gelatin; dil. (dilute) HCl, KCl, $HClO_4$, or $NaClO_4$	>0	—
In(III)	2 M HOAc, 2 M NH_4OAc, 0.01% gelatin	−0.71	3.70
	0.1 M KCl or HCl	−0.56	—
	1 M KCl	−0.60	—
	0.1 M KI	−0.53	—
	0.6 M H_2tart	−0.59	—
	1 M KOH or NaOH	−1.09	—
I^-	0.1 M KNO_3	(−0.03)	—

TABLE 3.6 (*Continued*)

Ion	Supporting Electrolyte	$E_{1/2}$ versus SCE	I
	A. Inorganic Ions (*Continued*)		
IO_3^-	1 M KCl, CsCl, LiCl, or NaCl	−1.16	—
	1 M HNO_3	0.00	—
	0.1 M biphthalate buffer, 0.1 M KCl, pH 3.2	−0.31	—
	0.1 M acetate buffer, 0.1 M KCl, pH 4.9	−0.50	—
	0.2 M phosphate buffer, pH 6.4	−0.79	—
	0.5 M $Na_2B_4O_7$, 0.2 M KNO_3, pH 9.2	−1.20	—
	0.1 M NaOH, 0.1 M KCl	−1.21	—
IO_4^-	0.16 M K_2SO_4, 1 M H_2SO_4	>0	—
		−0.12	—
	0.2 M HBO_2 + KOH, pH 10, 0.08% thymol	+0.02	—
K(I)	0.1 M Et_4NOH	−2.14	—
	0.1 M Et_4NOH in 50% ethanol	−2.10	1.70
Li(I)	0.1 M $(n\text{-}Bu)_4NOH$	−2.33	—
	0.1 M $(n\text{-}Bu)_4NOH$ in 50% ethanol	−2.31	1.19
Mn(II)	1 M NH_3, 1 M NH_4Cl, 0.004% gelatin	−1.66	—
	5 M $CaCl_2$	−1.45	—
	1 M KCl	−1.51	—
	1.5 M KCN	−1.33	—
	1 M NaF, pH 2.5–7, 0.01% gelatin	−1.55	3.93
	1 M KOH or NaOH	−1.70	—
	1.5 M NaOH, 3% mannitol	(−0.38)	−0.64
	0.25 M Na_2tart, 2 M NaOH	(−0.39)	−1.30
	1 M KSCN	−1.54	—
	0.3 M triethanolamine, 0.1 M KOH	(−0.5)	—
		−1.61	—

Mn(III)	0.4 M $K_4P_2O_7$, pH 2.3, 0.02% agar	>0	1.17
Mo(V)	0.2 M EDTA, pH 12	(−0.53)	—
Mo(VI)	0.03 M Na_2HPO_4, 0.9 M H_3cit, 0.1 M KCl, pH 2.8	−0.23	—
		−0.58	—
	0.1 M Na_3cit, pH 7	−1.11	—
	0.3 M HCl	−0.26	—
		−0.63	—
	2 M NH_4NO_3, 3 M HNO_3	−0.80	—
	2.3 M $HClO_4$, 0.008% gelatin	−0.24	—
	12 M H_2SO_4	0.00	—
		−0.13	1.24
	0.1 M H_2tart, pH 2.0	−0.22	—
		−0.22	—
	0.004 M EDTA, 0.1 M K_2SO_4, pH 2.5	−0.33	—
		−0.58	—
	0.05 M EDTA, pH 5.8	−0.80	1.47
	0.1 M EDTA, 0.1 M HOAc, 0.1 M NH_4OAc	−0.63	—
Na(I)	0.1 M Et_4NOH	−2.12	—
Nb(V)	0.1 M KCl, pH 2.6	−1.28	—
	0.3 M K_3cit, pH 6.8	−1.73	—
		−2.03	—
	3 M NH_4F	−1.90	—
	0.1 M KNO_3, pH 2.6	−1.03	—
	0.1 M $K_2C_2O_4$, pH 1–5.5	−1.53	—
	1 M K_2tart, pH 7.0	−2.00	—
Ni(II)	0.1 M NH_3, 0.1 M NH_4Cl	−0.92	—
	1 M NH_3, 1 M NH_4Cl	−1.10	3.56
	5 M $CaCl_2$	−0.56	—
	0.1 M KCl, 0.0003% sodium methyl red	−1.10	3.38
	0.1 M KCN, 0.1 M KCl	−1.42	—
	1 M KCN, 0.1 M KCl	−1.36	—

TABLE 3.6 (*Continued*)

A. Inorganic Ions (*Continued*)			
Ion	Supporting Electrolyte	$E_{1/2}$ versus SCE	I
	1 M NaF, 0.01% gelatin	−1.12	2.29
	12 M HCl	−0.80	—
	0.1 M KHphthalate	−1.14	—
	0.5 M pyridine, 1 M KCl, 0.01% gelatin	−0.78	—
	1 M KSCN, 0.01% gelatin	−0.68	3.59
	Saturated H_2tart	−1.05	—
	0.3 M triethanolamine, 0.7 M NH_3, 1 M NH_4Cl	−1.18	—
NH_2OH	1 M NaOH	(−0.43)	—
N_3^-	0.1 M KNO_3	(+0.25)	—
NO	Dilute HCl	−0.90	—
HNO_2	0.1 M H_2SO_4, 0.2 M Na_2SO_4	−0.98	—
	0.1 M KCl, 0.01 M HCl, 0.05–0.2 mM $UO_2(OAc)_2$	−1.00	7.45
NO_3^-	0.04 M $LaCl_3$	−1.58	—
	0.025 M $ZrOCl_2$, pH 1.7	−1.00	—
NH_4^+	Me_4NBr	−2.21	—
Me_4N^+	Me_4NBr	−2.67	—
$(n\text{-Bu})_4N^+$	Me_4NBr	−2.57	—
OH^-	0.1 M KNO_3	(+0.08)	—
H_2O_2	Phosphate–citrate buffer, pH 7	(+0.18)	—
		−1.00	—
	0.1 M NaOH	(−0.18)	—
		−1.00	—

O_2	0.1 M KNO_3, KCl, or most other common supporting electrolytes	−0.05	6.22
		−0.90	12.30
	7.3 M H_3PO_4	−0.23	—
		−0.64	—
	0.1 M NaOH	−0.18	—
		−1.00	—
Os(VI)	Saturated $Ca(OH)_2$	−0.40	—
		−1.16	—
	0.1 M Nagluconate, 1 M NaOH	−0.50	2.09
Os(VIII)	0.5 M acetate buffer, pH 4.7	>0	—
		+0.10	—
	Saturated $Ca(OH)_2$	>0	—
		−0.40	—
		−1.16	—
Pb(II)	2 M HOAc, 2 M NH_4OAc, 0.01% gelatin	−0.50	2.70
	5 M $CaCl_2$	−0.53	—
	0.1 M KCl, 0.005% gelatin	−0.40	3.85
	1 M KCl, 0.005% gelatin	−0.44	—
	0.25 M $(NH_4)_3$cit, pH 5	−0.77	—
	1 M Na_3cit, 0.1 M NaOH	−0.78	—
		−1.50	—
	1 M KCN	−0.72	—
	1 M NaF, pH 1–3, 0.01% gelatin	−0.41	4.08
	1 M HCl, 0.05% gelatin	−0.44	3.86
	12 M HCl	−0.90	—
	1 M NaOH, 0.005% gelatin	−0.76	3.40
	0.1 M KNO_3 or $NaNO_3$	−0.38	—
	1 M HNO_3, KNO_3, or $NaNO_3$	−0.40	3.67
	1 M $K_2C_2O_4$, pH 7–10.5	−0.58	—
	1 M $NaClO_4$ or $HClO_4$	−0.38	—

TABLE 3.6 (*Continued*)

A. Inorganic Ions (*Continued*)

Ion	Supporting Electrolyte	$E_{1/2}$ versus SCE	I
	7.3 M H_3PO_4	−0.53	—
	0.1 M KHphthalate	−0.40	—
	0.1 M $Na_4P_2O_7$, pH 10	−0.69	2.57
	0.5 M Na_2tart, pH 9	−0.58	2.40
	0.5 M Na_2tart, 0.01 M NaOH	−0.70	2.40
	0.3 M triethanolamine, 0.1 M KOH	−0.88	—
	0.3 M triethanolamine, 0.7 M NH_3, 1 M NH_4Cl	−0.56	—
	0.05 M EDTA, 1 M HOAc, 1 M NaOc, pH 4	−1.10	—
Pd(II)	1 M NH_3, 1 M NH_4Cl	−0.75	—
	1 M KCN	−1.77	—
	1 M monoethanolamine, 1 M KCl	−0.75	—
	0.1 M diethylamine, 1 M KCl	−0.74	—
	0.1 M ethylenediamine, 1 M KCl, 0.005% methyl red	−0.64	—
	1 M pyridine, 1 M KCl	−0.34	—
Pr(III)	0.1 M LiCl, 0.01% gelatin	−1.80	3.59
	0.1 M $N(CH_3)_4I$, 0.01% gelatin	−1.86	3.47
Ra(II)	Dil. KCl	−1.84	—
Rb(I)	0.1 M $[N(CH_3)_4]OH$	−2.03	—
Re(−I)	2.4 M HCl	(−0.17)	—
		(−0.34)	—
		(−0.47)	—
		−0.66	—

	1.2 M $HClO_4$	(−0.54)	—
		(−0.42)	—
		(−0.26)	—
		(+0.03)	—
Re(III)	2 M $HClO_4$	−0.28	—
		−0.46	—
Rh(III)	1 M pyridine and 1 M KCl or 1 M KBr	−0.41	—
	0.9 M KSCN	−0.39	—
$Rh(NH_3)_5Cl^{2+}$	1 M NH_4Cl	−0.93	—
	1 M NH_3, 1 M NH_4Cl	−0.93	—
	1 M KCN	−1.47	—
Ru(III)	0.2 M Nagluconate, pH 14	−0.67	1.00
Ru(IV)	1 M $HClO_4$	>0	0.91
		+0.20	1.53
		−0.34	2.89
S^{2-}	0.1 M KOH or NaOH	(−0.76)	—
SCN^-	0.1 M KNO_3	(+0.18)	—
$S_2O_3^{2-}$	0.1 M KNO_3	(−0.14)	—
$S_4O_6^{2-}$	1 M H_3PO_4 containing $(NH_4)HPO_4$, pH 1–8, 0.001% quinoline	−0.26	—
$S_2O_4^{2-}$	1 M NH_3, 0.5 M $(NH_4)_2HPO_4$, 0.01% gelatin	(−0.43)	4.09
SO_2	0.1 M HCl or HNO_3	−0.37	5.49
	Phthalate buffer, pH 3.0	−0.48	—
	Acetate buffer, pH 3.6	−0.54	—
	0.05 M phosphate buffer, pH 6.0, 0.1 M KNO_3	−1.23	—
SO_3^{2-}	0.1 M KNO_3	(+0.01)	—
Sb(III)	2 M HOAc, 2 M NH_4OAc, 0.01% gelatin	−0.40	—
		−0.59	4.20
	1 M Na_3Cit, 0.1 M NaOH	(−0.44)	—
		1.24	4.40

TABLE 3.6 (*Continued*)

A. Inorganic Ions (*Continued*)

Ion	Supporting Electrolyte	$E_{1/2}$ versus SCE	I
	1 M KCN	−1.09	—
	1 M HCl, 0.01% gelatin	−0.15	5.57
	12 M HCl	−0.52	—
		−0.66	—
	0.1 M NaOH, 0.003% thymolphthalein	(−0.37)	−2.93
		−1.07	5.90
	1 M KOH	(−0.45)	—
		−1.15	6.00
	1 M HNO_3, 0.01% gelatin	−0.30	5.10
	7.3 M H_3PO_4	−0.29	—
	1 M H_2tart, 1 M HCl	−0.14	3.66
	1 M Na_2tart, 0.1 M NaOH	(−0.30)	−2.60
		−1.30	3.50
Sb(V)	6 M HCl	>0	3.00
		−0.26	7.50
	12 M HCl	>0	—
		−0.48	—
		−0.57	—
Sc(III)	0.1 M LiCl or KCl, containing trace of HCl	−1.80	—
Se(−II)	0.05 M NH_3, 1 M NH_4Cl	(−0.84)	−4.90
		(0.00)	−6.00
	0.5 M H_3cit, pH 2.5	(−0.64)	−2.70
		(0.00)	−4.40
	0.5 M Na_2CO_3, pH 10.7	(−0.89)	−4.50

	1 M HCl	(−0.49)	−3.80
		(−0.10)	−5.10
	1 M NaOH	(−1.02)	−1.95
		(−0.94)	−3.78
Se(IV)	0.1 M NH_3, 0.1 M NH_4Cl	−1.64	11.00
	1 M NH_3, 1 M NH_4Cl	−1.53	11.02
	1 M HCl	>0	—
		−0.10	—
		−0.40	—
		−0.50	—
	1 M $(NH_4)_2$tart, 2 M NH_3	−1.53	10.20
Sm(III)	0.1 M Me_4NI, 0.0005 M H_2SO_4, 0.01% gelatin	−1.80	3.85
		−1.96	12.10
Sn(II)	2 M HOAc, 2 M NH_4OAc, 0.01% gelatin	(−0.16)	—
		−0.62	2.60
	0.5 M NaCl, 2 M $HClO_4$	−0.35	—
	1 M Na_3cit, 0.1 M NaOH	(−0.91)	—
		−1.12	—
	1 M NaF, pH 4–6, 0.01% gelatin	(−0.20)	−4.10
		−0.73	4.10
	12 M HCl, 0.002% triton X-100	−0.83	—
	1 M NaOH, 0.01% gelatin	(−0.73)	−3.45
		−1.22	3.45
	1 MNO_3, 0.01% gelatin	−0.44	4.02
	1 M $HClO_4$	(+0.14)	—
		−0.43	—
	7.3 M H_3PO_4	−0.58	—
	0.5 M H_2SO_4, 0.01% gelatin	−0.46	3.54
	0.4 M H_2tart, 0.1 M NaHtart, pH 2.3, 0.01% gelatin	−0.49	—
	0.4 Na_2tart, 0.1 M $NaClO_4$, pH 9.0, 0.01% gelatin	(−0.33)	—
		−0.92	—
	0.5 M Na_2tart, 0.1 M NaOH, 0.01% gelatin	(−0.71)	—
		−1.16	2.86

TABLE 3.6 (*Continued*)

A. Inorganic Ions (*Continued*)

Ion	Supporting Electrolyte	$E_{1/2}$ versus SCE	I
Sn(IV)	4 M NH_4Br, 0.005% gelatin	>0	—
		−0.50	6.52
	0.5 M NaCl, 2 M $HClO_4$	−0.47	—
	1 M Na_3cit 0.1 M NaOH	−1.22	—
Sr(II)	0.1 M Et_4NI	−2.11	3.46
Ta(V)	0.9 M HCl	−1.16	—
	0.5–1 M $K_2C_2O_4$, pH 0.5–3	−1.40	—
	0.1 M K_2tart, pH 3–5	−1.57	—
Te(−II)	0.1 M NH_3, 1 M NH_4Cl, 0.003% gelatin	(−1.10)	—
	0.5 M citrate buffer, pH 3.3, 0.03% gelatin	(−0.95)	—
	1 M HCl, 0.003% gelatin	(−0.73)	—
	1 M NaOH, 0.003% gelatin	(−1.20)	−3.50
		(−0.40)	—
Te(IV)	1 M NH_3, 1 M NH_4Cl, pH 9.4	−0.67	—
	0.5 M $NaBO_2$ or Na_2CO_3, pH 9.4	−0.88	—
	0.5 M H_3cit, pH 1.6	−0.05	—
		−0.40	5.72
	1 M NaF, pH 6.9, 0.01% gelatin	−0.89	6.70
	1 M NaOH, 0.003% gelatin	−1.10	—
		−1.19	9.75
	1 M $(NH_4)_2$tart, 2 M NH_3, pH 9.3	−0.70	6.40

Te(VI)	Acetate buffer, pH 5.6	−1.18	15.40
	NH_3–NH_4Cl buffer, pH 8.0, 0.0005% gelatin	−1.21	17.50
	carbonate buffer, pH 8.3	−1.37	16.60
	0.1 M KCl or $KClO_4$	−1.10	—
		−1.45	—
	0.5 M $(NH_4)_3$cit, pH 6.2, 0.003% gelatin	−1.19	—
	1 M Na_3cit, 0.1 M NaOH	−1.54	—
	1 M KCN	−1.36	—
	1 M NaF, pH 6.5–9.5, 0.01% gelatin	−1.50	2.35
	12 M HCl	>0	—
		−0.43	—
		−0.79	—
	1 M NaOH, 0.003% gelatin	−1.57	—
	Saturated $(NH_4)_2C_2O_4$ + NH_3, pH 8.0	−1.23	16.30
	1 M $(NH_4)_2$tart, pH 8.4	−1.38	13.00
Ti(III)	Saturated $CaCl_2$	(−0.12)	—
	0.01 M HCl	(−0.14)	—
	0.2 M $H_2C_2O_4$, pH 1	(−0.30)	−1.60
	Saturated H_2tart	(−0.44)	—
	0.1 M KSCN	(−0.46)	—
	EDTA, pH 1.0–2.5	(−0.22)	—
Ti(IV)	Saturated $CaCl_2$	−0.12	—
	0.2 M H_3cit	−0.37	—
	0.4 M Na_3cit, pH 5.5–6, 0.005% gelatin	−0.90	1.02
	0.1 M HCl, 0.005% gelatin	−0.81	1.56
	3.5 M lactic acid	−0.40	—
	0.2 M $H_2C_2O_4$, pH 0.5	−0.28	1.75
	Saturated phthalic acid	−0.93	—
	Saturated salicyclic acid	−0.35	—
	0.03 M H_2SO_4	−0.79	—
	Saturated H_2tart	−0.44	—

TABLE 3.6 (*Continued*)

Ion	Supporting Electrolyte	$E_{1/2}$ versus SCE	I
A. Inorganic Ions (*Continued*)			
	0.1 M KSCN	−0.46	—
	EDTA, pH 1.0–2.5	−0.22	—
	0.1 M EDTA, 2 M NaOAc	−0.53	—
Tl(I)	2 M HOAc, 2 M NH_4OAc, 0.01% gelatin	−0.47	2.30
	0.1 M NH_3, NH_4Cl, KCl, HCl, KOH, HNO_3, KNO_3, $HClO_4$, or $NaClO_4$	−0.46	2.70
	1 M NH_3, NH_4Cl, KCl, HCl, KOH, HNO_3, KNO_3, $HClO_4$, or $NaClO_4$	−0.48	—
	1 M Na_3cit, 0.1 M NaOH	−0.56	—
	0.1–1 M KCN	>0	—
	1 M NaF, pH 3.5–6.5, 0.01% gelatin	−0.50	2.67
	7.3 M H_3PO_4	−0.63	—
	0.2 M $Na_4P_2O_7$, 0.2 M KOH	−0.55	—
	17 M H_2SO_4	−0.98	—
	0.01 M EDTA, 1 M HOAc, 1 M NaOAc	−0.46	—
Tl(III)	0.6 M HCl	>0	3.83
		−0.45	5.72
	1 M $HClO_4$	>0	—
		−0.48	—
U(III)	1 M $HClO_4$	(−0.87)	−1.50
U(IV)	0.1 M $HClO_4$	−0.86	1.57
	1 M HCl	−0.89	—
U(V)	0.5 M $NaClO_4$, 0.01 M $HClO_4$	(−0.18)	−1.57

U(VI)	0.4 M HOAc, pH 2.7	−0.15	—
		−0.69	—
		−1.02	—
	2 M HOAc, 2 M NH_4OAc, 0.01% gelatin	−0.45	1.70
	0.5 M $(NH_4)_2CO_3$	−0.83	1.50
		−1.45	—
	1 M NH_3, 1 M NH_4Cl	−0.80	—
		−1.40	—
	1 M Na_2CO_3	−0.95	1.50
	0.1 M H_3cit, 0.1 M K_3Cit	−0.38	2.40
	1 M Na_3cit, 0.1 M NaOH	−0.98	—
	0.01 M HF	−0.21	2.48
	1 M NaF	−0.94	1.40
	0.1 M HCl	−0.18	1.54
		−0.94	—
	2 M HCl	−0.21	3.08
		−0.90	—
	12 M HCl	>0	—
		−0.63	—
	2 M hydroxylamine hydrochloride	−0.26	2.05
	0.5 M $H_2C_2O_4$	−0.13	3.20
	0.5 M $NaClO_4$, 0.01 M $HClO_4$	−0.18	1.57
	7.3 M H_3PO_4	−0.12	—
		−0.58	—
	0.05 M H_2SO_4	−0.22	2.00
		−0.90	2.35
		−1.06	—
	0.1 M Nagluconate, 0.1 M NH_4ClO_4, pH 11	−1.17	1.95
	0.1 M EDTA, 2 M NaOAc	−0.41	—

TABLE 3.6 (*Continued*)

Ion	Supporting Electrolyte	$E_{1/2}$ versus SCE	I
	A. Inorganic Ions (*Continued*)		
V(II)	1 M acetate buffer, pH 5.4	(−0.89)	−1.09
		(−0.11)	−3.36
	1 M KBr	(−0.50)	−2.03
	0.5 M $KHCO_3$, 0.5 M Na_2CO_3, pH 9.4	(−0.75)	−1.19
		(−0.18)	−4.16
	1 M Na_3cit, pH 7	(−1.17)	−0.87
	1 M KI	(−0.49)	−2.27
	1 M $K_2C_2O_4$, pH 6.5	(−1.09)	−1.43
	Saturated KHphthalate, pH 5.2	(−0.84)	−1.78
		(0.15)	−3.53
	1 M Na salicylate–salicylic acid buffer, pH 4.7	(−0.76)	−1.16
		(−0.15)	−3.52
	0.5 M H_2SO_4	(−0.50)	−1.74
	1 M Na_2tart, pH 6	(−0.17)	−1.07
	1 M NH_4SCN	(−0.47)	−2.04
	0.1 M EDTA, pH <8.3	(−1.27)	−1.10
V(III)	1 M acetate buffer, pH 5.4	−0.98	0.57
		−1.25	1.39
	1 M KBr, pH 2.5	−0.43	1.42
		−0.87	1.94
	0.5 M $KHCO_3$, 0.5 M Na_2CO_3, pH 9.4	(−0.34)	−2.80
	1 M KCN	−1.17	0.71
		−1.77	1.33
	1 M HCl or $HClO_4$, or 0.5 M H_2SO_4	−0.51	1.41

	1 M $K_2C_2O_4$, pH 3.5–6.5	-1.14	1.95
	Saturated KHphthalate, pH 5.2	(-0.10)	-1.13
		-0.88	1.22
	1 M Nasalicylate-salicylic acid buffer,	(-0.06)	-1.17
	pH 4.7	-0.97	0.26
		-1.21	1.26
	1 M NH_4SCN	-0.46	1.78
	0.1 M EDTA, pH 5–8.5	-1.27	1.20
V(IV)	1 M NH_3, 1 M NH_4Cl, 0.08 M Na_2SO_3	(-0.32)	-0.94
		-1.28	1.82
	0.5–1 M $KHCO_3$ saturated with CO_2	(-0.16)	-1.41
	1 M Na_3cit, 0.1 M NaOH	(-0.47)	—
		-1.76	—
	12 M HCl, 0.002% triton X-100	>0	—
		-0.62	—
		-0.75	—
	0.05 M H_2SO_4, 0.005% gelatin	-0.85	3.20
	0.1 M EDTA, pH 9.5	-1.25	2.20
	1.0 M $K_2C_2O_4$, pH 6	-1.31	4.32
V(V)	1 M NH_3, 1 M NH_4Cl, 0.005% gelatin	-0.96	1.60
		-1.26	4.72
	0.1 M HCl	>0	—
		-0.80	—
	1 M $K_2C_2O_4$, pH 5	>0	1.86
		-1.33	5.60
	7.3 M H_3PO_4	>0	—
		-0.54	—
		-0.91	—
	0.025 M H_2SO_4, 0.1 M KCl, 0.005% gelatin	>0	1.65
		-0.98	4.96

TABLE 3.6 (*Continued*)

A. Inorganic Ions (*Continued*)			
Ion	Supporting Electrolyte	$E_{1/2}$ versus SCE	I
	0.1 M H_2tart, pH 2.0	+0.40	—
		0.00	—
		−0.60	—
	0.1 M EDTA, pH 9.5	−1.22	3.29
W(III)	3 M HCl	(−0.65)	—
		(−0.45)	—
	12 M HCl	(−0.53)	—
	12 M HCl	(−0.56)	−1.10
W(V)	12 M HCl	−0.56	2.53
W(VI)	12 M HCl	>0	1.31
		−0.55	3.82
	10 M HCl	>0	1.42
		−0.60	4.33
	7.3 M H_3PO_4	−0.59	1.47
	0.1 M H_2tart, 5 M HCl	−0.33	—
		−0.68	4.40
Yb(III)	0.1 M NH_4Cl	−1.41	1.57
		−2.00	—
Zn(II)	2 M HOAc, 2 M NH_4OAc, 0.01% gelatin	−1.10	1.50
	1 M NH_3, 1 M NH_4Cl, 0.005% gelatin	−1.35	3.82
	1 M KCl, 0.0003% sodium methyl red	−1.00	3.42
	0.15 M K_3cit, pH <3	−1.06	—
	1 M Na_3cit, 0.1 M NaOH	−1.43	—
	1 M NaF, pH 4–6.5, 0.01% gelatin	−1.14	3.15

	1 M N_2H_4, 1 M $NaClO_4$, pH 8–9	−1.13	—
	1 M NaOH, 0.002% gelatin	−1.49	3.04
	0.01 M KNO_3	−0.99	—
	Nearly saturated $(NH_4)_2C_2O_4$	−1.30	—
	1 M $NaClO_4$	−1.00	—
	0.1 M KHphthalate	−1.01	—
	0.1 M pyridine, 0.1 M KCl, 0.01% gelatin	−1.02	—
	1 M pyridine, 0.1 M NaOH	−1.57	—
	Saturated H_2tart	−1.03	—
	0.5 M Na_2tart, pH 8.8, 0.01% gelatin	−1.15	2.30
	0.5 M Na_2tart, 0.1 M NaOH, 0.01% gelatin	−1.42	2.65
	0.1 M KSCN	−1.01	—
	0.3 M triethanolamine, 0.1 M KOH	−1.57	—
	0.3 M triethanolamine, 0.7 M NH_3 1 M NH_4Cl	−1.36	—

B. Organic Molecules

Compound	Supporting Electrode	$E_{1/2}$(V)	I
	Acids and Acid Derivates		
Acetic acid	0.1 M Et_4NClO_4/CH_3CN	−2.3	—
Acetylene dicarboxylic			
Acid	HCl–KCl (pH 0.5)	−0.56	7.1
Diethyl ester	HCl–KCl (pH 1.5)	−0.48	4.3
		−0.63	4.3
Acrylamide	0.05 M Me_4NI/30% EtOH	−1.91	3.49
Acrylic acid, ethyl ester	0.05 M Me_4NI/30% EtOH	−1.82	—
Acrylonitrile	0.05 M Me_4NI/30% EtOH	−1.96	—
Benzhydroxamic acid	0.05 M $(n\text{-}Bu)_4NCl$/90% EtOH	−2.18	4.6
Benzoic acid	0.1 M Et_4NClO_4/CH_3CN	−2.1	—
Fumaric acid	NH_3 buffer (pH 8)/10% EtOH	−1.57	4.1
Diethyl ester	NH_3 buffer (pH 8)/10% EtOH	−1.01	3.7

TABLE 3.6 (*Continued*)

Compound	Supporting Electrode	$E_{1/2}$(V)	I
	B. Organic Molecules (*Continued*)		
	Acids and Acid Derivates		
α-Ketoglutaric acid	0.7 M KCl + HCl (pH 2)	−0.63	2.71
Maleic acid	NH_3 buffer (pH 8)/10% EtOH	−1.35	3.6
Diethyl ester	NH_3 buffer (pH 8)/10% EtOH	−1.02	3.6
Oxalic acid	0.1 M Et_4NClO_4/CH_3CN	−1.6	2.06
Phthalic acid			
Diethyl ester	0.1 M Me_4NCl/75% EtOH	−1.87	4.64
Diphenyl ester	0.1 M Me_4NCl/75% EtOH	−1.65	4.47
	Carbonyls and Carbonyl Derivatives		
Acetaldehyde	0.06 M LiOH	−1.93	—
Acetone	0.1 M Bu_4NCl + 0.1 M Bu_4NOH/80% EtOH	−2.53	—
1,3-Diphenyl-	0.1 M Bu_4NCl + 0.1 M Bu_4NOH/80% EtOH	−2.10	—
Acetophenone oxime	0.1 M HCl–KCl/50% EtOH	−1.09	4.90
Benzaldehyde	0.1 M LiOH/50% EtOH	−1.51	—
Oxime	0.1 M HCl–KCl/50% EtOH	−0.77	4.60
tert-Butyl phenyl ketone	0.1 M LiOH/50% EtOH	−1.92	—
Butyrophenone	0.1 M LiOH/50% EtOH	−1.75	—
n-Caprylic aldehyde	0.05 M (*n*-Bu)$_4$NCl/90% EtOH	−2.35	—
Crotonaldehyde	0.2 M Me_4NOH/50% EtOH	−1.37	—
		−1.80	—
Dimethylglyoxime	0.1 M HCl/15% EtOH	−0.81	11.50
Formaldehyde	0.06 M LiOH	−1.75	—
Methyl vinyl ketone	0.1 M KCl	−1.42	—

Phthalaldehyde	OAc^- buffer (pH 5)/1.5% EtOH	−0.72	—
		−1.09	—
p-Quinone dioxime	0.1 M NaOH/55% EtOH	−1.12	—
	Halogen and α-Halocarbonyl Compounds		
Acetaldehyde, chloro-	NH_3 buffer (pH 8.5)	−1.03	—
Acetone		−1.67	—
Bromo-	OAc^- buffer (pH 4.5)	−0.34	3.30
Chloro-	OAc^- buffer (pH 4.5)	−1.15	2.90
Iodo-	OAc^- buffer (pH 4.5)	−0.14	3.30
Allyl bromide	Li_3cit/50% dioxane	−1.18	—
α-Chloro-	Li_3cit/50% dioxane	−0.88	—
Benzene			
Bromo-	0.05 M Et_4NBr/DMF	−2.24	4.20
Chloro-	0.05 M Et_4NBr/DMF	−2.55	—
Iodo-	0.05 M Et_4NBr/DMF	−1.64	—
Butyraldehyde			
α-Bromo-	0.1 M LiCl/50% dioxane	−0.48	—
		−1.91	—
α-Chloro-	0.1 M LiCl/50% dioxane	−1.33	—
		−1.90	—
Cyclohexane, bromo-	0.05 M Et_4NBr/DMF	−2.28	—
Cyclohexanone, 2-chloro-	0.1 M KCl	−1.45	1.85
Cyclopentanone, 2-chloro-	0.1 M KCl	−1.35	2.93
Phenacyl chloride	0.05 M $(n\text{-}Bu)_4NCl$/90% EtOH	−0.92	—
Propene, 1-bromo-	0.05 Et_4NBr/DMF	−2.50	—

TABLE 3.6 (*Continued*)

B. Organic Molecules (*Continued*)			
Compound	Supporting Electrode	$E_{1/2}$(V)	I
	Heterocyclic Compounds		
2-Acetothiophene	OAc^- buffer (pH 5)	−1.25	3.50
2,2′-Bipyridine	OAc^- buffer (pH 4.5), 0.1 M KCl	−1.14	—
		−1.25	—
2,4′-Bipyridine	OAc^- buffer (pH 4.5), 0.1 M KCl	−1.04	—
3,3′-Bipyridine	OAc^- buffer (pH 4.5), 0.1 M KCl	−1.55	—
Phthalimide	HCl–KCl/50% EtOH (pH 2)	−0.90	—
N-Propyl pyridinium bromide	0.05 M PO_4^{-3} buffer (pH 8)	−1.36	—
Pyrimidine	0.05 M PO_4^{-3} buffer (pH 7)	−1.30	3.50
Pyronin	PO_4^{-3} buffer/1% EtOH (pH 7)	−0.69	—
Quinoline	0.2 M Me_4NOH/50% EtOH	−1.50	—
Quinoxaline	0.05 M PO_4^{-3} buffer (pH 7)	−0.68	3.69
	Nitro and Related Compounds		
Azobenzene	0.01 M HCl + 0.02 M KCl/30% MetOH	−0.06	3.11
		−0.81	2.50
Azoxybenzene	0.01 M HCl + 0.02 M KCl/30% MeOH	−0.25	6.14
		−0.83	2.54
Benzenediazonium chloride	OAc^- buffer (pH 4)	−0.19	—
Cyclohexyl nitrate	0.5 M LiCl, 0.01% gelatin/11.25 wt. % EtOH	−0.63	3.06
Ethyl nitrate	0.5 M LiCl, 0.01% gelatin/11.25 wt. % EtOH	−0.82	3.39
Nitobenzene	OAc^- buffer/60% EtOH (pH 3)	−0.43	—
p-chloro-	OAc^- buffer/60% EtOH (pH 3)	−0.40	—
p-hydroxy-	OAc^- buffer/60% EtOH (pH 3)	−0.56	—

(tri)Nitroglycerin	0.1 M Me_4NCl/75% EtOH	−0.70	—
Nitrosocyclohexane dimer	0.1 M HCl–KCl/50% EtOH	−0.21	2.20
		−0.51	5.70
N-Nitrosodimethylamine	OAc^- buffer (pH 3.5)	−1.21	—
	Sulfur-Containing Compounds		
Cystine	NH_3 buffer (pH 9)	−0.70	—
		−1.30	—
Diethyl disulfide	0.025 M $(n\text{-Bu})_4NOH$/2-PrOH–MeOH–H_2O (2:2:1)	−1.78	—
Diphenyl disulfide	0.025 M $(n\text{-Bu})_4NOH$/2-PrOH-MeOH–H_2O (2:2:1)	−0.65	—
Diphenyl sulfone	0.1 M Me_4NBr/50% EtOH	−2.04	2.65
diphenyl sulfoxide	0.1 M Me_4NBr/50% EtOH	−2.07	2.54
Methyl phenyl sulfone	0.1 M Me_4NBr/50% EtOH	−2.14	2.55
Sodium diethyldithio phosphate	0.1 M $HClO_4$	(−0.06)	−2.00
Sodium formaldehyde sulfoxylate	0.1 M NaOH	(−0.42)	−3.84
thiourea	0.05 M H_2SO_4	(+0.04)	—
	Miscellaneous Compounds		
Benzoyl peroxide	0.3 M LiCl/C_6H_6–MeOH (1:1)	0.00	3.00
tert-Butyl hydroperoxide	0.3 M H_2SO_4/5% EtOH	−0.31	—
	0.3 M LiCl/C_6H_6–MeOH (1:1)	−0.96	5.80
Catechol	PO_4^{-3} buffer (pH 8)	(+0.10)	—
Chlorotriethyllead	1 M KCl	−0.68	—
Dichlorodiethyltin	1 M KCl	−0.57	—
Succinic acid peroxide	0.3 M LiCl/C_6H_6–MeOH (1:1)	−0.19	3.10

TABLE 3.6 (*Continued*)

B. Organic Molecules (*Continued*)			
Compound	Supporting Electrode	$E_{1/2}$(V)	I
	Unsaturated Hydrocarbons		
Acetylene, phenyl-	0.175 M (n-Bu)$_4$NI/75% dioxane	−2.37	—
Anthracene	0.175 M (n-Bu)$_4$NI/75% dioxane	−1.94	—
Azulene	0.175 M (n-Bu)$_4$NI/75% dioxane	−1.63	—
		−2.28	—
		−2.52	—
Naphthalene	0.175 M (n-Bu)$_4$NI/75% dioxane	−2.49	3.02
1-(1-Cyclohexenyl)	0.1 M (n-Bu)$_4$NI/75% dioxane	−2.42	—
		−2.48	—
1-(1-Cyclopentenyl)	0.1 M (n-Bu)$_4$NI/75% dioxane	−2.25	3.03
		−2.49	3.60
Phenanthrene	0.175 M (n-Bu)$_4$NI/75% dioxane	−2.44	—
		−2.67	—
Styrene	0.175 M (n-Bu)$_4$NI/75% dioxane	−2.34	—
β-Methyl-	0.175 M (n-Bu)$_4$NI/75% dioxane	−2.54	—

[a]Parentheses indicate an anodic wave, $I = 607\ n\ D^{1/2} = i_d/(m^{2/3}t^{1/6})$, based on average rather than maximum i_d values (multiplication by 7/6 will give appropriate I values for maximum diffusion currents). Negative I values indicate anodic current.

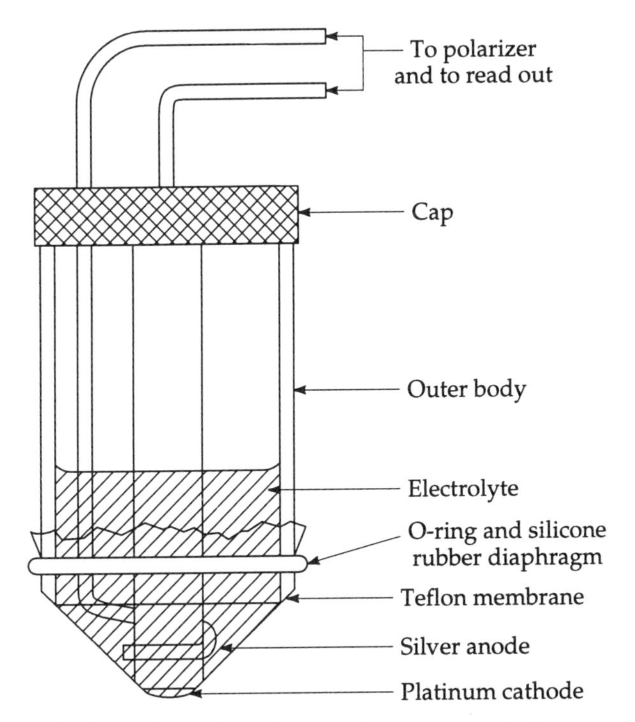

Figure 3.19 Clark electrode for the volammetric measurement of oxygen partial pressure.

sections the standard electrode potential, the dissociation constants of weak acids and basis, solubility products, and the formation constants of complex ions can be evaluated from polarographic half-wave potentials, if the electrode process is reversible. Furthermore, studies of half-wave potentials as a function of ligand concentration provide the means to determine the formula of a metal complex.

Although the use of voltammetric methods for the study of the electronic and molecular properties of organic and biological molecules was suggested many years ago, only recently has interest in this application developed among organic chemists and biochemists. One reason for this has been that the technology of voltammetry until recently has been such that meaningful thermodynamic and kinetic data were possible only for aqueous systems. From the organic chemist's viewpoint water is not an ideal solvent; nor does it represent a medium that is particularly relevant to organic processes. Thus, the application of voltammetric methods to physicochemical studies of organic systems has required the development of electronic instruments with three-electrode configurations that permit the use of the high-resistance solutions (characteristic of nonaqueous solvent systems). Under these conditions studies of the effects of substituents and of solvents on the half-wave potential of electroactive organic molecules are possible. Such studies provide a relative measurement of the free energies of removal (or of addition) of valence electrons, which can be correlated with molecular orbital calculations. Table 3.7 summarizes several examples of such applications, and indicates the general applicability of voltammetric methods for the generation of fundamental electronic data for structural studies.[68]

The techniques of voltage sweep and cyclic voltammetry provide the analytical and physicochemical capabilities of classical voltammetry and in addition provide the means for performing these measurements much more rapidly for a broader range of conditions. Cyclic voltammetry is particularly useful for the rapid assessment of thermodynamic reversibility, and for the evaluation of the stoichiometry for the electrode reaction.

TABLE 3.7 Half-Wave Reduction Potentials for Hydrocarbons in 75% Dioxane–H_2O and Energies of Lowest Vacant Molecular Orbitals

Hydrocarbon	$-m_{m^*} + 1$ E, Lowest Orbital	$-E_{1/2}$ versus SCE
Triphenylmethyl	0	1.05
1-Phenyl-6-biphenylenehexatriene	0.202	1.35
1-Phenyl-4-biphenylenebutadiene	0.251	1.46
Acenaphthylene	0.285	1.65
Tetracene	0.295	1.58
Perylene	0.347	1.67
1,2-Benzpyrene	0.365	1.85
Fluoranthene	0.371	1.77
1,4-Diphenylbutadiene	0.386	2.00
Azulene	0.400	1.64
Anthracene	0.414	1.96
Pyrene	0.445	2.11
1,2-Benzanthracene	0.452	2.00
1,2,5,6-Dibenzanthracene	0.474	2.03
4,5-Benzpyrene	0.497	2.00
Stilbene	0.504	2.16
Chrysene	0.520	2.30
2,2′-Binaphthyl	0.521	2.21
p-Quaterphenyl	0.536	2.20
Coronene	0.539	2.04
1,1-Diphenylethylene	0.565	2.25
p-Terphenyl	0.593	2.33
Phenanthrene	0.605	2.46
Naphthalene	0.618	2.50
Butadiene	0.618	2.63
Styrene	0.662	2.37
Triphenylene	0.684	2.49
Biphenyl	0.705	2.70

As outlined in the theoretical section of this chapter, controlled-potential methods have extensive application in the study of the kinetics and mechanisms of the electron-transfer reaction of electrochemical processes. Furthermore, associated reactions before and after the electron-transfer process are readily studied by controlled-potential methods. For a number of systems the rate constants for these associated chemical processes can be evaluated.

For processes that are sufficiently slow, polarographic, and voltammetric current–potential curves provide a measure of the heterogeneous rate constant and the transfer coefficient for the electrode reaction [see Eqs. (3.25) and (3.26)]. Although homogeneous chemical rate constants and mechanisms have been studied by these methods, there are much more convenient approaches that provide a quantitative measure of such associated processes. In particular, linear-sweep and cyclic voltammetry are ideally suited for the study of the kinetic phenomena of electrochemistry.[6] The kinetic parameters for the heterogeneous rate constants and the transfer coefficient can be evaluated from either single-sweep or cyclic voltammetry [Eqs. (3.38), (3.39), and (3.43)]. Cyclic voltammetry is especially well suited for the study of homogeneous post-electron-transfer processes. Because of the high scan rates associated with cyclic voltammetry, unstable intermediates can be observed and some measure of their lifetimes can be made. Voltammetric methods provide the means to elucidate reasonable mechanisms for the rate-controlling electron-transfer process as well as for any associated chemical processes that are rate-controlling. However, the overall electron stoichiometry frequently must be determined for the electrode reaction. This normally is accomplished by the use of controlled-potential coulometry, which provides a direct measure of the number of electrons per electroactive species. The latter knowledge is essential if a truly realistic mechanism is to be developed for the overall electrochemical process.

Both the ring–disk and thin-layer electrodes provide a convenient means for observing unstable intermediate products from electrochemical reactions. Quantitative evaluations of the lifetimes of these intermediates and of the products from such intermediates are readily evaluated by each of these methods.[52,53]

One of the most important, yet latent, applications of controlled-potential electrolysis is electrochemical synthesis. Although electrolysis has been used for more than a century to synthesize various metals from their salts, application to other types of chemical synthesis has been extremely limited. Before the advent of controlled-potential methods, the selectivity possible by classical electrolysis precluded fine control of the products. The only control was provided by appropriate selection of electrode material, solution acidity, and supporting electrolyte. By these means the effective electrode potential could be limited to minimize the electrolysis of the supporting electrolyte or the solvent. Today potentiostats and related controlled-potential-electrolysis instrumentation are commercially available that provide effective control of the potential of the working electrode to ± 1 mV, and a driving force of up to 100 V for currents of up to several amperes. Through such instrumentation electrochemical syn-

TABLE 3.8 Example of Practical Electrochemical Syntheses

Product	Reactant	Process	Ref.
3-Acetoxycyclohexene-1	Cyclohexene, acetic acid	Oxidative substitution	71
Adipionitrile	Acrylonitrile	Reductive coupling	72
Benzyl alcohol	Benzoic acid	Reduction	73,74
Chromium hexacarbonyl	Chromium acetylacetonate, carbon monoxide	Reduction	75
Cyanogen bromide	Hydrogen cyanide, ammonium bromide	Oxidation	76
Cyclohexadiene	Phthalic acid	Reduction	77
"Dewar benzene"	[2.2.0]Hexa-5-ene-2,3-dicarboxylic acid	Oxidation	78
Diethyl adipate	Ethyl acrylate	Reductive coupling	79
1,3-Dimethylcyclobutane	1,3-Dibromo-1,3-dimethylcyclobutane	Reductive cyclization	80
Dimethyl sebacate	Monomethyl adipate	Oxidation	81
Dimethyl 1,14-tetradecanodiate	Monomethyl suberate	Oxidative coupling	82
Polyacrylonitrile	Acrylonitrile	Oxidative polymerization	83
Polymethyl methacrylate	Methyl methacrylate	Reductive polymerization	84
Polystyrene	Styrene	Reductive polymerization	85
Propylene oxide	Propylene	Oxidation	86
Salicylaldehyde	Salicylic acid	Reduction	73,74,87
Sorbitol	Glucose	Reduction	88
Tetraethyl lead	Ethyl chloride, lead	Oxidation	89

thesis becomes possible for a wide range of solvent systems and supporting electrolytes.

Significant effort has been expended recently to develop practical syntheses based on electrochemistry. The review by Baizer and Lund[69] is the most extended up to date, and some of the more promising syntheses have been summarized.[70] Table 3.8 lists typical electroorganic syntheses.[71-89] Thus, the high degree of control possible through appropriate selection of the controlled potential, electrode material, supporting electrolyte, and pH permits highly efficient syntheses with few, if any, byproducts.

When electrochemical studies have established the ideal electrochemical conditions for a particular synthesis a simpler, more rapid and efficient chemical synthesis can be developed on the basis of the electrochemical knowledge. For example, if a given organic substance can be quantitatively reduced to give the desired product at a mercury-pool electrode held at potential of −0.900 V versus SCE, then an electrochemical synthesis could be accomplished with adequate instrumentation in approximately 20–30 min. However, these conditions could be duplicated chemically through the use of a cadmium amalgam in contact with an electrolyte solution that contains EDTA adjusted to pH 8. That this is true was first established by polarographic studies of cadmium ion in the presence of the specified supporting electrolyte. Such an example illustrates that the equivalent of crude potential control can be accomplished by the appropriate selection of the amalgamated material. A finer degree of potential control is then accomplished by appropriate selection of the supporting electrolyte ligand, and fine potential control is provided by adjustment of the electrolyte pH.

REFERENCES

1. Fick, A., *Pogg. Ann.* **1855,** *94*, 59.
2. Heyrovsky, J., *Chem. Listy* **1922,** *16*, 256.
3. Ilkovic, D., *Collect. Czech. Chem. Commun.* **1934,** *6*, 498.
4. Kemula, W.; Kublik, Z., *Roczniki Chem.* **1956,** *30*, 1005.
5. Kemula, W.; Kublik, Z., *Anal. Chim. Acta* **1958,** *18*, 104.
6. Nicholson, R. S.; Shain, I., *Anal. Chem.* **1964,** *36*, 706.
7. Nicholson, R. S., *Anal. Chem.* **1965,** *37*, 1351.
8. Eyring, H.; Glasstone, S.; Laidler, K. J., *J. Chem. Phys.* **1939,** *7*, 1053.
9. Kolthoff, I. M.; Lingane, J. J., *Polarography*, 2nd ed., Interscience Publishers, New York, 1952.
10. Meites, L., *Polarographic Techniques*, 2nd ed., Interscience Publishers, New York, 1965.
11. Heyrovsky, J.; Kuta, J., *Principles of Polarography*, Academic Press, New York, 1966.

12. Bond, A. M., *Modern Polarographic Methods in Analytical Chemistry*, Marcel Dekker, New York, 1980.
13. Bard, A. J.; Faulkner, L. R., *Electrochemical Methods*, Wiley, New York, 1980.
14. Galus, Z., *Fundamentals of Electrochemical Analysis*, 2nd ed., Ellis Horwood Ltd., Chichester, 1994.
15. Breyer, B.; Bauer, H. H., *Alternating Current Polarography and Tensammetry*, Vol. XIII, *Chemical Analysis*, Elving, P. J.; Kolthoff, I. M., eds., Interscience: New York, 1963.
16. Smith, D. E., *Crit. Rev. Anal. Chem.* **1971,** *2*, 247.
17. Kissinger, P. T., "Small-Amplitude and Related Controlled-Potential Techniques," in *Laboratory Techniques in Electroanalytical Chemistry*, Kissinger, P. T.; Heineman, W. R., eds., Marcel Dekker, New York, 1984, pp. 143–161.
18. Burge, D. E., *J. Chem. Ed.* **1970,** *47*, A81.
19. Osteryoung, J. G.; Osteryoung, R. A., *Am. Lab.* **1972,** *4(7)*, 8.
20. Osteryoung, J. G., *Acc. Chem. Res.* **1993,** *26*, 77.
21. Heineman, W. R.; Harry, B. M., Jr.; Wise, J. A.; Roston, D. A., "Electrochemical Preconcentration," in *Laboratory Techniques in Electroanalytical Chemistry*, Kissinger, P. T.; Heineman, W. R., eds., Marcel Dekker, New York, 1984, pp. 499–538.
22. Branina, Kh.; Neyman, E., *Electroanalytical Stripping Methods*, Vol. 126, *Chemical Analysis*, Winefordner, J. D., ed., Wiley, New York, 1993.
23. Bond, A. M.; Canterford, D. R., *Anal. Chem.* **1972,** *44*, 721.
24. Keller, H. E.; Osteryoung, R. A., *Anal. Chem.* **1971,** *43*, 342.
25. Pranz, E. P.; Osteryoung, R. A., *Anal. Chem.* **1965,** *37*, 1634.
26. Vassos, B. H.; Ewing, G. W., *Electroanalytical Chemistry*, Wiley, New York, 1983.
27. Matsuda, H.; Ayabe, Y., *Z. Elektrochem.* **1955,** *59*, 494.
28. Nicholson, R. S.; Shain, I., *Anal. Chem.* **1964,** *37*, 178.
29. Nicholson, R. S.; Shain, I., *Anal. Chem.* **1964,** *37*, 190.
30. Savéant, J. M.; Vianello, E., *Electrochim. Acta* **1963,** *8*, 905.
31. Savéant, J. M.; Vianello, E., *Electrochim. Acta* **1967,** *12*, 629.
32. Savéant, J. M., *Electrochim. Acta* **1967,** *12*, 753.
33. Savéant, J. M.; Vianello, E., *Electrochim. Acta* **1967,** *12*, 1545.
34. Mastragostino, M.; Nadjo, L.; Savéant, J. M., *Electrochim. Acta* **1968,** *13*, 721.
35. Nadjo, L.; Savéant, J. M., *J. Electroanal. Chem.* **1973,** *48*, 113.
36. Savéant, J. M.; Andrieux, C. P.; Nadjo, L., *J. Electroanal. Chem.* **1973,** *41*, 137.
37. Macdonald, D. D., *Transient Techniques in Electrochemistry*, Plenum Press, New York, 1977.
38. Greef, R.; Peat, R.; Peter, L. M.; Pletcher, D.; Robinson, J., *Instrumental Methods in Electrochemistry*, Ellis Horwood Ltd., Chichester, 1985.
39. Laviron, E., *J. Electroanal. Chem.* **1974,** *52*, 355.
40. Laviron, E., *J. Electroanal. Chem.* **1974,** *52*, 395.
41. Srinivasan, S.; Gileadi, E., *Electrochim. Acta* **1966,** *11*, 321.

42. Wopschall, R. H.; Shain, I., *Anal. Chem.* **1967,** *39*, 1514.
43. Wopschall, R. H.; Shain, I., *Anal. Chem.* **1967,** *39*, 1527.
44. Wopschall, R. H.; Shain, I., *Anal. Chem.* **1967,** *39*, 1535.
45. Wightman, R. M., *Anal. Chem.* **1981,** *53*, 1125A.
46. Cassidy, J.; Khoo, S. B.; Pons, S.; Fleischmann, M., *J. Phys. Chem.* **1985,** *89*, 8933.
47. Wightman, R. M.; Wipf, D. O., *Acc. Chem. Res.* **1990,** *23*, 64.
48. Fleischmann, M.; Pons, S.; Rolison, D. R.; Schmidt, P. P., eds., *Ultramicroelectrodes*, Datatech Systems, Morganton, N.C., 1987.
49. Bond, A. M.; Oldham, K. B.; Zoski, C. G., *Anal. Chim. Acta* **1989,** *216*, 177.
50. Cassidy, J. F.; Foley, M. B., *Chem. Br.* **1993,** *29*, 764.
51. Aoki, K., *Electroanal.* **1993,** *5*, 627.
52. Levich, V. G., *Physiochemical Hydrodynamics*, Prentice-Hall, Englewood Cliffs, N.J., 1963.
53. Albery, W. J.; Hitchman, M. L., *Ring Disc Electrode*, Oxford University Press, Oxford, 1971.
54. Pleskov, Yu. V.; Filinovskii, V. Yu, *The Rotating Disc Electrode*, Consultants Bureau, New York, 1976.
55. Cofré, P.; Sawyer, D. T., *Inorg. Chem.* **1986,** *25*, 2089.
56. Bruckenstein, S.; Napp, D. T., *J. Am. Chem. Soc.* **1968,** *90*, 6303.
57. Heineman, W. P.; Kissinger, P. T., "Large-Amplitude Controlled-Potential Techniques," in *Laboratory Techniques in Electroanalytical Chemistry*, Kissinger, P. T.; Heineman, W. R., eds., Marcel Dekker, New York, 1984, pp. 51–127.
58. Lingane, J. J., *Electroanalytical Chemistry*, 2nd ed., Interscience, New York, 1958.
59. Meites, L, "Controlled-Potential Electrolysis and Coulometry," in *Physical Methods of Chemistry*, 2nd ed., Vol. II, *Electrochemical Methods*, Rossiter, B. W.; Hamilton, J. F., eds., Wiley, New York, 1986, pp. 433–523.
60. Meites, L.; Zuman, P.; et al., eds., *CRC Handbook Series in Inorganic Electrochemistry*, Vols. 1–7, CRC Press, Boca Raton, Fla., 1980–1986.
61. Meites, L.; Zuman, P.; et al., eds., *Electrochemical Data*, Wiley, New York, 1974.
62. Meites, L.; Zuman, P.; et al., eds., *CRC Handbook Series in Organic Electrochemistry*, Vols. 1–4, CRC Press, Boca Raton, Fla., 1980–1986.
63. Bard, A. J.; Lund, H., eds., *Encyclopedia of Electrochemistry of the Elements*, Marcel Dekker, New York, 1973–1984.
64. Anderson, L. B.; Reilley, C. N., *J. Electroanal. Chem.* **1965,** *10*, 295.
65. Kolthoff, I. M.; Zuman, P., *Progress in Polarography*, Interscience Publishers, New York, 1962.
66. Clark, L. C., Jr., *Trans. Am. Soc. Artif. Intern. Organs* **1955,** *2*, 41.
67. Sawyer, D. T.; George, R. S.; Rhodes, R. C., *Anal. Chem.* **1959,** *31*, 2.
68. Mark, H. B., Jr., *Rec. Chem. Progr.* **1968,** *29*, 217.
69. Lund, H.; Baizer, M. M., eds., *Electroorganic Synthesis*, 3rd ed., Marcel Dekker, New York, 1991.

70. Shono, T., *Electroorganic Synthesis*, Academic Press, New York, 1991.
71. Shono, T.; Kosaka, T., *Tetrahedron Lett.* **1968,** 6207.
72. Prescott, J. H., *Chem. Eng.* **1965** (Nov. 8), 238.
73. Natarajan, K.; Udupa, K. S.; Subramanian, G. S.; Udapa, H. V. K., *Electrochem. Tech.* **1964,** *2*(5–6), 151.
74. Balakrishman, T. D.; et al., *Chem. Ind.* (London), **1970,** 1622.
75. Ercoli, R.; Guainazzi, M.; Silverstri, G., *Chem. Commun.* **1967,** 927.
76. Foreman, R. W.; Sprague, J. W., *Ind. Eng. Chem. Prod. Res. Develop.* **1963,** *2*, 303.
77. Condit, P. C., *Ind. Eng. Chem.* **1956,** *48*, 1252.
78. Radlick, P. H.; Kelm, R.; Spurlock, S.; Sims, J. J.; Tamelen van, E. E.; Whiteside, T., *Tetrahedron Lett.* **1968,** 5117.
79. Baizer, M. M., *Tetrahedron Lett.* **1963,** 973.
80. Rifi, M. R., *J. Am. Chem. Soc.* **1967,** *89*, 4442.
81. Kamneva, A. I.; Fioshin, Ya., M.; Kazakrova, L. I.; Itenberg, Sh. M., *Neftekhim.* **1962,** *2*, 550.
82. Anderson, J. D.; Baizer, M. M.; Petrovich, J. P., *J. Org. Chem.* **1966,** *31*, 3890.
83. Goldschmidt, S.; Stocke, E., *Ber.* **1952,** *85*, 630.
84. Wilson, C. L., *Rec. Chem. Progr.* **1949,** *10*, 25.
85. Bond, G. C., *Discuss. Faraday Soc.* **1966,** *10*, 25.
86. LeDuc, J. A. M. (to Pullman, Inc.), U.S. Patent 3, 342, 717 (September 19, 1967).
87. Udupa, K. S.; Subramanian, G. S.; Udapa, H. V. K., *Ind. Chem.* **1963** (May), 238.
88. Taylor, R. L., *Chem. Met. Eng.* **1937,** *44*, 588.
89. Bott, L. L., *Hydrocarbon Proc.* **1965,** *44*, 115.

CHAPTER 4

ELECTROCHEMICAL TITRATIONS AND CONTROLLED-CURRENT METHODS

4.1 INTRODUCTION

One of the most extensive applications of electrochemistry has been for endpoint detection in titrations. The latter continue to be important in analysis, in spite of their more tedious nature, because of the better precision and accuracy than is possible by direct electrochemical measurements. For example, in potentiometry a 0.25-mV error represents a 1% relative error in the concentration of the detected species.

Electrochemical endpoint detection methods provide a number of advantages over classical visual indicators. These methods can be used when visual methods of endpoint detection cannot be employed because of the presence of colored or clouded solutions and in the case of detection of several components in the same solution. They are more precise and accurate. In particular, such methods provide increased sensitivity and are often amenable to automation. Electrochemical methods of endpoint detection are applicable to most oxidation–reduction, acid–base, and precipitation titrations, and to many complexation titrations. The only necessary condition is that either the titrant or the species being titrated must give some type of electrochemical response that is indicative of the concentration of the species.

4.2 ENDPOINT DETECTION METHODS

Potentiometric Methods. By far the most common endpoint system for titrations is a potentiometric indicating electrode because of its simplicity and its universal applicability. It only requires a redox couple that gives some response that is indicative of either the titrant or the species to be titrated. In its simplest form the endpoint detection system consists of a reference electrode [usually the saturated calomel electrode (SCE)] plus an inert indicator electrode. An electric circuit used for potentiometric titrations is presented in Figure 4.1*a*. In the situation where both the titrant and the species to be titrated give reversible thermodynamic responses, the potential of the indicator is governed by either couple. An example of such a system is the titration of iron(II) by cerium(IV) with a platinum electrode as an "inert" indicator electrode:

$$Ce(OH_2)_6^{4+} + Fe(OH_2)_6^{2+} \rightarrow Fe(OH_2)_6^{3+} + Ce(OH_2)_6^{3+} \qquad (4.1)$$

For this system the indicator-electrode potential is represented by either half-reaction:

$$E_{ind} = E^{\circ}_{Fe(OH_2)_6^{3+}/Fe(OH_2)_6^{2+}} + \frac{0.059}{1} \log \frac{[Fe(OH_2)_6^{3+}]}{[Fe(OH_2)_6^{2+}]} \quad \text{at } 25°C \qquad (4.2)$$

$$E_{ind} = E^{\circ}_{Ce(OH_2)_6^{4+}/Ce(OH_2)_6^{3+}} + \frac{0.059}{1} \log \frac{[Ce(OH_2)_6^{4+}]}{[Ce(OH_2)_6^{3+}]} \quad \text{at } 25°C \qquad (4.3)$$

Prior to the endpoint the Nernst expression of Eq. (4.2) is most convenient for computing the indicator potential. In contrast, past the equivalence point Eq.

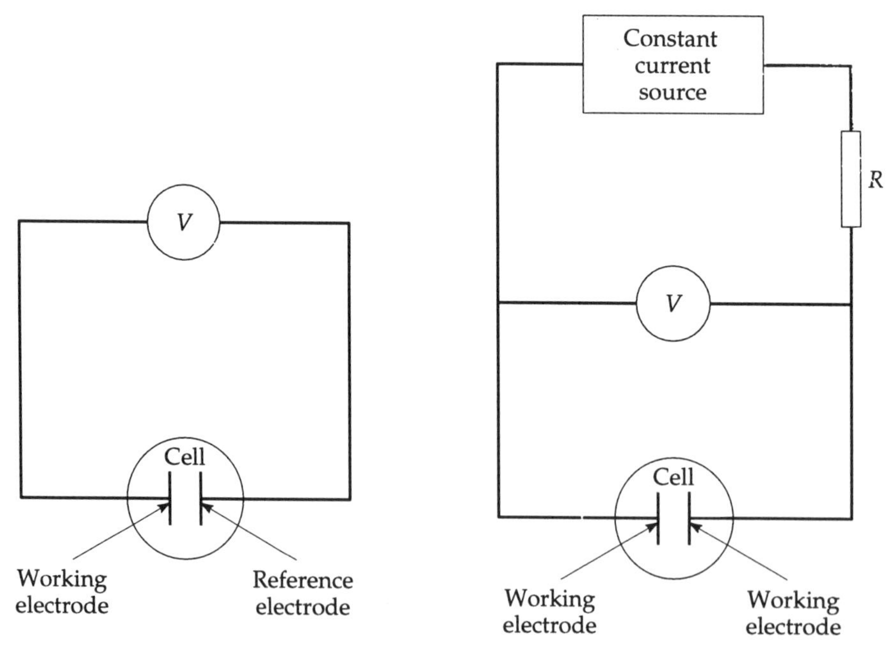

Figure 4.1 Electric circuit used for potentiometric titrations: (*a*) without polarizing current; (*b*) with polarizing current.

(4.3) is more convenient. To calculate the indicator potential at the equivalence point a combination of Eqs. (4.2) and (4.3) gives

$$2E_{ind} = E^{\circ}_{Fe(OH_2)_6^{3+}/Fe(OH_2)_6^{2+}} + E^{\circ}_{Ce(OH_2)_6^{4+}/Ce(OH_2)_6^{3+}} + \frac{0.059}{1}\log\frac{[Fe(OH_2)_6^{3+}][Ce(OH_2)_6^{4+}]}{[Fe(OH_2)_6^{2+}][Ce(OH_2)_6^{3+}]} \quad \text{at } 25°C \tag{4.4}$$

Consideration of Eq. (4.1) indicates that the conditions represented by

$$[Fe(OH_2)_6^{3+}]_{ep} = [Ce(OH_2)_6^{3+}]_{ep}; \quad [Fe(OH_2)_6^{2+}]_{ep} = [Ce(OH_2)_6^{4+}]_{ep} \tag{4.5}$$

prevail at the equivalence point (ep). Substitution of these quantities into Eq. (4.4) establishes that the indicator potential at the equivalence point is independent of the concentrations of the titrant and the reactant:

$$(E_{ind})_{ep} = \frac{E^{\circ}_{Fe(OH_2)_6^{3+}/Fe(OH_2)_6^{2+}} + E^{\circ}_{Ce(OH_2)_6^{4+}/Ce(OH_2)_6^{3+}}}{2} \tag{4.6}$$

This specific result can be generalized by the expression

$$(E_{ind})_{ep} = \frac{mE_1^{\circ} + nE_2^{\circ}}{m + n} \tag{4.7}$$

where m and n represent the number of electrons for the two half-reactions that make up the titration reaction.

Other examples of potentiometric titrations include acid–base titrations, in which an indicator electrode provides a response to hydronium ions, such as the glass electrode, quinhydrone electrode, or antimony electrode. In precipitation and complexation titrations the indicator electrode should provide the response to the active species in the solution. Thus, during the titration of chloride ions by silver nitrate, a silver electrode is an effective indicator electrode.

With any endpoint detection system several practical considerations are important for reliable results. For example, the indicator electrode should be placed in close proximity to the flow pattern from the burette, so that a degree of anticipation is provided to avoid overrunning the endpoint. Another important factor is that the indicator electrode be as inert and nonreactive as possible to avoid contamination and erratic response from attack by the titration solution. A third and frequently overlooked consideration is the makeup of the reference electrode and, in particular, its salt bridge. For example, a salt-bridge system that contains potassium chloride can cause extremely erratic behavior of any electrochemical system if the titrant solution contains perchlorate ion (because of the precipitation of potassium perchlorate at the salt-bridge titrant–solution interface). Likewise, a potassium chloride salt bridge in a potentiometric titration of either silver ion or mercurous ion can cause serious titration errors due to leakage of salt solution into the titration system.

Potentiometric titration curves normally are represented by a plot of the indicator-electrode potential as a function of volume of titrant, as indicated in Fig. 4.2. However, there are some advantages if the data are plotted as the first derivative of the indicator potential with respect to volume of titrant (or even as the second derivative). Such titration curves also are indicated in Figure 4.2, and illustrate that a more definite endpoint indication is provided by both differential curves than by the integrated form of the titration curve. Furthermore, titration by repetitive constant-volume increments allows the endpoint to be determined without a plot of the titration curve; the endpoint coincides with the condition when the differential potentiometric response per volume increment is a maximum. Likewise, the endpoint can be determined by using the second derivative; the latter has distinct advantages in that there is some indication of the approach of the endpoint as the second derivative approaches a positive maximum just prior to the equivalence point before passing through zero. Such a second-derivative response is particularly attractive for automated titration systems that stop at the equivalence point.

A specialized version of the potentiometric endpoint detection system is the use of two dissimilar metals as the electrode pair. For example, if a platinum

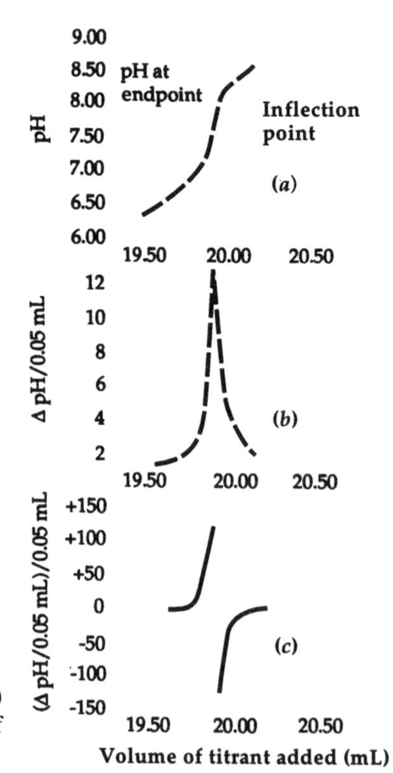

Figure 4.2 Potentiometric titration curves: (*a*) experimental titration data; (*b*) first derivative of curve *a*; (*c*) second derivative of curve *a*.

electrode is used in combination with a tungsten electrode, a sharp potential response frequently is observed that coincides with the equivalence point. This occurs not only with oxidation–reduction titrations but also with acid–base titrations. It comes about because of the generally more reversible behavior of platinum electrodes relative to tungsten electrodes. Although this is an empirical system, it offers the advantage of an extremely inert reference-electrode system. A related approach is to place the burette beneath the surface of the titrant solution and place a platinum wire inside the burette tip, and another platinum electrode in the titrant solution. This provides a differential response because the electrode inside the burrette maintains a reasonably constant potential and serves as the reference electrode.

The subject of potentiometric titrations has been exhaustively treated by the classic monograph by Kolthoff and Furman.[1] Although the second edition was published in 1931, it is still the definitive work. Unfortunately, it is no longer in print, but it is available in many chemical libraries. A complete and thorough discussion of the principles and theory of potentiometric titrations is provided, together with an extremely extensive summary of the many applications of potentiometric measurements.

Another specialized form of potentiometric endpoint detection is the use of dual-polarized electrodes, which consists of two metal pieces of electrode material, usually platinum, through which is imposed a small constant current, usually 2–10 μA. The scheme of the electric circuit for this kind of titration is presented in Figure 4.1*b*. The differential potential created by the imposition of the current is a function of the redox couples present in the titration solution. Examples of the resultant titration curve for three different systems are illustrated in Figure 4.3. In the case of two reversible couples, such as the titration of iron(II) with cerium(IV), curve *a* results in which there is little potential difference after initiation of the titration up to the equivalence point. The titration of arsenic(III) with iodine is representative of an irreversible couple that is titrated with a reversible system. Hence, prior to the equivalence point a large potential difference exists because the passage of current requires decomposition of the solvent for the cathode reaction (Figure 4.3*b*). Past the equivalence point the potential difference drops to zero because of the presence of both iodine and iodide ion. In contrast, when a reversible couple is titrated with an irreversible couple, the initial potential difference is equal to zero and the large potential difference appears after the equivalence point is reached.

The use of dual-polarized electrodes was first suggested more than 70 years ago;[2] the subject has been reviewed thoroughly by two more recent publications.[3,4] Almost all modern commercial pH meters have provision for imposing a polarizing current of either 5 or 10 μA to make possible measurements by dual-polarized electrode potentiometry. Such a provision is included because dual-polarized potentiometry is by far the most popular endpoint detection method for the Karl Fischer determination of water. For this titration a combination of reagents is used, including iodine; the response curve is similar to that of Figure 4.3*b*. In practice the response is many times more sensitive than

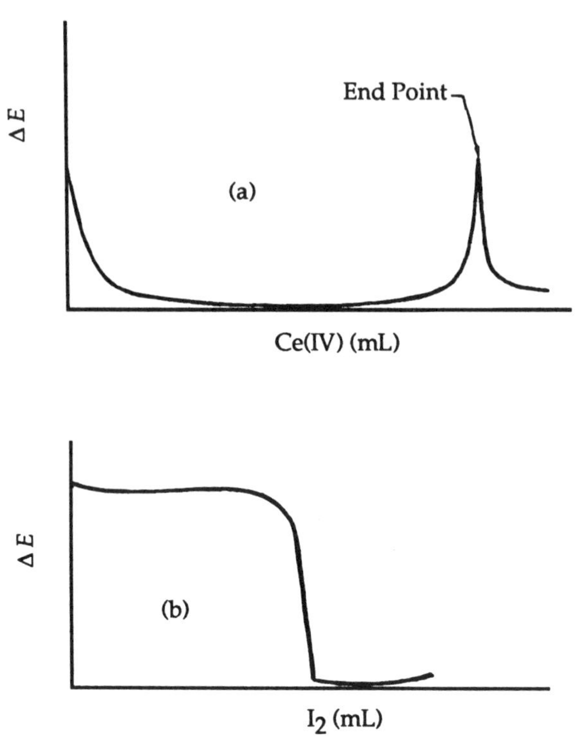

Figure 4.3 Dual-polarized electrode potentiometric titration curves: (*a*) titration of Fe(II) by Ce(IV); (*b*) titration of As(III) by I_2.

that obtained by a conventional potentiometric system. Without question a number of other potentiometric titrations would benefit from the use of the dual-polarized approach. This is to be recommended not only because of the increased sensitivity of the endpoint response that frequently occurs but also because of the simplicity of the electrode system (which avoids solution contamination and the other problems that frequently are associated with the reference electrode).

Amperometric Methods. Another form of electrochemical endpoint detection is the amperometric method. In its most common form, this consists of a polarizable microelectrode, usually the dropping-mercury electrode that is characteristic of polarography, in combination with a large nonpolarizable reference electrode. Figure 4.4 illustrates the scheme of an electric circuit used in this kind of titration. Thus, polarographic measurements are made of the titration system as a function of volume of titrant. Potential is applied across the indicator system such that it is on the diffusion-current plateau for either the titrant, the reactant, or both and the current intensity in the circuit is registered. If a dropping-mercury electrode is used, then all the technology and background

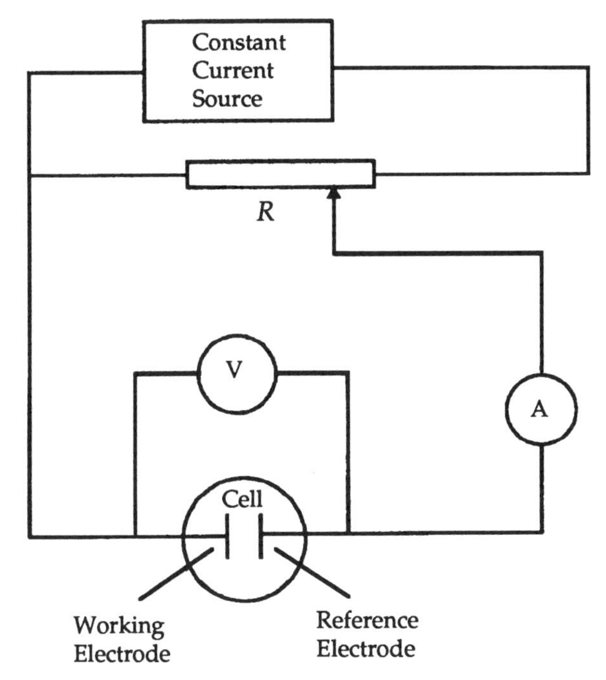

Figure 4.4 Instrumentation for amperometric titrations.

data from polarography may be used to set the proper conditions (supporting electrolyte and electrode potential) for an effective response that is characteristic of either the titrant or reactant concentration.

Figure 4.5 illustrates the titration curve that is obtained with an amperometric indicating system for the titration of lead ion with dichromate ion. If the applied potential is set on the plateau for the reduction of lead ion (approximately −0.5 V vs. SCE), curve *a* in Figure 4.5, will result. In contrast, if the applied potential is set at 0 V versus SCE, no current will flow until the point when excess chromate ion exists in the solution; curve *b* is indicative of the titration curve that would be obtained.

The amperometric approach to endpoint detection provides considerable latitude in the selection of the best conditions for the most specific and sensitive endpoint response. Furthermore, the response signal is directly proportional to the concentration of the observed species, whereas potentiometric responses are a logarithmic function of the concentration. Another attractive feature of amperometric endpoint detection is that the most important data are obtained prior to and after the equivalence point, whereas in potentiometric titrations the most important data occur at the equivalence point, which is the most unstable condition of the titration. With amperometric titrations an extrapolation of the straight-line portion of the curve, either prior to or after the equivalence point, to an intercept will provide an accurate measure of the equivalence point.

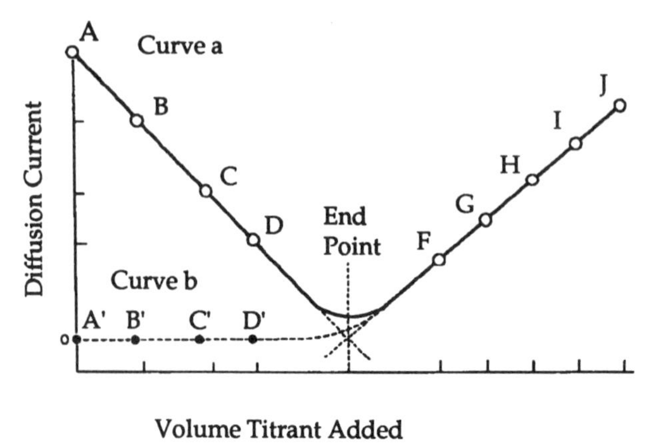

Figure 4.5 Amperometric titration curves for the titration of Pb(II) ions by $Cr_2O_7^{2-}$ ions at $E_{DME} = -0.8$ V versus SCE (curve *a*) and at $E_{DME} = 0.0$ V versus SCE (curve *b*).

To have as straight a line as possible, it is necessary to apply a dilution correction to the observed current. This can be simplified by using an extremely high concentration of titrant relative to the concentration of the species being titrated.

The entire subject of amperometric titrations has been reviewed in a number of monographs on electrochemistry;[4–6] a definitive work on this subject also has been published.[7] Because the amperometric titration method does not depend on one or more reversible couples associated with the titration reaction, it permits electrochemical detection of the endpoint for a number of systems that are not amenable to potentiometric detection. All that is required is that electrode conditions be adjusted such that either a titrant, a reactant, or a product from the reaction gives a polarographic diffusion current.

A specialized version of amperometric endpoint detection systems uses dual-polarized electrodes (often a pair of small platinum foils). In contrast to the dual-polarized potentiometric system, dual-polarized amperometric electrodes have a constant potential applied to them, such that one or the other of the electrodes gives a diffusion-controlled current response during some portion of the titration. The applied potential, which normally ranges from zero to several tenths of a volt, is ascertained either by consideration of polarographic half-wave potential data, or from experimental measurements. What is sought is an applied potential that is on the diffusion-current plateau for one or more couples involved in the titration. Figure 4.6 illustrates the current response that is obtained for a pair of reversible couples [titration of iron(II) with cerium(IV)]. The shape of the titration curve is determined by the changes of current intensity in the electric circuit. If the dual-electrode system is placed in the solution that contains iron(II) and a small voltage (e.g., ΔE equal to 50 mV) is applied, no current flow is detected in the circuit. If, however, the first portion of the titrant [cerium(IV)] is added, a portion of the iron(II) is oxidized to iron(III) and the

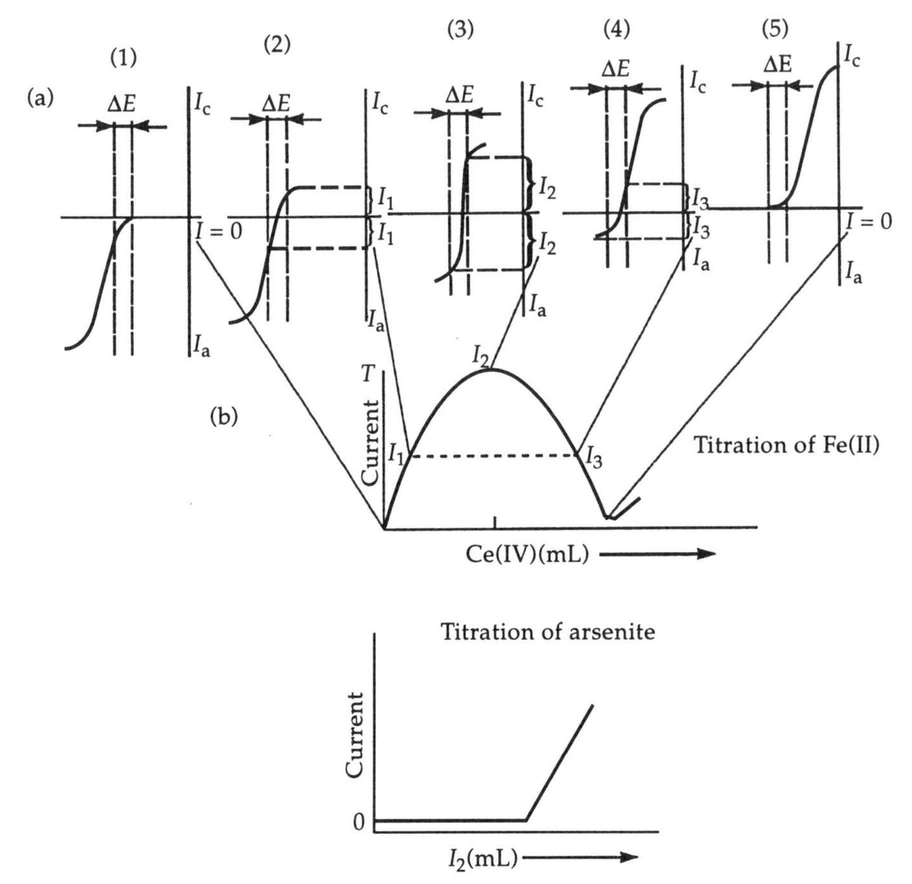

Figure 4.6 Dual-polarized electrode amperometric titration curves. Both curves result from the application of a 0.25-V potential across two identical platinum electrodes that are immersed in the titration solution.

reversible couple $Fe(OH_2)_6^{2+}/Fe(OH_2)_6^{3+}$ appears. The potential of anodic oxidation of $Fe(OH_2)_6^{2+}$ and the potential of cathodic reduction of $Fe(OH_2)_6^{3+}$ are almost identical. Therefore, the reduction of $Fe(OH_2)_6^{2+}$ takes place on the cathode and the oxidation of $Fe(OH_2)_6^{2+}$ occurs on the anode simultaneously, and a current is registered in the circuit. The maximum current is registered when 50% of the iron(II) has been titrated [when the concentrations of $Fe(OH_2)_6^{2+}$ and $Fe(OH_2)_6^{3+}$ are identical]. The further addition of the titrant causes the current to decrease because the concentration of $Fe(OH_2)_6^{2+}$ is decreased and the anodic current has to be equal to the cathodic current [the amount of $Fe(OH_2)_6^{2+}$ that is oxidized and the amount of $Fe(OH_2)_6^{3+}$ that is reduced have to be equal]. At the equivalence point the current is equal to zero because all the $Fe(OH_2)_6^{2+}$ has been oxidized and the anodic process cannot occur. The course of the titration curve after the equivalence point depends on the kind of titrant. When a titrant forms a reversible couple, the shape presented by curve *a* is observed; for an irreversible titrant this part of the titrant curve is a flat line. The titration of an irreversible couple with a reversible couple [arsenic(III) with iodine] gives the titration curve presented in Figure 4.6*b*. The latter curve also is characteristic of the response one obtains if a dual-polarized amperometric endpoint detection system is used for the Karl Fischer titration. The dual-polarized amperometric system is especially attractive because of the simplicity of the instrumentation. Only a microammeter plus a potentiometric voltage divider and a dry cell are necessary to produce a extremely sensitive and versatile endpoint detection system. The method has been known for many years but has not enjoyed the popularity that it deserves. Some of the older literature refers to the method as the "dead-stop" method.

Conductometric Methods. One of the oldest electrochemical detection methods is the conductometric monitoring of the ionic concentrations of species involved in a titration. The general approach is to make conductometric measurements during the course of the titration with a plot of the conductance as a function of volume of titrant. Figure 4.7*a* is representative of the kind of response that is obtained for the titration of hydrochloric acid with sodium hydroxide. The shape of the titration curve can be predicted by summing the ionic concentrations of the various species at any point during the course of the titration; the resulting summation will give the titration curve. Figure 4.7*b*

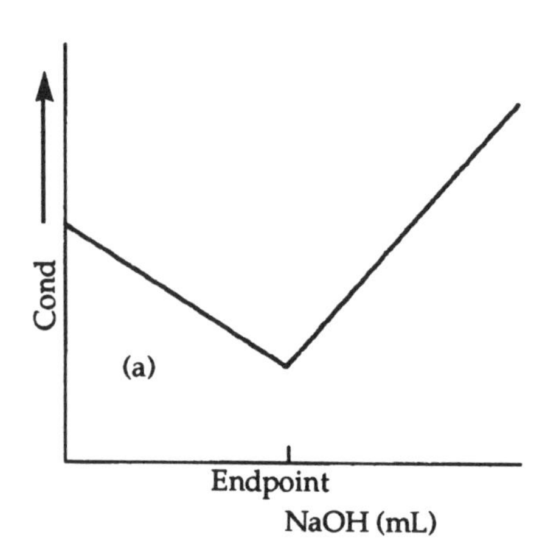

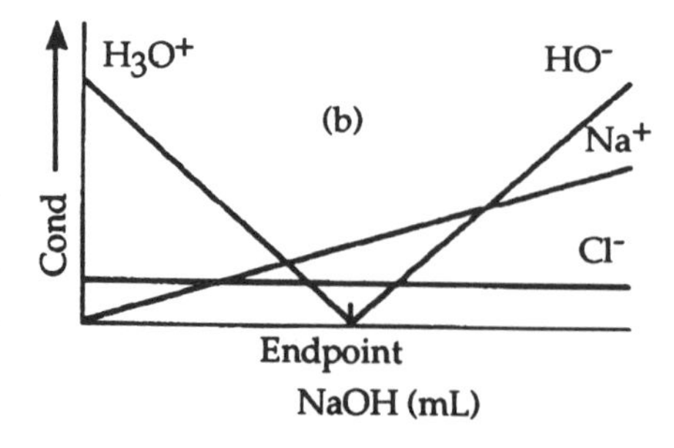

Figure 4.7 (*a*) Conductometric titration curve for the titration of HCl by NaOH. (*b*) Conductances of individual ions during the course of the titration; summation of these at any volume of titrant corresponds to the point on curve *a*.

indicates how these vary for the example titration. The titration itself can be represented by

$$[H_3O^+ + Cl^-] + [Na(OH_2)_4^+ + HO^-] \longrightarrow [Na(OH_2)_4^+ + Cl^-] + H_2O \tag{4.8}$$

The net effect is that prior to the equivalence point hydronium ion is replaced with sodium ion, while after the equivalence point sodium ion plus hydroxide ion are added to this solution. Because the equivalent conductance of hydronium ion is many times greater than sodium ion, there is a net decrease of the total solution conductance prior to the equivalence piont; after this point the conductance increases from the addition of sodium and hydroxide ion (the latter also has a large equivalent conductance).

As with amperometric titrations, to have straight-line portions of the titration curve dilution corrections must be made because the response is directly dependent on the concentration of the ionic species. Also, the important data are taken before and after the equivalence point rather than precisely at the equivalence point. The general conditions for effective conductometric measurements of solutions are discussed in Chapter 5 and are directly applicable when the system is used as the endpoint detection method. A particularly complete review of the subject has been presented.[8]

One factor that has caused conductometric titrations to have limited popularity has been the problem of electrode contamination and the resultant decrease in sensitivity and electrode stability. To overcome this problem, the general conductometric approach was extended from the conventional 1-kHz frequency range to a frequency of several megahertz. With the latter frequency the indicating electrodes can be placed on the outside of the titration vessel rather than immersed in the solution. This avoids electrode contamination as well as the need to have a highly activated platinized surface. A number of forms of the electrode configuration have been developed; both parallel plates and concentric plates give satisfactory responses. In general, high-frequency conductometric titrations have had limited application, in part because of the attendant difficulties associated with electronic circuits at the megahertz frequency range. A second limitation has been that the high-frequency conductometric response is not a simple linear function of ionic equivalent conductance, as is the case with conventional conductometric measurements. A particularly complete and authoritative discussion of the entire subject of high-frequency titrations has been presented.[9]

4.3 AUTOTITRATORS

One advantage of electrochemical endpoint detection is that it lends itself to automation through electronic circuitry. Not only is automation of a titration a labor saving advantage; it also eliminates human error and prejudice in the selection of the endpoint for a titration. The principles and some of the approaches to automated electrochemical titrations have been discussed in two monographs.[10,11]

Although all electrochemical endpoint detection methods (potentiometric, amperometric, and conductometric) can be a part of an automated titration system, to date the potentiometric method is the only one that has enjoyed much attention. Two approaches have been used to automate potentiometric titrations; one involves presetting the endpoint potential such that the titration is caused to proceed until the indicating system attains its preset point. Such instruments are commercially available and consist of a potentiometric pH meter that permits the endpoint potential to be preset and a solenoid activated stopcock assembly for delivery of titrant. An anticipatory circuit is provided such that as the endpoint potential is approached the rate of delivery is slowed to avoid overrunning the end of the titration. For reliable performance the burette tip must be placed such that the indicator electrode anticipates the endpoint concentration prior to its equilibrium attainment. This approach is convenient and simple in terms of instrumentation, but it demands an accurate prior knowledge of the proper potential to be set for the equivalence point. To gain maximum reliability the entire titration curve for a sample system must be determined so that the proper endpoint potential can be set. This can be accomplished by incrementally adjusting the setting for the endpoint potential and waiting for the titrator to attain that potential. A titration curve can be obtained by plotting potential settings versus volume of titrant delivered.

The second form of autotitrator for potentiometric titrations involves instrumentation that simultaneously delivers titrant and records the indicator potential. To obtain reliable and meaningful data, the rate of titrant delivery must be sufficiently slow to allow attainment of equilibrium. The placement of the electrodes must be such that the equilibrium concentration within the titration vessel is representatively sensed. Because potentiometric strip-chart recorders are the most convenient form of recording, tirant delivery normally must be at a constant rate (e.g., a motor-driven syringe). A convenient approach is to connect the drive assembly of the syringe to the chart-drive assembly for the recorder. If the latter is done, the rate of delivery of titrant can be adjusted such that the rate of potential change is made constant. Such provision will minimize the titration error and give as rapid a recording as is possible for a given system. Commercial versions of such instrumentation are available from many companies. Autotitrators that provide the entire titration curve give the maximum information concerning the sample system and its characteristics at the equivalence point. They assure that the proper point on the titration curve is selected as the endpoint. They also provide the maximum amount of reliable data for the evaluation of the dissociation constants of weak acids, weak bases, and complexed species.

In addition to these two versions of autotitrators, titrators have been developed that take advantage of the change in potential per increment of titrant added. Such instruments terminate the addition of titrant when the first derivative of potential relative to volume reaches a maximum, or when the second derivative of potential with respect to volume goes through zero (see Figure 4.2). The instrumentation based on this approach is commercially available; it has the advantage over a preset endpoint potential in that the specific sample system to be titrated actually determines when the endpoint will be selected (based on the second derivative of the titration curve). This system also must have a controlled rate of delivery of titrant, which is best accomplished through the use of a motor-driven syringe rather than from a conventional burette.

Whether a constant speed or a variable speed motor drive is used depends in part on the sophistication of the instrumentation and the nature of the anticipatory and detection section of the instrumentation. Because automated titration instrumentation frequently is needed for a specific problem, a great number of custom-fabricated assemblies have been developed.

4.4 pH-STATS

A related form of an automatic potentiometric titrator is instrumentation that permits the maintenance of the acidity or basicity of a solution over a period of time. Such devices are known as pH-stats, and find application in kinetic studies of hydrolysis reactions. The general approach is (by either manual or automatic means) to add either acid or base such that the pH in the solution is maintained constant over a period of time. Normally the amount of acid or base added as a function of time is sought in order that kinetic measurements may be made for the system. In its simplest form the acidity of the solution is monitored with a pH meter and controlled at a preselected value by the addition of acid or base from a burette; the quantity delivered as a function of time is recorded in a notebook. Obviously for the fast reactions this becomes difficult and dependent on the dexterity of the individual.

In general autotitrators that work with a preset endpoint lend themselves to application as pH-stats. All that is necessary is to record the volume of titrant added as a function of time. If a motor-driven syringe is used, this can be combined with a helical potentiometer and a strip-chart recorder to provide a volume–time curve; a manual version would include a digital register ganged to the motor-driven screw of the syringe. The digitizer can be read as a function of time and provides the necessary data for analysis of the kinetics of hydrolysis reaction.

pH-Stats have found their primary application in the study of hydrolysis rates such as

$$RCOOR' + HO^- \longrightarrow RCOO^- + R'OH \tag{4.9}$$

In the case of ester hydrolysis the rate law is expressed by the relation

$$\frac{-d[RCOOR']}{dt} = k[RCOOR'][HO^-] = \frac{-dm_{HO^-}}{V_t\,dt} \tag{4.10}$$

$$= \frac{C^\circ_{HO^-}}{V_t} = \left(\frac{dV_{HO^-}}{dt}\right)_{pH} = \frac{i}{V_t F} \tag{4.10a}$$

with the right-hand term representing the rate of disappearance of hydroxide ion from the sample solution (V_t) as a function of time (m_{HO^-}, moles of hydroxide ion; $C^{\circ}_{HO^-}$, concentration of hydroxide ion in burette). Equation (4.10a) indicates this rate of disappearance must be matched by an equivalent rate of addition of hydroxide if the pH of the solution is to be maintained. Furthermore, if the hydroxide ion were to be produced electrochemically in the solution, the right-hand term in Eq. (4.10a) would be indicative of the rate of hydroxide addition (in terms of current used to electrolyze the solution to provide hydroxide ion). Rearrangement of these equations gives a relation for the evaluation of hydrolysis rate constants

$$k = \left(\frac{dV_{HO^-}}{dt}\right)_{pH} \frac{C^{\circ}_{HO^-}}{[HO^-]m_{ester}} = \frac{i}{[HO^-]m_{ester}F} \tag{4.11}$$

where m_{ester} represents the moles of ester and F, the Faraday constant. As Eq. (4.11) indicates, pH-stats, whether using titrant delivery of hydroxide ion or electrochemical generation of hydroxide ion, provide an extremely convenient means for the evaluation of hydrolysis rate constants.[12]

Another application of pH-stats is to control solution pH without the use of a buffer system. Again, this can be accomplished by either a burette delivery system, a motor-driven syringe, or electrochemical generation of hydroxide ion or hydronium ion. This can be extremely useful for systems where all available buffer solutions interfere with the reaction of interest.

4.5 COULOMETRIC TITRATIONS

The coulometric method of titration is based on the electrochemical generation of the titrant and on Faraday's law of electrolysis, which equates equivalents of material to coulombs of electricity. The method was first discussed some 50 years ago[13] and has been investigated in detail in the intervening years. A particularly detailed discussion is presented by Lingane,[14] and a review and discussion are given by Deford and Miller in the treatise by Kolthoff and Elving.[15] The method is of sufficient interest and importance that it is reviewed biannually in *Analytical Chemistry*.[16]

The basic approach in coulometric titrations is to generate electrochemically (at constant current) a titrant in solution that subsequently reacts by a secondary chemical reaction with the species to be determined. For example, a large excess of cerium(III) is placed in the solution together with an iron(II) sample. When a constant current is applied, the cerium(III) is oxidized at the anode to produce cerium(IV), which subsequently reacts with the iron(II):

$$Ce(OH_2)_6^{3+} \longrightarrow Ce(OH_2)_6^{4+} + e^- \text{ (electrode reaction)} \tag{4.12}$$

$$Ce(OH_2)_6^{4+} + Fe(OH_2)_6^{2+} \longrightarrow Fe(OH_2)_6^{3+} + Ce(OH_2)_6^{3+} \text{ (secondary reaction)} \tag{4.13}$$

Should any iron(II) reach the anode, it also would be oxidized and thus not require the chemical reaction of Eq. (4.13) to bring about oxidation, but this would not in any way cause an error in the titration. This method is equivalent to the constant-rate addition of titrants from a burette. However, in place of a burette the titrant is electrochemically generated in the solution at a constant rate that is directly proportional to the constant current. For accurate results to be obtained the electrode reaction must occur with 100% current efficiency (i.e., without any side reactions that involve solvent or other materials that would not be effective in the secondary reaction). In the method of coulometric titrations the material that chemically reacts with the sample system is referred to as an *electrochemical intermediate* [the cerium(III)/cerium(IV) couple is the electrochemical intermediate for the titration of iron(II)]. Because one faraday of electrolysis current is equivalent to one gram-equivalent (g-equiv) of titrant, the coulometric titration method is extremely sensitive relative to conventional titration procedures. This becomes obvious when it is recognized that there are 96,485 coulombs (C) per faraday. Thus, 1 mA of current flowing for 1 second represents approximately 10^{-8} g-equiv of titrant.

In general, coulometric titrations use currents that range from 1 to 100 mA in magnitude. An essential part of a coulometric titration assembly is a reliable constant-current source with a range of accurately preset currents known with an error of less than 0.1%. Another crucial element in the instrumentation is a sensitive and rapid endpoint detection system. By its nature the coulometric titration method is a constant-rate-of-delivery system, which demands sensitive and rapid response if the endpoint is not to be overrun. Reference to the literature confirms the potentiometric, amperometric, and visual indicating systems have been applied to coulometric titrations. However, the potentiometric and the dual-polarized amperometric detection systems have enjoyed the widest application because of their selective and sensitive response. The final element of a coulometric titration system is an accurate timing device. This can be provided by either a precision stop-clock or an electronic counting system based on time. To summarize, a coulometric titration consists of passing a constant current through a solution until the endpoint is detected, at which time the period of electrolysis is recorded. The product of the current times the electrolysis time gives the number of coulombs, which is directly proportional to the number of equivalents of the sample substituent. For titration periods of 100 s or less, the timing device must be accurate to ± 0.1 s, if 0.1% accuracy is to be realized.

Coulometric titration procedures have been developed for a great number of oxidation–reduction, acid–base, precipitation, and complexation reactions. The sample systems as well as the electrochemical intemediates used for them are summarized in Table 4.1, and indicate the diversity and range of application for the method. An additional specialized form of coulometric titration involves the use of a spent Karl Fischer solution as the electrochemical intermediate for the determination of water at extremely low levels. For such a system the anode reaction regenerates iodine, which is the crucial component of the Karl Fischer titrant. This then reacts with the water in the sample system according to the

TABLE 4.1 Systems for Coulometric Titrations

Electrogenerated Reagent	Precursor	Determined Species
	Oxidizing Agents	
Ce(IV)	Ce(III), 1.5 M H_2SO_4	Fe(II), $Fe(CN)_6^{4-}$, Ti(III), As(III), U(IV), I^-, hydroquinone, phenols
Mn(III)	Mn(II), 3 M H_2SO_4	Fe(II), $C_2O_4^{2-}$, As(III), H_2O_2
Cl_2	Cl^-, HCl or H_2SO_4	As(III), I^-, unsaturated fatty acids
Br_2	Br^-, H_2SO_4	As(III), Sb(III), I^-, Tl(I), U(IV), NH_3, N_2H_4, U(IV), U(IV), phenols, 8-quinolinol, unsaturated hydrocarbons
I_3^-	I^-	As(III), $S_2O_3^{2-}$, H_2SeO_3, Sb(III), S^{2-}
	Reducing Agents	
$Fe(CN)_6^{4-}$	$Fe(CN)_6^{3-}$	Tl(I)
Ti(III)	$TiO(SO_4)$, 3.5 M H_2SO_4	Fe(III), Ce(IV), V(V), U(VI)
Sn(II)	$SnCl_4$, 3 M NaBr, HCl	Au(III), Ce(IV), Fe(III), I_2
Fe(II)EDTA	Fe(III)EDTA	Fe(III)
Fe(II)	Fe(III), 0.5 M H_2SO_4, H_3PO_4	Ce(IV), $Cr_2O_7^{2-}$, V(V), MnO_4^-
Cu(I)	Cu(II), 2 M HCl	$Cr_2O_7^{2-}$, IO_3^-, Br_2
U(IV)	$UO_2(SO_4)$, 0.15 M H_2SO_4	$Cr_2O_7^{2-}$, Ce(IV), Fe(III)
V(IV)	$NaVO_3$, 1 M H_2SO_4	

	Precipitating Agents	
Ag(I)	Ag, 0.5 M $HClO_4$	Cl^-, Br^-, I^-, $CH_3CS(NH_2)$, SCN^-, RSH
Hg(I)	Hg, 0.5 M $NaClO_4$, $HClO_4$	Cl^-, Br^-, I^-
$Fe(CN)_6^{4-}$	$Fe(CN)_6^{3-}$	Zn(II)
	Complexing Agents	
$\dot{H}(EDTA)^{3-}$	$HgNH_3(EDTA)^{2-}$, pH 8.5 (NH_4^+, NH_3)	Ca(II), Cu(II), Zn(II), Pb(II)
	Acid–Base Agents	
H_3O^+	H_2O, 0.5 M Na_2SO_4	Bases
HO^-	H_2O, 0.5 M Na_2SO_4	Acids
	Karl–Fischer Reagent	
I_2	''Spent'' Karl–Fischer reagent	H_2O

reaction

$$C_6H_5N \cdot I_2 + C_6H_5N \cdot SO_2 + C_6H_5N + CH_3OH + H_2O \longrightarrow$$
$$2\ C_6H_5N \cdot HI + C_6H_5N \cdot HSO_4CH_3 \qquad (4.14)$$

For this titration the dual-polarized amperometric endpoint detection system provides good sensitivity and rapid response.

The instrumentation for coulometric titrations consists of a galvanostat (a constant-current source), a cell equipped with an endpoint detector, and a timer.

The coulometric titration method lends itself to microscale analysis. However, accurate results require a good cell design. Efficient stirring is essential, and the placement of both the generator electrodes and the detection electrode is extremely important. Figure 4.8 illustrates a cell assembly that is satisfactory when a potentiometric endpoint detection system is used. Likewise, the assembly illustrated in Figure 4.9 is convenient for a dual-polarized amperometric endpoint detection system. In either cell a convenient means to introduce small-volume samples is desirable as well as provision to eliminate interfering gases from the solution (such as dioxygen).

To have an electrolysis current it is axiomatic that there be an anode as well as a cathode reaction. If the anode reaction is the one used to generate the

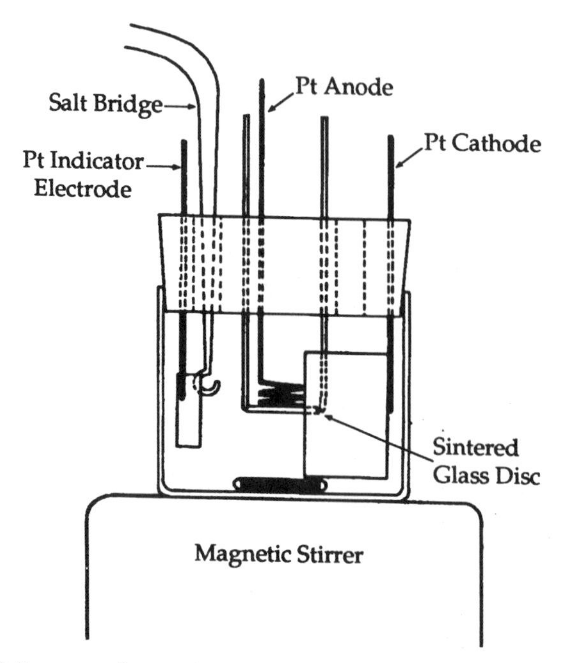

Figure 4.8 Cell system for coulometric titration by a platinum generator electrode and an isolated auxiliary electrode; system includes provision for potentiometric endpoint detection.

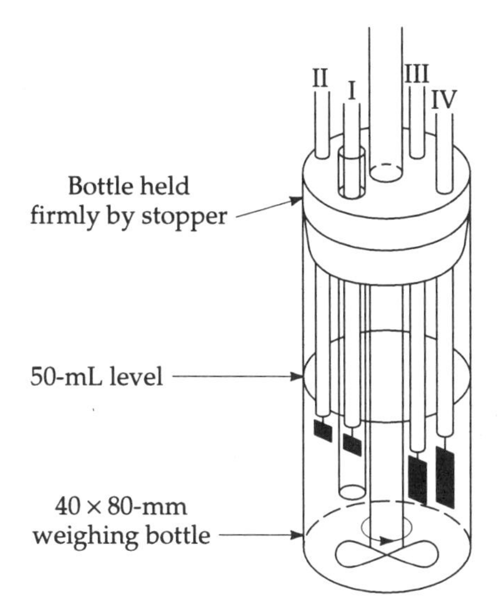

Figure 4.9 Coulometric titration cell with generator [II (generator anode, 0.7 × 0.7 cm)] and isolated auxiliary [I (generator cathode, 0.7 × 0.7 cm)] electrodes on the left side and a pair of identical platinum electrodes [III, IV (1.4 × 1.8 cm and 2.5 × 1.8 cm)] on the right for dual-polarized electrode amperometric endpoint detection.

titrant, then it is imperative that the products of the cathode not interfere with the titration. This can be accomplished in a number of ways, and the cells in Figures 4.8 and 4.9 indicate one approach; specifically, the electrode reaction that is not involved in the generation of titrant is isolated from the sample solution by a fritted disk at the end of a glass tube. Another approach is the use of ion-selective membranes or ion-exchange resins in the immediate proximity of the electrode.

In addition to the use of ion-exchange membranes to isolate the counter electrode from the working electrode, permeability-selective solid-state membranes can be used to generate coulometric titrants. For example, hydronium ion can be generated in water or nonaqueous solvents by oxidation of molecular hydrogen that has diffused through a palladium-membrane anode. Silver ion can be generated through a silver/silver sulfide membrane anode for the coulometric titration of chloride ion in strong oxidizing media (where a bare silver anode could not be used). In a similar fashion, fluoride ion can be generated at a europium-doped lanthanum fluoride membrane.

Because the generator electrodes must have a significant voltage applied across them to produce a constant current, the placement of the indicator electrodes (especially if a potentiometric detection system is to be used) is critical to avoid induced responses from the generator electrodes. Their placement should be adjusted such that both the indicator electrode and the reference electrode occupy positions on an equal potential contour. When dual-polarized amperometric electrodes are used, similar care is desirable in their placement to avoid interference from the electrolysis electrodes. These two considerations have prompted the use of visual or spectrophotometric endpoint detection in some applications of coulometric titrations.

Coulometric titrations have found their widest application where microanalysis titrations of high precision and accuracy are desired. Because small incremental additions of titrant are more convenient and precise through use of an electrolysis current than is possible through the use of a burette, coulometric titrations of small samples have the potential for enhanced accuracy. Furthermore, because the entire system (both titrant addition and detection) is electrochemical, coulometric titrations are especially amenable to automation. The technique is particularly useful for repetitive sample analysis, and has been extended to a number of industrial and commercial analytical devices. In particular, for the analysis of olefins and of sulfur compounds in petroleum streams, various forms of coulometric titrators are used.

Reference to Table 4.1 indicates that olefins can be determined by the electrochemical generation in situ of halogens. Bromine is effective for both olefins and sulfur compounds and is the basis for an automatic coulometric titrator for continuous analysis of petroleum streams.[17] The basic principle of this instrument is a potentiometric sensing system that monitors bromine concentration in a continuously introduced sample stream. The bromine in the solution reacts with the sample components and causes a decrease in the concentration of bromine. When this decrease is sensed by the potentiometric detection electrodes, the electrolysis current producing bromine adjusts itself to maintain the bromine concentration. Because the sample is introduced at a constant rate, the electrolysis current becomes directly proportional to the concentration of the sample component. Thus, the instrument records the electrolysis current as concentration of sample component and provides a continuous monitor for olefins or sulfur in petroleum streams.

Another example of the application of the principles of coulometric titrations to a continuous on-stream analyzer is the moisture analyzer developed by Keidel.[18] It illustrates one of the outstanding advantages of the coulometric generation of a titrant; namely, an intermediate is produced as a titrant that would not be available in standard solutions. The principle of the moisture analyzer is to place a phosphoric acid solution between two closely spaced platinum electrodes (helically wound in a glass tube). When current is passed between the two electrodes, the water in the phosphoric acid is electrolyzed

$$2\ H_3PO_4 \xrightarrow[4e^-]{\text{electrolysis}} 2\ H_2 + O_2 + 2\ HPO_3 \tag{4.15}$$

to yield hydrogen, oxygen, and HPO_3. The latter species, even at high applied potential, is a sufficiently poor conductor of electricity that virtually no current passes between the electrodes. If a moisture-containing gas stream is passed through this cell at a constant flow rate, any water in the stream will react with HPO_3 to produce phosphoric acid

$$HPO_3 + H_2O \longrightarrow H_3PO_4 \tag{4.16}$$

which is a good conductor of electricity and will allow the reaction represented by Eq. (4.15) to occur. Hence, the electrolysis current that is observed is directly proportional to the moisture content of the sample stream. For a flow rate of 100 mL min^{-1} at atmospheric pressure, a current of 13.2 mA results for each part per million (ppm) of water in the gas phase. The general features of this form of automated coulometric titration have been discussed,[19] as well as its extension to the determination of moisture in organic liquids.[20]

Another example of the use of the coulometric titration method as a continuous monitor is its application as a gas-chromatographic detector for the analysis of pesticides and sulfur–halogen compounds. The approach is to burn the effluent from a gas-chromatographic column in a hydrogen flame to convert any halogen compound to hydrogen halides and any sulfur compound to H_2S. These products are carried in the gas-flow path into a coulometric titration cell with a silver anode as the generator electrode and a silver indicator electrode for a potentiometric detection system. The indicating circuit is set such that the silver ion concentration in the solution is controlled to a preset level. Any halide ion or H_2S introduced into the cell solution will react with the silver ion and cause its concentration to decrease. When this is detected, the electrolysis current is activated electronically and produces silver ion to maintain its preset level. The resulting electrolysis current is monitored on a strip-chart recorder as a function of time and gives elution peaks that appear identical to the elution peaks observed with a conventional gas-chromatographic detection system. This system has proved extremely useful for the determination of minute amounts of halogen- and sulfur-containing compounds in the presence of large excesses of other extraneous materials.

Both the automatic coulometric titration of petroleum streams and the continuous monitoring of pesticides and sulfur–halogen compounds indicate that the coulometric titrator method is amenable to the automatic maintenance of the concentration of a component in a solution system. A manual version of this approach has been used to study the kinetics of hydrogenation of olefins as well as to determine the rate of hydrolysis of esters.[12] The latter system is a pH-stat that is based on the principles of coulometric titrations. Equations (4.9)–(4.11) indicate how this approach is applied to the evaluation of the rate constants for ester hydrolysis. A similar approach could be used to develop procedures for kinetic studies that involve most of the electrochemical intermediates summarized in Table 4.1. The coulometric titration method provides a convenient means to extend the range of systems that can be subjected to kinetic study in solution.

4.6 CONTROLLED-CURRENT METHODS

The principle of controlled current electrolysis has been known since the beginning of this century.[21] However, the utilization of this form of electrochemistry remained dormant for 50 years until three groups of investigators illustrated its many advantages for analytical and physicochemical measurements.[22–24] Several works describe this technique in detail,[25–27] and other reviews summarize many applications of chronopotentiometry to chemical problems.[28–30]

The basic form of controlled current electrolysis is called chronopotentiometry. As the name implies, while a constant current is passed between a pair of electrodes immersed in a quiescent solution, the potential of the measuring electrode (''working electrode'') is monitored as a function of time. The basic form of instrumentation, illustrated in Figure 4.10, indicates that the potential of the working electrode normally is measured relative to a non-current-carrying reference electrode (to minimize errors from the *iR* drop between the other two electrodes). Figure 4.11 illustrates the potential–time curve (''chronopotentiogram'') for the reduction of iron(III) at a platinum electrode. At the beginning of the electrolysis the potential changes very little with time because of the ''buffered'' (or poised) condition, whereby both iron(II) and iron(III) are in the vicinity of the working electrode. However, as the electrolysis continues the iron(III) is depleted, which forces the electrode potential to shift sharply in a negative direction until some other species can be reduced to maintain the constant current. As indicated in Figure 4.11, if other reducible ions are absent in the sample solution, then either hydronium ions or water will be reduced.

The point at which the sample ion is depleted in the vicinity of the working electrode is called the transition time τ; this quantity is related to a number of variables including the sample-ion concentration. In 1901 Sand[21] derived the equation that describes the functional dependence of the transition time for a constant-current electrolysis of a diffusion-controlled process

$$i\tau^{1/2} = \frac{\pi^{1/2} nFAD^{1/2} C^{\mathrm{b}}}{2} \tag{4.17}$$

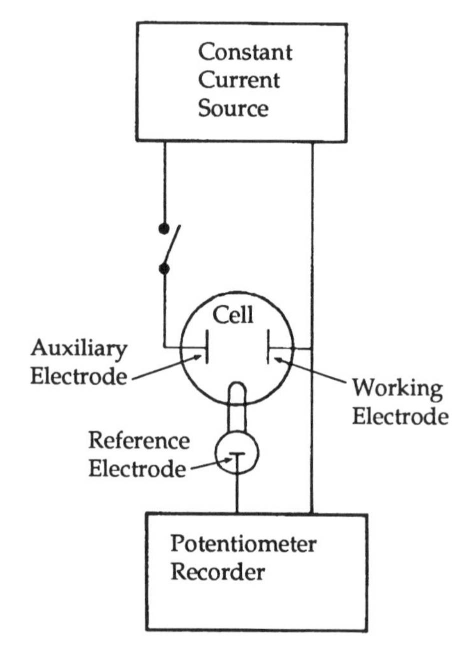

Figure 4.10 Instrumentation for chronopotentiometry.

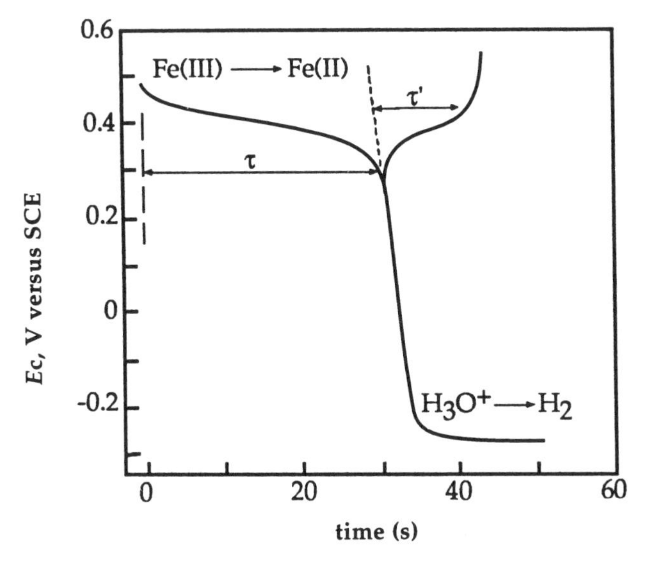

Figure 4.11 Chronopotentiogram for the reduction of Fe(III) at a platinum electrode. τ represents the transition time for the reduction process and τ' the transition time on current reversal for the oxidation of the reduction product.

where i = electrolysis current
n = number of electrons in the electrolysis reaction
F = faraday
A = area of the working electrode
D = diffusion coefficient of the electroactive species
C = concentration of the electroactive species

This relation can be expanded for a two-step electrolysis involving two sample species. Because the first species continues to diffuse to the electrode even after the first transition time, the expression is more complex than a simple additive relation:

$$i(\tau_1 + \tau_2)^{1/2} = \frac{\pi^{1/2}FA}{2}(n_1 D_1^{1/2} C_1^{\mathrm{b}} + n_2 D_2^{1/2} C_2^{\mathrm{b}}) \tag{4.18}$$

The relation for the second transition time can be obtained by subtracting Eq. (4.17) from Eq. (4.18) to give

$$i\tau_2^{1/2} = \frac{\pi^{1/2}FA}{2}(2n_1 n_2 D_2^{1/2} D_2^{1/2} C_1^{\mathrm{b}} C_2^{\mathrm{b}} + n_2^2 D_2 C_2^{\mathrm{b}2})^{1/2} \tag{4.19}$$

For the special condition where C_1^{b} equals C_2^{b} and D_1 equals D_2 (or for consecutive electrolysis by two steps), the ratio of transition times simplifies to

$$\frac{\tau_2}{\tau_1} = \frac{2n_2}{n_1} + \left(\frac{n_2}{n_1}\right)^2 \tag{4.20}$$

Thus, for 2 one-electron steps the second transition time is three times as long as the first. For a one-electron step followed by a two-electron step the ratio increases to 8. In contrast to polarography, chronopotentiometry offers enhanced analytical sensitivity for a second (or third) species. Advantage of this characteristic can be taken by adding a known amount of a species electrolyzed prior to the unknown species; in some cases this can provide a tenfold increase in sensitivity.

Reference to Figure 4.11 indicates that reversal of the current at (or before) the transition time will cause the product species to be electrolyzed until it is depleted from the vicinity of the electrode. If both the primary and reverse processes are diffusion controlled, then the transition time for the reverse process τ' is related to the forward process by

$$\tau' = \frac{t}{3} \; (t \leq \tau) \tag{4.21}$$

Thus, current-reversal chronopotentiometry is useful for characterizing the electrode process as well as for ascertaining the nature of the product species.

If the chronopotentiometric reaction involves a thermodynamically reversible process without complications, Eq. (4.17) may be used to relate concentrations to transition times in the Nernst expression. For a system where both reactant and product are soluble species (such as in Figure 4.11), the chronopotentiogram is described by the relations

$$E = E_{\tau/4} + \frac{RT}{nF} \ln \frac{\tau^{1/2} - t^{1/2}}{t^{1/2}} \tag{4.22}$$

$$E = E_{\tau/4} + \frac{0.059}{n} \log \frac{\tau^{1/2} - t^{1/2}}{t^{1/2}} \quad \text{at } 25°\text{C} \tag{4.22a}$$

where $t^{1/2}$ is the time of electrolysis and $E_{\tau/4}$ is the quarter-wave potential (the point on the curve where the logarithmic term becomes zero); equivalent to the half-wave potential in polarography. For a process where the product species is insoluble, Eq. (4.22) takes the form

$$E = E_{\tau/4} + \frac{0.059}{n} \log (\tau^{1/2} - t^{1/2}) \quad \text{at } 25°\text{C} \tag{4.23}$$

In current-reversal chronopotentiometry the reverse wave of a reversible process is described by the relation

$$E = E_{\tau/4} + \frac{0.059}{n} \log \frac{\tau^{1/2} - [(\tau + t')^{1/2} - 2t'^{1/2}]}{(\tau + t')^{1/2} - 2t'^{1/2}} \tag{4.24}$$

where t' is the reverse electrolysis time. Also, for the reverse wave

$$E_{0.222\tau'} = E_{\tau/4} \tag{4.25}$$

When the electrolysis process is irreversible (in a thermodynamic sense) but still limited by linear diffusion, the potential–time relationship takes a form that includes the heterogeneous electron-transfer kinetic parameters. For a cathodic process the relation is

$$E = \frac{0.059}{\alpha n_a} \log \frac{nFAC^{\mathrm{b}}k_c}{i} + \frac{0.059}{\alpha n_a} \log \frac{\tau^{1/2} - t^{1/2}}{\tau^{1/2}} \quad \text{at 25 °C} \tag{4.26}$$

where E = electrode potential versus the NHE
α = cathodic transfer coefficient (a kinetic symmetry parameter)
k_{c} = heterogeneous cathodic electron-transfer rate constant
n_a = number of electrons in the rate-controlling step

and the other terms have their usual meaning. For an anodic process the relation is

$$E = \frac{-0.059}{(1 - \alpha)n_a} \log \frac{nFAC^{\mathrm{b}}k_{\mathrm{a}}}{i} - \frac{0.059}{(1 - \alpha)n_a} \log \frac{\tau^{1/2} - t^{1/2}}{\tau^{1/2}} \quad \text{at 25°C} \tag{4.27}$$

where $(1 - \alpha)$ is the anodic transfer coefficient and k_{a} the heterogeneous anodic electron-transfer rate constant. Thus k_{c} and k_{a} represent the rate of electron transfer at 0.000 V versus NHE for unit concentration and unit electrode area. If the formal thermodynamic potential ($E°'$) for the process is known (or can be evaluated), then k_{c} and k_{a} may be converted to a simplified rate constant k_{s}, which is a measure of the rate of electron transfer at the formal potential:

$$k_{\mathrm{s}} = k_{\mathrm{c}} \exp \frac{-\alpha n_a FE'}{RT} = k_{\mathrm{a}} \exp \frac{(1 - \alpha)n_a FE'}{RT} \tag{4.28}$$

Consideration of Eqs. (4.26) and (4.27) indicates that chronopotentiometry can be used to evaluate the kinetic parameters for electron-transfer reactions. By plotting the electrode potential E versus log $[(\tau^{1/2} - t^{1/2})/\tau^{1/2}]$, one obtains a linear curve whose slope permits evaluation of αn_a [or $(1 - \alpha)n_a$]. Extrapolation of such a plot to the point where $t = 0$ causes the logarithmic terms of Eqs. (4.26) and (4.27) to go to zero and thereby permits k_{c} and k_{a} to be evaluated from the values of $E_{t=0}$. Plots of $E_{t=0}$ versus log C^{b} and log $1/i$ provide another means of evaluating αn_a.

The chronopotentiometric wave for current reversal of a cathodic process also may be evaluated using the relation

$$E = \frac{-0.059}{(1-\alpha)n_a} \log \frac{2k_a}{\pi^{1/2} D_{red}^{1/2}} - \frac{0.059}{(1-\alpha)n_a} \log [(\tau + t')^{1/2} - 2t'^{1/2}] \tag{4.29}$$

where D_{red} is the diffusion coefficient of the reduced species and t' is the time of the reverse electrolysis.

Table 4.2 summarizes a number of diagnostic criteria for analysis of chronopotentiometric potential–time curves that are complicated by various kinetic schemes.[31]

Chronopotentiometry is especially useful for the study of pre-electron-transfer steps that occur prior to the electron-transfer reaction. These can be generalized by the expression

$$Y \underset{k_b}{\overset{k_f}{\rightleftharpoons}} ox \xrightarrow{ne^-} red; \qquad K = \frac{k_f}{k_b} \tag{4.30}$$

where Y is a nonelectroactive species that is in slow equilibrium with ox, which is electroactive. As a result of this complication, Eq. (4.17) is modified to

$$i\tau_k^{1/2} = \frac{\pi^{1/2} nFAC^b D^{1/2}}{2} - \frac{\pi^{1/2} i}{2AK(k_f + k_b)^{1/2}} \operatorname{erf} [(k_f + k_b)^{1/2} \tau_k^{1/2}] \tag{4.31}$$

where erf represents the error function. If the argument of erf is greater than 2, erf approaches unity. Then

$$i\tau_k^{1/2} = \frac{\pi^{1/2} nFAC^b D^{1/2}}{2} - \frac{\pi^{1/2} i}{2AK(k_f + k_b)^{1/2}} \tag{4.32}$$

or

$$\tau_k^{1/2} = \tau_d^{1/2} - \frac{\pi^{1/2}}{2K(k_f + k_b)^{1/2}} \tag{4.33}$$

This approach can be expanded to include a process of the form

$$pY \underset{k_b}{\overset{k_f}{\rightleftharpoons}} ox \xrightarrow{ne^-} red; \qquad K = \frac{k_f}{k_b} \tag{4.34}$$

Then

$$\tau_k^{1/2} = \tau_d^{1/2} - \frac{\pi^{1/2} nFAD^{1/2}}{2pi} \left[\frac{i}{nFAD^{1/2} K} \operatorname{erf} (k_b \tau)^{1/2} \right]^{1/p} \tag{4.35}$$

TABLE 4.2 Diagnostic Criteria of Chronopotentiograms for Various Kinetic Schemes

Kinetic Scheme	Linear Log Plot	Slope Log Plot	$\partial E_{1/4}/\partial \log i$	$\partial E_{1/4}/\partial \log C$	τ_r/τ_f[a]
$ox \overset{\text{rapid}}{\rightleftarrows} red$	$(\tau^{1/2} - t^{1/2})t^{1/2}$	RT/nF	0	0	$\frac{1}{3}$
$ox \overset{\text{slow}}{\rightleftarrows} red$	None		0 to $-RT/\alpha nF$	0 to $+ RT/\alpha nF$	$\frac{1}{3}$
$ox \longrightarrow red$	$\tau^{1/2} - t^{1/2}$	$RT/\alpha nF$	$-RT/\alpha nF$	$+RT/\alpha nF$	$\frac{1}{3}$ or 0
$ox \overset{\text{rapid}}{\rightleftarrows} red$(insoluble)	$\tau^{1/2} - t^{1/2}$	RT/nF	0	$+RT/nF$	1
$ox \overset{\text{slow}}{\rightleftarrows} red$(insoluble)	None		$-RT/nF$ to $-RT/\alpha nF$	RT/nF to $RT/\alpha nF$	1
$ox \longrightarrow red$(insoluble)	$\tau^{1/2} - t^{1/2}$	$RT/\alpha nF$	$-RT/\alpha nF$	$+RT/\alpha n$	1 or 0
$ox \overset{\text{rapid}}{\rightleftarrows} red \overset{\text{slow}}{\rightleftarrows} Y$	None		0 to $-RT/nF$	0 to $+RT/nF$	0–$\frac{1}{3}$
$ox \overset{\text{rapid}}{\rightleftarrows} red \overset{\text{rapid}}{\rightleftarrows} Y$	$\tau^{1/2} - t^{1/2}$	RT/nF	$-RT/nF$	$+RT/nF$	0
$ox \overset{\text{rapid}}{\rightleftarrows} red \overset{\text{slow}}{\longrightarrow} Y$	None		0 to $-RT/nF$	0 to $+RT/nF$	0–$\frac{1}{3}$
$m\ ox \overset{\text{rapid}}{\rightleftarrows} p\ red$	$(\tau^{1/2} - t^{1/2})^m/t^{p/2}$	RT/nF	0	$(m - p)\ RT/nF$	$\frac{1}{3}$
$2\ ox \overset{\text{rapid}}{\rightleftarrows} 2\ red \overset{\text{slow}}{\rightleftarrows} Y$	None		0 to $-RT/3nF$	$-RT/nF$ to $+RT/nF$	0–$\frac{1}{3}$
$2\ ox \overset{\text{rapid}}{\rightleftarrows} 2\ red \overset{\text{rapid}}{\longrightarrow} Y$	$\tau^{1/2} - t^{1/2}$	RT/nF	$-RT/3nF$	$+RT/nF$	0
$2\ ox \overset{\text{rapid}}{\rightleftarrows} 2\ red \overset{\text{slow}}{\longrightarrow} Y$	None		0 to $-RT/3nF$	$-RT/nF$ to $+RT/nF$	0–$\frac{1}{3}$

[a]Subscripts r and f denote reverse and forward, respectively.

For the condition where $(k_b t) > 2$

$$\tau_k^{1/2} = \tau_d^{1/2} - \frac{\pi^{1/2}}{2p}\left(\frac{nFAD^{1/2}}{i}\right)^{1-1/p}(k_b^{1/2}K)^{-1/p} \tag{4.36}$$

and

$$i\tau_k^{1/2} = i\tau_d^{1/2} - \frac{\pi^{1/2}}{2p}(nFAD^{1/2})^{(1-1/p)}(i^{1/p})(k_b^{1/2}K)^{-1/p} \tag{4.37}$$

A diagnostic approach for the analysis of various pre-electron-transfer chemical reactions[32] is summarized in Table 4.3. Thus, the variation of the first and second derivatives of the quantity $i\tau^{1/2}$, as a function of current, indicates the type of pre-electron-transfer chemical process. Two experimental plots will provide sufficient data to use the criteria of this table; $i\tau^{1/2}$ versus i at constant C and $\tau^{1/2}$ versus C at constant i.

Through the use of current-reversal chronopotentiometry, reaction sequences of the type

$$ox \overset{ne-}{\rightleftarrows} red \overset{k_f}{\longrightarrow} Y \tag{4.38}$$

can be investigated. However, the functions are complex, and the use of a graphical expression is much more convenient.[33] Figure 4.12 indicates the variation of the ratio of τ_r/t_1, as a function of $k_f t_1$, where t_1 represents the time of forward electroysis and τ_r represents the reverse transition time. Thus, for any value of the ratio, the quantity $k_f t_1$ can be evaluated. By plotting of $k_f t_1$ versus t_1, the slope of the linear curve is equal to k_f of Eq. (4.38).

Constant current electrolysis also is applicable to thin-layer cells such that

TABLE 4.3 Diagnostic Criteria of Chronopotentiograms for Various Reaction Mechanisms

Mechanism	$\partial i\tau^{1/2}/\partial i$	$\partial^2 i\tau^{1/2}/\partial i^2$	$C\tau = 0^a$	$(\partial\tau/\partial T)^b$
$ox \longrightarrow red$	0	0	0	+
$Y \rightleftarrows ox \longrightarrow red$	−	0.	+	++
$p\,Y \rightleftarrows ox \longrightarrow red$	−	+	+	++
$Y \rightleftarrows p\,ox;\ ox \longrightarrow red$	−	−	+	++
ox (absorbed) $\longrightarrow red$	+	Variable	−	−

^{a}Obtained by extrapolation of a plot of τ versus C to $\tau = 0$.
^{b}Temperature coefficient of τ.

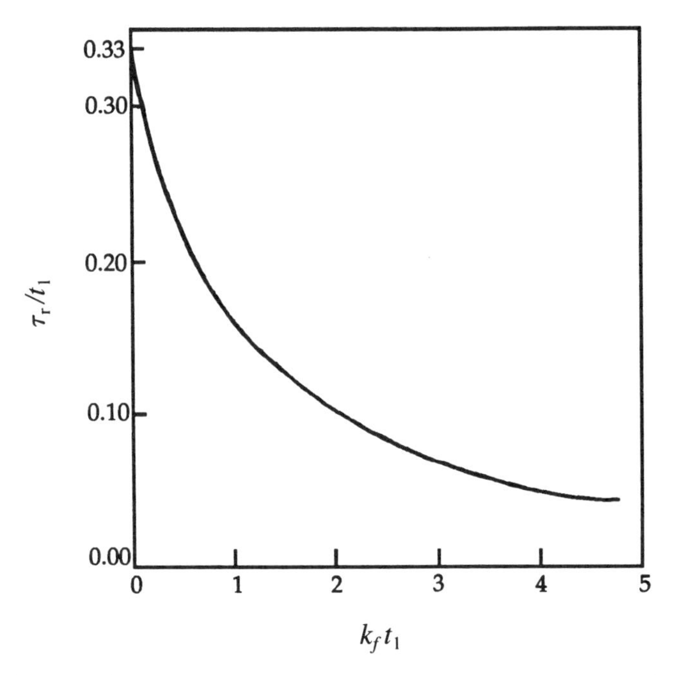

Figure 4.12 Variation of reverse transition time τ_r with the time of forward electrolysis t_1. The quantity k_f represents the unimolecular homogeneous rate constant for the conversion of the product of the forward electrolysis to a nonelectroactive species (in terms of the reverse electrolysis).

Eq. (4.17) becomes

$$i\tau = nFAC^{b}d \tag{4.39}$$

where d is the thickness of the electrode spacing. Clearly, such a system provides a convenient and rapid means of evaluating the number of electrons n for an unknown process. The reverse transition time should be the same as that for the forward process, unless there are postelectrochemical complications.

A number of embellishments of controlled-current electrolysis have been proposed for specialized systems, but discussion of these is beyond the scope of this chapter. Those interested in pursuing this material are referred to discussions of programmed-current techniques,[34] single-step and double-step galvanostatic methods,[35] and chronocoulometry.[36]

As indicated at the beginning of this chapter, the methodology of chronopotentiometry is straightforward. Some care is necessary to ensure that the electrode configuration and placement are such as to cause the electrolysis to be limited by semiinfinite linear diffusion. Chapter 5 indicates a number of indicator-electrode designs that are appropriate for such performance. As in any diffusion-controlled process, absence of vibration of the electrolysis cell is essential to obtain meaningful quantitative data. In general, electrolysis times from 0.1 s to about 30 s represent the convenient and useful quantitative range for conventional instrumentation and laboratory conditions.

Although there has been much debate about the proper method of measuring the transition time τ in chronopotentiometry, most workers now agree on the procedure outlined in Figure 4.11. For electrolysis times of less than 0.1 s, specialized recording of the potential–time curve is necessary, usually with an oscilloscope. When the transition time is less than several milliseconds, correction must be made for the current–time integral required to charge the double layer. This problem has led to the development of the double-pulse galvanostatic method.[35]

REFERENCES

1. Kolthoff, I. M.; Furman, N. H., *Potentiometric Titrations*, 2nd ed., Wiley, New York, 1931.
2. Willard, H. H.; Fenwick, F., *J. Am. Chem. Soc.* **1922,** *44*, 2516.
3. Reilley, C. N.; Cooke, W. D.; Furman, N. H., *Anal. Chem.* **1951,** *23*, 1223.
4. Reilley, C. N.; Murray, R. W., "Introduction to Electrochemical Techniques," in *Treatise on Analytical Chemistry*, Part I, Vol. 4, Kolthoff, I. M.; Elving, P. J., eds., Interscience Publishers, New York, 1963, pp. 2181–2187.
5. Lingane, J. J., *Electroanalytical Chemistry*, 2nd ed., Interscience Publishers, New York, 1958, pp. 267–295.
6. Delahay, P., *New Intrumental Methods in Electrochemistry*, Interscience Publishers, New York, 1954, Chap. 10.
7. Stock, J. T., *Amperometric Titrations*, Interscience Publishers, New York, 1965.
8. Loveland, J. W., "Conductometry and Oscillometry," in *Treatise on Analytical Chemistry*, Part I, Vol. 4, Kolthoff, I. M.; Elving, P. J., eds., Interscience Publishers, New York, 1963, pp. 2604–2618, 2624–2625.
9. Reilley, C. N., "High-Frequency Methods," in *New Instrumental Methods in Electrochemistry*, Delahay, P., ed., Interscience Publishers, New York, 1954, Chapter 15.
10. Lingane, J. J., *Electroanalytical Chemistry*, 2nd ed., Interscience Publishers, New York, 1958, Chap. VIII.
11. Furman, N. H., "Potentiometry," in *Treatise on Analytical Chemistry*, Part I, Vol. 4, Kolthoff, I. M.; Elving, P. J., eds., Interscience Publishers, New York, 1963, pp. 2283–2286.
12. Farrington, P. S.; Sawyer, D. T., *J. Am. Chem. Soc.* **1956,** *78*, 5536.
13. Szebelledy, L.; Somoggi, Z., *Z. Anal. Chem.* **1938,** *112*, 313, 323, 332, 385, 391, 395, 400.
14. Lingane, J. J., *Electroanalytical Chemistry*, 2nd ed., Interscience Publishers, New York, 1958, pp. 484–616.

15. DeFord, D. D.; Miller, J. W., "Coulometric Analysis," in *Treatise on Analytical Chemistry*, Part I, Vol. 4, Kolthoff, I. M.; Elving, P. J., eds., Interscience Publishers, New York, 1963, Chap. 49.
16. Bard, A. J., *Anal. Chem.* **1968,** *40*, 64R; ibid., **1966,** *38*, 88R; ibid., **1964,** *36*, 70R; ibid., **1962,** *34*, 57R.
17. Shaffer, P. A., Jr.; Briglio, A., Jr.; Brockman, J. A., Jr., *Anal. Chem.* **1948,** *20*, 1008.
18. Keidel, F. A., *Anal. Chem.* **1959,** *31*, 2043.
19. Czuha, M., Jr.; Gardiner, K. W.; Sawyer, D. T., *J. Electroanal. Chem.* **1962,** *4*, 51.
20. Cole, L. G.; Czuha, M., Jr.; Mosley, R. W.; Sawyer, D. T., *Anal. Chem.* **1959,** *31*, 2048.
21. Sand, H. J. S., *Phil. Mag.* **1902,** *1*, 45.
22. Gierst, L.; Juliard, A., *J. Phys. Chem.* **1953,** *57*, 701.
23. Delahay, P., *New Instrumental Methods in Electrochemistry*, Interscience Publishers, New York, 1954, Chap. 8.
24. Lingane, J. J., *Electroanalytical Chemistry*, 2nd ed., Interscience Publishers, New York, 1958, Chap. XXII.
25. Bard, A. J.; Faulkner, L. R., *Electrochemical Methods*, Wiley, New York, 1980, Chap. 7.
26. Heineman, W. P.; Kissinger, P. T., "Large-Amplitude Controlled-Current Techniques," in *Laboratory Techniques in Electroanalytical Chemistry*, Kissinger, P. T.; Heineman, W. R., eds., Marcel Dekker, New York, 1984, pp. 129–142.
27. Murray, R. W., "Chronoamperometry, Chronocoulometry, and Chronopotentiometry," in *Physical Methods of Chemistry*, 2nd ed., Vol. II, *Electrochemical Methods*, Rossiter, B. W.; Hamilton, J. F., eds., Wiley, New York, 1986, pp. 525–589.
28. Lingane, P. J., *CRC Crit. Rev. Anal. Chem.*, **1971,** *1*, 587.
29. Macdonald, D. D., *Transient Techniques in Electrochemistry*, Plenum Press, New York, 1977, Chap. 5.
30. Galus, Z., *Fundamentals of Electrochemical Analysis*, 2nd ed., Ellis Horwood Ltd., Chicester, 1994.
31. Reinmuth, W. H., *Anal. Chem.* **1960,** *32*, 1514.
32. Reinmuth, W. H., *Anal. Chem.* **1961,** *33*, 322.
33. Testa, A. C.; Reinmuth, W. H., *Anal. Chem.* **1960,** *32*, 1512.
34. Murray, R. W.; Reilley, C. N., *J. Electroanal. Chem.* **1962,** *3*, 64.
35. Birke, R. L.; Roe, D. K., *Anal. Chem.* **1965,** *37*, 450–455.
36. Anson, F. C., *Anal. Chem.* **1966,** *38*, 54.

CHAPTER 5

INDICATOR ELECTRODES

5.1 MEASUREMENT OF ELECTRODE POTENTIALS

The potential of a single electrode or half-cell cannot be directly measured in a simple way. Any potential measuring device such as a voltmeter or electrometer has two metallic terminals that must be connected across the two points between which the potential difference is measured. One terminal can be connected to the metallic electrode, but the other terminal must make connection to the solution through a wire as illustrated in Figure 5.1. The immersion of the connecting wire in the solution creates a second metal–solution interface whose potential difference will be included in the measurement. A fuller discussion of this point is available.[1]

Because the second interface created by immersing a connecting wire into the solution would have an erratic and ill-defined potential, the potential of a working electrode must be measured with respect to some reference electrode whose potential is stable and reproducible. This measurement will therefore include at least two single-electrode or half-cell potentials. Although this measurement is simple to make experimentally, its interpretation can be complicated by the presence of junction potentials between solutions of different composition, by resistive (*iR*) drops in cells in which a net current is flowing, and by internal polarization of the electrodes caused by net chemical changes produced by the passage of a current.

Figure 5.2 represents the two most common configurations used to make potential measurements. The two-electrode configuration (Figure 5.2*a*) is a system in which the cell current passes through both the working electrode and the reference electrode. In potentiometric measurements with glass electrodes or other specific ion electrodes of high internal impedance, a potential-mea-

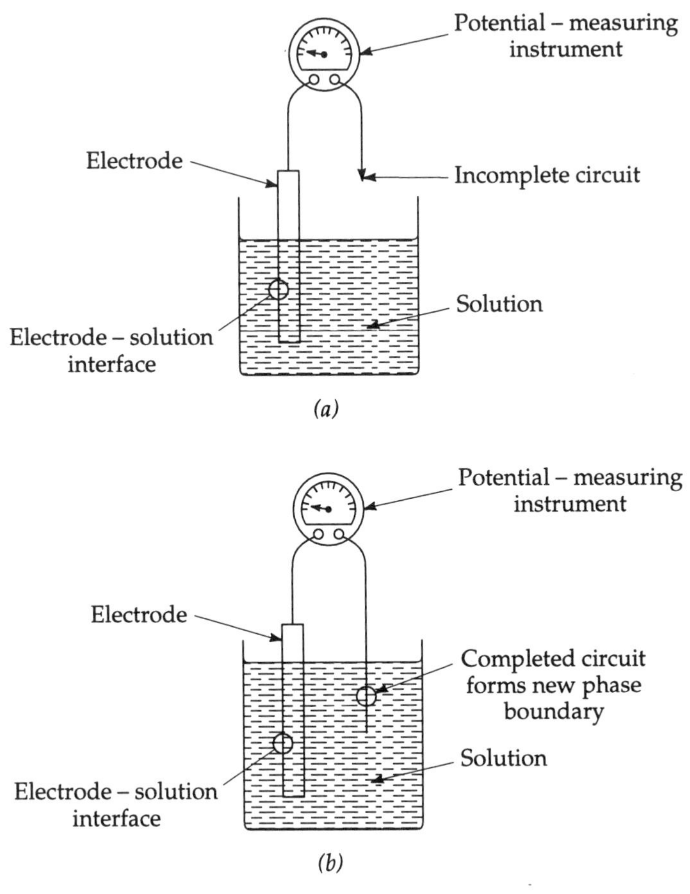

Figure 5.1 Cell and circuit elements for the measurement of electrode potentials. The upper system (*a*) illustrates the dilemma of attempts to measure single-electrode potentials.

suring device is necessary with a correspondingly high input impedance. Under these conditions, the net current flowing in the cell is very small (10^{-10}–10^{-15} A), and the internal resistive drop in the cell is negligible.

A two-electrode configuration also can be used in a voltammetric or polarographic cell in which the current is measured as a function of the applied potential. In this case the working-electrode potential will be less than the applied potential because of the iR drop in the cell. In addition, the current passing through the reference electrode may cause its potential to deviate from its equilibrium (zero-current) value, due to changes in concentration of the electroactive species at the metal–solution interface. Both of these effects act to reduce the potential of the working electrode:

$$E_{\text{working electrode}} = E_{\text{cell}} - iR_{\text{cell}} - E_{\text{polarization}} \tag{5.1}$$

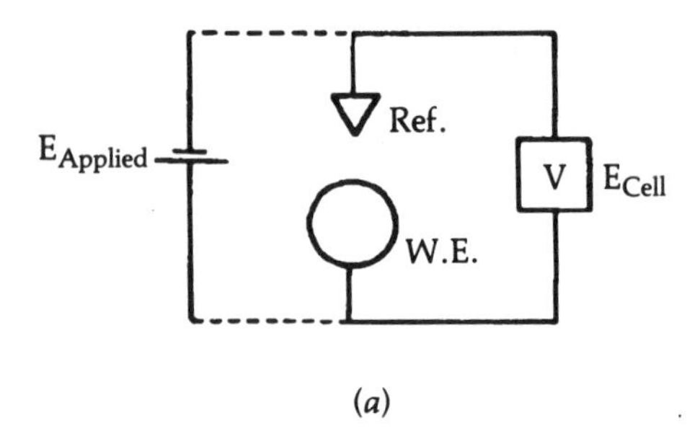

(*a*)

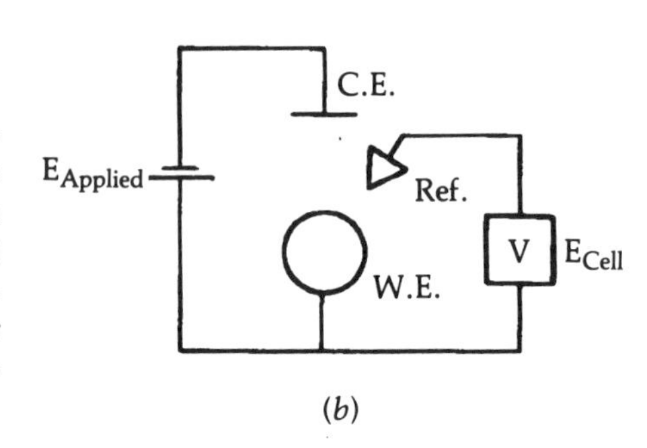

(*b*)

Figure 5.2 Circuits for the measurement of cell potentials: (*a*) two-electrode system with current passing through both working and reference electrodes; (*b*) three-electrode system with current passing through the working and counter electrodes but not the reference electrode.

To avoid serious errors, the cell current and internal cell resistance must be kept as small as possible, and the reference electrode must be designed to have low internal resistance and a metal–solution interface of sufficient area to minimize internal polarization. Under ordinary polarographic conditions (10-μA current and 1000-Ω internal cell resistance) the error amounts to 10 mV.

To minimize errors in voltammetric work, the three-electrode configuration (Figure 5.2*b*) is commonly used. The cell current flows between the working electrode and the counter or auxiliary electrode, while the potential of the working electrode is measured with respect to the reference electrode using a high-impedance measuring device. This avoids internal polarization of the reference electrode and compensates for the major portion of the *iR* drop in the cell.

Junction Potentials. The reference electrode is often isolated from the working-electrode compartment by a salt bridge or by a Luggin capillary so that one or more junctions exist in the cell. A potential difference will arise between two different ionic solutions in contact because of the differential mobility of the ions across the junction. This is illustrated in Figure 5.3, where HCl of concentration C_{HCl} and KCl of concentration C_{KCl} form a junction. The HCl and KCl will diffuse in opposite directions across the junction from the region of their greater activity (concentration) to the one of lower activity. Because the H_3O^+ ion has a greater mobility than does either $K(H_2O)_4^+$ or Cl^- ion, it diffuses more rapidly across the junction. This separation of charge (the liquid-junction potential, E_j) decelerates the H_3O^+ ions and accelerates the

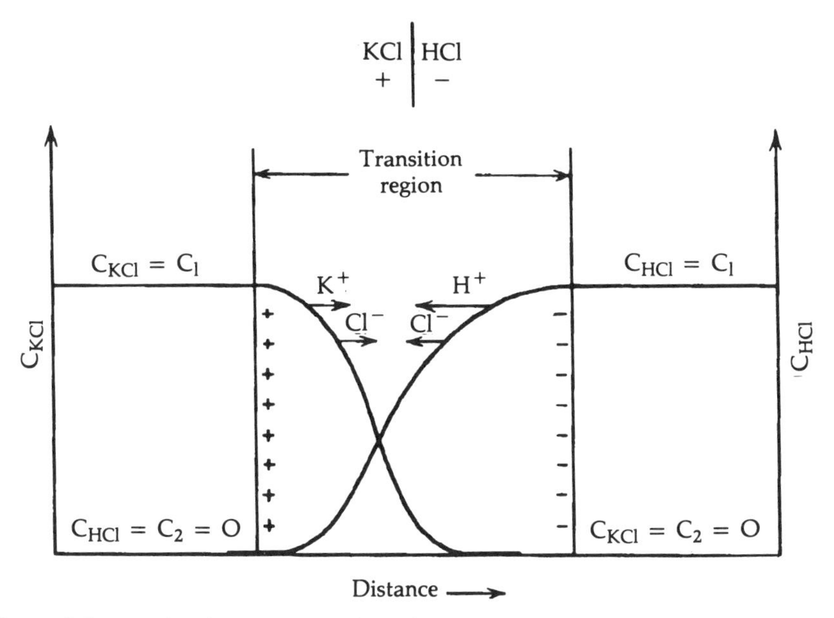

Figure 5.3 Junction between two electrolyte solutions, one containing KCl and the other HCl. The concentration profiles indicate the higher mobility of H^+ ions relative to K^+ ions.

$K(H_2O)_4^+$ ions until their velocities are equal. If the junction is constructed so that a stable diffusion geometry is established, a steady-state potential will be established that is approximately constant with time.

A junction potential also exists across the interface between electrolyte solutions of the same concentration in two different solvents. Because of this, potential measurements in two different solvents cannot be related even though the same reference electrode is used for both solvents. The dilemma can be resolved only by making extrathermodynamic assumptions because of the impossibility of making a measurement of a single-electrode potential. A number of systems have been proposed as the basis for a suitable reference electrode scale to relate directly potential measurements in one solvent to those made in another solvent. These include $Rb(H_2O)_6^+/Rb$,[2] ferricinium$^+$/ferrocene,[3] and (o-phenanthroline)$_3$Fe(III)/(o-phenanthroline)$_3$Fe(II).[4] At present there is no generally accepted solution; it is one of several interrelated problems that have been succinctly stated and critically reviewed by Popovych.[5,6] These interrelated problems include (1) the establishment of a single-solvent-independent scale for pH and for other ion activities, (2) the formulation of a solvent-independent standard potential series with a single reference point $E^\circ_{H_3O^+/H_2}$ = 0 V in water only, and (3) the evaluation of liquid-junction potentials at aqueous–nonaqueous interfaces.

The direct potentiometric measurement of a junction potential is not possible because of the impossibility of directly measuring a single-electrode potential.

Therefore the fraction of the total cell potential due to the junction potential cannot be unambiguously assigned. However, it is possible to estimate junction potentials indirectly or to make calculations based on assumptions about the geometry and distribution of ions in the region of the junction. For a junction between two dilute solutions of the same univalent electrolyte (concentrations C_1 and C_2), the liquid-junction potential is described by

$$E_j = \frac{-RT}{F} \frac{\sum u(C_2 - C_1)}{\sum uz(C_2 - C_1)} \ln \frac{\sum uzC_2}{\sum uzC_1} \tag{5.2}$$

where u is the ionic mobility ($cm^2\ s^{-1}\ V^{-1}$) and z is the charge of the ion. Both u and z are taken as positive for cations and negative for anions, and the activity coefficients are assumed to be unity, which is the Henderson equation[7] for liquid-junction potential.

If the charges on the anion and cation of the electrolyte are equal, Eq. (5.2) reduces to

$$E_j = \frac{-RT}{zF} \frac{(u_+ - u_-)}{(u_+ + u_-)} \ln \frac{C_2}{C_1} \tag{5.3}$$

where z and the ionic mobilities of the cation and anion (u_+ and u_-, respectively) are not taken as positive. This equation predicts that as the mobilities of the cation and anion approach the same value, the junction potential will approach zero. The reasonableness of junction potentials calculated from Eq. (5.3) can be examined by considering the following cell with liquid junction:

$$\mathrm{Ag}|\underset{E_1}{\mathrm{AgCl}}|M\mathrm{Cl}(C_1)\underset{E_j}{\|}M\mathrm{Cl}(C_2)|\underset{E_2}{\mathrm{AgCl}}|\mathrm{Ag} \tag{5.4}$$

Defining E_{cell} as the potential of the right-hand electrode E_2, measured with respect to the left-hand electrode E_1, we may write E_{cell} as

$$E_{cell} = E_2 - E_1 + E_j \tag{5.5}$$

The reversible half-cell potentials E_1 and E_2 may be calculated from the Nernst equation for the half-reaction

$$\mathrm{AgCl(s)} + e^- \longrightarrow \mathrm{Ag(s)} + \mathrm{Cl}^-$$

$$E = E^\circ_{\mathrm{AgCl/Ag}} - \frac{RT}{F} \ln a_{\mathrm{Cl}^-} \tag{5.6}$$

so that E_{cell} will be given by

$$E_{cell} = \frac{-RT}{F} \ln \frac{(a_{\mathrm{Cl}^-})_2}{(a_{\mathrm{Cl}^-})_1} + E_j \tag{5.7}$$

In dilute solutions the activity of the chloride ion may be calculated as

$$a_{Cl^-} = \gamma_{\pm} C_{Cl^-} \quad (5.8)$$

where $\gamma_{\pm}$ is the mean ionic activity coefficient in the electrolyte and can be measured in a thermodynamically rigorous fashion in cells without liquid junction.

To calculate the junction potential, values of the mobilities [or of the transport numbers (t)] are necessary. Because the junction involves the same electrolyte on both sides of the junction the expressions

$$t_+ = \frac{u_+}{(u_+ + u_-)} \quad \text{and} \quad t_- = \frac{u_-}{(u_+ + u_-)} \quad (5.9)$$

are valid and $t_+ + t_- = 1$, where t_+ is the fraction of current carried by the cation and t_- is the fraction carried by the anion. Substituting the transport numbers for mobilities in Eq. (5.3), the junction potential can be expressed as

$$E_j = \frac{-RT}{zF}(2\,t_+ - 1)\ln\frac{C_2}{C_1} \quad (5.10)$$

Transport numbers[8] can be used to calculate the junction potentials from Eq. (5.10). Table 5.1 gives the experimental E_{cell} values and the values of E_{cell} calculated from Eqs. (5.2), (5.8), and (5.10). The reasonable agreement indicates that junction potentials for dilute solutions of 1 : 1 electrolytes calculated from Eq. (5.3) or (5.10) are credible.

Cells with Liquid Junctions and Elimination of Junction Potentials. When electrochemical cells are employed to obtain thermodynamic data, high accuracy ($\pm$0.05 mV) requires the use of cells that are free from liquid junction (in the sense that the construction of the cell does not involve bringing into contact two or more distinctly different electrolyte solutions). Otherwise, the previously discussed uncertainties in the calculation of liquid-junction potentials will limit the accuracy of the data.

The difference between cells with and without liquid junctions is one of degree rather than of kind. The cell used to establish the mean ionic activity coefficients of HCl

$$(Pt)|H_2,\ H_3O^+Cl^-|AgCl|Ag \quad (5.11)$$

really contains two or more solutions. Because silver ion is reduced by hydrogen, in practice the half-reactions are separated and should, strictly, be represented as follows:

$$(Pt)|H_2,\ H_3O^+Cl^-(\text{satd. with } H_2)\|H_3O^+Cl^- (\text{satd. with AgCl})|AgCl|Ag \quad (5.12)$$

TABLE 5.1 Computed Potentials of the Liquid Junction in Cells of the Type $Ag|AgCl|MCl(C_1)\|MCl(C_2)|AgCl|Ag$

Concentrations (M)								
C_1	C_2	Electrolyte	$\gamma_\pm$ at C_1	$\gamma_\pm$ at C_2	t_+	E_j Calculated Eq. (5.10) (mV)	E_{cell} Calculated Eq. (5.7) (mV)	E_{cell} Observed (mV)
0.005	0.01	NaCl	0.928	0.904	0.392	3.84	−13.3	−13.4
		KCl	0.927	0.902	0.490	0.36	−16.7	−16.8
		HCl	0.928	0.906	0.825	−11.57	−28.8	−28.3
0.005	0.04	NaCl	0.835	0.928	0.391	11.64	−39.1	−39.6
		KCl	0.832	0.927	0.490	1.07	−49.6	−49.6
		HCl	0.844	0.928	0.826	−34.83	−85.8	−84.2
0.04	0.06	NaCl	0.812	0.835	0.388	2.33	−7.4	−7.6
		KCl	0.807	0.832	0.490	0.21	−9.4	−9.6
		HCl	0.824	0.844	0.829	−6.85	−16.7	−16.3

Source: Ref. 8, pp. 162, 164, 226.

In addition, a solution of pure hydrochloric acid might be present between the two saturated solutions. In this particular case, the considerations outlined in previous sections would lead us to believe that the liquid-junction potential in such a cell is negligible. If, however, hydrogen and silver chloride were more soluble, the resulting liquid-junction potential might not be negligible.

Several types of cells of necessity involve liquid junctions. Examples[9] include (1) half-cells of the second kind

$$M(\mathrm{s})|MX(\mathrm{s})|X^-(\mathrm{solution}) \qquad (5.13)$$

where the anion, X^-, is protonated in acidic solution; and the salt, MX, reacts in basic solution; for example

$$\mathrm{Hg(liq)|Hg_2CrO_4(s)|CrO_4^{2-}} \qquad (5.14)$$

Such electrodes cannot be investigated in cells of the type

$$\mathrm{(Pt)|H_2(g),\ H_3O^+}X^-|MX(\mathrm{s})|M \qquad (5.15)$$

or

$$\mathrm{Hg|HgO(s)|K(H_2O)_4^+OH^-,\ K(H_2O)_4^+}X^-|MX(\mathrm{s})|M \qquad (5.16)$$

because X^- is protonated in the cell of Eq. (5.15) and MX(s) reacts with hydroxide ion in the cell of Eq. (5.18); and (2) electrodes for which both the oxidized and reduced forms of the half-cell reaction are soluble. For example, the cell

$$\mathrm{(Au)|K_4Fe^{II}(CN)_6,\ K_3Fe^{III}(CN)_6,\ KCl|AgCl(s)|Ag} \qquad (5.17)$$

is not usable because the reaction

$$\mathrm{Fe^{II}(CN)_6^{4-} + AgCl \longrightarrow Ag + Cl^- + Fe^{III}(CN)_6^{3-}} \qquad (5.18)$$

occurs spontaneously in the solution.

Various attempts have been made to circumvent these problems and to eliminate junction potentials, including (1) extrapolation procedures designed to eliminate the difference between the compositions of the two solutions in the appropriate limit, (2) separation of the two solutions by means of a double-junction salt bridge, (3) the use of double cells with dilute alkali metal amalgam connectors, and (4) the use of glass or other types of ion-specific electrodes as "bridging" reference electrodes.

The extrapolation procedures used for cells with liquid junction are time-consuming, and the method is not entirely free of theoretical pitfalls.[9] Because salt bridges usually involve double junctions, an important distinction needs to be made betwen the behavior of single-junction and double-junction salt bridges.[15] In a single junction the residual junction potential can be reduced, in principle, to a low value by using a large concentration of an equitransferent electrolyte on one side of the junction. However, the use of equitransferent solutions (containing mixtures of KCl and KNO_3) in a double-junction bridge actually increases the residual junction potential between 0.1 M KCl and 0.1 M HCl. Thus, although the ions of the concentrated solution may contribute the major portion of the two junction potentials at either end of the bridge, the residual full-bridge potential is determined mainly by the dilute solutions at either end of the bridge.[15] A properly designed double-junction salt bridge can reduce but not completely eliminate the junction potential. This conclusion results from studies with the cell:

$$\mathrm{Hg|Hg_2Cl_2|HCl(0.1\ M)||KCl(0.1\ M)|Hg_2Cl_2|Hg} \tag{5.19}$$

In the absence of a bridge the cell potential (attributable mainly to the junction potential) is about 27 mV. When a KCl bridge is inserted, the cell potential decreases with increasing concentration of the KCl bridge electrolyte as shown in Figure 5.4. For a concentration of 3.5–4.0 M KCl in the bridge, the residual junction potential has decreased to about 1–2 mV.

Design of junction devices. Considerable effort has been applied to the design of junction devices for reference electrodes and salt bridges. For highest stability and reproducibility, the junction should be formed within a cylindrical tube by one of the methods described in Ref. 10 or 11. Such junctions can be

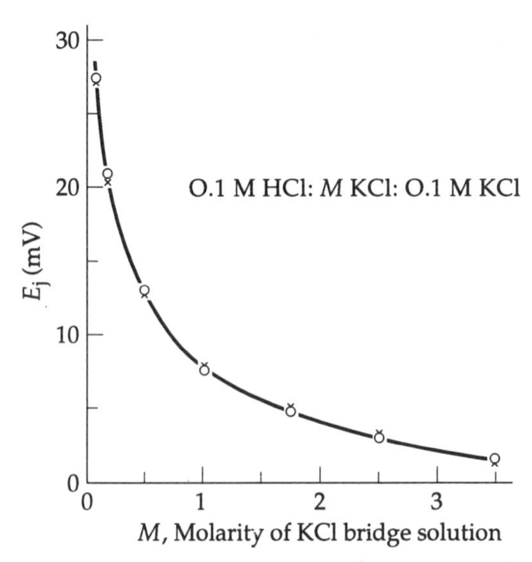

Figure 5.4 Junction potential between a 0.1 M HCl solution and a 0.1 M KCl solution as a function of the concentration of KCl in the connecting salt bridge.

categorized as the "constrained diffusion" type with cylindrical symmetry. However, these rather elaborate methods of forming junctions are not popular for routine use because of their inconvenience.

Most of the commercially manufactured electrodes have junctions that can be classified as "continuous impeded-flow" types; a number of different junction devices are illustrated in Figure 5.5. Their relative merits are discussed in the following section.

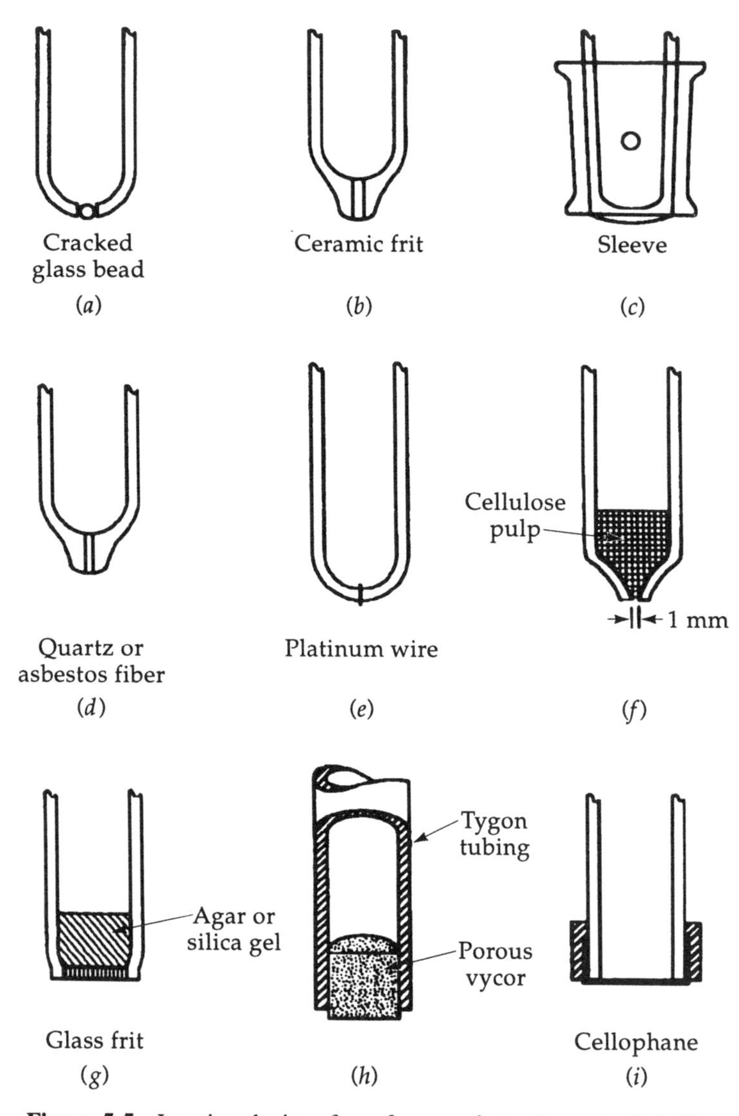

Figure 5.5 Junction devices for reference electrodes and salt bridge.

Flow rate of junctions. The flow rate of a junction is of concern to the user for two reasons: (1) a high flow rate may tend to contaminate the sample with the filling electrolyte and (2) too low a flow rate may lead to erratic junction potentials or a tendency of the junction to clog if the sample contains colloidal material. Some typical flow rates and recommended applications for the different junction styles are summarized in Table 5.2.

The electrolyte of the impeded-flow junctions should stream into the test solution at a constant rate by a single leakage path, preferably with cylindrical symmetry. The presence of several separated openings through which the bridge solution may flow at varying rates is not conducive to the establishment of a reproducible steady state. A constant and substantial flow rate is probably the most critical factor for obtaining reproducible junction potentials. For pH measurements of ordinary precision (± 0.05 pH unit) a reproducibility of about ± 3 mV is tolerable, and almost any of the junctions described will work satisfactorily provided the junction is not clogged.

When specific ion electrodes are employed, an error of less than ± 1 to 2% of the activity of the ion is desirable. This requires that the potential measurement have an error of less than 0.1 to 0.2 mV; the reproducibility of the junction potential becomes a critical factor in the measurement. The sleeve-type junction

TABLE 5.2 Typical Liquid-Junction Characteristics

Junction	Approximate Resistance (Ω) 3.5 M Aqueous KCl	Electrolyte Flow Rate (μL h^{-1})	Comments
Soft glass in Pyrex	1000	3–30	General use; avoid precipitates and colloids
Quartz in Pyrex	1000	3–30	General use; avoid precipitates and colloids
Platinum wire	1000	3–30	General use; avoid strong oxidizing and reducing agents, precipitates, and colloids
Ceramic frit	1000	3–30	Can be used in strong bases; not recommended for precipitates and colloids
Asbestos fiber	1000–3000	10–200	General use; avoid precipitates and colloids
Carborundum frit	100–500	100–500	General use; not recommended for low-ionic-strength solutions
Ground sleeve	100	100–500	General use; recommended for slurries and colloids; not recommended for strongly acid solutions

is recommended for specific-ion-electrode measurements. This junction has a relatively high flow rate that tends to produce a flow velocity sufficient to overcome backdiffusion of sample ions into the region of the junction. In addition, the ability to flush the junction makes restoration convenient should it become contaminated.

Choice of electrolyte for salt bridges and reference electrodes. Many of the difficulties encountered in potentiometric measurements can be attributed to erratic or drifting junction potentials caused by clogged junctions. Certain elementary rules should be observed in choosing the filling solution for a salt bridge or reference electrode, particularly when they will be used in organic solvents or solutions that are only partially aqueous.

1. The electrolyte should not react with any species in the sample. For example, a reference electrode solution that contains silver chloride or mercurous chloride should not be used in contact with a solution containing sulfide ion because metallic sulfides may precipitate in the junction and clog it. The formation of soluble reaction products (such as a complex ion) may alter the mobility of a species in the bridge electrolyte, upsetting a desirable equitransferent condition. Sodium perchlorate is often used in aqueous solutions to maintain a high ionic strength. If it is contact with a low-flow-rate junction that contains a high concentration of potassium chloride, potassium perchlorate will precipitate in the junction and clog it.

2. In an analytical measurement that employs an ion-specific-electrode, the filling electrolyte should not contain the ion being measured in the sample, or an ion that produces a significant interference at the sensing electrode. This rule is particularly important in trace determinations on small-volume samples where contamination of the sample by the filling solution may be significant.

3. The ionic strength of the filling solution should be at least tenfold that of the sample so that the bridge solution will largely determine the junction potential and "swamp out" the effect of the sample ions.

4. The filling solution should be equitransferent. If the rates at which positive and negative charge diffuse into the sample are nearly equal, a minimum junction potential results. The best choice is usually a 1 : 1 electrolyte with the ionic mobilities (refer to Tables 7.7 and 7.8) of the cation and anion nearly the same. An aqueous KCl solution fulfills this condition and is commonly used as a bridge electrolyte, when $K(H_2O)_4^+$ and Cl^- ions do not pose problems. Solutions that contain high levels of strong acids or bases are particularly troublesome because the high mobilities of H_3O^+ and HO^- ions make it impossible to "swamp out" their effect on the junction potential. If the level of H_3O^+ or HO^- ions is approximately constant in the test solution, the junction potential can be minimized by use of a filling solution with about the same level of acid or base.

5. In a junction formed between different solvents, the electrolyte must be soluble in both solvents. Potassium chloride is insoluble in many commonly

used organic solvents such as acetronitrile and dimethyl sulfoxide. Therefore, aqueous reference electrodes that employ 3–4 M KCl cannot be used without an intervening salt bridge and an electrolyte soluble in both solvents. Quaternary ammonium perchlorate or halide salts commonly are used as bridge electrolytes in organic solvents. Tetraethylammonium picrate or tetra-*n*-butylammonium tetraphenylborate ($NBu_4^+BPh_4^-$) has been suggested as bridge electrolytes because they are nearly equitransferent in many organic solvents.

Some Practical Considerations in the Use of Salt Bridges. Salt bridges are most commonly used to diminish or stabilize the junction potential between solutions of different composition and to minimize cross-contamination between solutions. For example, in working with nonaqueous solvents an aqueous reference electrode often is used that is isolated from the test solution by a salt bridge that contains the organic solvent. However, this practice cannot be recommended, except on the grounds of convenience, because there is no way at present to relate thermodynamically potentials in different solvents to the same aqueous reference-electrode potential; furthermore, there is a risk of contamination of the nonaqueous solvent by water.

Salt bridges (particularly those that employ saturated KCl) also have been widely used in the conversion of polarographic and other potentials from one reference-electrode scale to another. This conversion is commonly made by direct comparison of the two reference (ref) electrodes in the cell

$$\text{ref electrode 1}|\text{KCl salt bridge}|\text{ref electrode 2} \qquad (5.20)$$

However, there is convincing experimental evidence that this can introduce substantial errors because the residual junction potential errors are not negligible.[32] The reference electrodes should be connected to one another through the sample solution in the cell:

$$\text{ref electrode 1}|\text{sample solution}|\text{ref electrode 2} \qquad (5.21)$$

The potential difference of the cell of Eq. (5.21) will more closely represent the potential difference of the two reference-electrode scales.

With voltammetric techniques like polarography or coulometry a salt bridge often is used to isolate the working- and counter-electrode compartments. Because the cell current flows through the salt bridge, the resistance of the salt bridge must be kept as low as possible to minimize *iR* drop. Many of the junction devices shown in Figure 5.5 are more suited for potentiometric work where current flow is negligible. For voltammetric work, a larger junction area is needed to obtain low resistance. This is accomplished by the use of glass-frit or porous separators. Several designs are illustrated in Figure 5.6, and salt bridges that are incorporated into cell designs are described later in this chapter.

Gels are sometimes used in conjunction with fritted tubes because they lower flow rates while maintaining a low resistance. Aqueous solutions that are not

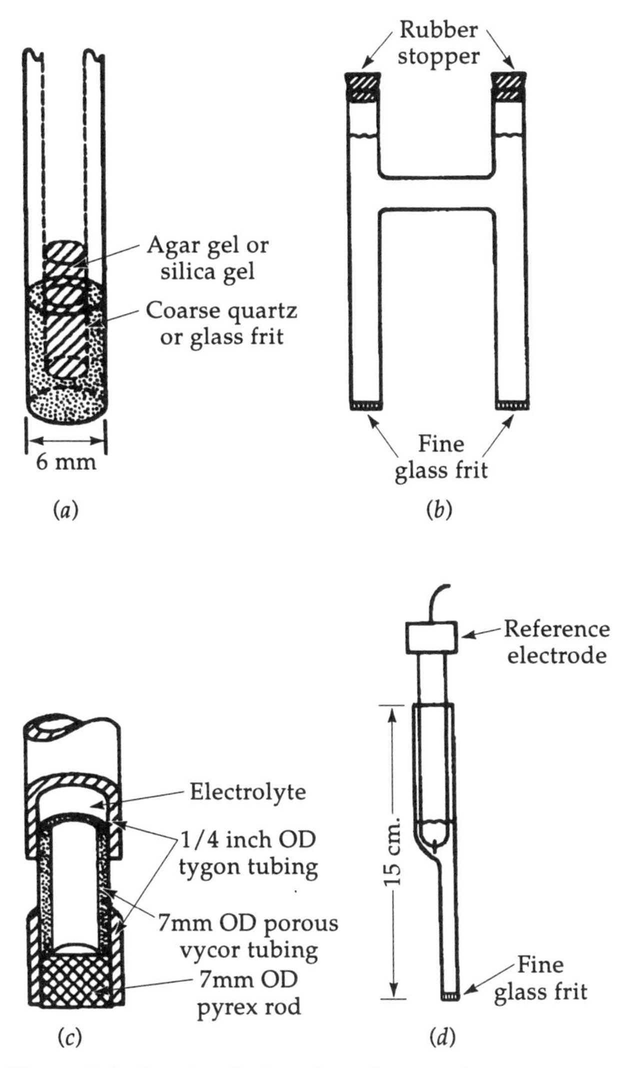

Figure 5.6 Junction devices for voltammetric measurements.

strongly acidic or basic are readily gelled with a 4% solution of agar. The agar is first dissolved in hot water, then the salt is added, and the hot fluid gel is drawn or pipetted into the salt bridge where it is allowed to harden. A convenient device for filling the cross-tube of an H-type cell is shown in Figure 5.7. Once prepared, an agar bridge should not be allowed to dry out and should be stored in contact with the bridge electrolyte when not in use. Agar will not gel properly in many organic solvents, and agar bridges are not recommended for connecting aqueous and organic solvent solutions. Five grams of methyl cellulose in 100 mL of 0.5 M $LiClO_4$ in pyridine gives a colorless, transparent,

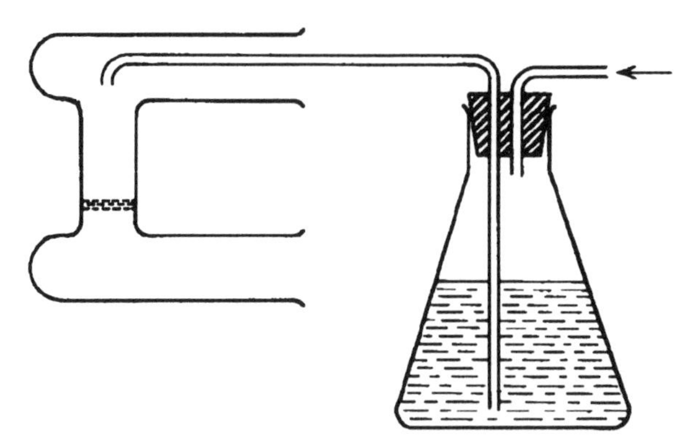

Figure 5.7 System for filing H-cell salt bridge with agar electrolyte.

conductive gel;[13] this material may be suitable for gelling other organic solvents.

The use of a 7% fumed-silica gel with silica frit bridges of the style shown in Figure 5.6*a* has been recommended.[14] Fumed silica forms stable gels with water and a number of organic solvents, but the aqueous gel is not stable above pH 8.

5.2 REFERENCE ELECTRODES

Properties of the Ideal Reference Electrode. An ideal reference electrode should show the following properties: (1) it should be reversible and obey the Nernst equation with respect to some species in the electrolyte; (2) its potential should be stable with time; (3) its potential should return to its initial value after small currents are passed through the electrode (no hysteresis); (4) if it is an electrode of the second kind (e.g., Ag/AgCl), the solid phase must not be appreciably soluble in the electrolyte; and (5) it should show low hysteresis with temperature cycling.

Because the flow of electric current always involves the transport of matter in solution and chemical transformations at the solution–electrode interface, local behavior can only be approached. It can be approximated, however, by a reference electrode whose potential is controlled by a well-defined electron-transfer process in which the essential solid phases are present in an adequate amount and the solution constituents are present at sufficiently high concentrations. The electron transfer is a dynamic process, occurring even when no net current flows; and the larger the anodic and cathodic components of this exchange current, the more nearly reversible and nonpolarizable the reference electrode will be. A large exchange current increases the slope of the current–potential curve so that the potential of the electrode is more nearly independent of the current. The current–potential curves (polarization curves) are frequently used to characterize the reversibility of reference electrodes.

Reference electrodes can be classified into several types: (1) electrodes of the first kind; a metallic or soluble phase in equilibrium with its ion

$$H_3O^+|H_2(Pt);\ \text{ferrocinium}|\text{ferrocene};\ Ag(OH_2)_2^+|Ag;$$

$$\text{amalgam electrodes of the type } M\ (H_2O)_n^+|M(Hg)$$

(2) electrodes of the second kind; a metallic phase in equilibrium with its sparingly soluble metal salt

$$AgCl|Ag;\ Hg_2Cl_2|Hg;\ Hg_2SO_4|Hg;\ HgO|Hg$$

and (3) miscellaneous; glass electrodes, ion-specific electrodes, electrodes of the third kind.

Reference Electrodes for Use in Aqueous Solutions. The extensive studies of reference electrodes have been critically surveyed.[15–17]

The hydrogen electrode. The hydrogen electrode is discussed first because it is the primary reference electrode used to define an internationally accepted scale of standard potentials in aqueous solution. By convention, the potential of an electrode half-reaction that is measured with respect to the normal hydrogen electrode (NHE; also written as SHE, standard hydrogen electrode) is defined as the electrode potential of the half reaction. This convention amounts to an arbitrary assignment for the standard potential of the hydrogen electrode as zero at all temperatures. Thus, there is in effect a separate scale of electrode potentials at each temperature level.

The hydrogen electrode also is one of the most reproducible electrodes that is available; properly prepared electrodes will show bias potentials of less than 10 μV when compared with one another.

The hydrogen electrode commonly consists of a platinum foil, the surface of which is able to catalyze the reaction

$$2H_3O^+ + 2e^- \longrightarrow H_2 + H_2O \tag{5.22}$$

where H_3O^+ and H_2 are at unit activity. Platinum black is the most useful catalyst and will function either dispersed in the solution[17] or deposited on the surface of a bright platinum electrode. Platinum black usually is deposited from a 1–3% solution of chloroplatinic acid (H_2PtCl_6) that contains a small amount of lead acetate. The lead acetate is not essential, but it is desirable, because electrodes prepared with lead acetate have a longer life and are less susceptible to poisoning. The electrode is poisoned by traces of sulfide and cyanide, but when this occurs, the electrode potential changes so markedly that malfunction is obvious. Poisons apparently act by displacing adsorbed hydrogen.[18] Drifting potentials are observed when species such as benzoic acid and nitrophenols are present that are readily hydrogenated.

Two forms of the hydrogen electrode are shown in Figure 5.8.[19] In the cell of Figure 5.8*a* the electrodes are supported in a stopper and a flexible polyvinyl chloride tube is used to make the connection to the second half-cell. The cell of Figure 5-8*b* is equipped with a presaturator and a hydrogen bypass and is designed to be immersed completely in a thermostatted water or oil bath. Although the half-cell of Figure 5.8*b* shows the platinum–glass seal immersed in the solution, a 2–3-cm "stalk" of platinum wire is recommended so that the glass–metal seal can be kept out of solution; a strained glass–metal junction immersed in solution can give rise to spurious potentials. To avoid this problem, cover with glass the whole region of the spot weld between the platinum foil and wire as shown in Figure 5.9.[20] The connection in the contact tube should be made by spot-welding or silver-soldering copper wire to the platinum or by use of Wood's metal (a low-melting alloy). The use of mercury is considered objectionable by many authorities because of the possibility that it might work its way through the seal.

Before platinization, electrodes should be cleaned by brief immersion in a solution prepared by combining 3 volumes of 12 M hydrochloric acid with 1 volume of 16 M nitric acid and 4 volumes of water (50% aqua regia). This mixture also can be employed to strip the platinum black from used electrodes. The electrodes are then washed in 16 M nitric acid, rinsed in water, cathodized

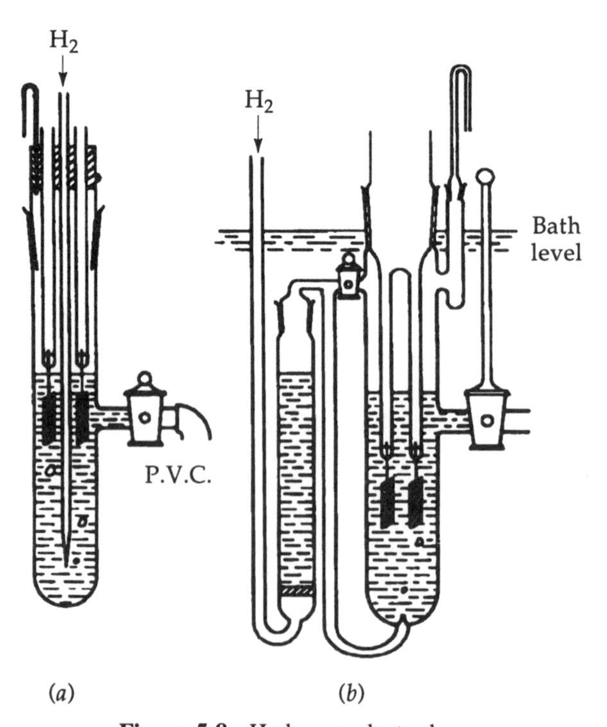

Figure 5.8 Hydrogen electrodes.

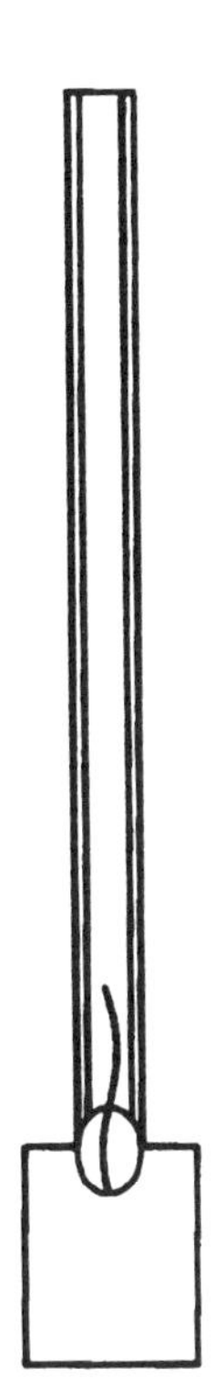

Figure 5.9 Platinum electrode sealed in "soft" glass with the wire-foil spot weld covered by glass.

in 0.01 M sulfuric acid, and given a final thorough rinse in distilled water. The cathodization step is designed to remove surface oxides and should be carried out immediately before platinization.

Platinizing procedures have been reviewed;[21] a solution of 0.072 M (3.5%) chloroplatinic acid plus 1.3×10^{-4} M (0.005%) lead acetate, at a current density of 30 mA cm^{-2} for up to 10 min, has been recommended. A deposition time of 5 min should be adequate for hydrogen and for conductance electrodes; smaller deposits speed equilibration and reduce adsorption from the electrolyte. Good stirring is essential, and no gas should be evolved at the platinum cathode. The chlorine evolved at the anode can easily be prevented from interacting with the cathode by employing a salt bridge between two beakers or an H-type cell with a glass-frit separator.

The surface coat of platinum black should appear to be a uniform, jet-black compact deposit. After platinization the electrode should be washed and stored in water. If dry electrodes are exposed to air for a prolonged period, their catalytic activity is lost and they must be replatinized before use.

The active surface of the platinized foil electrode should be covered completely by the solution when in use. It should be supplied with pure hydrogen gas at the rate of one to two bubbles per second from a jet about 1 mm in diameter. To avoid changes in concentration, the hydrogen can be passed through a saturator that contains water or cell solution before it enters the cell.

Electrolytic grade hydrogen (available commercially in cylinders) is a convenient source of gas. It typically contains 0.15–0.20% of oxygen and 0.03–0.05% nitrogen. The potential of a hydrogen electrode will be biased by the presence of dissolved oxygen in the solution because the oxygen is readily reduced at the catalytically active surface.[22] Therefore, oxygen should be excluded from the cell during measurements and should be removed from the hydrogen gas supply by one of the methods described in Chapter 6. The palladium–membrane electrolytic generators provide a convenient source of dry, oxygen-free hydrogen and may be less expensive than cylinders in the long run.

The partial pressure of hydrogen in most experiments will not be exactly 1 atm. In aqueous solutions that contain no volatile solutes such as ammonia or carbon dioxide, the partial pressure of hydrogen gas at the electrode surface is obtained by subtracting the vapor pressure of water and adding the depth effect[23] to the barometric pressure. The solubility of hydrogen is determined by the depth of immersion of the jet through which the gas enters; this excess pressure slightly supersaturates the solution. An empirical expression for the partial pressure of hydrogen in the cell is given by

$$P_{H_2} = P_{\text{barometric}} - P_{H_2O} + 0.030\,h \tag{5.23}$$

where all pressures are expressed in millimeters of Hg and h is the depth of immersion of the jet in millimeters.

The palladium–hydrogen electrode. Substrates other than platinum have been used for the preparation of hydrogen electrodes. Although a palladium electrode is normally a poor substitute for platinum because the catalytic activity of palladium is lower and it decays faster, the capacity of palladium to dissolve up to 30 volumes of hydrogen allows the preparation of a unique kind of hydrogen electrode that can be used in solutions that contain no dissolved hydrogen. A palladium–hydrogen micro-reference electrode suited for use in fast potentiostatic current measurements has been described[24] that uses a palladium wire welded to platinum and charged with hydrogen electrolytically. The exposed area of the palladium wire can be as small as 10^{-4} cm^2, and it can be placed within 0.002 cm of the working electrode. The normal Luggin capillary cannot be placed this close to the electrode surface without ''screening'' the electrode, which results in a nonuniform current density at the electrode surface. The palladium micro-reference electrode gives a steady potential of +50 mV with respect to a hydrogen electrode in the same solution and is stable for about 24 h. The electrode is easily renewed by recharging it electrolytically with hydrogen. The potentials of the palladium–hydride reference electrode with respect to the platinum–hydrogen electrode between 25 and 195°C have been determined.[25]

The silver–silver chloride reference electrode. Next to the hydrogen electrode, the silver–silver chloride electrode is probably the most reproducible and reliable reference electrode, and it is certainly one of the most convenient electrodes to construct and use:[26]

$$AgCl(s) + e^{-} \longrightarrow Ag(s) + Cl^{-} \qquad (5.24)$$

The solubility of silver chloride in water is about 10^{-5} M at 25°C, which sets a lower limit on the use of the electrode as an ion-specific electrode for chloride ion. The solubility in saturated KCl solution increases to about 6×10^{-3} M, due to the formation of soluble complexes of the type $AgCl_2^-$. For this reason the saturated KCl electrolyte must be presaturated with silver chloride; otherwise the electrode becomes stripped of its AgCl coating.

Because of this tendency to form anionic complexes, from the point of view of establishing satisfactory reference electrodes of the second kind, the most significant constant is not necessarily the solubility product constant, but the equilibrium constant for the reaction

$$AgCl(s) + Cl^{-} = AgCl_2^{-} \qquad (5.25)$$

In the half-cell of Eq. (5.24), the concentration of $AgCl_2^-$ must be small compared to that of Cl^-, or a liquid-junction potential will result because the mobilities of $AgCl_2^-$ and Cl^- are not the same. Thus, for a reference electrode of the second kind to be effective in cells without appreciable junction potentials, the equilibrium constant for the reaction of Eq. (5.25) must be smaller than unity (preferably <0.1). In water, methanol, formamide, and *N*-methylformamide, this criterion is met, but in most organic solvents the equilibrium constant for the reaction of Eq. (5.25) ranges from 30 to 100. The silver chloride electrode is not recommended for general use in organic solvents.[27]

The potential of the silver–silver chloride electrode is sensitive to traces of bromide in the solution used to deposit AgCl. The presence of 0.01 mole percent (mol %) of bromide in a KCl electrolyte is sufficient to alter the potential of electrodes immersed in the solution by 0.1–0.2 mV.[28] The potentials are not greatly affected by traces of iodide or cyanide. Light of ordinary intensities does not have a marked effect on the potential of the electrodes, but exposure to direct sunlight should be avoided.

Of the several methods for preparation of silver/silver chloride electrodes, three are widely used:

1. A silver wire is made the anode in a 0.1 M HCl or KCl solution and coated with AgCl. Electrodes prepared in this way are suitable for potentiometric titrations or routine voltammetric use. Mechanical strains in the silver wire introduced in the drawing process, and impurities in the silver or in the chloride electrolyte generally cause the bias potentials of such electrodes to be rather high (± 5 mV), but they are reasonably stable with time. To coat the silver electrode with AgCl, it is first cleaned with 3 M nitric acid, washed thoroughly with water, and put in an H-type cell with a glass-frit separator, or the cathode is separated from the silver anode by a salt bridge. The silver is

coated with silver chloride at a current density of about 0.4 mA cm^{-2} for about 30 min, using 0.1 M HCl as the electrolyte. The electrodes are thoroughly washed and allowed to age for 1–2 days. The color of the fresh electrodes is variable, ranging from a sepia to a pale tan or brown. After washing, the color will usually change to a pink, grayish-pink, or plum color.

2. The electrolytic type of AgCl electrode is formed by the electrodeposition of silver on a platinum wire, foil, or gauze. The surface of the carefully washed deposit is then converted to silver chloride by electrolysis in a chloride solution as described above. Electrodes prepared in this fashion from pure solutions have a lower bias potential than does the preceding type. The silver-plating is carried out by first thoroughly cleaning the platinum electrodes as described in the preceding section on their preparation for platinization. They are then immersed in a solution that contains 10 g L^{-1} of pure $KAg(CN)_2$ (freed from excess cyanide by the addition of dilute $AgNO_3$ if necessary) and plated with silver for approximately 6 h at 0.4 mA cm^{-2}. The anode and cathode compartments should be separated by a salt bridge or glass frit and adequate stirring should be provided. The silver plate obtained should be snowy white and velvet-like in appearance, and should wet uniformly. After soaking in concentrated NH_3/H_2O for 1–6 h, it is washed with water, frequently over a period of 1–2 days. The electrodes are then coated with AgCl as described above.

3. In the thermal-electrolytic-type AgCl electrode, silver is formed by heating a paste of silver oxide coated on a spiral of platinum wire. Part of the spongy silver mass in converted to AgCl by the electrolysis process described above. Because the silver is porous, the electrodes tend to be rather sluggish, but they can be prepared with very small bias potentials (± 20 μV).

Silver–silver chloride reference electrodes are available from all the principal manufacturers of ion-specific and pH electrodes. They usually are furnished with a saturated KCl (saturated with AgCl) filling solution, but some manufacturers will supply the electrodes in a ''dry'' condition so that the user may add the electrolyte. This can be an advantage because a saturated KCl filling electrolyte is a source of difficulty in laboratories where the temperature fluctuates or goes much below 25°C. For the latter condition potassium chloride will precipitate in the junction and will have to be dissolved with dilute KCl or water and flushed. For this reason most authorities recommend the use of 3.5 M KCl (saturated with AgCl) for use in the silver chloride reference electrode. Junction potentials are not seriously altered, so that pH measurements are unaffected.

Relating a reference electrode to the standard hydrogen electrode. In using any reference electrode other than the hydrogen electrode there is the problem of knowing what potential on the hydrogen scale to ascribe to it. Direct comparison by means of the usual salt bridge or liquid junction presents two problems. First there is no direct way to determine the necessary single ion activities. Second, the liquid-junction potential must be evaluated. There are two ways to deal with this. For the typical electrode system

Ag|AgCl|KCl(m)‖liquid junctions or salt bridge‖test solution|test electrode

$$E^{\circ\prime} = E_{AgCl/Ag} \tag{5.26}$$
$$E^{\circ\prime} + E_j = E_{AgCl/Ag} + E_{liquid junction}$$

The potential $E^{\circ\prime}$ may be calculated from the known value of $E_{AgCl/Ag}$ by use of the equations:

$$E^{\circ\prime} = E_{AgCl/Ag} = E^{\circ}_{AgCl/Ag^-} - \frac{RT}{F} \ln a_{Cl^-} \tag{5.27}$$

$$= E^{\circ}_{AgCl/Ag^-} - \frac{RT}{F} \ln m_{Cl^-}\gamma_{Cl^-} \tag{5.28}$$

where m is the molality. The unknown term γ_{Cl^-} may be approximated by equating it to the corresponding mean ionic activity coefficient $\gamma_{\pm}$, for the particular KCl solution concerned or by calculating the activity coefficient from the extended Debye–Hückel equation.[29] The value of ($E^{\circ\prime} + E_f$) may then be determined by adding to $E^{\circ\prime}$ the calculated liquid-junction potential, provided all the parameters for its calculation are known or can be approximated.

Alternatively, the silver–silver chloride half-cell can be directly compared with a hydrogen electrode, as, for example, in the cell

$$\text{(Pt)}|H_2, \text{HCl}(m) \text{ or NBS buffer}\|\text{KCl}(m)|\text{AgCl}|\text{Ag} \tag{5.29}$$

($E^{\circ\prime} + E_j$) is then given by the observed cell potential minus that due to the hydrogen electrode, calculated from

$$E_H = \frac{RT}{F} \ln a_{H_3O^+} = \frac{RT}{F} \ln m_{H_3O^+}\gamma_{H^+}$$
$$= -\frac{2.3RT}{F} \text{pH} \tag{5.30}$$

For this calculation some arbitrary assumption must be made, such as $\gamma_{H_3O^+} = \gamma_{\pm}$; the value of $\gamma_{\pm}$ can be estimated from the Debye–Hückel equation. Alternatively, the known pH of an NBS primary standard buffer can be used in Eq. (5.30). This latter procedure probably is the most practical and, if the pH of the solutions lies between 2 and 12 and they contain only simple ions in concentrations less than 0.2 M, good constancy of E_j is found. This value of ($E^{\circ\prime} + E_j$) consequently relates specifically to solutions that are similarly constituted.

Values of ($E^{\circ\prime} + E_j$) for silver chloride and calomel reference electrodes are listed in Table 5.3. For reasons described previously, the unsaturated 3.5 M KCl electrode appears to be the best choice for routine work.

Most commercially available silver chloride (or calomel) reference elec-

TABLE 5.3 Standard Potentials ($E^{\circ\prime} + E_j$) and Temperature Coefficients for Cells of the Type (Pt)/H_2, H_3O^+ (a = 1)‖KCl/MCl(satd.)/M

MCl/M	KCl	$E^{\circ\prime} + E_j$ (V at °C)							$\frac{d(E^{\circ\prime} + E_j)}{dt}$ (mV deg^{-1} at 25°C)
		10	15	20	25	30	35	40	
AgCl/Ag	3.5 M (at 25°C)	0.215	0.212	0.208	0.205	0.201	0.197	0.193	−0.73
	Satd.	0.214	0.209	0.204	0.199	0.194	0.189	0.184	−1.01
Hg_2Cl_2/Hg	0.1 M (at 25°C)	0.336	0.336	0.336	0.336	0.335	0.334	0.334	−0.08
	1.0 M (at 25°C)	0.287	—	0.284	0.283	0.282	—	0.278	−0.29
	3.5 M (at 25°C)	0.256	0.254	0.252	0.250	0.248	0.246	0.244	−0.39
	Satd.	0.254	0.251	0.248	0.244	0.241	0.238	0.234	−0.67

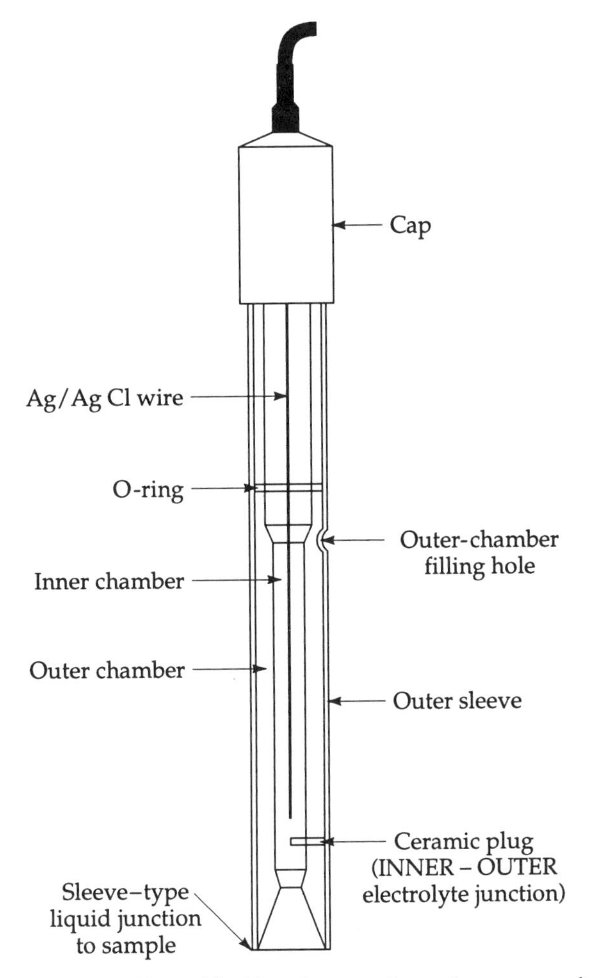

Figure 5.10 Silver–silver chloride reference electrode, commercial design.

trodes (see Figure 5.10) are not designed to pass substantial currents and should not be used for two-electrode voltammetry (where the current passes through the reference electrode). It is best to make your own silver chloride reference electrode for this purpose; a recommended design is shown in Figure 5.11.[53] This reference electrode has a low internal resistance and sufficient surface area ($\sim 10\ cm^2$) to pass currents up to 20 μA without significant change of potential.

Figure 5.12 illustrates a small silver chloride reference electrode used by the authors that employs a dual-glass junction (soft glass in Pyrex) with a low leak rate. This electrode is suitable for three-electrode voltammetry where the cell current does not pass through the reference electrode. The choice of filling electrolyte will depend on the solvent system used, although the authors have used an aqueous 1 M $(CH_3)_4N^+Cl^-$ filling solution connected to organic solvent

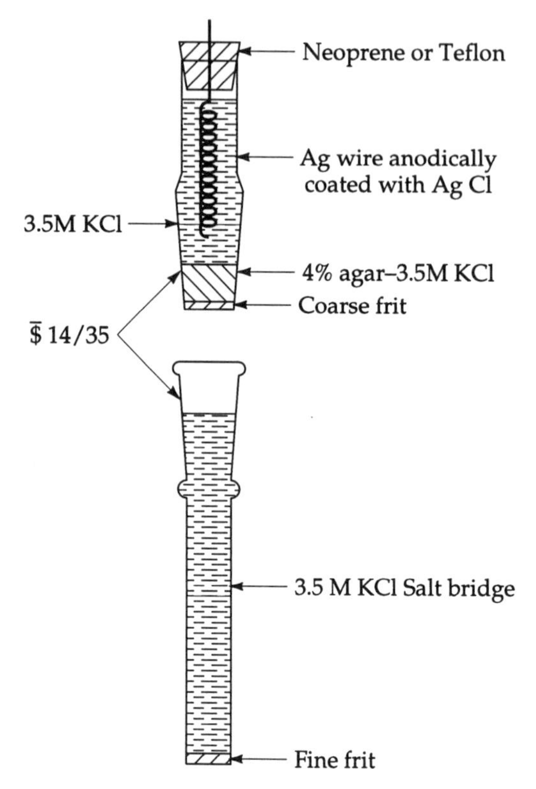

Figure 5.11 Design for a "home-made" silver–silver chloride reference electrode with low resistance.

solutions by a Luggin capillary bridge (see Chapter 6) that is filled with the sample solution. When used with organic solvents such as acetronitrile, dimethyl sulfoxide, or dimethylformamide, the junction potentials certainly are not negligible, but they are reproducible to ± 5 mV or less. Although convenient, this practice is not recommended where it is necessary to exclude water completely.

The mercury–mercurous chloride (calomel) electrode. The calomel electrode was used extensively as a chloride electrode, but it has been all but abandoned for this purpose in favor of the silver chloride electrode. The fixed-potential saturated or 3.5 M KCl calomel electrode always has been popular for use with glass electrodes in pH measurements and in polarographic work; most of the vast compilations of aqueous polarographic half-wave potentials were referred to the aqueous saturated calomel electrode (SCE).

The electrode is based on the half-reaction

$$Hg_2Cl_2(s) + 2\,e^- \longrightarrow 2\,Hg(l) + 2\,Cl^- \qquad (5.31)$$

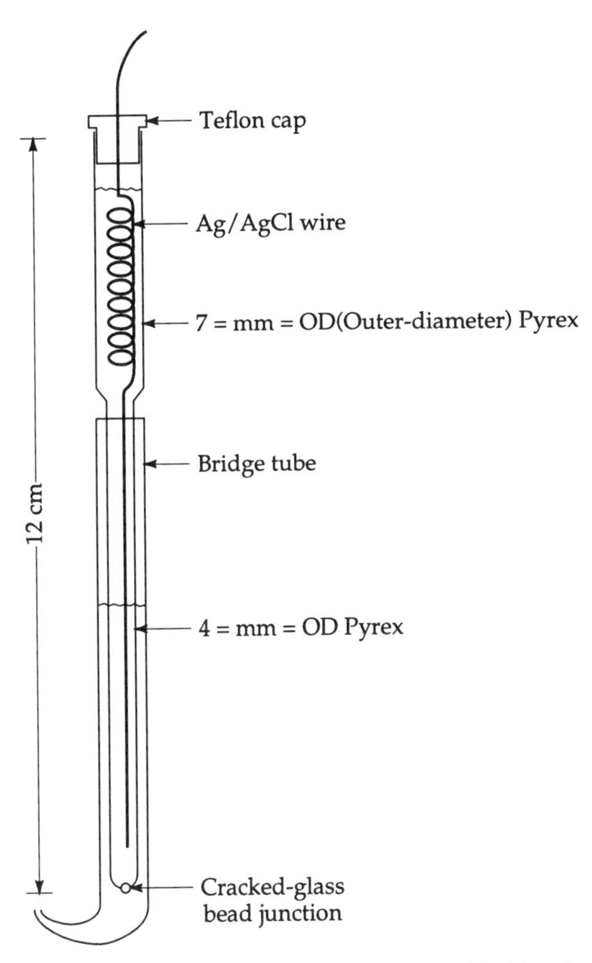

Figure 5.12 Design for a small, low-leakage silver–silver chloride reference electrode and salt bridge.

and the "classical" electrode is prepared by grinding calomel (Hg_2Cl_2), mercury, and a little potassium chloride solution together and placing the resultant slurry in a layer about 1 cm thick on the surface of the mercury contained in a clean test tube. External contact to the mercury is usually made with a platinum wire sealed into soft glass. Several forms of the calomel reference electrode are shown in Figure 5.13, and they are available from suppliers of pH and reference electrodes.

The preparation and properties of the calomel electrode have been reviewed;[30] superior reproducibility is claimed for a calomel electrode prepared by shaking dry, finely divided (0.1–0.5 μm) calomel with pure mercury. The calomel spreads over the mercury to form a "pearly" skin, and some of this skin is introduced on to a mercury surface where it spreads immediately. So-

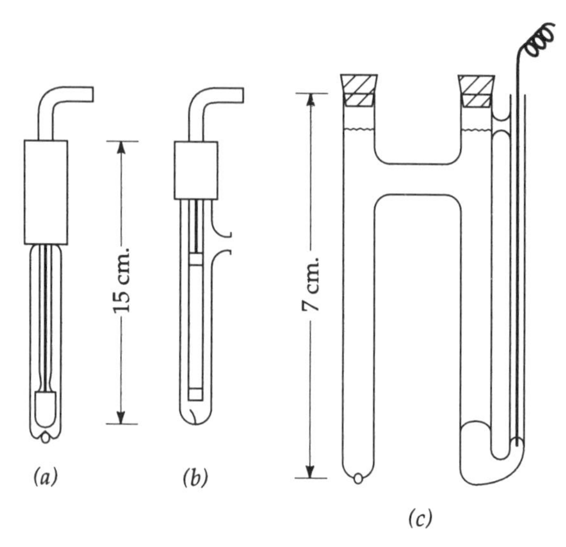

Figure 5.13 Calomel reference electrodes: (*a*, *b*) commercial designs with cracked-glass and fiber junctions, respectively; (*c*) home-made design with cracked-glass junction on the bottom of left tube.

lution is then added carefully to this ''skin'' electrode to give an electrode with substantial improvement in performance over that of paste electrodes.

To prevent aqueous solution from creeping down between the mercury and the glass wall of the cell, the cell walls should be treated with silicone preparations to render the glass hydrophobic. Because this adversely affects the platinum contacts sealed into the glass, a design should be used that allows the mercury to form its own connection by filling a long-bore capillary tube; a removable platinum wire contact can then be used at the remote end of this tube, as in Figure 5.13.

To determine the potential of the SCE on the hydrogen scale, the same problems arise as discussed for the silver–silver chloride reference electrode, and $E^{\circ\prime}$ and $(E^{\circ\prime} + E_j)$ are determined in the same manner. The recommended values[31] are tabulated for the calomel electrode in Table 5.3.

Although extensive data are available for the calomel electrode, the silver–silver chloride electrode appears to be superior to it because of its ease of preparation and use, lower sensitivity to the presence of oxygen, and small temperature hysteresis.

The mercury–mercuric oxide electrode. The mercury–mercuric oxide electrode is uniquely well behaved among metal–metal oxide electrodes.[32] The potential of the cell

$$(\mathrm{Pt})\mathrm{H_2}|\mathrm{NaOH(aq)}|\mathrm{HgO}|\mathrm{Hg} \qquad (5.32)$$

in agreement with theory, is constant within a mean deviation of ±0.06 mV for sodium hydroxide molalities between 0.001 and 0.3 mol kg^{-1}. The $E°$ of the cell is 0.926 V at 25°C. Several other related fixed potential electrodes have been described,[33] which in combination with a SCE (potential assumed to be 0.2446 V vs. the hydrogen electrode at 25°C) and a KCl salt bridge yield the following values:

$$Hg|HgO|Ba(OH)_2(\text{satd.}): \quad E = 0.1462 \text{ V vs. NHE at } 25°\text{C}$$

$$Hg|HgO|Ca(OH)_2(\text{satd.}): \quad E = 0.1923 \text{ V vs. NHE at } 25°\text{C}$$

This electrode, which is available commercially, is well suited for use in alkaline solutions.

The mercury–mercurous sulfate electrode. Several commercial suppliers offer the mercury–mercurous sulfate electrode with a saturated potassium sulfate electrolyte. The potential ($E°' + E_j$) of this electrode system is 0.658 V on the hydrogen scale at 22°C.[34] The electrode constitutes one-half of the Weston standard cell,[35] an international secondary voltage standard, and is outstanding in reproducibility,[36] in spite of the slight tendency of mercurous sulfate to hydrolyze and its rather high solubility.

Because special preparative procedures are not necessary (except for the preparation of mercurous sulfate electrolytically), this reference electrode is recommended for use in place of the silver chloride or calomel electrodes when chloride ion must be rigorously excluded.

Quasi-reference electrodes. Precise potentiometric measurements require reference electrodes that are highly reproducible, but there are many applications where this is less essential. For example, in routine analytical polarographic or voltammetric measurements, the accurate measurement of the current is more important than the precise measurement of potential. For these purposes a simple quasi-reference electrode is suitable. In a halide-containing supporting electrolyte, a silver wire or mercury pool will adopt a reasonably steady potential that is reproducible to within ±10–20 mV.

A platinum wire also will function satisfactorily as a quasi-reference electrode[37] in solutions of high resistance or in potentiostatic current measurements. Figure 5.14 illustrates a platinum wire for use in a high-resistance medium;[38] it minimizes the uncompensated resistance in the cell and replaces a bulky reference electrode and Luggin capillary. Wire quasi-reference electrodes also are used to measure fast electrode reactions by analysis of the system response to a current or potential step. Here the use of a standard reference electrode with its relative high resistance results in a long rise time (damped pulse form) as well as damped oscillations. In some experiments, both fast response for meaningful information to be gathered in the first few microseconds and a long-term stable reference potential are required. The use of a dual-reference-electrode system is recommended for this purpose[39] and is il-

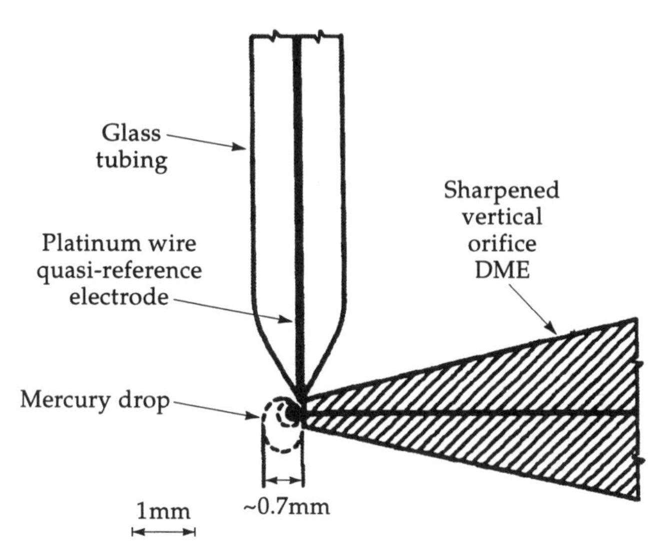

Figure 5.14 Platinum-wire quasi–reference electrode in combination with a dropping-mercury electrode.

lustrated by Figure 5.15. At relatively short times (or high frequencies) the potentiostat responds to the potential at the wire reference electrode, to allow fast-pulse conditions to be met. At longer times, including DC conditions, only the standard reference electrode, chosen for steady-state reference potential stability, is controlling. The values of R and C must be chosen to match the desired frequency response and input impedance of the potentiostat; the original paper should be consulted for details.

For potentiometric titrations, the potential changes at the endpoint are relatively large (100–200 mV) and a simple quasi–reference electrode may be quite adequate, even if it has an uncertainty or drift of a few millivolts. Also, for redox titrations that involve strong oxidizing agents with a platinum or noble-metal indicator electrode, a silver wire or mercury pool cannot be used because they are easily oxidized. A platinum quasi–reference electrode cannot be used because its potential will shift with that of the indicator electrode. For such a situation, the glass electrode will serve as a highly satifactory quasi–reference electrode if the hydronium ion concentration remains essentially constant. A pH meter may be used to measure the potential changes, with the platinum indicator electrode connected into the terminal for the reference electrode in pH measurements and the glass electrode in its normal terminal.

Temperature effects. The ordinary glass electrode–reference electrode pair that is used for pH measurements is not well suited to measurements far removed from room temperature. This is because the electrodes are immersed only partially, with the tips of the electrodes in the solution and the tops of the electrodes at ambient temperature. This creates a thermal gradient in the body

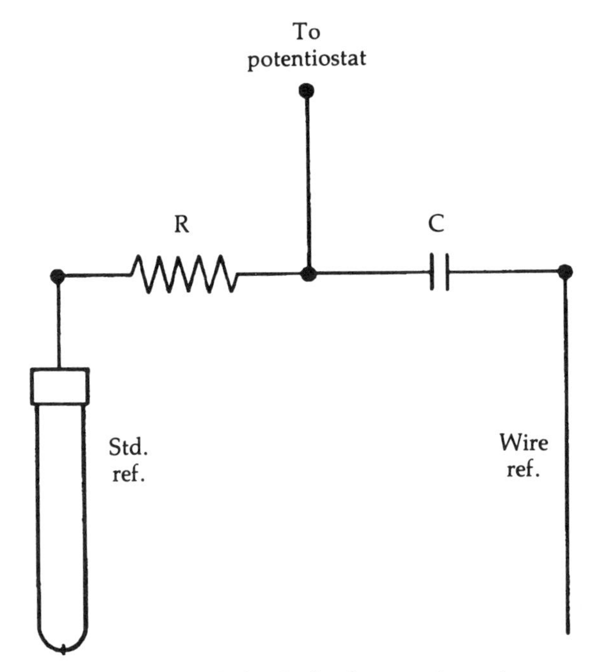

Figure 5.15 Circuit for dual-reference-electrode system.

of the electrode, which causes errors due to Seebeck (thermocouple) effects at junctions between dissimilar metals in the body of the electrodes. Whenever possible, the electrodes should be operated in an isothermal environment; thermal gradients should be limited to the external connecting leads, which should be made of the same metal (e.g., copper).

Other difficulties may arise if the two electrodes are located in different thermal environments with the temperature of one electrode varying while the other remains constant. Although the true temperature coefficients of the potentials of single electrodes cannot be obtained by thermodynamic means, an indication of the changes in electrode potential can be obtained by observing the change in cell potential when nonisothermal conditions are created deliberately. Although these values have no thermodynamic meaning, they indicate the magnitude of errors that result from unavoidable thermal gradients in a cell. In Table 5.3 the change of the reference-electrode potentials with temperature, $d(E^\circ + E_j)/dT$, are tabulated relative to the hydrogen electrode at 25°C. The size of the temperature coefficients indicates that temperature control that is better than ± 0.1°C is necessary for precise work.

Reference Electrodes for Use in Polar Aprotic Solvents. The increased use of polar aprotic solvents for electrochemical studies has inspired a search for suitable reference electrodes. Although the description of an aprotic solvent is somewhat ambiguous (see Chapter 6), we include in this class those solvents

TABLE 5.4 Reference Electrodes for Use in Dipolar Aprotic Solvents[a]

Reference Electrode	Acetonitrile	Propylene Carbonate	Dimethyl Formamide	Dimethyl Sulfoxide
H_3O^+/H_2	S	S	S	NR
Glass (pH)	S		S	S
Glass (cationic)	S	S	S	
$Ag(solv)^+/Ag$	S	S	S	S
AgCl/Ag	NR	NR	NR	NR
Hg_2Cl_2/Hg	NR	NR	NR	NR
$Li(OH_2)_4^+/Li$		S	S	S
$Li(OH_2)_4^+/Li(Hg)$		S		S
TlCl/Tl(Hg)				S
$Fe(Cp)_2^+/Fe(Cp)_2(Pt)$	S	S	S	S
I_3^-/I^-		S		
Aqueous SCE + salt bridge	S	S	S	S
Aqueous AgCl/Ag + salt bridge	S	S	S	S
$CdCl_2/Cd(Hg)$			S	S

[a]Abbreviations: Cp = cyclopentadienyl; S = satisfactory stability and reproducibility; NR = not recommended (unstable or not reproducible.)
Source: Ref. 27.

with dielectric constants greater than 30, which show weak acidic character and are difficult to oxidize or to reduce. Table 5.4 outlines reference electrode systems for use in such solvents.[27,40] Several monographs[41–44] contain discussions of nonaqueous reference electrodes. These address the problem of measuring electrode potentials in nonaqueous solvents and the comparison of potential scales in different solvents, and aspects of pH measurements and acidity functions.

Until recently, the most popular reference half-cell for potentiometric titrations, polarography, and even kinetic studies has been the saturated aqueous calomel electrode (SCE), connected by means of a nonaqueous salt bridge (e.g., Et_4NClO_4) to the electrolyte under study. The choice of this particular bridge electrolyte in conjunction with the SCE is not a good one because potassium perchlorate and potassium chloride have a limited solubility in many aprotic solvents. The junction is readily clogged, which leads to erratic junction potentials. For these practical reasons, a calomel or silver–silver chloride reference electrode with an aqueous lithium chloride or quaternary ammonium chloride fill solution is preferable if an aqueous electrode is used.

The hydrogen electrode. The hydrogen electrode has been used successfully in the formamides, anhydrous acetone, propylene carbonate, tetrahydrofuran, 1,2,3,3-tetramethylguanidine, and acetronitrile (although there are conflicting reports about the reproducibility of the hydrogen electrode in acetronitrile). In

Me_2SO solutions platinized platinum electrodes apparently decompose the solvent by catalytic reduction.

The calomel electrode. Calomel and other mercurous halides disproportionate in a number of organic solvents, and attempts to use the calomel electrode in polar aprotic solvents, have, for the most part, been unsuccessful. For this reason, it is not advisable to replace the aqueous electrolyte of an ordinary calomel reference electrode with an electrolyte dissolved in an aprotic solvent.

The silver–silver chloride electrode. The silver chloride reference electrode is not generally suitable as an electrode of the second kind because of the large solubility of AgCl in many aprotic solvents from formation of anionic complexes with chloride ion. In many cases the silver chloride solubility will essentially be that of the added chloride. This contributes significantly to the junction potential in cells with liquid junction and makes the electrode unsuitable for precise potentiometric work.

The silver–silver ion electrode. One of the most satisfactory and widely used electrodes is the silver–silver ion electrode, which appears to be reversible in all aprotic solvents except those that are oxidized by silver ion. The electrode is easily made by putting a silver wire (or silver-plated platinum) in a solution of 0.001–0.01 M $AgNO_3$ or $AgClO_4$. The polarizability of these electrodes indicates that if they are to be used in voltammetric work, it should be with a three-electrode circuit (see Figure 5.2*b*) so that the cell current does not pass through the reference electrode.

Although $AgNO_3$ is nearly equitransferent in acetronitrile,[45] junction potentials can be minimized further by use of a salt bridge that contains the same nonaqueous supporting electrolyte as the sample solution. By this approach, electrolytic connection is made with a compartment containing a silver electrode in the same electrolyte as the salt bridge, but with the addition of 0.001 to 0.01 M $AgNO_3$ (see Figures 5.16 and 5.17). This type of reference electrode requires more complex glassware than the conventional reference electrode does, and the supporting electrolyte must not contain impurities that react with $Ag(OH_2)_4^+$, but the extra effort yields lower junction potentials and greater reproducibility of potential measurements.

The difficulties that can arise with aqueous–nonaqueous liquid junctions are illustrated by the values reported for the cell

$$\text{Aqueous SCE}|\text{junction or salt bridge}|0.01\text{ M AgNO}_3|\text{Ag (acetonitrile)} \quad (5.33)$$

These range from 0.253 V for 0.1 M $NaClO_4$ in acetronitrile in the salt bridge to 0.300 V for a single junction with no salt bridge.[46]

Silver–silver ion electrodes have been employed to study liquid-junction potentials between electrolyte solutions in different solvents by use of the cell

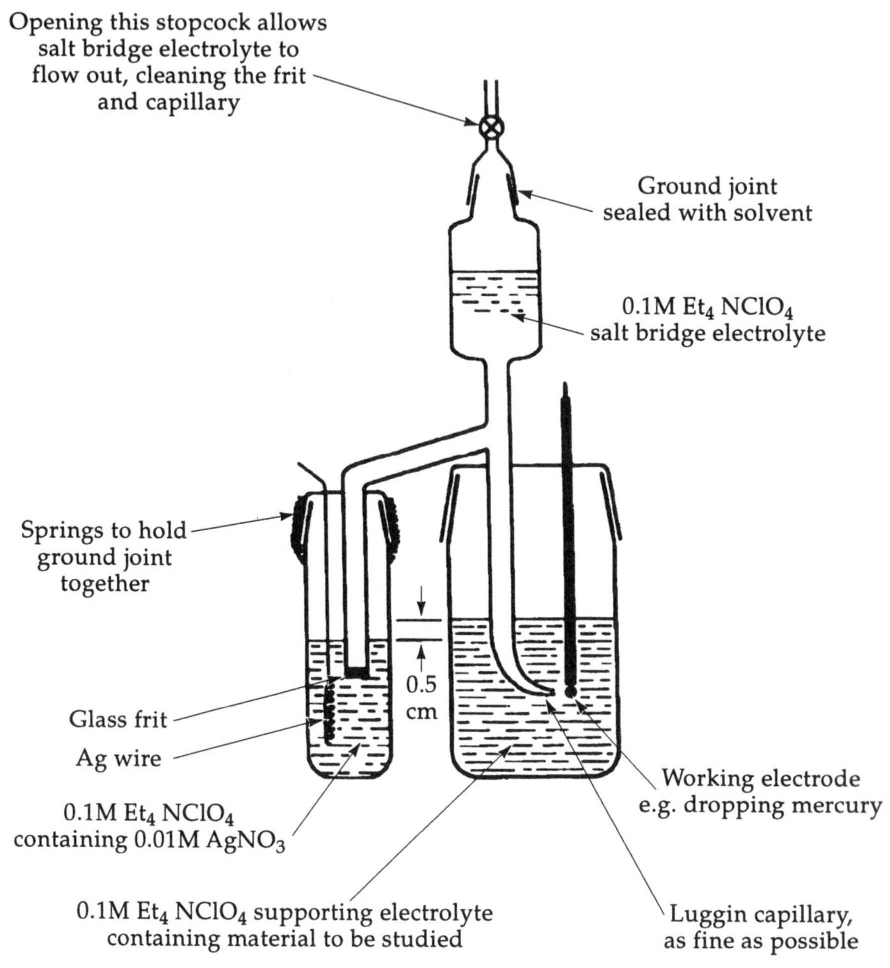

Figure 5.16 Silver–silver ion reference electrode and salt bridge for use with non-aqueous solvents. Design allows junction potentials to be minimized.

$$\mathrm{Ag|AgClO_4\,(0.01\,M)(acetronitrile,\,S1)|TEA\,Pic\,(0.1\,M)\,(bridge\,solvent,\,S3)|}$$
$$\mathrm{AgClO_4\,(0.01\,M)|Ag\,(solvent\,S2)} \tag{5.34}$$

in which TEA Pic is tetraethylammonium pictrate.[12] When S2 is dimethyl sulfoxide (Me_2SO), and the bridge solvent (S3) is varied over a representative group of solvents, the potentials are independent of the bridge solvent to within 5 mV. However, this is not true when the bridge solvent is formamide. The situation when the bridge solvent is water has not been tested because of the low solubility of TEA Pic in water. Furthermore, when S2 is formamide, water, or methanol, rather than Me_2SO, the potential of the cell varies by up to 100 mV as the bridge solvent S3 is changed.

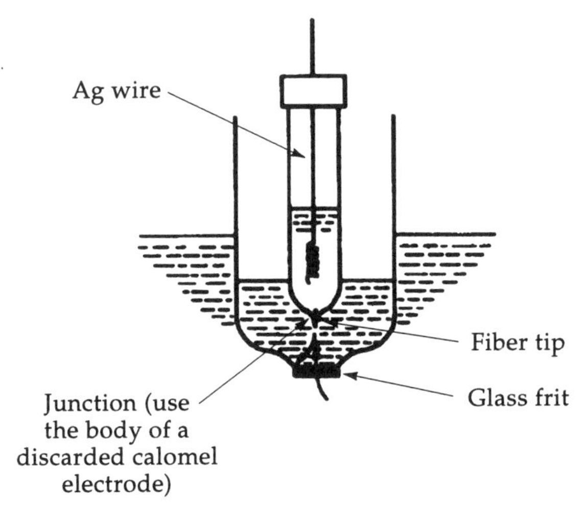

Figure 5.17 Design for simplified silver–silver ion reference electrode and salt bridge for the use with nonaqueous solvents.

There is a linear correlation between the observed cell potential and mutual heats of solution of the various solvents in the cell. Junction potentials of up to 100 mV have been attributed to the free-energy changes associated with the transport of solvent molecules across the junction; this is in addition to the junction potential due to the passage of ions across the junction. The liquid-junction potentials due to different solvents can be reduced by careful selection of bridge solvents and electrolytes. To minimize the liquid-junction potentials between two different solvents [as in the cell of Eq. (5.34)], the bridge solvent must not interact strongly (small mutual heats of solution) with either of the other solvents, and the transport numbers of the cation and anion of the bridge electrolyte should be equal. Either tetraethylammonium picrate or tetrabutylammonium tetraphenylborate are reasonable choices as bridge electrolytes, but the latter is preferred for voltammetric work because the picrate ion is reduced easily.

The ferrocene/ferricinium ion electrode. For nonaqueous electrochemistry, IUPAC recommends[47] the use of the ferrocene/ferricinium ion [$Fe^{II}(Cp)_2$/$Fe^{II}(Cp)_2^+$, HCp = cyclopentadiene] couple as an internal standard. The couple has been chosen because its potential is largely independent of the solvent ($E°$ = +0.40 V vs. NHE in water;[3] and +0.69, +0.72, +0.76, and +0.68 V vs. NHE in MeCN, DMF, py, and Me_2SO, respectively).[48] The ferricinium ion is unstable in some organic solvents because of decomposition.[49,50] Recently the use of bis(pentamethylcyclopentadienyl) iron(II) has been proposed to avoid the problem.[51] The $Fe^{II}(Cp)_2$/$Fe^{III}(Cp)_2^+$ couple cannot be used as an internal standard for some systems due to overlapping waves.[52] In these cases other compounds such as tris-(1,10-phenanthroline)iron(II),[4] cobaltocene,[3] or a variety of aromatic compounds[53] can be used to provide a virtual continuum of reduction potentials.

Other reference electrodes for use in polar aprotic solvents. Emphasis has been given to the use of the silver–silver ion reference electrode because it is almost universally applicable, and because standardization on the use of one reference electrode system simplifies the comparison of data between different workers. However, a number of other reference electrodes have been used (see Table 5.4), particularly those that have resulted from the vast amount of battery research. These include the $Li/Li(solv)^+$ and other alkali metal electrodes that function reversibly in Me_2SO, propylene carbonate, and hexamethylphosphoramide. The thallium–thallous halide electrodes of the second kind also function reversibly in Me_2SO and propylene carbonate. The cadmium amalgam–cadmium chloride reference electrode also functions reversibly in dimethylformamide and may be a useful substitute for the silver–silver ion reference electrode, which may be unstable in dimethyformamide.[54]

A number of redox couples at platinum electrodes (with both species soluble in the solvent) have been important to the establishment of a relation between potential scales in different solvents. Although they are useful in the comparison of redox potentials, they appear to have no advantages over the silver–silver ion system in the preparation of practical reference electrodes.

Stability of reference electrodes prepared in organic solvents. Reference electrodes in organic solvents ordinarily are not as stable with time as those with aqueous electrolytes. Most aprotic organic solvents undergo slow chemical change from reactions of impurities, hydrolysis of the solvent by traces of water, or reaction with a component of the half-cell reaction. The stability of reference electrodes should be checked frequently by comparison with fresh reference electrodes prepared with recently purified solvents and electrolytes.

Reference Electrodes for Use in Nonpolar Solvents. Solvents such as dichloromethane (not highly polar) present special problems. Their low dielectric constants promote extensive ion association, and cell resistances tend to be large. For this reason they are often used in mixtures with more polar solvents. Because dichloromethane and other nonpolar solvents are not miscible with water, use of an aqueous reference electrode with such solvents is not practical unless a salt bridge with some mutually miscible solvent is used. A better approach is to use a reference electrode of known reliability prepared in a solvent miscible with dichloromethane or to use the reference electrode based on the half-cell in dichloromethane.[88]

$$Ag|Ag_3I_4^-Bu_4N^+ \qquad (Bu = n\text{-butyl}) \tag{5.35}$$

Reference Electrodes For Use in Fused-Salt Systems. The kinds of investigations that require reliable reference electrodes in molten salts duplicate those in aqueous electrolytes. Specifically, these include (1) thermodynamic studies, such as the determination of the half-cell potentials of various electrodes relative to a particular reference electrode, or determination of thermodynamic data for molten-salt mixtures; (2) kinetic studies, such as the measurement of exchange currents (important in battery- and fuel-cell technology; (3) mass transport and energy output studies, such as determination of transference numbers or measurement of thermoelectric power in pure or mixed electrolytes; and (d) the study of electroactive chemical species in molten salts by use of voltammetric techniques (which may, of course, be used also for purposes of chemical analysis).

The literature on fused-salt systems is too vast to survey here, but good access to the literature is provided in two reviews.[55,56] Both contain useful information on solvent systems and suitable reference electrodes.

Alkali metal chloride and nitrate eutectic mixtures that melt in the range 200–500°C are widely used as fused-salt solvents, because they have a large liquid range from their melting to their boiling points, and because they can be used at temperatures below the softening point of glass.

The silver–silver ion electrode. Of the reversible metal electrodes, silver has been most often employed. There is only one stable oxidation state of silver; above 300°C there is no danger of oxide formation because Ag_2O is unstable.[57] The metal has no observable tendency to dissolve in molten silver salts and is highly reversible in mixed chloride and nitrate eutectics. The Ag(I) ion can be introduced into the melt by either adding silver nitrate to a nitrate melt (AgCl to a chloride melt) or by anodizing a silver electrode. The potentials of silver nitrate concentration cells show ideal thermodynamic behavior up to 0.5 mol % in $(Na,K)NO_3$ eutectic and in $NaNO_3$.[58]

Silver metal usually is employed in the form of wires or foils. In contact with a melt that contains Ag(I), silver metal continuously recrystallizes, so that a wire of small diameter may eventually be converted to a fragile string of loosely joined crystals. The rate of the process depends on temperature, and also appears to proceed more rapidly at a kinked or otherwise strained point in the wire.

The simplest reference electrode may be prepared by dipping a pure clean silver wire in a melt contained in a glass (or silica) tube. At higher temperatures, there is a possibility that glass will exchange sodium ions for silver ions; where this is a problem, silica is preferred over glass. The junction to the solution may be made by an asbestos fiber pinch-sealed into the bottom of the tube, by means of a porous ceramic plug sealed into the bottom of the tube, or through a thin glass bulb (see Figure 5.18).

Silver is rather difficult to seal into glass, and it is easer to seal a platinum wire into glass and then spot weld a sturdy silver wire to the platinum as shown in Figure 5.18. Corning No. 1720 glass is used to make the junction between the reference electrode compartment and the melt that contains the working electrode. This provides a barrier through which the transport of Ag(I) is insigificant, yet it has a resistance of less than 3000 Ω at 700°C.[59]

Although a number of other reference electrodes are capable of reproducible

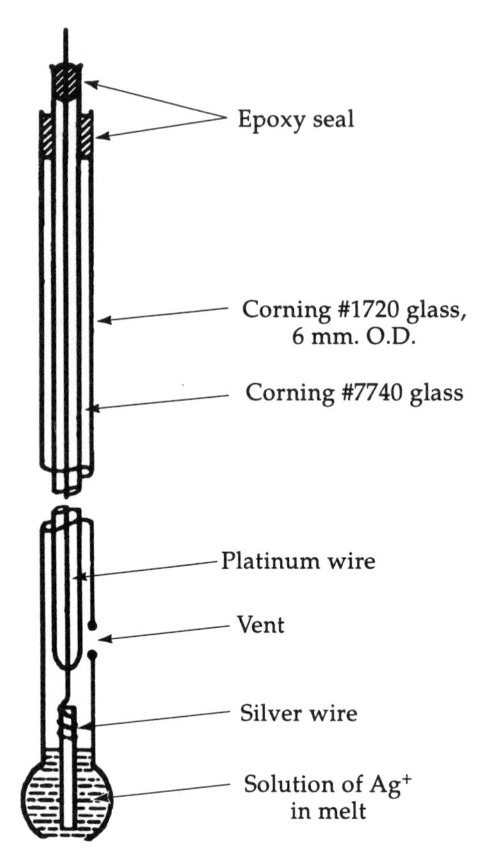

Figure 5.18 Design for silver–silver ion reference electrode for use in molten-salt systems.

behavior in molten salts, including the $Pt/PtCl_2$ electrode used in chloride melts,[60] none of these appear to be as widely applicable as the silver–silver ion electrode.

5.3 VOLTAMMETRIC INDICATOR ELECTRODES

Electrode Materials and Their Electrochemical Behavior. There is abundant evidence that the rate of electron transfer across an electrode–solution interface is dependent on the physical and chemical properties of the electrode material. The term *electrocatalysis* has been coined for this effect, and studies of the oxidation of hydrocarbons[61] and the reduction of water and hydronium ion[62] have provided ample evidence for its existence.

Table 5.5 indicates the range of exchange current densities that have been observed for the hydrogen evolution reaction on various metals. Note that the

TABLE 5.5 The Exchange Current Density j_0 for the Hydrogen Evolution Reaction in 1 M H_2SO_4

Metal	$-\log j_0$ (A/cm^2)
Palladium	3.0
Platinum	3.1
Rhodium	3.6
Iridium	3.7
Nickel	5.2
Gold	5.4
Tungsten	5.9
Niobium	6.8
Titanium	8.2
Cadmium	10.8
Manganese	10.9
Thallium	11.0
Lead	12.0
Mercury	12.3

value for mercury is about 10 orders of magnitude smaller than that for platinum. This difference is consistent with the fact that mercury is a much more useful electrode material for the study of cathodic processes than is platinum or other noble metals.

Electrode reactions are analogous to any heterogeneous chemical reaction where reaction takes place on a catalytic surface, with one important difference.[37] The heterogeneous chemical reaction does not involve a net charge transfer across the interface and is potential independent, while the electrode reaction involves a net charge transfer across the interface, and therefore the reaction rate is potential-dependent. In effect, the activation energy of the electrode reaction can be controlled by varying the potential.

The net current density, described by the Buther–Volmer equation [see Chapter 1, Eq. (1.61)]

$$j = j_0\left[\exp\frac{(1-\alpha)n_aF\eta}{RT} - \exp\frac{-\alpha n_aF\eta}{RT}\right] \tag{5.36}$$

is proportional to the moles of electrons transferred per square centimeter of electrode surface, and is a measure of the rate of the electrode reaction.

The voltage "window" or range of accessible potentials is limited on the negative side by reduction of the supporting electrolyte or solvent. In protic solvents, the limiting reaction will usually be hydrogen evolution. Therefore, for an electrode to be useful well into the negative potential region, it must have a low exchange current density and high overpotential for reduction of hydronium ions. On the positive side, the potential range will be limited by

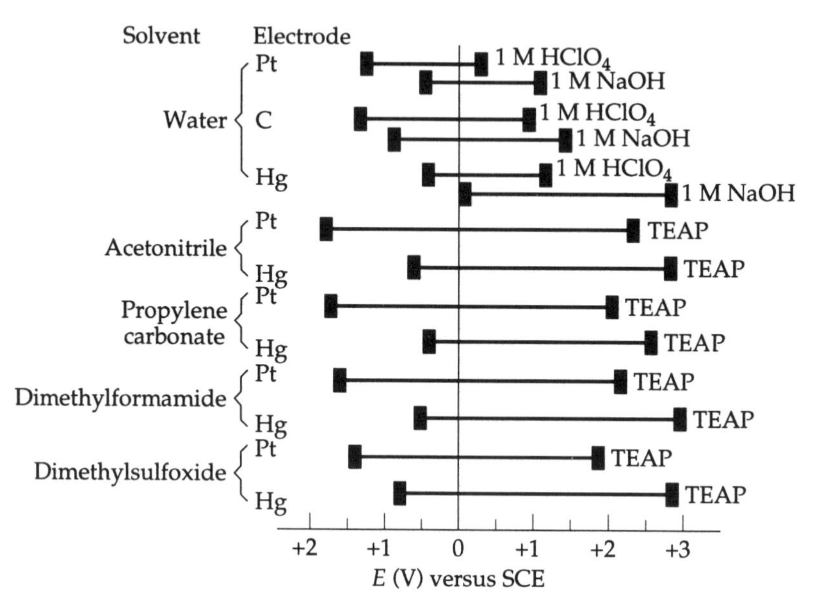

Figure 5.19 Voltage limits for various electrode materials in several solvents (TEAP = tetroethylammonium perchlorate).

oxidation of supporting electrolyte or solvent, by oxidation of the electrode material to form soluble metal ions or metal oxides, or by formation of molecular and chemisorbed oxygen in water or other oxygen containing solvents.

Figure 5.19 summarizes the positive and negative voltage limits for some commonly used electrode materials in several solvents. Wherever possible, the data for a particular solvent has been referred to a single reference electrode. Absolute values of the electrode potential for different solvent systems cannot be directly compared, however, because they are often referred to different reference electrodes and because of the uncertainty in our knowledge of junction potentials between different solvent systems.

The most commonly used electrode materials are mercury; various forms of carbon, gold, platinum, and other noble metals; carbide, borides; and conducting tin oxide films on glass.

Mercury (general properties). Because mercury is liquid down to −39°C, it can be used in dropping, streaming, or pool configurations that are impossible with solid electrodes. It also can be coated or plated in thin layers on other metals, which then show properties similar to mercury electrodes. The dropping or streaming mercury electrodes have the special advantage of providing a continuously renewed surface, which helps to minimize the effects from adsorption of solution impurities or from fouling of the electrode surface by films produced in the electrode reaction. In addition, the surface is smooth and continuous and does not require the pretreatment and polishing that is common with solid electrodes.

The advantages of the liquid surface and large overpotential for hydrogen evolution make mercury the material of choice for cathodic processes, unless the use of mercury is specifically contraindicated by some incompatibility with the system. Incompatibility can arise from strong specific absorption, as with some sulfur-containing compounds, or in high-temperature systems such as fused salts because of the low boiling point of mercury (356.6°C).

Cathodic limits on mercury. In aqueous or other protic solvents the reduction of hydronium ion or solvent generally will limit the negative potential range. The nature of some electrode reactions at highly negative potentials on mercury has been examined.[63] For example, $K(OH_2)_4^+$ and $Na(OH_2)_4^+$ ions are reduced reversibly in aqueous solutions, but the process is accompanied by a parallel irreversible reaction due to an amalgam dissolution reaction of the alkali metal with water that produces hydrogen.

In dipolar aprotic solvents, the availability of hydronium ions is much lower and consequently the cathodic limit is extended. Reversible or nearly reversible waves can be readily observed for the reduction of Group I and some Group II metal ions.[64,65]

In fused-salt systems, reduction of traces of water (if present) or of the metal cations ordinarily will be the cathodic limiting process.

Anodic limits on mercury. Mercury is readily oxidized, particularly in the presence of anions that precipitate or complex mercury(I) or mercury(II) ions, such as the halides, cyanide, thiosulfate, hydroxide, or thiocyanate. For this reason, mercury is seldom used to study anodic processes except for those subtances that are easily oxidized, for example, Cr(II), Cu(I), and Fe(II). Under carefully controlled conditions, mercury can be coated with a thin layer of mercury(I) chloride such that it does not interfere with electron transfer in the oxidation of a number of organic compounds, particularly amines.[66]

Platinum, gold, and other noble metals (Pd, Rh, Ir). Platinum and gold are the most commonly used metallic solid electrodes. These metals are readily obtained in high purity, are easy to machine, and can be fabricated readily into a variety of geometric configurations—wires, rods, flat sheets, and woven gauzes. They are resistant to oxidation, but are not totally inert as often assumed in early work.

Negative potential limit. Both platinum and palladium have extremely small overpotentials for hydrogen evolution, which is the basis for their use in the construction of reversible hydrogen electrodes. Gold has a significantly larger overpotential, but it is much smaller than for mercury. Platinum also adsorbs hydrogen readily and the amount of adsorbed hydrogen can be used to estimate the true surface area of the platinum. Palladium dissolves hydrogen readily in the bulk metal. Although this can be an asset in the construction of a palladium–hydrogen reference electrode, it makes palladium unsuited for use in most voltammetric work.

Gold does not appreciably adsorb hydrogen and this factor together with its larger overpotential for hydrogen evolution makes gold the metal of choice for the study of cathodic processes. Curiously, platinum is used more often than are all other metals combined, probably because of tradition and the fact that gold is much more difficult to seal into glass than is platinum.

Positive potential limit. At sufficiently positive potentials, all the noble metals form an oxide film or layer of oxygen in aqueous solution with a fairly well-defined stoichiometry that can be used to estimate true surface areas. The exact nature of this oxygen layer has been the subject of much investigation and some controversy, but a consensus seems to be emerging[67] that the oxygen film consists of chemisorbed oxygen with nucleation and growth of an oxide phase under severe anodic conditions. The voltammetric behavior that is observed on platinum, palladium, rhodium, and gold in 1 M sulfuric acid is shown in Figure 5.20. The characteristic hydrogen adsorption–desorption peaks that appear on the first three metals are almost absent on gold. (Palladium was not taken into the hydrogen evolution region because of its capacity to dissolve hydrogen, which gives it almost unlimited capacity at ordinary voltammetric current levels.)

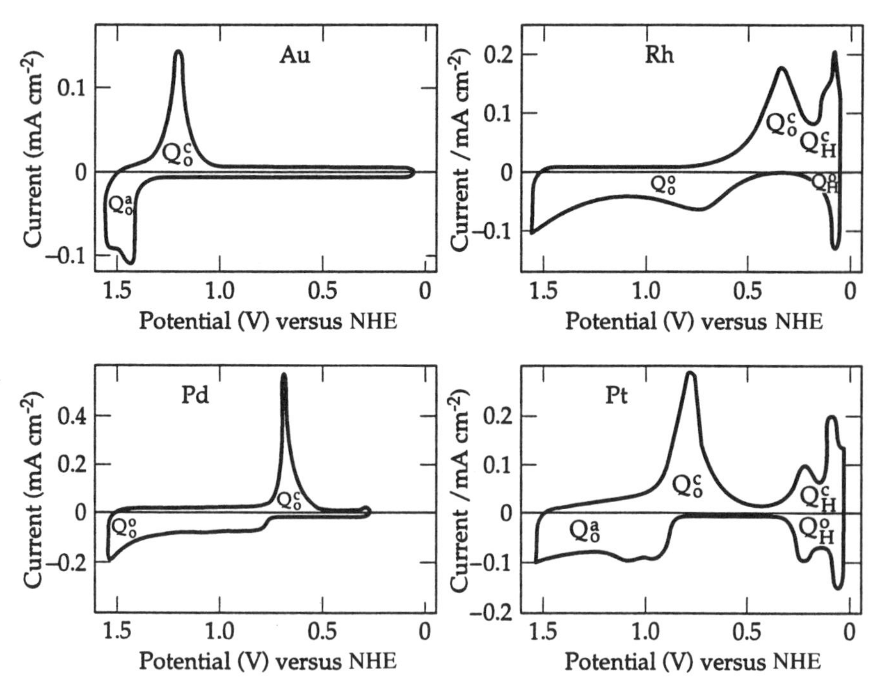

Figure 5.20 Cyclic voltammograms for platinum, palladium, rhodium, and gold electrodes in 1 M H_2SO_4: Q^c_o, cathodic current due to oxide reduction; Q^a_o, anodic current due to oxide formation; Q^c_H, cathodic current due to H_2 formation; Q^a_H, anodic current due to H_2 oxidation.

The potential cycling illustrated by Figure 5.20 is a commonly used pretreatment procedure for attainment of a reproducible "active" surface. Less widely known is the fact that in aqueous solution this cycling procedure causes the dissolution of appreciable quantities of metal. The discrepancy between the integrated anodic and cathodic oxygen adsorption–desorption peaks has been shown to be due to dissolution of the metal. Typical values are given in Ref. 68 and indicate that platinum and gold dissolve to a much lesser extent than do palladium and rhodium.

Gold is more readily oxidized than platinum in the presence of complexing anions such as the halides or cyanide. The effect of chloride can be seen in Figure 5.21.[69] The enhancement of the peaks probably is due to the process

$$Au(s) + 4\,Cl^- \longrightarrow AuCl_4^- + 3\,e^- \tag{5.37}$$

For this reason, gold electrodes should not be used at highly positive potentials in solutions that contain halides or other complexing ions.

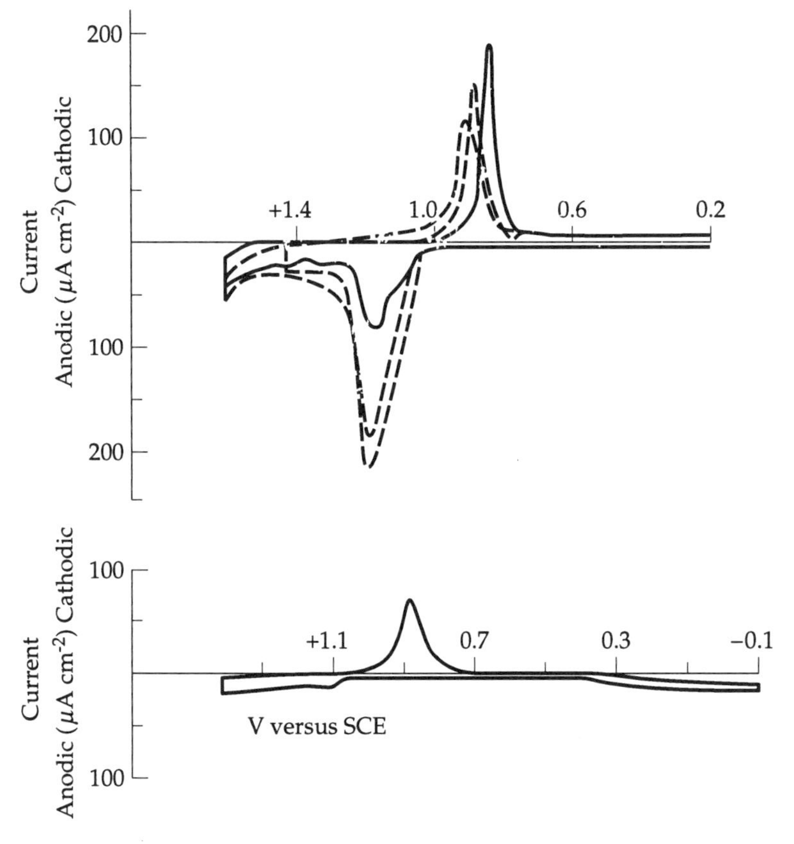

Figure 5.21 Cyclic voltammograms for a gold electrode: upper curve for successive scans in the presence of chloride ion; lower curve for sulfate electrolyte.

In polar aprotic solvents that are largely free of water, the formation of a chemisorbed oxygen layer is less pronounced and polished platinum shows a positive limit that is larger than any of the other commonly used electrode materials. Gold is almost as good, but it cannot be used with many complexing anions. On the negative side, gold is superior to other metals (except mercury) because of its larger overpotential for hydrogen evolution.

Carbon (general properties). Several different forms of carbon have been used to make satisfactory electrodes, including spectroscopic-grade graphite (usually impregnated with ceresin or paraffin wax), pyrolytic graphite (a high-density, highly oriented form of graphite), carbon paste (spectroscopic-grade graphite mulled in sufficient Nujol, bromonaphthalene, or bromobenzene to form a stiff paste), graphite dispersed in epoxy resin or silicone rubber, and vitreous or "glassy" carbon.

Vitreous or glassy carbon. The use of glassy carbon as an electrode material was first suggested in 1962.[70-72] Glassy carbon is an electrically conductive material, highly resistant to chemical attack and gas-impermeable, and obtainable in a pure state. It has many properties in common with pyrolytic graphite, but does not need to be oriented as does pyrolytic graphite. Several groups[73-76] have described uses of glassy carbon. Some of its cited advantages relative to platinum are (1) low cost, (2) pretreatment by polishing with metallographic paper, (3) larger overpotential for production of hydrogen and dissolved oxygen, and (4) increased reversibility for several redox couples and reactions that involve subsequent proton transfer. Relative to platinum the disadvantages are (1) a high residual current in 1 M sulfuric acid and (2) surface roughening as a result of recrystallization at high current densities.

Glassy carbon has a unique advantage in the determination of trace metals by stripping voltammetry. In one procedure[76] the trace metals are deposited with mercury after addition of mercuric nitrate to the supporting electrolyte. As a result, the trace metals are plated out into an extremely thin layer of mercury, and in the subsequent stripping step a single well-resolved peak is obtained for each metal. After a determination, the surface can be renewed by wiping with a cellulose tissue to remove all the mercury.

Wax-impregnated graphite. Spectroscopic-grade graphite is porous, and penetration of the electrode by solution or by oxygen makes it unsuitable for use in voltammetry. Impregnation of the graphite under vacuum by molten ceresin wax or by paraffin gives much more satisfactory reproducibility. The surface is easily renewed by light sanding with fine sandpaper. The reproducibility in aqueous solution is improved by dipping the electrode in a surface-active agent (0.001% Triton X-100), which increases the wettability of the electrode surface by the solution.[77]

Pyrolytic graphite. Pyrolytic graphite is produced by pyrolysis of hydrocarbons under reduced pressure to give a deposit of highly ordered carbon crystallites on a substrate maintained at 1000–2500°C. The graphite so formed is anisotropic with the planes of the hexagonal graphite rings parallel to the surface of the substrate. Pyrolytic graphite is highly impervious to liquids and gases, inert to chemical attack, and free of entrapped gases and metallic contaminants. Although relatively thin samples of pyrolytic graphite can be prepared, most of the data that have been reported were obtained with commercially available pyrolytic graphite in a thick coherent form prepared at about 2000°C. In fabricating an electrode, care must be taken to ensure that only the planes parallel to the *ab* crystallographic axes are exposed. The edges of the planes are more vulnerable to penetration by solution, such as graphite behaves in the formation of intercalation compounds by diffusion of atoms between the planes of the graphite. The surface of pyrolytic graphite is renewed by light sanding or by cleavage. The fabrication and use of a pyrolytic graphite electrode for voltammetry in aqueous solution have been discussed;[78] the behavior of a home-made pyrolytic carbon film electrode has been compared with that of wax-impregnated graphite.[79] The reproducibility is as good and the carbon-film electrode exhibits nearly reversible behavior for the hexacyanoferrate(III/II) couple.

At the low current densities (0.03–0.3 $\mu A\ cm^{-2}$) that are used in anodic stripping voltammetry, the reproducibility of pyrolytic graphite electrodes is improved by impregnation with ceresin wax.[80] At higher current densities (30 $\mu A\ cm^{-2}$), wax impregnation is not necessary. Experience indicates that the wax-impregnated pyrolytic graphite electrode gives superior reproducibility relative to wax-impregnated spectroscopic graphite.

Because the residual current largely is a function of the differential capacitance of the electrode in solution, information about the latter is of practical significance. A differential capacitance study of the basal plane of stress-annealed pyrolytic graphite indicates a surprisingly low minimum value of 3 $\mu F\ cm^{-2}$, which is attributed to a space-charge component within the graphite.[81] In contrast, for the edge orientation of pyrolytic graphite, the minimum capacitance value is 50 $\mu F\ cm^{-2}$; the value for glassy carbon is 13 $\mu F\ cm^{-2}$. The capacitance associated with the compact double layer should have a value of 15–20 $\mu F\ cm^{-2}$ in the absence of surface states on the basis of the values obtained with metal electrodes.

Dispersed-graphite electrodes. A number of carbon electrodes have been prepared by dispersing graphite in ceresin wax,[82] epoxy polyers,[83] silicone rubber,[84] or Nujol.[85] The Nujol dispersion forms a thick paste and has been widely used as the carbon-paste electrode (CPE).[86] It is easy to prepare, shows fair reproducibility ($\pm 5\%$), and has a low residual current in the anodic region with a good anodic potential range. In this respect it is superior to platinum and gold because it shows none of the troublesome oxide-film formation. In the negative region it shows a persistent residual current that has been attributed to the presence of oxygen dissolved in the paste or adsorbed on the surface of the graphite particles. The surface is renewed by removing about 3 mm of the paste and replacing it with fresh paste.

The CPE prepared in the usual way tends to disintegrate in nonaqueous solvents. The addition of sodium lauryl sulfate to the paste prevents wetting of the graphite by acetronitrile, nitromethane, or propylene carbonate; this makes the electrode more suitable for use in nonaqueous solvents.[87]

A dispersion of graphite in ceresin wax has been suggested as a way to prepare an electrode suitable for use in nonaqueous solvents. The hot paste is tamped into a Teflon tube and allowed to solidify. No pretreatment is necessary, and the surface can be renewed by wiping with cellulose tissue.

The use of a compressed acetylene-black electrode for coulometric analysis of adsorbed organic compounds has been proposed.[88] A 1.2-cm-diameter glass tube that is closed with a frit is mounted in a rubber stopper and inserted into a vacuum flask. Next, about 0.1 g of acetylene black is added and tamped firmly; then a test solution containing about 0.5 mg of sample is added and drawn through the acetylene black at a controlled flow rate such that the compound to be determined is adsorbed quantitatively. A graphite collector electrode, which carries the current to the compressed acetylene black, is then introduced and held in position with a weight. A small amount of electrolyte is added above the collector, and several drops are drawn through the acetylene black. The electrode then is transferred to an electrolysis cell and a constant-current electrolysis is performed (typically with a 2-mA current); the potential between the graphite electrode and a calomel reference electrode is recorded. The potential–time curve usually shows a substantial break that indicates the coulometric endpoint. The utility of this technique appears to be greatest for semimicroanalysis of electroactive organic compounds.

The carbon cloth electrode. When heated to high temperatures, certain woven hydrocarbon polymer fabrics can be converted to carbon or graphite cloth that appears to have a voltage range in various solvents similar to that of other carbon electrodes. The use of this material for anodic and cathodic work in aqueous and nonaqueous media has been reviewed.[89] The electrochemical area of such electrodes is about three times their geometric area. The material has good mechanical properties and withstands current densities of 100 mA cm^{-2}. At current densities tenfold higher, the resistance of the cloth, normally just a few ohms per square centimeter, leads to uneven current distribution and poor potential control. This material appears to have its greatest utility in preparative-scale electrosynthesis; it is not particularly suited to routine electroanalytical analysis. The cloth is sufficiently inexpensive that one can simply cut out a piece of the size needed and discard if after the electrolysis.

Metal and semiconductor materials (borides, carbides, nitrides, and silicides). Tin oxide-coated glass has been used as an electrode material in electrochemical spectroscopy.[90] By doping of the tin oxide with antimony, an *n*-type semiconductor is formed. The surface is chemically inert and is transparent in the visible region of the spectrum. However, it is more useful for its optical transparency than as an electrode material.

The current–voltage curves of semiconductor-type electrodes deviate from the behavior of metals, mainly because of the internal resistance and nonuniform internal potential distribution of the electrode. Unlike solution resistance, it is not possible to compensate electronically for this deviation by positive feedback.

A comparative study of a number of metal and semiconductor materials as voltammetric electrodes has been made.[91] All the electrodes were examined in a similar manner and their characteristics recorded in terms of (1) available positive potential limit versus SCE; (2) extent and reproducibility of residual currents, and (3) the reproducibility of E_{peak} and i_{peak} using the hexacyanoferrate(II) ion as a model compound for the study of one-electron reversible oxidation. The results for a number of electrodes are summarized in Table 5.6.

The results obtained indicate that with the possible exception of chromium and tungsten carbide, none of the materials studied is comparable to vitreous carbon. There appears to be no reason why chromium cannot be used in place of platinum for measurements in the near-positive potential range; the voltammograms that are obtained are well defined and reproducible. Despite having

TABLE 5.6 Anodic Potential Range of Electrode and Peak Potential for Oxidation of $Fe^{II}(CN)_6^{4-}$

	Anodic Potential Range in Aqueous Solution (V vs. SCE)[a]			Peak Potential for Oxidation of 10^{-3} M $Fe^{II}(CN)_6^{4-}$	
Electrode Material	pH 1.0	pH 4.2	pH 10.0	pH 1.0	pH 4.2
Pt	1.1	1.1	0.9	0.35	0.23
C (vitreous)	1.3	1.4	0.95	0.33	0.28
C (pyrolytic graphite)	1.3	1.3	0.9	0.35	0.27
Pt + 10% Rh	1.4	1.1	0.6		0.26
Pd	1.2	1.0	0.15		0.47
Os	0.55	0.55	0.4	0.40	0.27
Ir	0.6	0.5–0.8	0.55		0.25
Re	0.6	0.4	0.1		0.37
Zr	1.5	1.3		NW	NW
Mo	0.2	−0.2	−0.1	NW	NW
Nb	NW[b]	NW	NW	NW	NW
Ta	NW	1.4	NW	NW	NW
Cr	0.9	0.7	0.35	0.34	0.27
Ti	1.6–1.9			0.45	NW
B_4C_3	1.3	0.9	0.8	0.35	0.27
TaC	0.5	0.4	0.1	0.45	
WC	0.95	0.8	0.7	0.47	

[a]Anodic limit is defined as the potential at which the current becomes equal to one-half of the value of the peak current for the oxidation of 10^{-3} M $Fe^{II}(CN)_6^{4-}$.
[b]NW = poorly defined voltammogram.
Source: Ref. 91.

the largest positive potential limit, titanium is of little use as an electrode material; voltammograms are irreproducible and seriously influenced by the previous history of the electrode. Zirconium, tantalum, niobium, and molybdenum exhibit classic features of passivation; oxidation waves for Fe(II) are not obtained and the residual-current curves have significant slopes and are erratic.

Lead has been much used as a cathode material in organic electrosynthesis;[92] it has a high hydrogen overvoltage and is easy to work mechanically. However, it does not appear to be as useful as vitreous carbon or platinum for general voltammetric work.

Measurement of Electrode Area. Because of surface roughness, the real or true surface area of a solid electrode is greater than the projected or geometric area. However, if the electrode is polished to a smooth surface finish, this will be of no consequence in most voltammetric work. The depth of the depleted region around the electrode surface (the diffusion-layer thickness) is substantially larger than the characteristic dimensions of surface roughness for electrolysis times that are greater than 1 s. [The diffusion-layer thickness may be crudely approximated by the term $(Dt)^{1/2}$, where D is the diffusion coefficient ($cm^2\ s^{-1}$) and t is the time.]

However, in an adsorption measurement (where surface coverage is limited to a monolayer) or where the surface is rough (e.g., a platinized platinum electrode), an approximation to the true surface area may be needed. For a platinum electrode this can be obtained by measuring the amount of charge required to deposit a monolayer of adsorbed hydrogen atoms. Integration of the current–time curve for the hydrogen adsorption peaks (H_{C1} and H_{C2} of Figure 5.22) gives the integral shown as Q_H in Figure 5.23. The integration is carried out from about +0.4 to +0.05 V versus a hydrogen electrode in 0.5 M H_2SO_4. The total number of coulombs is corrected for charging of the double layer. The double-layer charging current is assumed to be approximately constant and equal to the current measured at the minimum (about +0.5 V vs. NHE in Figure 5.22) of the double-layer charging region. A monolayer of hydrogen corresponds to about 120 $\mu C\ cm^{-2}$ of real surface area (after correction for double-layer charging). The area also can be estimated by use of a constant-current method, or by measurement of the coulombs required to oxidize a monolayer of adsorbed hydrogen (H_{A1}, H_{A2}, H_{A3} of Figure 5.22), or by measurement of the coulombs required to deposit a monolayer of oxygen (O_{A1}, O_{A2}, O_{A3} of Figure 5.22). The accuracy of these surface area measurements is such that the error limits probably are $\pm$10–20%. The relative merits of the methods have been discussed.[21,93]

The area of a polished electrode (taken to be the projected or geometric area in most voltammetric experiments at times >1 s) usually is measured directly or electrochemically. If the electrode is of regular geometry, such as a disk, sphere, or wire of uniform diameter, its characteristic dimensions can be measured by use of a micrometer, optical comparator, or traveling microscope and the area calculated.

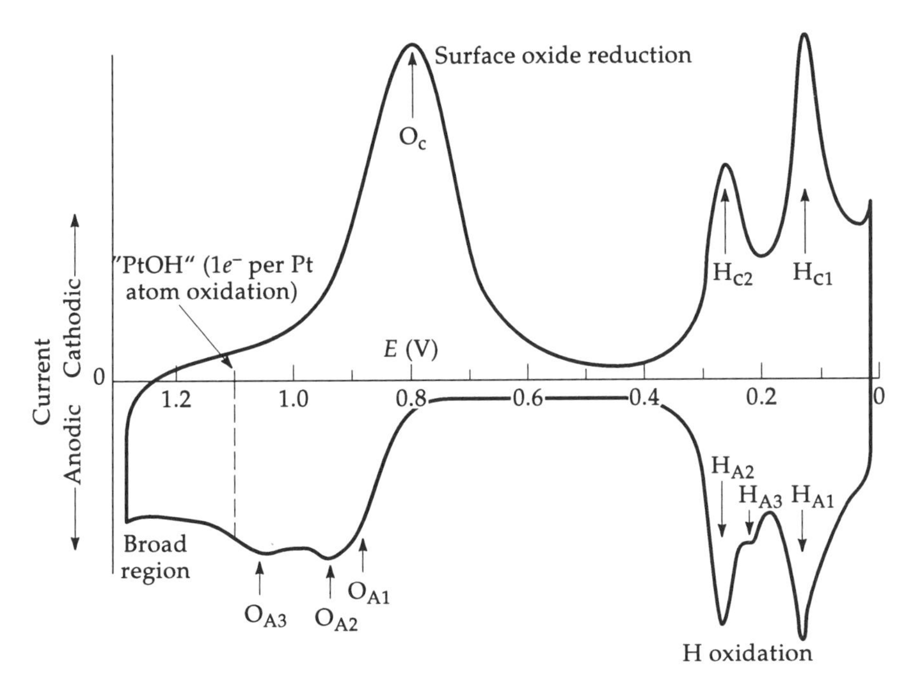

Figure 5.22 Cyclic voltammogram for a platinum electrode. Regions of oxide formation and reduction, as well as formation of H_2 (and atomic hydrogen) and its oxidation, are indicated.

In practice two methods are used for stationary planar electrodes in quiescent solution: chronoamperometry and chronopotentiometry. By use of an electroactive species whose concentration, diffusion coefficient, and n value are known, the electrode area can be calculated from the experimental data. In chronoamperometry, the potential is stepped from a value where no reaction takes place to a value that ensures that the concentration of reactant species will be maintained at essentially zero concentration at the electrode surface. Under conditions of linear diffusion to a planar electrode the current is given by the Cottrell equation [Chapter 3, Eq. (3.6)]:

$$i = nFAC^{\mathrm{b}}\left(\frac{D}{\pi t}\right)^{1/2} \tag{5.38}$$

The product $it^{1/2}$ should remain essentially constant at a shielded planar electrode for electrolysis periods from 1 to 30 s or more. By use of a servorecorder accurate data usually can be obtained for times between 10 and 30 s.

In the chronopotentiometric method the transition time is measured for constant current and the electrode area is calculated from the relation [Chapter 4, Eq. (4.17)]:

$$i\tau^{1/2} = \frac{\pi^{1/2}nFAD^{1/2}C^{\mathrm{b}}}{2} \tag{5.39}$$

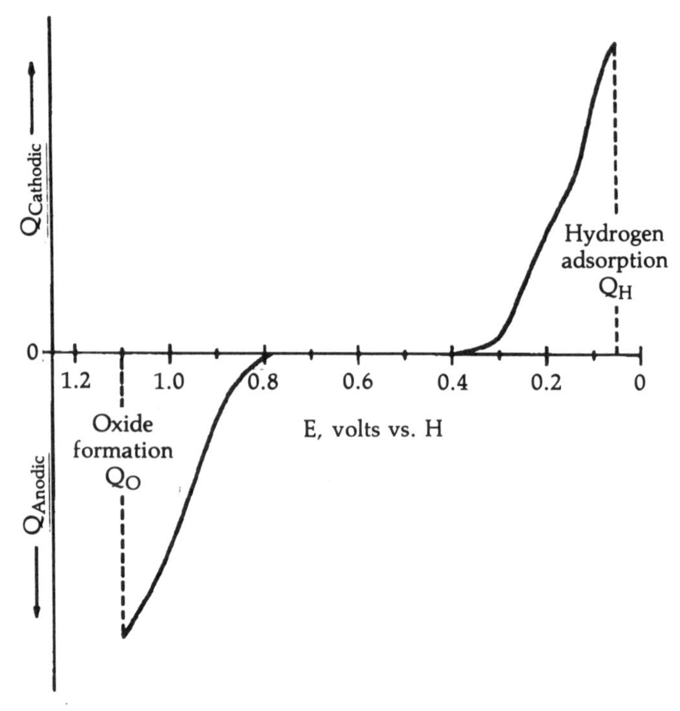

Figure 5.23 Current–time integral as a function of potential for a platinum electrode in 0.5 M H_2SO_4.

This equation is valid for only semiinfinite linear diffusion to a shielded planar electrode; the effects of electrode geometry (upward or downward diffusion) and electrode shielding on the transition time have been discussed.[94]

The step-potential method has an advantage over the constant-current method because the double-layer charging takes place at the beginning of the measurement and afterward is negligible. Potential changes in the constant-current method cause the flow of double-layer charging current to change during the entire experiment, particularly at the transition time.[94]

Both of these methods may be applied to nonplanar electrodes if the results are obtained at electrolysis times sufficiently short that the diffusion layer remains thin in comparison to the radius of curvature of the nonplanar electrode surface. For example, the spherical hanging-mercurcy-drop electrode provides chronoamperometric data that deviate less than 1–2% from the linear-diffusion Cottrell equation out to times of about 1 s. With solid wire electrodes of cylindrical geometry, similar conclusions apply, but at short times surface roughness effects yields a real surface area that is larger than the geometric area.

Table 5.7 gives values for diffusion coefficients that can be used for determination of electrode areas.[95] These values apply only to the specified temperature and supporting electrolyte composition. A critical evaluation of the

TABLE 5.7 Selected Diffusion Coefficients at 25°C in Water and Acetonitrile

Substance	Concentration (mM)	Solvent[a] Medium	$D \times 10^5$ ($cm^2\ s^{-1}$)
$Ag(OH_2)_4^+$	2.5–4.0	0.1 M KNO_3	1.55 ± 0.02
$Fe^{III}(CN)_6^{3-}$	4.0	0.1 M KCl	0.762 ± 0.01
$Fe^{III}(CN)_6^{3-}$	4.0	1.0 M KCl	0.763 ± 0.01
$Fe^{II}(CN)_6^{4-}$	4.0	0.1 M KCl	0.650 ± 0.02
$Fe^{II}(CN)_6^{4-}$	4.0	1.0 M KCl	0.632 ± 0.02
$Cd(OH_2)_6^{2+}$	2.5	0.1 M KNO_3	0.69 ± 0.02
$Cd(OH_2)_6^{2+}$	2.5	1.0 M KNO_3	0.68 ± 0.02
$Cd(OH_2)_6^{2+}$	2.5	0.1 M KCl	0.716 ± 0.01
$Cd(OH_2)_6^{2+}$		0.1 M KCl	0.70 ± 0.013[b]
$Cd(OH_2)_6^{2+}$	2.5	1.0 M KCl	0.80 ± 0.02
Cd metal		Mercury	1.6[c]
Aniline	1–2	0.1 M TEAP[d] in acetonitrile	2.60 ± 0.15
Naphthalene	1–2	0.1 M TEAP in acetonitrile	2.68 ± 0.1
Biphenyl	1–2	0.1 M TEAP in acetonitrile	2.33 ± 0.2
9,10-Dihydro -9,10-dimethylphenazine	1–2	0.1 M TEAP in acetonitrile	2.00 ± 0.05
Perylene	1–2	0.1 M TEAP in acetonitrile	2.59 ± 0.15
9,10-Diphenylanthracene	1–2	0.1 M TEAP in acetonitrile	1.97 ± 0.1

Source: Ref. 95 unless otherwise indicated.
[a]Solvent is water unless otherwise indicated.
[b]Weighted "best value"; Macero and Rulfs, *J. Electroanal. Chem.*, **1964,** *7*, 328.
[c]Stevens and Shain, *Anal. Chem.*, **1966,** *38*, 865.
[d]TEAP = tetraethylammonium perchlorate; average *D* calculated from data given by Bacon and Adams, *Anal. Chem.*, **1970,** *42*, 524.

measurement of electrode surface areas by the chronoamperometric method has been presented.[96]

Electrode Pretreatment. There is ample evidence that the rate of electron transfer at a solid electrode is sensitive to the surface state and previous history of the electrode. An electrode surface that is not clean usually will manifest itself in a voltage-sweep experiment to give a decrease in the peak current and a shift in the peak potential. Various pretreatment methods have been employed to clean or "activate" the surface of electrodes; the process is intended to produce an enhancement of the reversibility of the reaction (i.e., produce a greater rate of electron transfer).[97] This activation or cleaning process may function in two ways: by removing adsorbed materials that inhibit electron transfer and by altering the microstructure of the electrode surface.

The pretreatment process ordinarily begins by polishing the electrode surface until it is smooth and bright. Techniques similar to those used in the preparation of metallographic specimens can be used; the electrode is polished with successively finer grades of alumina, silicon carbide, or diamond dust. Simple polishing is often adequate for work in dipolar aprotic solvents, whose adsorption may be less of a problem than in aqueous solutions. After thorough rinsing and drying with cellulose tissue, the electrode is ready for use. The polishing ordinarily is repeated daily or before every experiment.

More stringent cleaning is used in aqueous solutions. Adams recommends a chemical pretreatment with "cleaning solution" (chromic–sulfuric acid mixture) to oxidize the surface heavily, followed by thorough rinsing.[98] Others recommend treatment with hot nitric acid (platinum and gold) or aqua regia (platinum only) to accomplish the same thing—the oxidation of the electrode surface to a reproducible state. After rinsing, the electrode is placed in the test solution and held at a potential that will reduce the oxide layer. This generally produces a clean and reproducible surface.

Sometimes the same result can be attained by heavily anodizing (e.g., 1 s at 10–100 mA cm^{-2}) the electrode surface to remove adsorbed material. The oxide layer then is reduced at a potential sufficient to remove the oxide layer but not reduce hydronium ion or solvent. The pretreatment cycle is repeated before every experiment, usually in the test solution. For voltammetric work in aqueous solution, chemical pretreatment and electrochemical cycling is preferred to simple polishing or in addition to polishing.

The current–potential behavior shown in Figure 5.22 is claimed to be characteristic of a clean platinum surface in a clean test solution,[99] and can be used as a criterion of solution and electrode cleanliness in aqueous 0.5 M H_2SO_4. The presence of organic material generally will cause a decrease in the hydrogen adsorption peaks and the appearance of new peaks.

Less is known about electron-transfer kinetics on carbon electrodes, but one study indicates that electron transfer in the Fe(II)/Fe(III) system is slower on carbon electrodes than on platinum; the order of rate constants is k_s(Pt) > k_s (glassy carbon) > k_s (carbon paste).[100] The maximum rates are observed after treatment with chromic acid, except for the reduction of iodate ion where simple polishing gives the highest rate. The potentiometric response of some carbon electrodes after various pretreatments has been studied.[101] In addition, the surfaces were examined by infrared spectroscopy, scanning electron microscopy, and direct titration of surface groups. The results indicate that the glassy carbon electrode has a negligible concentration of surface acid–base groups and comes closest to an ideal inert redox electrode.

Construction and Mass-Transport Properties of Voltammetric Electrodes

Mercury electrodes. Mercury is a widely used electrode material for the study of cathodic processes because of its high overpotential for hydronium ion reduction. Its clean liquid surface eliminates the problems that are associated with solid electrodes (which must be polished or cleaned by various chemical pretreatments).

The presence of noble metals in mercury will lead to a reduced overpotential for the evolution of hydrogen and should be reduced to trace levels (<0.1 ppm). This is usually accomplished by vacuum distillation or by distillation in the presence of a stream of air. Base metals should be removed by chemical treatment before distillation, and the still that is employed should be reserved for purifying mercury for electrochemical use.

Pure mercury should retain a bright mirror surface indefinitely in contact with dry air and should not leave a ring if stored in a glass vessel. In fact, the appearance of the mercury is a more sensitive test for base metals than is emission spectroscopy. Their presence at a concentration of 1 ppm is indicated by either the formation of a film or "tailing" of the mercury when it is shaken in a glass flask or rolled around in a clean porcelain dish. A simple test for purity consists of placing a few millimeters of mercury in a clean stoppered vial or flask along with about three times the volume of pure distilled water. When shaken vigorously, pure mercurcy will form a fairly stable foam that disappears gradually within 5–15 s. In contrast, mercury that contains a substantial amount of base metals does not foam at all, and mercury that contains mere traces of base metals will form a foam that is stable for 1 or 2 s and then collapses suddenly. For instance, an amalgam of 1 mg of copper per 1 kg of mercury will fail to give stable foam when shaken with distilled water. The water that is used for the test must not contain any organic material, and the container for the test must be scrupulously clean.

Although the preceding methods are satisfactory for purifying small quantities of mercury, electrolytic methods are also effective and convenient, and provide a purity equal to the purest reagent-grade mercury.[145]

Mercury is toxic, and every effort should be made to contain spills by placing trays under the apparatus. If a spill occurs, it should be immediately cleaned up with a suction apparatus or a "mercury magnet," which is a spiral of copper wire that has been amalgamated. Dry sweeping should be avoided because it disperses the mercury. It is almost inevitable that tiny droplets of mercury will become scattered about in a polarographic laboratory, and these are almost impossible to clean up. The best insurance against mercury poisoning is a good ventilation system that changes the air several times per hour. Work with mercury should never be carried out in a closed, unventilated room.

Mercury most often is used in the form of a dropping electrode (Figure 5.24), a hanging-mercury drop (Figure 5.25), or a mercury pool (Figure 5.26). It also may be plated on platinum to give an electrode with properties intermediate between those of platinum and mercury.

The dropping-mercury electrode. The usefulness of the dropping-mercury electrode (DME) for analytical voltammetry was discovered by Heyrovsky, and the history of this discovery has been recounted.[102] The DME usually is prepared from a 10–20-cm length of glass-capillary tubing with an approxi-

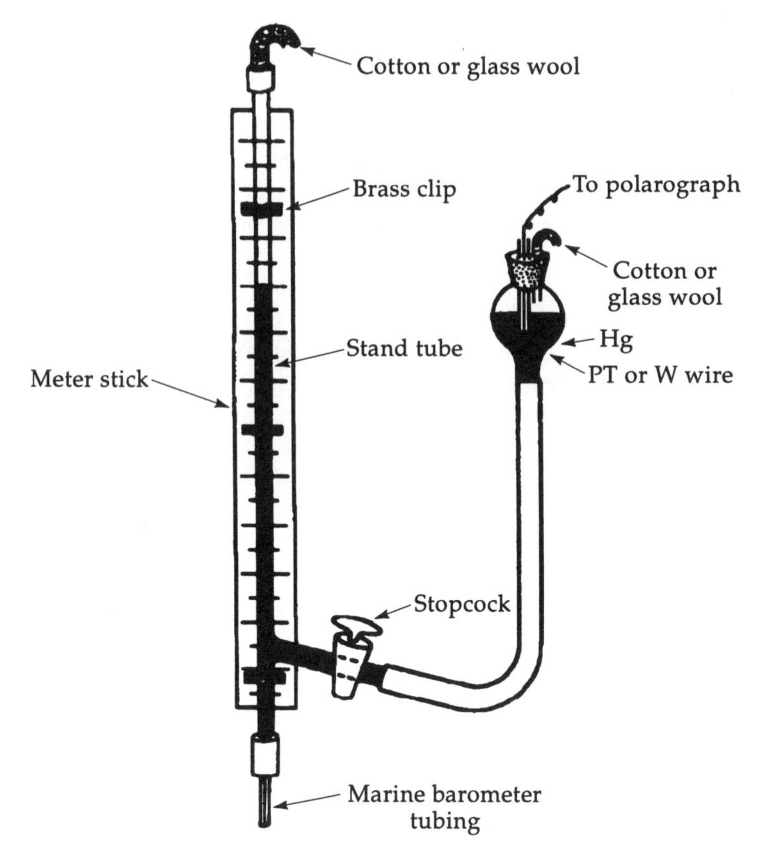

Figure 5.24 Dropping-mercury-electrode assembly.

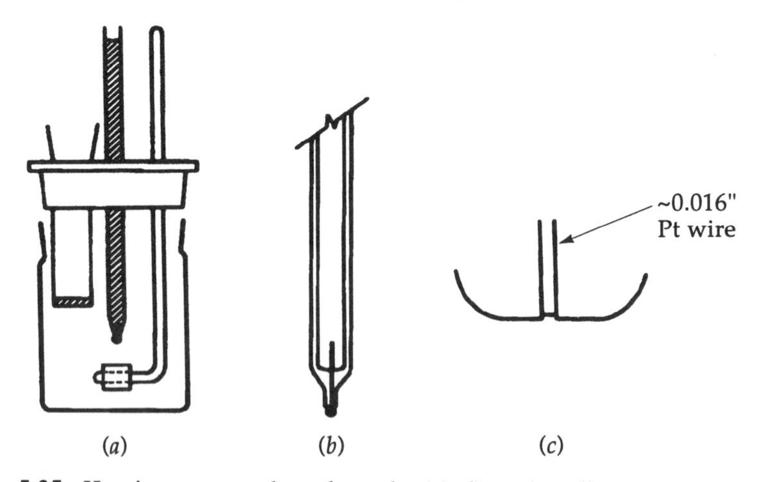

Figure 5.25 Hanging-mercury-drop electrode: (*a*) electrode–cell assembly with scoop to transfer mercury drops from DME to the electrode; (*b*, *c*) details of construction for amalgamated platinum electrode to which mercury drops are attached.

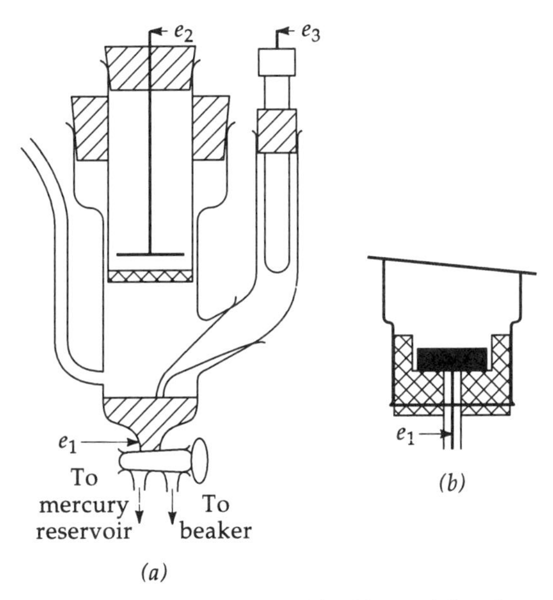

Figure 5.26 Mercury-pool electrodes: (*a*) cell with provision for pool replacement; (*b*) plastic cup for pool (e_1, working electrode; e_2, auxiliary electrode; e_3, reference electrode).

mately 0.05-mm internal diameter. (Corning marine barometer tubing frequently has the requisite diameter.) The length of the capillary is selected so that a convenient head of mercury above the capillary will provide a drop time of 3–10 s; such capillaries can be purchased. Figure 5.24 illustrates a convenient arrangement for the DME. The meter stick adjacent to the stand tube is used to measure the height of the mercury column. The connections to the reservoir and capillary tubing can be made with neoprene rubber or Tygon tubing. They usually are wired on for safety.

In cutting the capillary to length, care must be taken to ensure that the end of the capillary is perpendicular to the capillary and that there are no rough edges or cracks around the orifice. This can be verified by inspection under a low-power microscope.

To prevent clogging of the capillary by the test solution, a flow of mercury should always be started before the capillary is immersed in the solution. At the end of the experiment, the DME is removed and thoroughly rinsed with solvent and distilled water while mercury is still flowing. When the water has evaporated, the mercury reservoir is lowered until the head of mercury above the capillary is below the sidearm. The capillary can be stored dry or in a test tube filled with mercury.

A number of variations of the basic DME have been devised. The vertical-orific capillary, formed by bending a capillary into a right angle and cutting off the capillary near the bend, is claimed to eliminate problems of current maxima and depletion effects (growth of a drop in a solution depleted of electroactive material by the preceding drop), and to provide greater uniformity from drop to drop.[103] The fabrication of a Teflon DME for use in solutions of HF has been described.[104] A number of mechanical devices have been developed that regulate the drop time by dislodging the drop at precisely controlled intervals (usually by a sharp rap on the capillary).[105,106] A regulated drop time often is used in experiments to synchronize current measurements with drop growth.

The hanging-mercury-drop electrode. A number of techniques require a stationary electrode of fixed size—for example, chronocoulometry, anodic stripping voltammetry, and voltammetry with positive feedback iR compensation. This need is fulfilled, while still retaining the desirable properties of a mercury electrode, by the hanging-mercury-drop electrode (HMDE) shown in Figure 5.25. The electrode can be fabricated by sealing a small-diameter platinum wire in soft glass and etching the platinum wire back from the end in aqua regia. The wire then is amalgamated and one or more mercury drops are attached to its end by means of a scoop (Figure 5.25*a*). The latter is used to catch drops from a DME and transfer them to the HMDE.[107] Kemula[108] uses a different style of hanging drop in which the drop hangs from a glass capillary tube connected to a reservoir containing mercury. A threaded screw advancing into the reservoir displaces mercury to form a drop on the end of the capillary.

The mercury-pool electrode. Mercury pools of sufficient diameter to approach a planar configuration obey the equations derived for linear diffusion to a planar electrode. This has certain theoretical advantages because of the large number of equations that have been derived for the planar electrode geometry, especially in terms of constant-current chronopotentiometry and linear-potential sweep chronoamperometry.

Compared to the DME, the mercury pool has an enhanced analytical sensitivity; the ratio of the voltammetric peak current to the residual (charging) current is approximately 10 times as great for a mercury pool.[109,110] In an effort to eliminate the rounding at the edge of the mercury pool that is created by the meniscus, three types of cells have been devised that use platinum rings to flatten the mercury pool.[111] Although these are successful in providing a flatter surface, the exposure of platinum to the solution may decrease the overpotential for hydrogen evolution. Figure 5.26 illustrates two mercury-pool electrode systems.

Mercury-plated electrodes. Platinum can be plated with mercury to form an electrode with many of the properties of a mercury surface. This is done by cleaning a platinum electrode in nitric acid, connecting it as the cathode in perchloric acid solution over a pool of mercury, and then dipping it into the mercury while it is still connected in the circuit. The method works well for plating a planar platinum inlay electrode to give a flat mercury-coated surface, but it can be used for electrodes of any shape.[112]

There is much evidence that mercury forms intermetallic compounds with platinum at the mercury–platinum interface. Apparently these do not diffuse to the surface of the mercury so that it retains many of the desired properties of mercury. Hoewver, there is a slight reduction in the overpotential for hydrogen evolution at a plated electrode in comparison to a pure mercury electrode.[113]

Solid electrodes in unstirred solution. The most useful solid electrodes are platinum, gold, vitreous carbon, and carbon paste. The preferred configuration for theoretical work is a flat planar surface sealed in glass with epoxy or snugly fitted into a Teflon shroud. The electrodes ordinarily are unshielded, although shielded designs have been described.[94]

Platinum electrodes of smaller diameter may be made by sealing platinum wire into soft glass (Corning Code 0120 or 0088). The technique for doing this is shown by the sequence of drawings in Figure 5.27. A thin capillary or soft glass is first fused onto the platinum wire (avoid overheating, which forms bubbles in the glass or melts the platinum); the glass bead that is formed on the wire is then sealed into a neck-down soft-glass tube. A planar voltammetric electrode may be made by cutting off the wire and polishing the platinum surface flush with the glass tube. Or a small sphere may be melted on the end of the wire to give a nearly spherical electrode. Glass that contains lead must be worked in an oxidizing flame to avoid reducing the lead.[114]

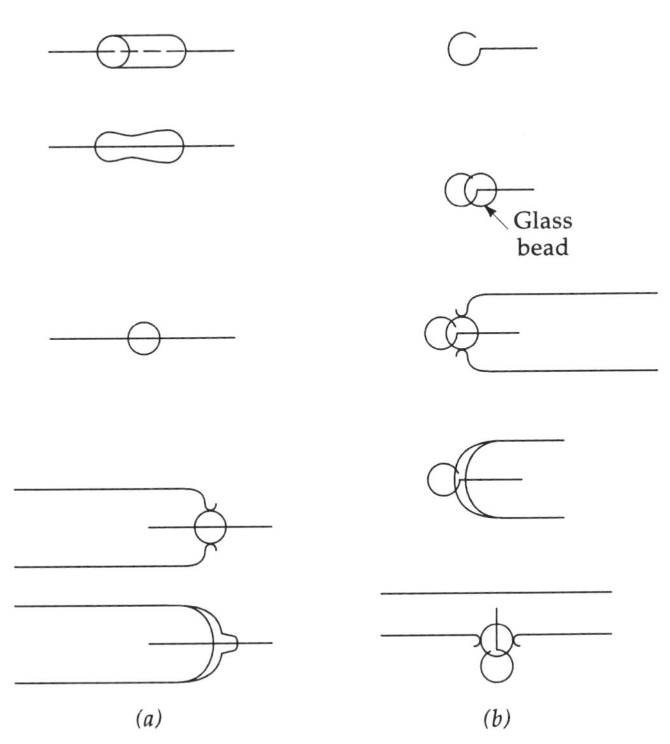

Figure 5.27 Techniques for sealing platinum in soft-glass tubing: (*a*) cylindrical wire electrode; (*b*) wire-loop electrode.

Most other metals cannot be sealed into glass as easily as platinum because there is not a sufficiently good match between the coefficients of expansion of the glass and the metal (Table 5.8). For such metals, an electrode of planar configuration can be made by forcing a cylindrical billet of the metal into a slightly undersized hole that has been drilled in a Teflon rod. Shrinkable Teflon tubing also is available, which contracts about the metal wire or billet when it is heated with a hot-air dryer. This usually produces a tight fitting seal that is impervious to aqueous solutions.

When the electrode material is too soft or fragile to force into an undersized hole in Teflon, the holder shown in Figure 5.28 can be used. The electrode is inserted into the Teflon collet first. When tightly threaded onto the trunk, the Kel-F chuck squeezes the slip-fit Teflon collet to make a tight seal.[115]

Thin-layer electrodes, which most commonly are made of platinum, require care and precision in fabrication (as shown in Figure 5.29). The spacing between the electrode and the precision capillary must be reproducible and is maintained by the precisely machined ridges or steps on the platinum rod. The cell is filled and emptied by controlling the pressure above the electrode with the upper Teflon stopcock. Several cycles of filling and emptying the cell are used before each experiment. The theoretical equations that apply to mass transport in a thin-layer cell have been reviewed.[116–118] The primary advantage of the thin-layer cell is that it confines the reactant within 10^{-3} cm of the electrode surface so that each reactant particle has immediate access to the electrode surface. This simplifies the mass transport in the cell to the extent that the equations that govern thin-layer electrodes are simple combinations of Faraday's law and the Nernst equation for a reversible system. Surface active

TABLE 5.8 Physical Properties of Selected Metals and Glasses

	Lattice Structure[a]	Density, g/cm^3 (20°C)	Melting or Softening Point (°C)	Linear Coefficient of Expansion (0–100°C) $\times 10^{-7}$
Ruthenium	HCP	12.45	2310	91
Osmium	HCP	22.61	3050	61
Rhodium	FCC	12.41	1960	83
Iridium	FCC	22.65	2443	68
Palladium	FCC	12.02	1552	111
Platinum	FCC	21.45	1769	91
Silver	FCC	10.5	961	188
Gold	FCC	19.3	1063	143
Pyrex (Corning No. 7740)		2.23	820	32.5[b]
Soft glass, soda-lime glass (Corning 0088 flint glass)		2.47	700	92[b]
Vycor (Corning 7900)		2.18	1500	8[b]

[a]HCP = hexagonal close-packed; FCC = face-centered cubic (cubic close-packed).
[b]0–300°C

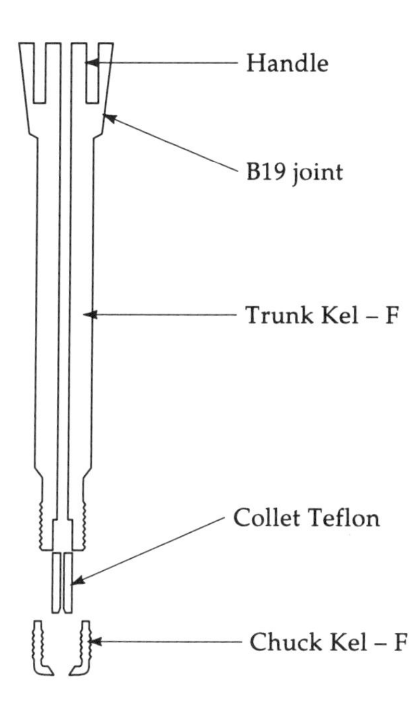

Figure 5.28 Holder for soft or fragile electrode materials.

impurities are virtually excluded from the electrode surface unless they are present at high concentrations. Because the solution volume is so small, a low concentration of surface-active material, even if it collects at the electrode surface, covers only a small fraction of the surface. For these reasons thin-layer cells promise to combine electrochemical techniques with electron and photon spectroscopy. The latter is necessary for studies of the relationship between the electrochemical behavior and the atomic and electronic structure of the electrode surface.[118]

A simple thin-layer cell that is made by threading a platinum wire through a Teflon capillary has been described,[119] as has a micrometer type thin-layer cell for estimating n values in electrochemical reactions (Figure 3.15).[120]

Carbon electrodes. The vitreous carbon electrode usually is fabricated by sealing a plug or disk, but from a thick sheet of glassy carbon, into a glass tube with epoxy cement. The surface of the electrode is then polished until it is bright and smooth. No further treatment is necessary.

The carbon-paste electrode is made by mulling together 15 g of graphite and 9 mL of Nujol until the entire mixture appears uniformly wetted and has the consistency of a stiff paste. The paste is packed into a Teflon cup like that shown in Figure 3.14*b* and smoothed with a spatula. A more elaborate device is shown in Figure 5.30.[121] The carbon paste is packed into the cylindrical cavity; a new surface can be produced by advancing the threaded rod and slicing off the old surface with a thin wire stretched across a bow (pictured in Figure

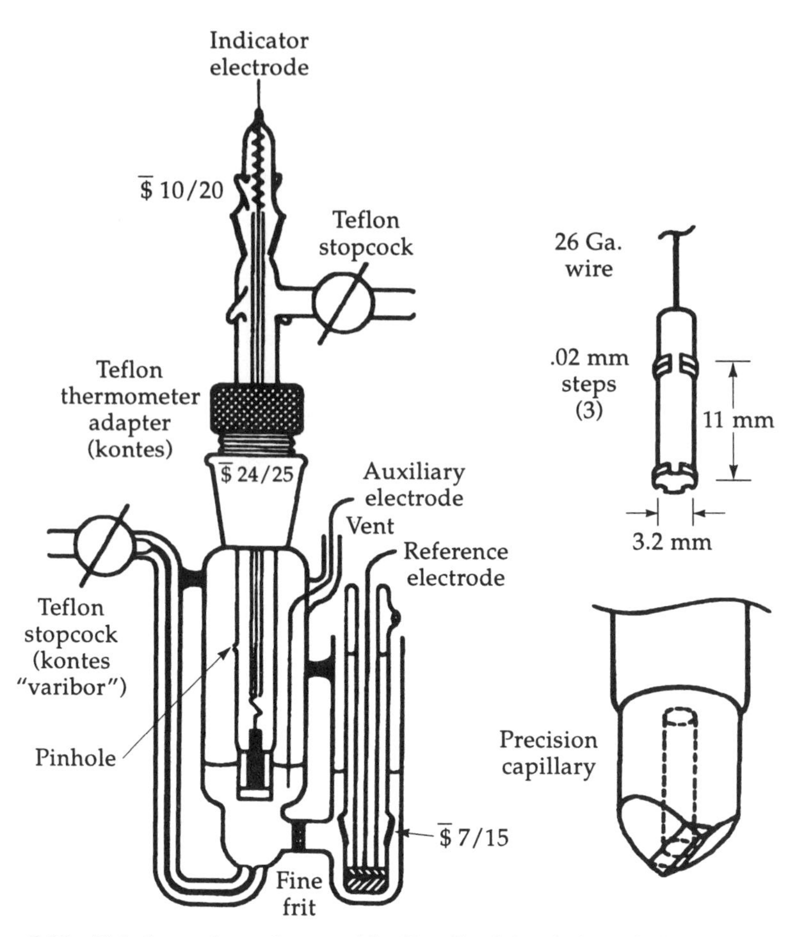

Figure 5.29 Thin-layer electrode assembly. Details of the platinum indicator electrode shown in upper right and of the precision ground-glass capillary in lower right.

5.30*b*). This saves the labor of digging out and repacking the carbon paste when a fresh electrode surface is desired. The thin-wire cutter also appears to produce a more even and reproducible surface.

Hydrodynamic and stirred-solution electrodes. Certain advantages result when the electrode is moved past the solution or vice versa. The increased mass transport increases the current and often increases the sensitivity (although not necessarily the signal-to-noise ratio). In addition, hydrodynamic electrodes such as the rotated platinum electrode and rotated-disk electrode exhibit a current–potential behavior similar to that of the DME. That is, they give the familiar plateau when the current is limited by mass transport to the electrode surface and the current is proportional to the solution concentration of the electroactive species.

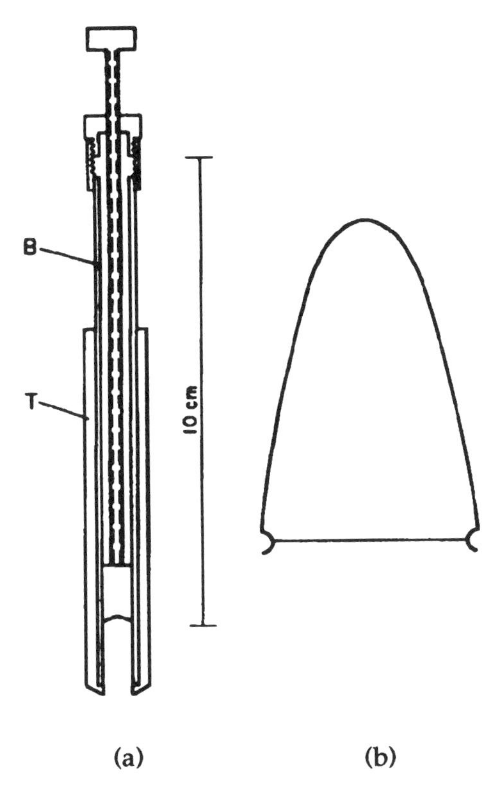

Figure 5.30 Carbon-paste electrode with threaded screw to advance paste: (*a*) electrode assembly with machined cylinder channel (T) and threaded piston (B); (*b*) bow with stretched wire for slicing off old surface after the electrode plug is advanced.

The vibrating mercury electrode. A DME that is connected by a shaft to an eccentric that is driven at about 200 Hz produces a drop time of the order of 5 ms.[122] A number of advantages are claimed over the conventional dropping-mercury electrode: (1) the suppression of maxima without the addition of surfactants; (2) the elimination or minimization of kinetic and catalytic currents, and (3) the elimination of stirring effects on the shape of current–voltage curves. The principal drawback to the electrode is that the surface-area increase during a drop life is about 10 times that of the DME for its normal drop time. Therefore the charging current is increased by tenfold to produce a steeply increasing residual-current baseline.

The rotated platinum electrode. Three designs of rotated platinum electrodes are shown in Figure 5.31. The first design (Figure 5.31*a*) produces the smoothest current–voltage curves because the flow is not as turbulent around the electrode surface as for the electrode of Figure 5.31*b*. The electrode of Figure 5.31*c* must be rotated smoothly with little eccentricity or wobble to obtain reproducible currents; this method is not recommended by the authors.

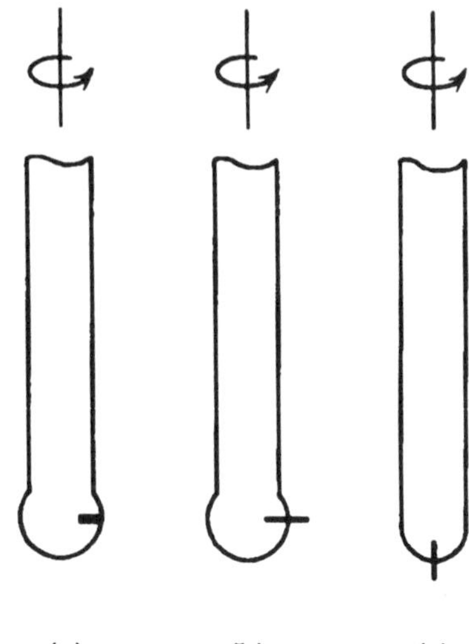

Figure 5.31 Rotated platinum electrodes. (a) (b) (c)

The electrodes usually are rotated at about 600 rpm. Contact to the platinum wire is made internally by filling the electrode with mercury. A stationary wire dips into the mercury pool at the top to make external contact to the potential-control circuitry. The platinum–glass seals are prone to crack, which causes erratic currents that are associated with the leaking of mercury to the electrode surface. Once cracked, the electrodes are not easily repaired and should be discarded. Table 5.9 indicates the dependence of the current on the rotational speed of the electrode.

The rotated disk and ring–disk electrodes that are shown in Figures 5.32 and 5.33 are the most useful of the hydrodynamic electrodes, from both practical and theoretical viewpoints. With such electrodes, the equations for mass

TABLE 5.9 Limiting Currents at Electrodes in Stirred Solution

Electrode	Expressions[a] for the Limiting Current
Rotated platinum electrode (RPE)	$i_L = knA(R)^{0.3\text{–}0.6}C^b$
Rotated disk electrode (RDE)	$i_L = 1.88 \times 10^5 nD^{2/3}v^{-1/6}r^2\omega^{1/2}C^b$
Platinum tubular electrode (PtTE)	$i_L = 5.31 \times 10^5 nD^{2/3}L^{2/3}V_f^{2/3}C^b$
Multiple-mesh flow-through electrode (MMFE)	$i_L = kNV_f^{1/3}C^b$

[a]Terms: i_L = limiting current (A); n = number of electrons transferred per mole in the oveall reaction; D = diffusion coefficient (cm^2 s); R = rotational speed (rps); ω = angular velocity (rad s^{-1}, $\omega = 2\pi R$); r = disk radius (cm); A = electrode area (cm^2); v = kinematic viscosity (cm^2 s^{-1}) (viscosity/density); L = length of tubular electrode (cm); V_f = solution flow rate (cm^3 s^{-1}); N = number of meshes in the multiple mesh electrode; C^b = concentration (mol cm^{-3}); k = constant.

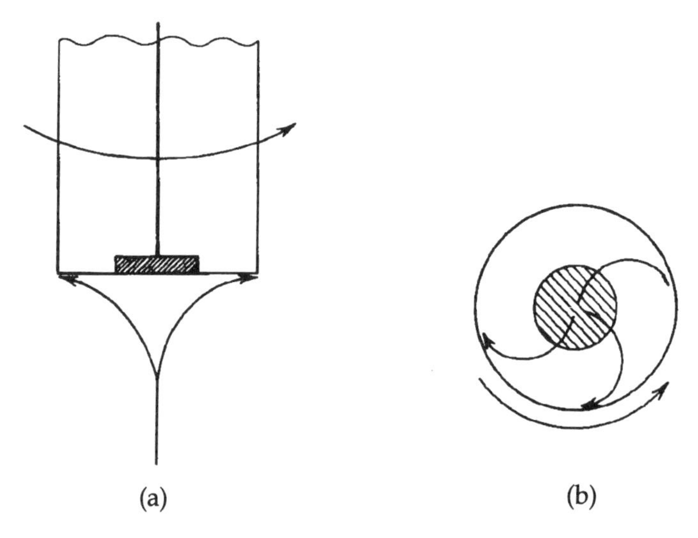

Figure 5.32 Rotated-disk electrode.

Figure 5.33 Rotated ring–disk electrode.

transport are in close agreement with experimental results. Furthermore, the ring–disk electrode allows the products that are produced at the disk to be detected as they are swept past the ring. The ring and the disk are insulated from one another so that their potentials can be controlled independently.

The construction of these electrodes requires precise machining of the disk, the Teflon shroud, and the spacers. Ring–disk electrodes now are available commercially (Pine Instrument Co., Grove City, Penn.), and most workers will prefer to buy them rather than to attempt their fabrication. The expression for the limiting current at a disk electrode is shown in Table 5.9.

Flow electrodes. Rather than move the electrode past the solution, the sample solution can be flowed past a stationary electrode. The tubular platinum electrode (Figure 5.34) and the gold micromesh flow-through electrode (Figure 5.35) are both ingenious attempts to produce electrodes that are useful for the measurement of electroactive materials in a continuously flowing stream. Ap-

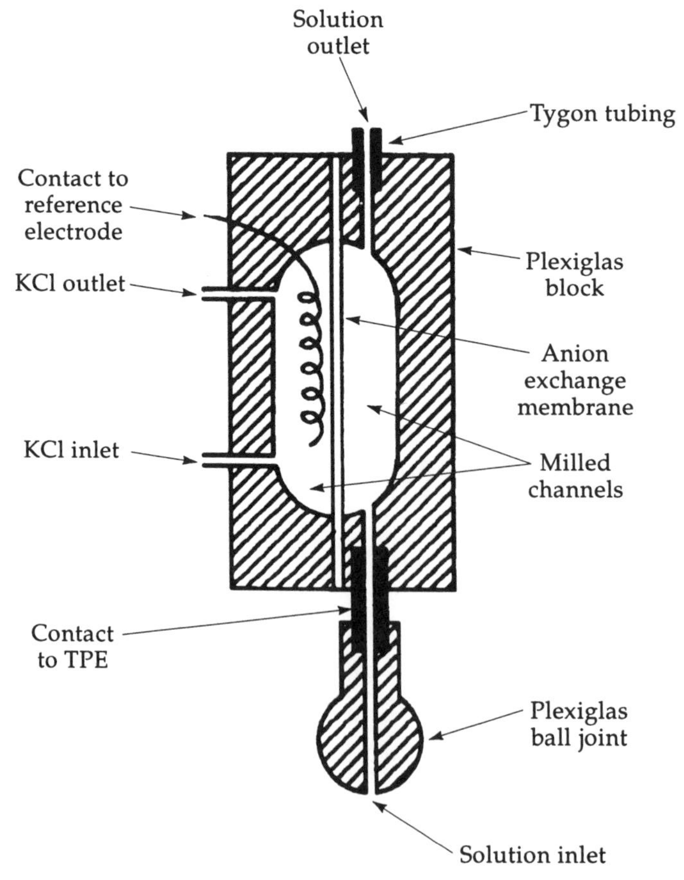

Figure 5.34 Tubular platinum electrode (TPE) for flowing sample solutions.

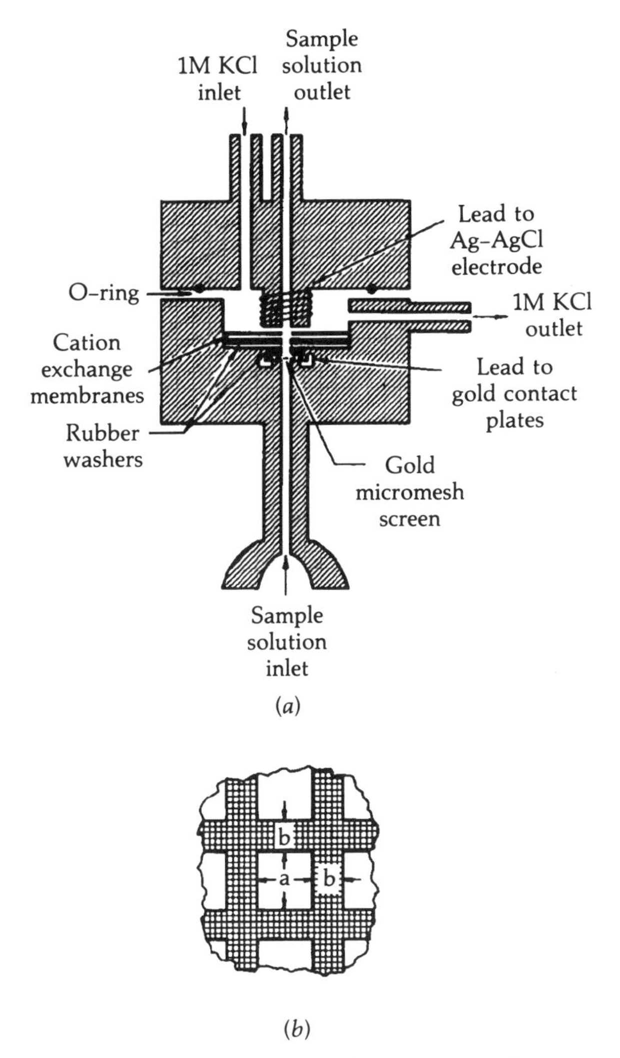

Figure 5.35 Gold micromesh flow-through electrode: (*a*) cell system; (*b*) detail of gold micromesh screen.

plications of these electrodes to the analysis of flowing streams have been described,[123,124] and their mass-transport properties are summarized in Table 5.9. The manner of their construction is fairly obvious from the figures; the references can be consulted for more detail.

The flow electrode of Figure 5.36 is designed for external generation of a coulometric intermediate. This is most useful in situations where the sample contains some species that would interfere with the electrode reaction that generates the coulometric reactant. A substantial flow is required through the electrode to sweep the products of the reaction into the sample, and a design

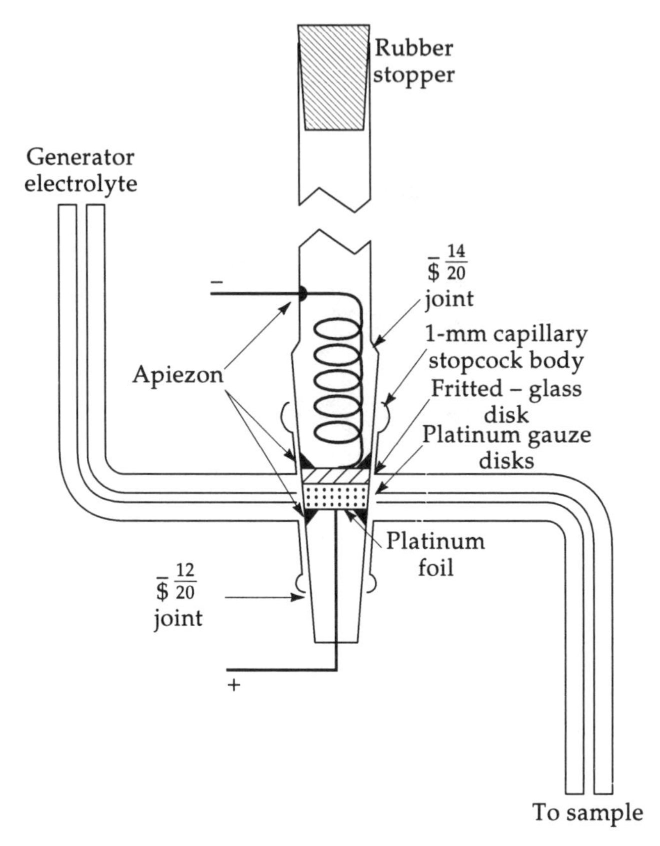

Figure 5.36 Flow cell for external coulometric generation of a titrant.

that is claimed to be an improvement over that shown in Figure 5.36 has been described.[125]

5.4 OPTICALLY TRANSPARENT ELECTRODES

Several methodologies have been developed to permit spectral observations by use of optically transparent electrodes. An internal-reflectance spectrophotometry (IRS) cell system is depicted in Figure 5.37. Such measurements can be made by use of electrodes that are composed of solid germanium,[126,127] tin oxide,[128,129] or metal films coated on glass.[130,131] The electromagnetic radiation strikes the electrode–solution interface at an angle that exceeds the critical angle and penetrates the solution to a depth of less than one wavelength of the light. Therefore, only a thin layer of solution next to the metal-film electrode interacts with the beam of light. Hence this type of experiment provides a means of studying processes that occur at or near the electrode, with little or not inter-

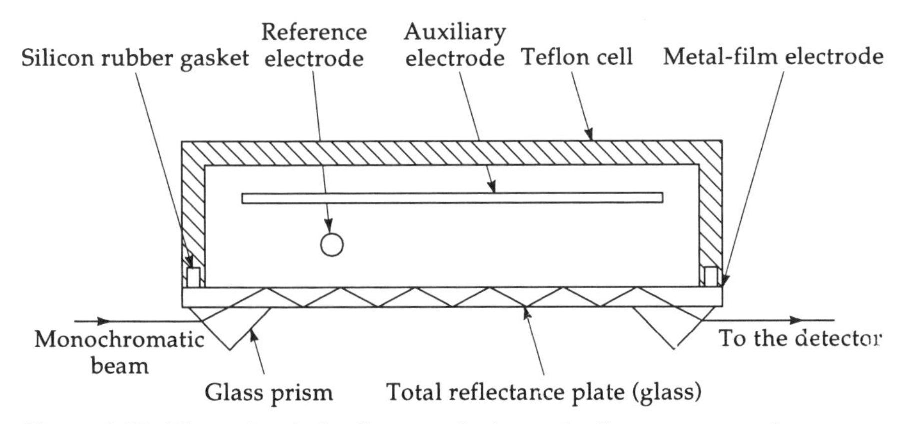

Figure 5.37 Electrochemical cell system for internal reflectance spectrophotometry.

ference from bulk-solution components. The provision for multiple reflections enhances the sensitivity of the measurement.

A second technique is the measurement of spectral transmission through optically transparent tin oxide coated on glass or quartz;[132, 133] through thin metal films of platinum, gold, or mercury on glass;[134, 135] or through gold minigrid electrodes that are produced from thin gold foils by a photoetching process.[136, 137] These electrodes are not difficult to fabricate and provide a way to measure the spectral properties of the products of an electrode reaction. The construction of optically transparent platinum and gold electrodes by standard vapor-deposition techniques has been discussed.[138] These same electrodes also can be plated with thin films of mercury ($\sim 10^{-6}$ cm) to provide a 0.4-V extension in the negative direction without seriously degrading the optical properties.

Figure 5.38 illustrates a rotated disc electrode that is surrounded by a concentric shell of light wires sealed in epoxy cement. Products produced electrochemically at the disc electrode are swept past the light wires, and their spectrum is recorded by slowly scanning with the monochromator.[139]

Figure 5.39 depicts a rotated electrode system that reverses the roles of the light and the electrode from that of preceding assembly. Here an intense beam of light shines through the transparent disk to produce a photochemical reaction. The products of the photochemical reaction are swept past the disk electrode, where they may be monitored for their electroactivity.[140]

5.5 MERCURY INDICATOR ELECTRODES

Although mercury seldom is used as an indicator electrode in redox titrations (because it is so readily oxidized), it is used extensively for potentiometric titrations with complexing agents such as ethylenediaminetetraacetic acid

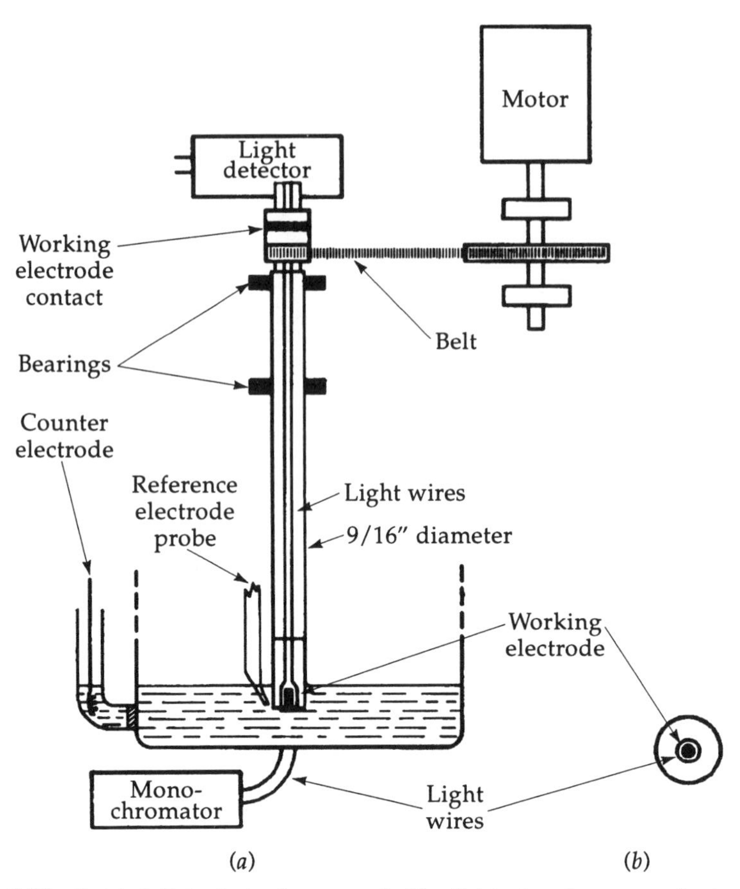

Figure 5.38 Rotated-disk electrode surrounded by light wires for spectroelectrochemical measurements.

(EDTA). In this application the $M(OH_2)_n^{2+}$ cation that is to be titrated forms a soluble EDTA complex that is appreciably less stable than the HgEDTA complex.[141] The mercury electrode is constructed in a form like that of Figure 5.40*a* or *b*, and is used in conjunction with a saturated calomel reference electrode. The electrode is immersed in a solution that contains the ion to be titrated $[M(OH_2)_n^{2+}]$, and a few drops of 0.01 M HgY^{2-} are added (where Y^{4-} represents the EDTA anion). This establishes the potential of the mercury electrode according to the half-cell reaction

$$Hg^{II}(OH_2)_6^{2+} + 2\,e^- \rightleftharpoons Hg + 6\,H_2O \tag{5.40}$$

so that E_{cell} is given by

$$E_{cell} = E^{\circ}_{Hg} - E_{SCE} + \frac{2.3RT}{2F} \log\,[Hg(OH_2)_6^{2+}] \tag{5.41}$$

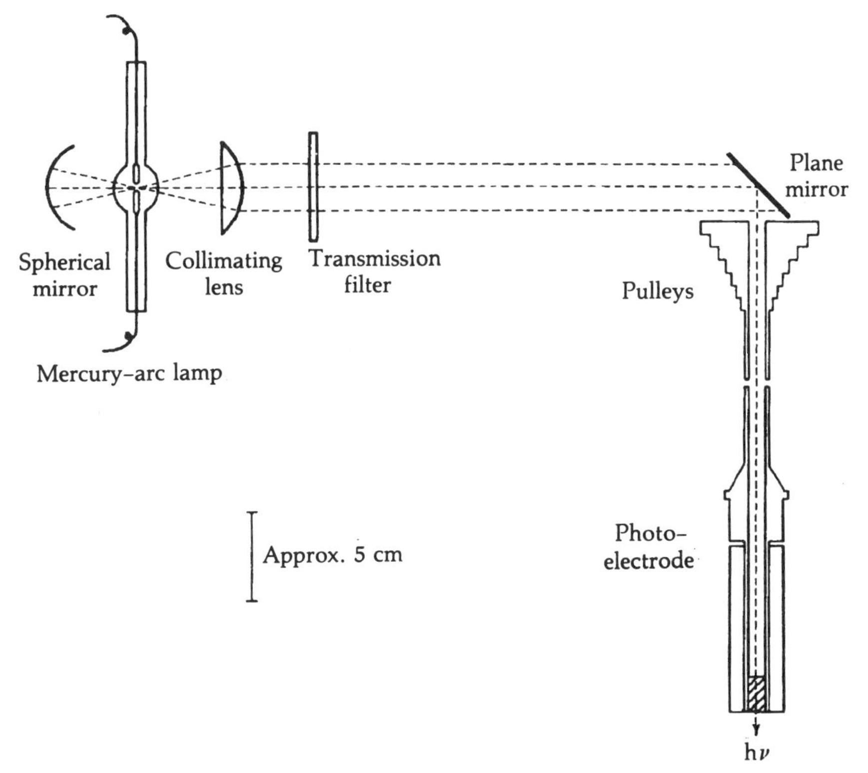

Figure 5.39 Transparent rotated-disk electrode system for producing and electrochemically monitoring photochemical reaction products.

Both the $Hg(OH_2)_6^{2+}$ and $M^{II}(OH_2)_n^{2+}$ ions form complexes with EDTA by the general reaction

$$M^{II}(OH_2)_n^{2+} + Y^{4-} = MY^{2-} + nH_2O \tag{5.42}$$

with equilibrium formation constants

$$K_{Hg} = \frac{[HgY^{2-}]}{[Hg(OH_2)_6^{2+}][Y^{4-}]} \tag{5.43}$$

$$K_M = \frac{[MY^{2-}]}{[M^{II}(OH_2)_n^{2+}][Y^{4-}]}, \qquad K_{Hg} >> K_M \tag{5.44}$$

Substitution of Eqs. (5.43) and (5.44) in Eq. (5.41) yields

$$E_{cell} = E^\circ_{Hg} - E_{SCE} + \frac{2.3RT}{2F} \log \frac{[HgY^{2-}]K_M M(OH_2)_n^{2+}]}{K_{Hg}[MY^{2-}]} \tag{5.45}$$

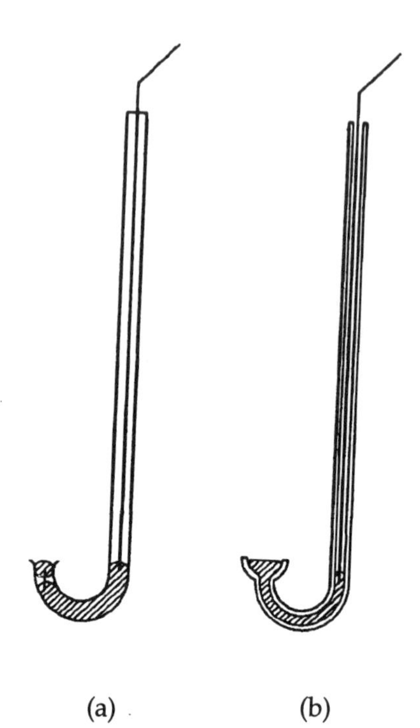

Figure 5.40 Potentiometric mercury indicator electrodes.

During the course of the titration the concentration of HgY^{2-} remains essentially constant because it is so much more stable than the MY^{2-} complex. The E_{cell} is determined mainly by the ratio $[M^{II}(OH_2)_n^{2+}]/[MY^{2-}]$, which changes slowly in the middle of the titration but rapidly near the equivalence point as the concentration of $M^{II}(OH_2)_n^{2+}$ drops to a small value. This gives a sharp potential change that signals the endpoint. The method is general and can be applied to most cations that form soluble EDTA complexes that are appreciably less stable than the mercury(II)-EDTA complex.

5.6 SOLID INDICATOR ELECTRODES

The number of reversible metal–metal ion electrodes is limited so that the accurate direct potentiometric measurement of the activity of a metal ion with an electrode of the same metal usually is not feasible, except perhaps with the $Ag/Ag^{I}(OH_2)_4^{+}$ system. However, a number of metal ion–metal half-reactions are sufficiently reversible to give a satisfactory potentiometric titration with a precipitation ion or complexing agent. These couples include $Cu^{II}(OH_2)_6^{2+}/Cu$, $Pb^{II}(OH_2)_4^{2+}/Pb$, $Cd^{II}(OH_2)_6^{2+}/Cd$, and $Zn^{II}(OH_2)_6^{2+}/Zn$. However, all these metals can be determined by EDTA titration and the mercury electrode that is described in the preceding section.

Platinum and gold often are used as inert redox indicator electrodes in titrations where both the oxidized and reduced species are soluble in solution [e.g., Ce(IV)/Ce(III), Fe(III)/Fe(II), $Cr_2O_7^{2-}$/Cr(III), and I_2/I^-]. Here the electrode functions as an inert substrate that can mediate electron transfer and respond to the potential determined by the relative concentrations of the soluble redox species. Platinum electrodes are used most widely because the metal is easily sealed into soft glass to make a convenient electrode, and electron transfer reactions seem to be as fast on platinum as any other substrate. When used in this manner, the electrode should have a reasonably large surface area (1–2 cm^2). This can be achieved by spot-welding a platinum foil to a short length of sturdy platinum wire that has been sealed into glass tubing. If preferred, a tightly wound spiral of platinum wire can be used to obtain a reasonable surface area. A larger surface area renders the electrode more resistant to polarization, so that redox measurements can be made with a low-impedance voltmeter or potentiometer.

pH-Sensitive Solid Indicator Electrodes. Although platinum and iridium under certain conditions will respond to changes in pH, they have not been widely used for this purpose and do not appear to be as reliable as the antimony electrode. The latter apparently functions as a pH-sensitive electrode according to the reaction

$$Sb_2O_3(s) + 6\,H_3O^+ + 6\,e^- \longrightarrow 2\,Sb(s) + 3\,H_2O \qquad (5.46)$$

The electrode can be made by casting pure antimony in a glass tube, then cracking off the glass tube and remounting the antimony stick in a glass or plastic tube with epoxy cement. The electrode is not as reproducible as a glass electrode, but gives close to a 59-mV per pH unit response at 25°C, and it can be used in slurries or solutions that contain fluoride ion (which would ruin glass electrodes). The electrode is available commercially in a form that is intended for industrial applications. It also has been used in a micro form for pH titrations of nanoliter samples of biological fluids. In this application the antimony that is contained in a glass capillary is drawn out with a microelectrode puller into a 5-μm-diameter tip that can be immersed in a 5-nL sample under oil.[142]

5.7 SELECTIVE-ION ELECTRODES

During the past 20 years great progress has been made in the development and applications of ion-selective electrodes. The electrodes show good selectivity and sensitivity for the measurement of ion activities, and indirectly, the concentration of enzymes, enzyme substrates, and neutral gaseous molecules like CO_2 and NH_3. Selective-ion electrodes can be classified roughly into four types as shown in Figure 5.41. These are the glass electrodes, which can be made with good selectivity for H_3O^+ and other cations (Figure 5.41*a*); liquid ion-

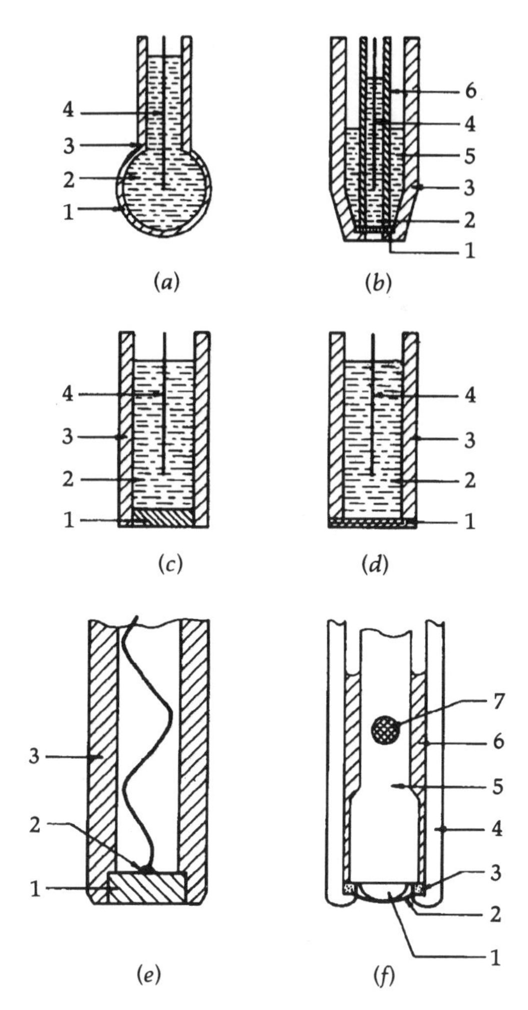

Figure 5.41 Selective-ion electrodes: (*a*) glass membrane; (*b*) liquid ion exchange; (*c*) homogeneous solid membrane; (*d*) heterogeneous solid membrane; (*e*) solid membrane without reference electrode; (*f*) gas-permeable membrane; 1, sensing electrode; 2, electrolyte, 2(*e*) ohmic contact, 2(*f*) gas-permeable membrane; 3, membrane support; 4, reference electrode, 4(*f*) outer electrode body, 5(*b*) liquid ion exchanger; 5(*f*) electrode body; 6(*b*) reference electrode body, 6(*f*) electrolyte; 7, liquid junction.

exchanger electrodes, which can be made selective for both cations and anions (Figure 5.41*b*); homogeneous and heterogeneous solid membrane electrodes, which can be made selective for both cations and anions (Figure 5.41*c–e*); and permeable membrane electrodes, which allow diffusion of gaseous molecules (CO_2 and NH_3) that by hydrolysis or some other reaction produce HO^- or H_3O^+ ions. The latter are sensed by a solid-state electrode inside the membrane. Table 5.10 lists a number of commercially available electrodes by type.

TABLE 5.10 Commercial Ion-Selective Electrodes

Electrode Designated for	Type	pH Range	Principal Interferences
Ammonia	Permeable membrane		Volatile amines
Bromide	Solid state	0–14	CN^-, I^-, S^{2-}
Cadmium	Solid state	1–14	Ag(I), Hg(II), Cu(II), Fe(II), Pb(II)
Calcium	Liquid ion exchanger	5.5–11	Zn(II), Fe(II), Cu(II), Ni(II)
Chloride	Solid state	0–14	S^{2-}, Br^-, I^-, CN^-
Chloride	Liquid ion exchange	2–10	I^-, NO_3^-, Br^-, HCO_3^-, SO_4^{2-}, F^-
Cupric	Solid state	0–14	S^{2-}, Ag(I), Hg(II), Fe(III)
Cyanide	Solid state	0–14	S^{2-}, I^-
Fluoride	Solid state	0–8.5	HO^-
Fluoroborate	Liquid ion exchanger	2–12	NO_3^-, Br^-, OAc^-
Iodide	Solid state	0–14	S^{2-}
Lead	Solid state	2–14	Ag(I); Hg(II); Cu(II); high levels of Cd(II), Fe(III)
Nitrate	Liquid ion exchanger	2–12	ClO_4^-, I^-, Br^-, NO_2^-, Cl^-
Perchlorate	Liquid ion exchanger	4–10	I^-, NO_3^-, Br^-
pH (H^+)	Glass	0–14	Na(I), at high pH
Potassium	Liquid ion exchanger		Cs(I), NH_4^+, H_3O^+, Ag(I)
Silver/sulfide	Solid state	0–14	Hg(II)
Silver	Glass	4–8	H_3O^+
Sodium	Solid state		Ag(I), H_3O^+
Sodium	Glass	3–12	Ag(I), H_3O^+, Li(I), K(I)
Sulfur dioxide	Permeable membrane	1.7	HF, acetic acid, HCl
Thiocyanate	Solid state		S^{2-}, CN^-, $S_2O_3^{2-}$, Cl^-, HO^-
Water hardness (divalent cation)	Liquid ion exchanger	5.5–11	Zn(II), Fe(II), Cu(II), Ni(II), Ba(II)

The theory and application of selective-ion electrodes have been extensively reviewed.[143–151] One of the interesting sidelights is the fact that the internal reference electrode may be replaced by an apparent ohmic contact in many instances, as illustrated by Figure 5.41*e* for the solid membrane electrode. Thus the glass electrode can be filled with mercury in place of the internal reference electrode,[152] or a gold contact that is plated over with copper can be used.[153] Likewise, a selective-ion electrode for calcium ion has been described that is coated on a platinum electrode;[154] the contact appears to be mainly ohmic.

The simplest method of measurement with ion-selective electrodes is direct potentiometry by use of the Nernst equation. However, this makes extreme demands on the reproducibility of the junction potential, and there is the problem of variation of activity with ionic strength. Concentration-cell techniques have proved to be very precise, especially in terms of null-point potentiometry.

The usual procedure involves adjustment of the composition of one of the half-cell solutions to match the other. This is indicated by a potential null between identical indicator electrodes that are specific for the ion. The method has been employed to measure fluoride ion at the level of 0.4–190 ng in volumes that range from 5 to 100 μL; the relative error is less than 1%.

Calibration techniques that use buffers to adjust the ionic strength and pH of the solution are effective for certain kinds of samples. For example, the detection of fluoride in public water supplies often is carried out by dilution of both standards and samples with a buffer that contains acetic acid, sodium chloride, and sodium citrate (with the pH adjusted to pH 5.0–5.5 by use of sodium hydroxide). This buffer performs three functions: (1) it fixes the ionic strength of the standards and samples to the same level, principally determined by the buffer; (2) the solution is buffered in a region where HO^- ion does not interfere; and (3) any Fe(III) or Al(III) ions are complexed by citrate to release the fluoride ion that is bound by these ions.

The electrodes also can be used effectively to determine the endpoints in potentiometric titrations, a technique that generally provides better accuracy than direct potentiometry.

Undoubtedly the area of greatest potential application is in the field of clinical chemistry and the study of biological systems. Electrodes can be fabricated that can penetrate into volumes as small as a single cell.[155] The potential of

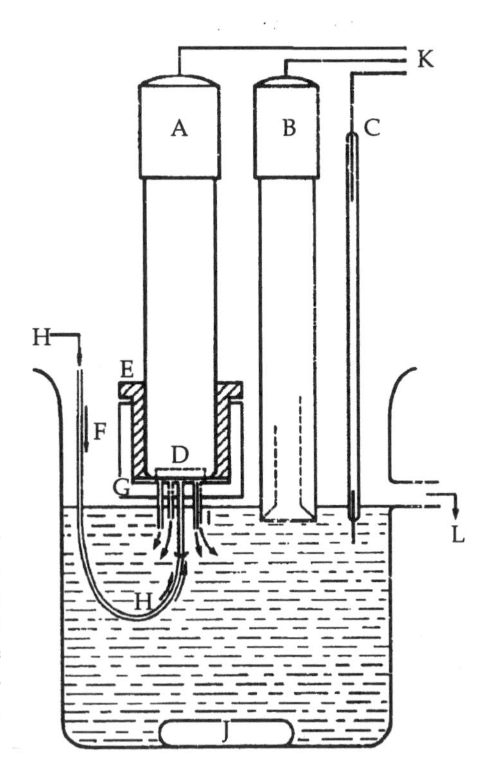

Figure 5.42 Flow cell for a selective-ion electrode: *A*, sensor electrode; *B*, reference electrode; *C*, solution ground; *D*, sensing membrane; *E*, Teflon sleeve; *F*, Plexiglas cap; *G*, washer; *H*, sample inlet flow; *I*, sample outlet flow; *J*, magnetic stirring bar; *K*, potentiometer; *L*, solution outlet.

TABLE 5.11 Biological Materials That Can Be Assayed by Ion-Specific Electrodes

Enzyme	Substrate	Sensing Electrode
β-glucosidase	Amygdalin	CN^-
Rhodanese	$S_2O_3^{2-}/CN^-$	CN^-
β-Cyanoalanine synthase	L-Cysteine/CN^-	S^{2-}/CN^-
α-Chymotrypsin	Diphenylcarbamyl fluoride	F^-
Urease	Urea	Glass/antibiotic
Deaminase enzymes		Glass/antibiotic
Cholinesterase	Acetylcholine	Acetylcholine
Peroxidase	HOOH	I^-
Catalase	HOOH	I^-
Glucose oxidase	β-D-glucose	I^-
Hypoxanthine oxidase	Hypoxanthine	I^-
Uricase	Uric acid	I^-

Source: Ref. 157.

the electrodes for continuous analysis of enzymes has been demonstrated by use of flow cells of the type shown in Figure 5.42.[156] Table 5.11 lists a few of the enzymes that can be determined by use of selective-ion electrodes.[157]

REFERENCES

1. Bockris, J. O'M., Reddy, A. K. N., *Modern Electrochemistry*, Vol. 2, Plenum Press, New York, 1970, pp. 644–678.
2. Pleskov, V. A., *Usp. Khim.* **1947,** *16*, 254.
3. Koepp, H. M.; Wendt, H.; Strehlow, H., *Z. Elektrochem.* **1960,** *64*, 483.
4. Nelson, I. V.,; Iwamoto, R. T., *Anal. Chem.* **1961,** *33*, 1975.
5 Popovych, O.; Dill, A. J., *Anal. Chem.* **1969,** *41*, 456.
6. Popovych, O., *Crit Rev. Anal. Chem.* **1970,** *1*, 73.
7. Henderson, P., *Z. Physik. Chem.* **1907,** *59*, 118; **1908,** 63, 325.
8. MacInnes, D. A., *The Principles of Electrochemistry*, Dover Publications, New York, 1961, pp. 85, 226.
9. Rock, P. A., *J. Chem. Educ.* **1970,** *47*, 683.
10. Finkelstein, N. P.; Verdier, E. T., *Trans. Faraday Soc.* **1957,** *53*, 1618.
11. Guggenheim, E. A., *J. Am. Chem. Soc.* **1930,** *52*, 1315; *J. Phys. Chem.* **1930,** *36*, 1758.
12. Cox, B. G.; Parker, A. J.; Waghorne, W. E., *J. Am. Chem. Soc.* **1973,** *95*, 1010.
13. Turner, W. R.; Elving, P. J., *Anal. Chem.* **1965,** *37*, 467.
14. Clem, R. G.; Jakob, F.; Anderberg, D., *Anal. Chem.* **1971,** *43*, 292.
15. Ives, D. J. G.; Janz, G. J., eds., *Reference Electrodes*, Academic Press, New York, 1961.

16. Bates, R. G., *Determination of pH*, 2nd ed., Wiley-Interscience, New York, 1973, Chap. 10.
17. Hills, G. J.; Ives, D. J. G., in *Reference Electrodes*, Ives, D. J. G.; Janz, G. J., eds., Academic Press, New York, 1961, Chap. 2, p. 102.
18. Franklin, T. C.; Naito, M.; Itoh, T.; McClelland, D. H., *J. Electroanal. Chem.* **1970,** *27*, 303.
19. Ref 17, p. 100.
20. Ref 16, p. 290.
21. Feltham, A. M.; Spiro, M., *Chem. Rev.* **1971,** *71*, 177.
22. Ref 16, p. 294; Ref 17, p. 92.
23. Ref 16, p. 282; Ref 17, p. 95.
24. Fleischmann, M.; Hiddleston, J. N., *J. Sci. Instrum. Series 2.* **1968,** *1*, 667.
25. Dobson, J. V., *J. Electroanal. Chem.* **1972,** *35*, 129.
26. Janz, G. J., in *Reference Electrodes*, Ives, D. J. G.; Janz, G. J., eds., Academic Press, New York, 1961, Chap. 4.
27. Butler, J. N., in *Advances in Electrochemistry and Electrochemical Engineering*, Vol. 7, Delahay, P., ed., Interscience Publishers, New York, 1970, pp. 106–114.
28. Pinching, G. D.; Bates, R. G., *J. Res. Natl. Bur. Stand.* **1946,** *37*, 311; Ref 16, p. 329.
29. Ref. 16, p. 48.
30. Ref. 17, Chap. 3.
31. Ref. 17, pp. 160–161.
32. Ives, F. J. G., Ref. 17, Chap. 7.
33. Samuelson, G. J.; Brown, D. J., *J. Am. Chem. Soc.* **1935,** *57*, 2711.
34. Mattock, G., *pH Measurement and Titration*, Heywood, London, 1961, p. 153.
35. Ref 16, p. 395.
36. Ref 16, p. 398.
37. Fisher, D. J.; Belew, W. L.; Kelley, M. T., in *Polarography 1964*, Vol. 2, Hills, G. J., ed., Interscience Publishers, New York, 1966, p. 1043.
38. Belew, W. L.; Fisher, D. J.; Kelley, M. T.; Dean, J. A., *Chem. Instrum.* **1970,** *2*, 297.
39. Herrmann, C. C.; Perrault, G. P.; Pilla, A. A. *Anal. Chem.* **1968,** *40*, 1173.
40. Hills, G. J., Ref 17, p. 433.
41. Mann, C. K.; Barnes, K. K., *Electrochemical Reactions in Nonaqueous Systems*, Marcel Dekker, New York, 1970, Chap. 1.
42. Strehlow, H., in *The Chemistry of Non-aqueous Solvents*, Vol. II, Lagowski, J. J., ed., Academic Press, New York, 1967, Chap. 4.
43. Bates, R. G., ibid., Chap. 3.
44. Bates, R. G., in *Solute-Solvent Interactions*, Coetzee, J. F.; Ritchie, C. D., eds., Marcel Dekker, New York, 1969, Chap. 2.
45. Ward, W., quoted in Ref 27, p. 115.
46. Larson, R. C.; Iwamoto, R. T.; Adams, R. N., *Anal. Chim. Acta.* **1967,** *25*, 371; quoted in Ref 27, p. 135.

47. Gritzner, G.; Kuta, J., *Pure Appl. Chem.* **1984,** *56*, 461.
48. Barrette, W. C., Jr.; Johnson, H. W., Jr.; Sawyer, D. T., *Anal. Chem.* **1984,** *56*, 1890.
49. Alexander, R.; Parker, A. J.; Sharp, J. H.; Waghorne, W. E., *J. Am. Chem. Soc.* **1972,** *94*, 1148.
50. Diggle, J. W.; Parker, A. J., *Electrochim. Acta.* **1973,** *18*, 975.
51. Bashkin, J. K.; Kinlen, P. J., *Inorg. Chem.* **1990,** *29*, 4507.
52. Gagné, R. R.; Koval, C. A.; Lisensky, G. C., *Inorg. Chem.* **1980,** *19*, 2855.
53. Bauer, D.; Beck, J. P., *Bull. Soc. Chim. Fr.* **1973,** 1252.
54. Ref 41, p. 18.
55. Laity, R. W., in Ref 17, Chap. 12.
56. Janz, G. J., ed., *Molten Salts Handbook*, Academic Press, New York, 1967.
57. Senderoff, S.; Brenner, A., *J. Electrochem. Soc.*, **1954,** *101*, 31; quoted in Ref 55, p. 585.
58. Boxall, L. G.; Johnson, K. E., *Anal. Chem.* **1968,** *40*, 831.
59. Caton, R. D., Jr.; Wolfe, C. R., *Anal. Chem.* **1971,** *43*, 660.
60. Laitinen, H. A.; Ferguson, W. S., *Anal. Chem.* **1957,** *29*, 4.
61. Ref. 1, pp. 1156–1170.
62. Trasatti, S., *J. Electroanal. Chem.* **1972,** *39*, 163.
63. Reeves, R. M.; Sluyters-Rehbach, M.; Sluyters, J. H., *J. Electroanal. Chem.* **1972,** *34*, 55, 69; **1972,** *36*, 101, 287.
64. Butler, J. N., *J. Electroanal. Chem.* **1967,** 14, 89.
65. Headridge, J. B.; Ashraf, M.; Dodds, H. L. H., *J. Electroanal. Chem.* **1968,** *16*, 116.
66. Kuwana, T.; Adams, R. N., *J. Am. Chem. Soc.* **1957,** *79*, 3609; *Anal. Chim. Acta.* **1959,** *20*, 51, 60.
67. Rand, D. A. J.; Woods, R., *J. Electroanal. Chem.* **1971,** *31*, 29.
68. Rand, D. A. J.; Woods, R., *J. Electroanal. Chem.* **1972,** *35*, 209.
69. Gaur, J. N.; Schmid, G. M., *J. Electroanal. Chem.* **1970,** *24*, 279.
70. Yamada, S.; Sato, H., *Nature.* **1962,** *193*, 261.
71. Plock, C. E., *J. Electroanal. Chem.* **1968,** *18*, 289.
72. Zittel, H. E.; Miller, F. J., *Anal. Chem.* **1965,** *37*, 200.
73. Jennings, V. J.; Forster, T. E.; Williams, J., *Analyst.* **1970,** *95*, 718.
74. Kopanica, M.; Vydra, F., *J. Electroanal. Chem.* **1971,** *31*, 175.
75. Cauquis, G.; Serve, D., *J. Electroanal. Chem.* **1972,** *34*, Appl. 1.
76. Florence, T. M., *J. Electroanal. Chem.* **1970,** *27*, 273.
77. Elving, P. J.; Smith, D. L., *Anal. Chem.* **1960,** *32*, 1849.
78. Miller, F. J.; Zittel, H. E., *Anal. Chem.* **1963,** *35*, 1866.
79. Beilby, A. L.; Brooks, W., Jr.; Lawrence, G. L., *Anal. Chem.* **1964,** *36*, 22.
80. Eisner, U.; Mark, H. B., Jr., *J. Electroanal. Chem.* **1970,** *24*, 345.
81. Randin, J. P.; Yeager, E., *J. Electroanal. Chem.* **1972,** *36*, 257.
82. Covington, J. R.; Lacoste, R. J., *Anal. Chem.* **1965,** *37*, 420.
83. Swofford, H. S., Jr.; Carman III, R. L., *Anal. Chem.* **1966,** *38*, 966.

84. Pungor, E.; Szepesuary, E.; Havas, J., *Anal. Letters.* **1968,** *1*, 213.
85. Adams, R. N., *Anal. Chem.* **1958,** *30*, 1576.
86. Adams, R. N., *Electrochemistry at Solid Electrodes*, Marcel Dekker, New York, 1969, p. 26.
87. Marcoux, L. S.; Prater, K. B.; Prater, B. G.; Adams, R. N., *Anal. Chem.* **1965,** *37*, 1446.
88. Voorhies, J. D.; Davis, S. M., *Anal. Chem.* **1960,** *32*, 1855.
89. Hand, R.; Carpenter, A. K.; O'Brien, C. J.; Nelson, R. F., *J. Electrochem. Soc.*, **1972,** *119*, 74.
90. Strojek, J. W.; Kuwana, T., *J. Electroanal. Chem.* **1968,** *16*, 471.
91. Alder, J. F.; Fleet, B.; Kane, P. O., *J. Electroanal. Chem.* **1971,** *30*, 427.
92. Baizer, M. M.; Lund, H., eds., *Organic Electrochemistry*, 3rd ed., Marcel Dekker, New York, 1991.
93. Biegler, T.; Rand, D. A. J.; Woods, R., *J. Electroanal. Chem.* **1971,** *29*, 269.
94. Bard, A. J., *Anal. Chem.* **1961,** *33*, 11.
95. von Stackelberg, M.; Pilgram, M., Toome, W., *Z. Elektrochem.* **1953,** *57*, 342.
96. Ref 86, pp. 45–61.
97. French, W. G.; Kuwana, T., *J. Phys. Chem.* **1964,** *68*, 1279.
98. Ref 86, pp. 206–208.
99. Angerstein-Kozlowska, H.; Conway, B. E.; Sharp, W. B. A., *J. Electroanal. Chem.* **1973,** *43*, 9.
100. Taylor, R. J.; Humffray, A. A., *J. Electroanal. Chem.* **1973,** *42*, 347.
101. Majer, V.; Vesely, J.; Stulik, K., *J. Electroanal. Chem.* **1973,** *45*, 113.
102. Koryta, J., *J. Chem. Educ.* **1972,** *49*, 183.
103. Smoler, I., *Coll. Czech. Chem. Commun.* **1954,** *19*, 238.
104. Bond, A. M.; O'Donnell, T. A.; Waugh, A. B., *J. Electroanal. Chem.* **1972,** *39*, 137.
105. Belew, W. L.; Fisher, D. J.; Jones, H. C.; Kelley, M. T., *Anal. Chem.*, **1969,** *41*, 779.
106. Means, D. K.; Mark, H. B., Jr., *Anal. lett.* **1971,** *4*, 23.
107. Demars, R. D.; Shain, I., *Anal. Chem.* **1957,** *29*, 1825.
108. Kemula, W., in *Advances in Polarography*, Vol. 1, Longmuir, I. S., ed., Pergamon Press, London, 1960, pp. 105–143.
109. Streuli, C. A.; Cooke, W. D., *Anal. Chem.* **1953,** *25*, 1691.
110. Streuli, C. A.; Cooke, W. D., *Anal. Chem.* **1954,** *26*, 963.
111. Kuempel, J. R.; Schaap, W. B., *Anal. Chem.* **1966,** *38*, 664.
112. Ramaley, L.; Brubaker, R. L.; Enke, C. G., *Anal. Chem.* **1963,** *35*, 1088.
113. Hassan, M. Z., Untereckler, D. F.,; Bruckenstein, S., *J. Electroanal. Chem.* **1973,** *42*, 161.
114. Wheeler, E. L., *Scientific Glassblowing*, Interscience Publishers, New York, 1958, p. 14.
115. Capon, A.; Parsons, R., *J. Electroanal. Chem.* **1972,** *39*, 275.
116. Hubbard, A. T.; Anson, F. C., in *Electroanalytical Chemistry*, Bard, A. J.; ed., Marcel Dekker, New York, 1970, p. 129.

117. Reilley, C. N., *Rev. Pure Appl. Chem.* **1968,** *18*, 137.
118. Hubbard, A. T., *Crit. Rev. Anal. Chem.* **1973,** *3*, 201.
119. Shaeffer, J. C.; Peters, D. G., *Anal. Chem.* **1970,** *42*, 430.
120. McClure, J. E.; Maricle, D. L., *Anal. Chem.* **1967,** *39*, 236.
121. Lindquist, J., *J. Electroanal. Chem.* **1968,** *18*, 204.
122. Cover, R. E.; Connery, J. G., *Anal. Chem.* **1969,** *41*, 918.
123. Blaedel, W. J.; Boyer, S. L., *Anal. Chem.* **1971,** *43*, 1538.
124. Blaedel, W. J.; Boyer, S. L., *Anal. Chem.* **1973,** *45*, 258.
125. Kesler, R. B., *Anal. Chem.* **1963,** *35*, 963.
126. Mark, H. B., Jr.; Pons, B. S., *Anal. Chem.* **1966,** *38*, 110.
127. Tallent, D. R.; Evans, D., *Anal. Chem.* **1969,** *41*, 835.
128. Hansen, W. N.; Kuwana, T.; Osteryoung, R. A., *Anal. Chem.* **1966,** *38*, 1810.
129. Winograd, N.; Kuwana, T., *J. Am. Chem. Soc.* **1970,** *92*, 224.
130. Pons, B. S.; Mattson, J. S.; Winstrom, L. O.; Mark, H. B., Jr., *Anal. Chem.* **1967,** *39*, 685.
131. Gottesfeld, S.; Ariel, M., *J. Electroanal. Chem.* **1972,** *34*, 327.
132. Kuwana, T.; Darlington, R. K.; Leedy, D. W., *Anal. Chem.* **1964,** *36*, 2023.
133. Osa, T.; Kuwana, T., *J. Electroanal. Chem.* **1969,** *22*, 389.
134. Yildiz, A.; Kissinger, P. T.; Reilley, C. N., *Anal. Chem.* **1968,** *40*, 1018.
135. Heineman, W. R.; Kuwana, T., *Anal. Chem.* **1971,** *43*, 1075.
136. Heineman, W. R.; Burnett, J. N.; Murray, R. W., *Anal. Chem.* **1968,** *40*, 1974.
137. Petek, M.; Neal, T. E.; Murray, R. W., *Anal. Chem.* **1971,** *43*, 1069.
138. Von Benken, W.; Kuwana, T., *Anal. Chem.* **1970,** *42*, 1114.
139. McClure, J. E., *Anal. Chem.* **1970,** *42*, 551.
140. Johnson, D. C.; Resnick, E. W., *Anal. Chem.* **1972,** *44*, 637.
141. Reilley, C. N.; Schmid, R. W., *Anal. Chem.* **1958,** *30*, 947.
142. Karlmark, B., *Anal. Biochem.* **1973,** *52*, 69.
143. Rechnitz, G. A., *Accts. Chem. Res.* **1970,** *3*, 69.
144. Eisenman, G., ed., *Glass Electrodes for Hydrogen and Other Cations*, Marcel Dekker, New York, 1967.
145. Freiser, H., ed., *Ion-Selective Electrodes in Analytical Chemistry*, Plenum Press, New York, 1978.
146. Vesely, J.; Weiss, D.; Stulik, K., *Analysis with Ion-Selective Electrodes*, Wiley, New York, 1978.
147. Cammann, K., *Working with Ion-Selective Electrodes. Chemical Laboratory Practice*, Springer Verlag, Berlin, 1979.
148. Bailey, P. L., *Analysis with Ion-Selective Electrodes*, 2nd ed., Heyden, London, 1980.
149. Koryta, J.; Stulik, K., *Ion-Selective Electrodes*, 2nd ed., Cambridge University Press, Cambridge, 1983.
150. Ammann, D., *Ion-Selective Microelectrodes*, Springer Verlag, Berlin, 1986.
151. Umerawa, Y., ed., *CRC Handbook on Ion-Selective Electrodes. Selectivity Coefficients*, CRC Press, Boca Raton, Fla., 1990.

152. Ritchie, C. D.; Uschold, R. E., *J. Am. Chem. Soc.* **1967,** *89*, 1721.
153. Ref 144, p. 446.
154. Cattrall, R. W.; Freiser, H., *Anal. Chem.* **1971,** *43*, 1905.
155. Ref 144, Chap. 18.
156. Llenado, R. A.; Rechnitz, G. A., *Anal. Chem.* **1973,** *45*, 826.
157. Rechnitz, G. A., *Research/Development*, **1973,** *24*, 18.

CHAPTER 6

ELECTROCHEMICAL CELLS AND INSTRUMENTATION

6.1 ELECTROCHEMICAL CELLS: INTRODUCTION

General Requirements. Electrochemical techniques are used under a variety of conditions. No matter what the purpose of the procedure, the design and construction of a suitable cell must meet certain basic requirements related to the desired level of precision in the measurement, optimum electrode geometry, the chemical reactivity of the system, the working temperature and pressure of the system, the scale of the system (micro or macro), the need to exclude contaminants from the laboratory atmosphere, and reasonable convenience in changing solutions and cleaning the cell between measurements. The cell also may need to be designed to be compatible with auxiliary equipment such as rotated electrodes, spectroscopic systems, or systems for controlling temperature and pressure.

There are several factors to consider in optimizing cell geometry. In laboratory-scale experiments the working solution volume may range from 10 μL of a dilute solution (as in a thin-layer cell) to a solution volume as large as a liter (as in preparative electrochemistry on the mole scale). Currents may range from the submicroampere region to several amperes. The specific resistance of the cell electrolyte may be as low as 2 Ω-cm (1 M aqueous HCl) or as large as 10^5 Ω-cm (0.1 M sodium acetate in glacial acetic acid).

Effects of solution resistance. The electrode geometry becomes a crucial factor whenever the ohmic (*iR*) drop in a cell becomes large. First, the ohmic drop imposes a natural limit on the current that can pass through the cell because the product of the total cell resistance and the current cannot exceed the output voltage of the potential source. This limit may be encountered in solvent systems of high specific resistance or in solutions of moderate resistance if the current is large, as in preparative electrochemistry. Second, the ohmic drop results in an error in the measured potential of the working electrode using either two-electrode or three-electrode circuits (see Figure 5.2).

Ohmic drop in two-electrode circuits. In a simple two-electrode cell of the type often employed in the polarography of aqueous solutions, the cell current passes through both the working electrode and the reference electrode. As discussed in Chapter 5, it is possible to design a reference electrode of low resistance and sufficiently large area so that its potential is not seriously affected by microampere polarographic currents. However, if accurate half-wave potential measurements are to be made, the product of the total cell resistance and the current must not exceed 1–2 mV. This requires that the cell resistance not exceed 200 Ω for currents of 10 μA. The approximate cell resistance can be conveniently measured with a simple AC conductivity bridge. The value of the measured resistance will depend on the surface area of the electrodes, so the measurement is best made with the actual electrodes to be used. Because an electrochemical cell does not behave as a pure resistance, a DC ohmmeter should not be used because the cell potential will cause a false reading.

Ohmic drop in three-electrode circuits. In modern coulometry and voltammetry the use of a potentiostat and a three-electrode configuration is the routine practice. The three electrodes are usually called the *working, reference,* and *counter* (or *auxiliary*) electrodes (see Figure 5.2). The cell current passes between the working electrode immersed in the test solution and the counter electrode, which may be in the test solution but is usually isolated from it by a single- or double-junction glass frit.

The reference electrode normally is located in a salt bridge and connected to the solution by a Luggin capillary whose tip is placed close to the working electrode. This minimizes the error in the measured potential associated with the iR drop in the solution between the tip of the Luggin capillary and the working electrode. If the cell currents are large, the solution resistance high, or fast potential changes necessary, careful attention must be paid to optimizing the electrode geometry. To achieve the greatest accuracy of potential control, the characteristics of the potentiostat and electrolysis cell must be carefully matched as components of a feedback control system.

The external leads from the potentiostat to the electrodes may also contribute significant resistance and capacitance that must be taken into account if the cell currents are large and if fast response is desired. Most metallic working electrodes will have very low resistance, but a typical dropping-mercury electrode (DME) may have a resistance as large as 100 Ω because the mercury-filled lumen of the capillary is so small (~0.005-cm diameter). This resistance makes a contribution to the total cell resistance and to the uncompensated resistance in a three-electrode circuit.

The extent to which a three-electrode circuit can compensate for ohmic drop in the cell has been the subject of a number of investigations and is well understood in principle, although exact equations may not be available for every electrode geometry. The effects are illustrated by reference to Figure 6.1, which shows a DME and the tip of a Luggin capillary that makes contact with the reference electrode. The DME may either be the normal blunt-tip glass capillary (dashed lines of Figure 6.1), or a tapered capillary may be used to minimize the shielding effect that distorts the symmetry of the current flux at the mercury drop.

Experimental measurements[1,2] of the uncompensated solution resistance indicate that it changes rapidly with distance in the immediate vicinity of the mercury drop, and attains an approximately constant limiting value at a distance greater than about 0.5 cm. The latter corresponds to a distance roughly 10 times the maximum radius of the mercury drop. This is illustrated in Figure 6.2, where the potential of a reference electrode in a movable Luggin capillary probe at different distances from the mercury drop has been measured with respect to an identical reference electrode located several centimeters away from the drop on the side opposite the counter electrode. An important feature

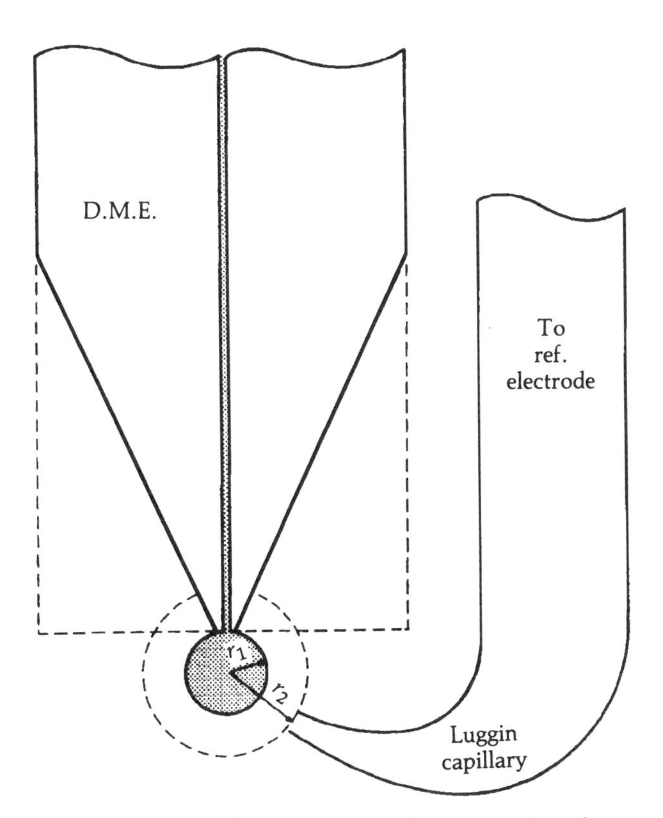

Figure 6.1 Luggin capillary and its placement relative to a dropping-mercury electrode.

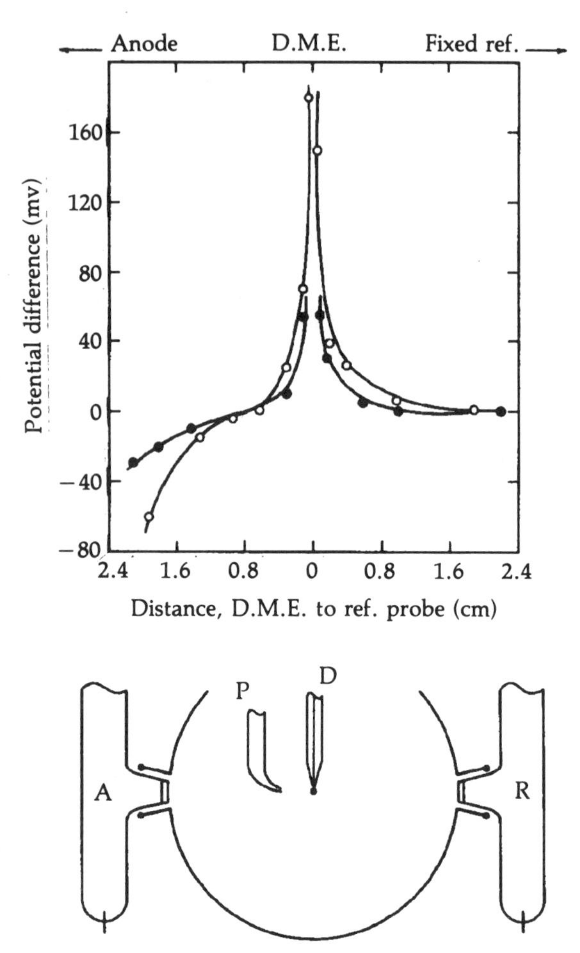

Figure 6.2 Potential difference between a reference electrode in a movable Luggin capillary probe (P) and an identical fixed-reference electrode (R) placed opposite to the counter electrode (A). The dropping-mercury electrode (D) is placed between the fixed-reference electrode and the counter electrode; the probe electrode is on side of DME opposite the fixed-reference electrode.

to note is the nearly spherical symmetry of the gradients at points close to the DME. (The slight gradients on the side of the DME in the direction of the counter electrode indicate that the symmetry probably would be seriously distorted if the counter electrode were placed closer than 1 cm from the DME, or if the cell diameter were smaller than 1 cm.)

The ohmic iR drop at the DME is that expected for a spherically symmetric radial current flux between the surface of an inner sphere of radius r_1 centimeters (mercury drop in this case) and the surface of an (imaginary) outer sphere of radius r_2 (dashed circle of Figure 6.1 at the tip of the Luggin capillary). Using the model of concentric spherical electrodes of radii r_1 and r_2

separated by a medium of specific resistance ρ Ω-cm), the expression for the resistance of the solution occupying the volume between the two spheres, assuming a spherically symmetrical current flow, is given by

$$R = \frac{\rho}{2\pi}\left(\frac{1}{r_1} - \frac{1}{r_2}\right) \tag{6.1}$$

This equation apparently was first derived by Ilkovic,[3] and Nemec[4] has derived an equivalent form of the equation. In the case under consideration, the solution iR drop "seen" at the tip of the Luggin capillary will be that between r_1 (the surface of the mercury drop) and r_2 (the distance from the center of the mercury drop to the tip of the Luggin capillary). If the glass capillary has a blunt end, the shielding of the drop requires the introduction of a correction factor. An empirical factor of $\frac{3}{2}$ has been suggested[5] and a value of $\frac{16}{9}$ has been proposed on theoretical grounds.[6] Assuming this approximation and that the tip of the Luggin capillary is more than 0.5 cm from the drop surface, the uncompensated solution resistance R, of a blunt-end DME is given as

$$R = \frac{3}{2}\frac{\rho}{4\pi r_1} \qquad (r_2 >> r_1) \tag{6.2}$$

For $r_1 = 0.05$ cm, and a potential error not to exceed 2 mV, the specific resistance of the electrolyte must be less than 200 Ω-cm for currents of 10 μA. This approximate calculation indicates that the errors in potential due to iR drop will be 200 mV when the specific resistance of the electrolyte is 10^4 Ω-cm.

The solution iR drop at the DME will also be time-dependent because r_1, the drop radius, is a function of time. For this reason a stationary hanging-mercury-drop electrode (HMDE) is to be preferred or the vertical orifice (Smoler) DME can be used (see Figure 5.14). The tip of a platinium-wire quasi–reference electrode can be placed as close as 0.1 drop diameter (about 0.003 cm) because the drop grows in the downward direction.[7] This gives nearly complete compensation in an electrolyte with a specific resistance of 15,000 Ω-cm for a cell with total resistance of about 10^5 Ω. The effect of the polargrams of placing the quasi–reference electrode at different distances from the electrode surface is shown in Figure 6.3.

Optimum geometry in voltammetry with microelectrodes. In electrolytes of specific resistance of 100-Ω cm or less, the geometry is not very critical. The working electrode should be placed between the counter and reference electrodes and is ordinarily an appreciable distance (>1 cm) from both. At small, nearly spherical electrodes, all points on the surface of the electrode will be essentially equidistant from the counter electrode and the current density will be almost uniform over the surfaces of the microelectrode.

The reference electrode is usually connected to a high-impedance imput so

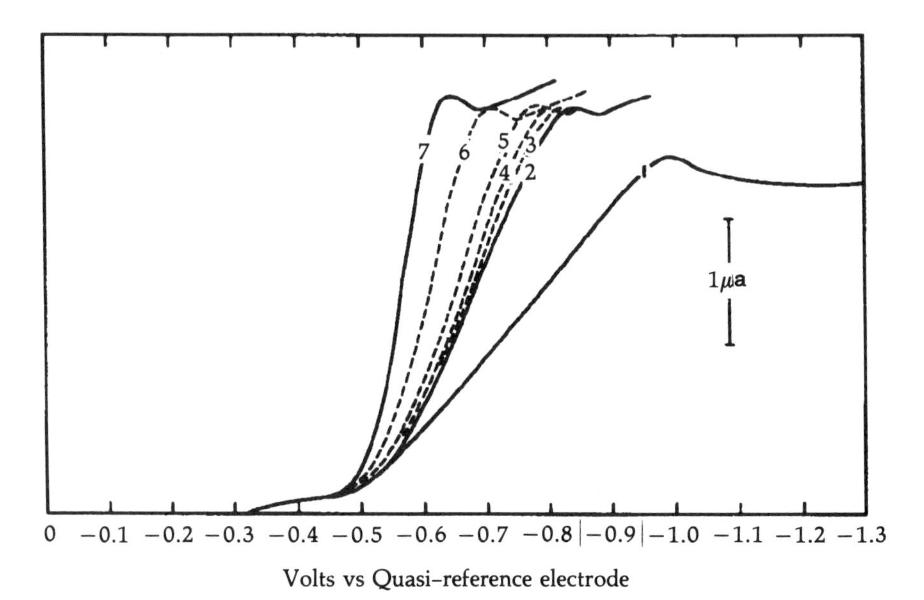

Figure 6.3 Polarograms as a function of reference-electrode placement in a solution system with a specific resistance of 15,000 Ω-cm: (1) two-electrode polarogram; (2–6) quasi-reference electrode placed increasingly close to DME; (7) quasi-reference electrode less than 0.1 drop radius from DME.

that the current and iR drop through the Luggin capillary and reference electrode are negligible. The Luggin capillary and reference electrode may have a rather high resistance (10–100 kΩ) unless fast response is desired, in which case the reference electrode must have a low resistance.

As the specific resistance of the solution increases, the geometry becomes more important, as has been shown in the previous sections. The uncompensated resistance will remain large unless the tip of the reference electrode is located very close to the working-electrode surface. This tip must be quite small; otherwise the current density will be nonuniform over the electrode surface because of distortion of the equipotential lines by the top of the reference electrode.

Electrode geometry in controlled-potential electrolysis. When fast response and accuracy of potential control are desired, considerable attention must be paid to the design of the cell–potentiostat system, and several papers have discussed the critical parameters and made recommendations for optimum cell design.[8–11] In general, to achieve stability and an optimum potentiostat rise time for a fast potential change, the total cell impedance should be as small as possible, and the uncompensated resistance should be adjusted to an optimum (nonzero) value that depends on the characteristics of the cell and potentiostat.[9,12] The electrode geometry also should provide for a low-resistance reference electrode and a uniform current distribution over the surface of the working electrode, which requires that the equipotential lines in the solution near the electrode surface be parallel to the electrode surface.

A cell system has been designed for the potentiostatic transient investigation of fast electrode reactions.[9] The main novel feature of the cell is the elimination of the classical Luggin capillary, as shown in Figure 6.4. The design provides a low-ohmic-resistance reference electrode with low stray capaitances.

In controlled potential coulometry, fast response is seldom necessary and some design compromises can be made for the sake of convenience. Here, the most important thing is to arrange the counter and working electrodes so that a uniform current density will be obtained. If this is not possible, the positioning of the reference-electrode tip becomes further restricted. This can be understood by reference to Figure 6.5, which shows a cross section of an idealized potential and current distribution on the inside of a cylindrical gauze working electrode with an eccentric, cylindrical counter electrode, under concentration polarization conditions. As in all current-distribution systems, the current and potential lines are orthogonal, and the current distribution is indicated by the convention of spacing the current lines to indicate current flux per unit of area (the more closely spaced the current lines, the higher the current density). The metal of the gauze electrode is assumed to be an equipotential, and the boundary (for mathematical purposes) is the adjacent layer of solution (the electrical double layer). The potential that is represented by the dashed contour lines at the

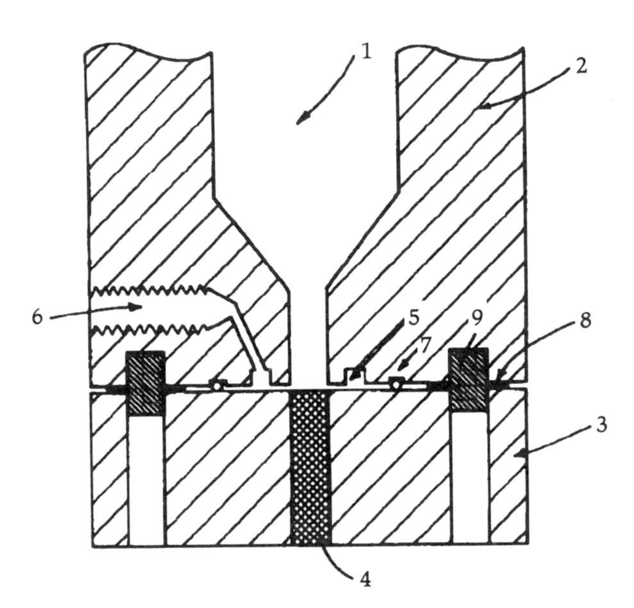

Figure 6.4 Cell design for the study of fast electrode reactions. System provides low-resistance reference electrode and low stray capacitances: (1) counter-electrode chamber; (2) Kel-F top; (3) Teflon bottom; (4) working electrode; (5) reference-electrode groove; (6) reference-electrode connection; (7) Viton O-ring; (8) stainless-steel spacer; (9) stainless-steel locating pin.

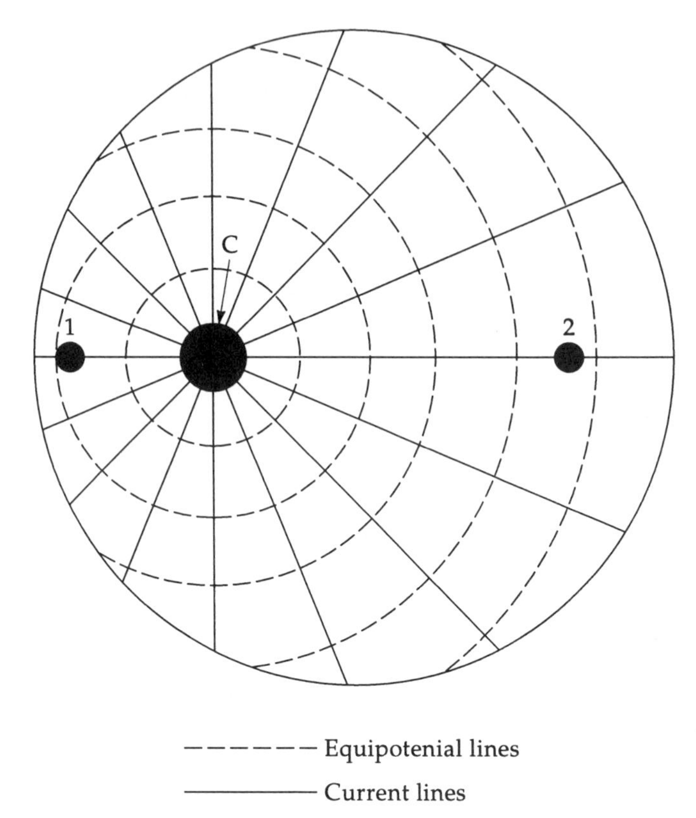

Figure 6.5 Idealized potential and current distribution diagram for the inside of a cylindrical gauze electrode and an eccentric interior cylindrical counter electrode (C). Positions 1 and 2 represent other placements of the counter electrode.

electrode surface is equivalent to the usual electrochemical potential difference between the metal and this adjacent layer of solution, which is measured with respect to some reference half-cell. The terms *electrode potential* or *potential of the surface of the electrode* refer to this potential difference at specific points along the surface of the electrode.

When polarization occurs at an electrode with nonideal geometry (e.g., when the current is limited by rate of electron transfer or by mass transport), there is a gradient in potential in the solution adjacement to the electrode, and associated with this is a tangential as well as normal component of the current at the electrode surface.[13] This causes the equipotential lines to intersect the electrode and the current lines to enter the electrode at angles other than 90°. (In the absence of polarization, or in a polarized electrode with ideal geometry, the equipotential lines would be parallel to the electrode surface, and the current lines would intersect the electrode at an angle of 90°.)

Under the effects of electrode polarization, proper placement of the reference electrode in the cylindrical gauze cell is dictated by the potential distribution.

If the reference electrode is placed at location 2, the electrode potential exceeds the control potential in the region of location 1 (i.e., it is more positive in the case of an oxidation or more negative in the case of a reduction). To avoid exceeding the control potenital, the reference electrode tip must be positioned on an equipotential that does not intersect the working electrode (such as location 1). There will inevitably be some uncompensated *iR* drop in the solution, and to minimize uncompensated *iR* drop errors, the tip should be located on an equipotential that approaches the working electrode closely. If the distance from the working electrode is made greater, more uncompensated *iR* drop is introduced, but there is no inherent potential control difficulty in this arrangement, and the larger *iR* drop error is not detrimental in most coulometric procedures. Simply stated, the best practice is to *locate the reference electrode tip on the line of minimum separation between the counter electrode and the working electrode.*[8] This will achieve satisfactory potential control (for analytical purposes) for all types of geometric arrangements.

Materials for the Construction of Cells and Electrodes. Materials used for the construction of cells should be usable over a wide temperature range, dimensionally stable, impervious to aqueous and organic solvents and reagents, durable, and easy to machine or fabricate. Another highly desirable property is for the material to be transparent so that the electrodes and the solution can be observed.

Glass. Glass comes closest to fulfilling all the requirements mentioned above and is the most widely used material. It is attacked by hydrofluoric acid, concentrated alkalies, and basic fused salts but has outstanding chemical resistance to almost all other chemical reagents. The chemical and physical properties of the most commonly used laboratory glasses are summarized in Table 6.1. Borosilicate glasses are the easiest to work with and can be used at temperatures up to 600°C. Borosilicate fritted-glass disks also are widely used as cathode–anode separators or to isolate reference-electrode and working-electrode compartments. The pore sizes for glass frits of standard porosities are listed in Table 6.2.

The large variety of commercially available ground-glass joints, glass frits, and stopcocks, and the services of a glassblower make it easy to fabricate cells for special purposes, and a selected few of these are described in a later section. Most companies that sell electrochemical instrumentation offer a number of cells of different styles, as well as electrodes and other accessories.

A number of plastics are used in electrochemical cells, and some of their pertinent physical and chemical properties are described in the following paragraphs and listed in Table 6.3.

Teflon. The Teflon TFE polymer is prepared from tetrafluoroethylene by Du Pont; it is white and nearly opaque except in thin sheets. Teflon FEP is a copolymer of tetrafluoroethylene and hexafluoropropylene, which is more readily molded and is translucent. The FEP polymer has more limited chemical

TABLE 6.1 Chemical Composition and Thermal Properties of Common Laboratory Glasses

Component	Soda-Lime "Soft Glass" Standard Flint	Pyrex Corning No. 7740	Vycor Corning No. 7900
	Chemical Composition (%)[a]		
SiO_2	71–73.5	80.5	96
B_2O_3	—	12.9	
Na_2O	14–17	3.8	
K_2O	0–1.5	0.4	
CaO	5–6	—	
MgO	3.5–4.5	—	
Al_2O_3	—	2.2	
Fe_2O_3	0.006–0.1	—	
	Thermal Properties[b]		
Linear coefficient of expansion (°C, 0–300°C) × 10^{-7}	92	32.5	8
Softening point (°C)	700	820	1500
Annealing point (°C)	521	565	910
Strain point (°C)	480	515	820

[a]Wheeler, E. L., *Scientific Glassblowing*, Interscience Publishers, New York, 1958, p. 12.
[b]*Laboratory Glass Blowing with Corning's Glasses*, Corning Glass Works, Corning, N.Y., 1961.

resistance, particularly to halogenated solvents and a lower working-temperature limit of 205°C. The TFE polymer is the most chemically resistant plastic available. It is inert, even at elevated temperatures, to practically all chemicals except fused alkali metals, chlorine trifluoride, and fluorine at elevated temperature and pressure. Teflon tape and molded stirring bars have been reported to react vigorously with sodium–potassium alloy at room temperature.[14]

TABLE 6.2 Porosity of Fritted Glass

Porosity	Type	Nominal Pore Size (μm)
Coarse	C[a]	40–60
Medium	M[b]	10–15
Fine	F[c]	4–5.5
Very fine	VF	2–2.5
Ultrafine	UF	0.9–1.4

[a]Type C: filtering coarse materials; mercury filtration.
[b]Type M: gas dispersion; salt bridges
[c]Type F: salt bridges.

TABLE 6.3 Physical Properties and Chemical Resistance of Plastics

	Teflon TFE	Teflon FEP	Kel-F 81	Methyl Methacrylate	Polyamide (Nylon–Zytel)	Poly-styrene	Poly-propylene	Linear Polyethylene	Conventional Polyethylene
Physical Properties									
Temperature limit (°C) (continuous)	250	200	140	50	140	70	130	110	80
Density ($g\ cm^{-3}$)	2.15	2.15	2.13	1.19	1.14	1.05	0.90	0.94–0.96	0.91–0.93
Linear coefficient of expansion (°C × 10^{-5})	10	8.3–10.5	4.8		10–15	6–8		11–13	16–18
Dielectric constant	2.2	2.2	2.5	3.5	4.0–7.6	2.5		2.3	2.3
Dielectric strength ($V\ mil^{-1}$)	400–600	500–600	400–600	400	385	500–700		450–500	460–700
Water absorption (% 24 h^{-1})	0.01	0.01	0.01	0.3	1.5	0.03–0.05	0.02	0.02	0.03
Relative O_2 permeability	0.6	0.6				0.11	0.11	0.08	0.40
Thermal conductivity [cal/(cm^2)(s)/°C/cm × 10^{-4}]	6.0	6.0	6.3		5.5	2.9		12	8
Chemical Resistance									
Acids, inorganic		E[a]	E	X	X	NR	E	E	E
Acids, organic		E	E	X	X	G	E	E	E
Alcohols		E	E			G	E	E	E
Aldehydes		E	E			NR	G	G	G
Amines		E	E			G	G	G	G
Bases		E	E			E	E	E	E
Esters		E	G[b]	NR[c]		NR	E	E	E
Ethers		E	G			X	G	G	G
Glycols		E	E			NR	E	E	E
Hydrocarbons, aliphatic		E	E		E	NR	G	G	G
Hydrocarbons, aromatic		E	X[d]	NR		NR	G	G	G
Hydrocarbons, halogenated		E	X	NR		NR	G	G	G
Ketones		E	G	NR		NR	G	G	G

[a]Excellent—no effect to 1 year.
[b]Good—no effect at 24 h.
[c]Not recommended—contact causes damage.
[d]Fair—short exposures causes some effect at room temperature.

Teflon is relatively soft, and very light finishing cuts must be used to machine it to a smooth finish and precise tolerances. Because it will flow under pressure, it frequently is used to fabricate electrodes by force-fitting rods or wires into a slightly undersized hole that has been drilled in the Teflon, or by using heat-shrinkable Teflon tubing. Using either of these techniques tight seals can be made between the Teflon and the electrode material, with little danger that solution will creep between the Teflon and the electrode. Seals to Teflon TFE are impractical with ordinary adhesives, although special adhesive kits are available for use with Teflon FEP. Although Teflon TFE is used at temperatures up to 300°C as insulation for wiring, it may give off traces of the monomer when heated much above 250°C.

Kel-F. This is a chlorotrifluoroethylene polymer, which has neglible water-vapor transmission and is nonflammable. Kel-F is suitable for use from −200 to +200°C, and is unaffected by concentrated alkalies and strong acids, including aqua regia. It is widely used in handling liquid hydrofluoric acid and is resistant to most organic solvents. Kel-F is swelled slightly by highly halogenated compounds and some aromatics. It is more rigid than Teflon and easier to machine with good dimensional stability. The material can be molded under heat and pressure, and this technique has been used to seal electrodes.[15]

Nylon. The various polyamides are referred to generically as *nylon* and are characterized by toughness and good abrasion resistance. These materials are readily machined and molded and can be used to 150°C. Although nylon will withstand boiling water, it absorbs water with resultant swelling and loss of flexural strength. Nylon is resistant to weak acids, weak and strong alkalies, petroleum oils, and many common solvents. It is attacked by strong acids, oxidizing agents, phenols, and formic acid. Because of its tendency to absorb liquids, nylon is not recommended for uses where it will have prolonged contact with solutions.

Acrylates. Polymethylmethacrylate (Lucite, Plexiglas, Perspex) is easily machined and is widely used with dilute aqueous solutions. Acrylates are resistant to nonoxidizing acids and weak alkalies, but are attacked by concentrated oxidizing acids and strong alkalies. They will withstand petroleum oils and most alcohols, but are generally unsuitable for use in contact with organic solvents. They are dissolved by ketones, esters, and aromatic and chlorinated hydrocarbons. They are thermoplastic and cannot be used continuously above 75°C. The acrylates are perfectly clear and transparent and often are used to make shields or inert atmosphere enclosures. The materials burn slowly when ignited.

Epoxy resins. The epoxy resins possess excellent adhesive properties and are often used to seal electrode materials into glass and plastic tubes (except Teflon). They will generally withstand temperatures to 200°C and are resistant to weak acids and alkalies and to organic solvents generally. They are affected by strong alkalies and attacked by certain strong acids.

Polyolefins. Polyethylene (conventional and linear) and polypropylene have excellent chemical resistance and are readily molded and machined, although they are rather soft. Conventional polyethylene adheres well to metals, and polyethylene tubing can be readily sealed around metal rods and wires to make simple electrodes suitable for use at temperatures below 60°C. The material is resistant to mineral acids and bases (except concentrated sulfuric and perchloric acids) and most organic solvents except halogenated or aromatic hydrocarbons.

Polyvinyl chloride. This material (PVC) is most commonly encountered in the form of flexible tubing (Tygon and other trade names). It has good transparency and chemical resistance, and is most commonly used for short gas connections between the cell and the gas supply. Its heat resistance varies with composition, but most formulations should not be used above 80°C. It is resistant to salt solutions, alkalies, weak acids, alcohols, and aliphatic hydrocarbons. PVC is subject to attack by strong acids and oxidizing agents, and dissolves in ketones and esters; it also is attacked by aromatic hydrocarbons. The tubing frequently is compounded with plasticizers, such as dioctylphthalate, and these may leach out and be a source of contamination. Although PVC tubing often is used for gases, it has a high gas permeability. Hence at modest flow rates over moderate distances considerable oxygen contamination of a nitrogen or argon supply occurs when PVC tubing is used.

Changes in Solution Composition Caused by Structural Materials. Materials used for the construction of electrodes and cells may alter the composition of solutions by either surface adsorption of solutes or dissolution of adventitious impurities in the structural material. Several studies have indicated that polyethylene, polypropylene, Teflon, stainless steel, borosilicate glass (Pyrex), Vycor, quartz, "soft glass," and silicone-coated surfaces all adsorb appreciable quantities of certain ions. For most of the elements investigated, the total adsorption on plastic or glassware is negligible at millimolar concentrations of the solute. However, in work at trace levels ($<10^{-5}$ M), such as those encountered in the sampling and analysis of natural waters, considerable attention must be paid to the choice of materials used in sampling containers and the various chemical procedures. The adsorption from dilute aqueous solutions on glass and plastic surfaces of Cs, Ce(III), ^{90}Sr–Y, ^{140}Ba–La, ^{95}Zr–Nb, ^{106}Ru–Rh, and I^- at the tracer level (10^{-10} M) has been investigated.[16] The results indicate that for most of the elements studied the use of borosilicate glassware is preferable to the use of polypropylene, but that for Cs, Ru, and Zr there is a lower loss on polypropylene surfaces. The protection of laboratory glassware with a silicone containing hydrophobic agent results in some decrease in adsorption, but probably not enough to justify the trouble and expense.

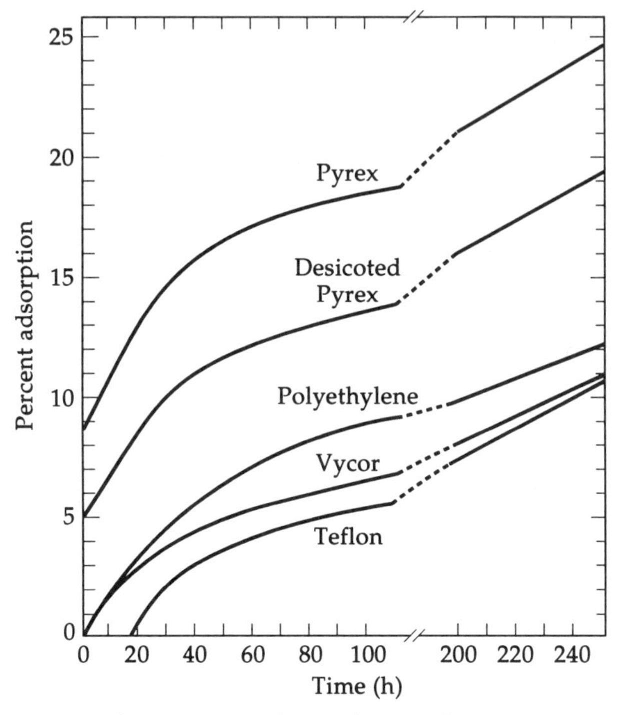

Figure 6.6 Adsorption of silver ion as a function of time.

The adsorption of trace silver ion (2×10^{-6} M) has been studied by use of a silver ion-selective electrode; the results are shown in Figure 6.6. For periods of up to a day, silver losses on Teflon were less than 2%; at the end of 30 days the losses (28%) were least for Vycor glass. The study indicates that none of the materials is suitable for long-term storage of solutions containing low levels of silver unless a complexing ligand such as thiosulfate is present.

Mercury is particularly liable to loss on polyethylene. The loss of mercury at 0.05 ppm concentration (added as mercuric chloride) is such that in 4 h 95% is lost from filtered creek-water samples and 40% is lost from distilled-water samples, both stored in polyethylene. This loss rate is so rapid that some mechanism other than adsorption may be responsible; one suggestion is that traces of reducing agents react with Hg(II) to form small amounts of Hg(I) that disproportionate to Hg(II) and Hg. The mercury formed is presumed to be lost by volatilization.

A wide variety of solvents, reagents, and structural materials encountered normally in the trace-element analysis of seawater have been analyzed for trace-element impurities by neutron activation analysis and gamma (γ)-ray spectrometry.[17] Some of the results obtained for 10 trace elements are shown in Table 6.4 and indicate that many substances contain high impurity levels of various elements. Particular note should be made of the high concentrations of zinc in

TABLE 6.4 Trace-Element Concentrations in Structural Materials

Sample	Source	Concentration (ppb)									
		Zn	Fe	Sb	Co	Cr	Sc	Cs	Ag	Cu	Hf
Construction materials											
Teflon	*a*	9.3	35	0.4	1.7	<30	<0.004	<0.01	<0.3	22	NM
Plexiglas	*b*	<10	<140	<0.01	<0.05	<10	<0.002	<0.06	<0.03	<9.5	NM
Polyvinyl chloride	*c*	7120	270,000	2690	45	2	4.5	<1	<5	630	NM
Surgical rubber tubing	*d*	3×10^6	UM	<100	<30	UM	<8	<100	1240	<6	NM
	e	5×10^6	UM			UM					
Neoprene rubber	*f*	1.8×10^7	UM	290	2300	UM	3090	UM	<1000	NM[s]	NM
Nylon (block)	*g*	UM[r]	UM	UM	1.4×10^6	UM	UM	UM	UM	UM	UM
Polyethylene tubing	*h*	55	7.4	9000	140	254	11	<100	<300	NM	<100
Container and related material											
Borosilicate glass	*i*	730	280,000	2900	81	UM	106	<100	<0.001	NM	597
Vycor glass	*i*	UM	UM	1.1×10^6	UM	UM	UM	UM	UM	UM	UM
Quartz tubing	*j*	1.5	395	0.05	0.44	6.5	0.03	1.1	0.05	2	<0.005
Quartz tubing	*k*	<1	NM	<0.01	12	2.5	0.39	<0.1	<0.01	0.04	<0.005
Quartz tubing	*l*	UM	UM	1940	0.89	UM	UM	1390	<0.1	0.09	<0.01
Quartz tubing	*m*	21	NM	43	0.64	230	0.18	0.30	<0.1	0.05	27
Quartz tubing	*n*	33	NM	38	1.1	602	0.16	<0.1	<0.1	0.03	26
Polyethylene	*o*	28	10,400	0.18	0.07	76	0.008	<0.05	<0.1	6.6	<0.5
Polyethylene	*p*	25	10,600	0.83	0.31	19	0.36	<0.15	<0.1	15	<0.5
Kimwipe tissue	*q*	48,800	1000	16	24	500	14	<0.1	~0.8	NM	NM

[a] Du Pont.
[b] Manufacturer unknown.
[c] Manufacturer unknown.
[d] KentLatex Products, Inc.
[e] Rubber Latex Products, Inc.
[f] Atlantic India Rubber Works.
[g] Manufacturer unknown.
[h] Interstate Plastics Company.
[i] Corning Glass Company.
[j] "Spectrosil," Thermal American Fused Quartz Company.
[k] "Suprasil," Engelhard Industries, Inc.
[l] Quartz Products.
[m] United States Quartz.
[n] General Electric Company.
[o] Nalgene, No. 6250 container.
[p] The Chemical Rubber Co., 2/5 dram polyethylene, flip-top vial.
[q] Kimberly-Clark.
[r] UM: unable to measure because of interfering radionuclides.
[s] Not measured.

neoprene and natural rubber tubing and in Kimwipe cellulose tissue; zinc, antimony, and iron in polyvinyl chloride tubing; antimony in one source of polyethylene tubing; iron in one source of polyethylene bottles; iron and antimony in borosilicate glass; a large concentration of antimony in a sample of Vycor glass; and cobalt in a sample of nylon. Teflon appeared to be relatively free of trace-metal impurities, and the purest material of those studied was a sample of Plexiglas.

Some samples of plastic tubing contain plasticizers (particularly PVC tubing), and fillers and mold-release agents often are employed. In contact with a solution these materials may be leached out and introduce contaminates to the sample.

The Maintenance of an Inert Atmosphere. Because oxygen is chemically reactive with many substances and electrochemically reducible, electrochemical work requires exclusion of oxygen. This is generally done by purging the system with a flow of inert gas (nitrogen, argon, helium, etc.), by enclosing the system in an inert-atmosphere glovebox, or by using a vacuum line. A properly designed vacuum-line system is considered the most effective way of rigorously excluding oxygen and water, but it is possible to maintain an atmosphere below 1 ppm of oxygen and water in a glove box, and a glovebox allows the use of simpler cells and more convenient manipulation.

For polarographic and voltammetric work where the presence of oxygen interferes merely because of its electrochemical reducibility, purging the solution and cell with a flow of inert gas is entirely satisfactory and is capable of reducing the oxygen concentration in the cell atmosphere to something between 20 and 200 ppm. However, certain species react rapidly with oxygen, and an inert gas containing 20 ppm of oxygen may be intolerable. For example, a 50-mL solution of 10^{-4} M Cr(II) that is purged with nitrogen that contains 20 ppm oxygen at a flow rate of 200 mL min^{-1} will be completely oxidized in about 15 min, assuming that all the oxygen flowing into the system reacts with the Cr(II) ion.

The choice of inert gas to some extent depends on local costs and availability. In addition to the cost of the gas, transportation and demurrage costs of gas cylinders can be substantial. Nitrogen is most often used because it is cheapest and is widely available in a "prepurified" grade that contains not more than 20 ppm of dioxygen. However, nitrogen reacts with lithium at room temperature, forms molecular complexes with certain transition metals,[18] and may be reactive in fused-salt systems at high temperature. In these circumstances the use of argon or helium is preferred. Helium can be obtained in very high purity by diffusion through quartz[19] or by use of absorption columns.[20] Hydrogen also can be obtained in an extremely pure state by electrolysis and/or diffusion through palladium–silver alloys, but its flammability, low density, and potential reactivity with some systems make it less suitable for routine use. The use of a gas that is denser than air, such as argon, is preferred so that the solution will be covered with a blanket of inert gas.

Residual oxygen can be removed from nitrogen or other inert gases by bubbling through aqueous solutions containing vanadous ion or chromous ion that are continuously regenerated through contact with amalgamated zinc; another convenient recipe uses vanadous ion.[21] Gases treated by bubbling through aqueous solutions will not be suitable for use with nonaqueous solvents unless there is provision to dry the gas, and the continual carryover of water will require an inconveniently large drying capacity.

Perhaps the most efficient and convenient methods for providing a supply of dry oxygen-free inert gas employs BTS catalyst (BASF) in conjunction with molecular sieves. The catalyst is a pelleted form of finely divided copper on an inert support and is used in a system like that shown in Figure 6.7. The nitrogen passes through molecular sieves, BTS catalyst, and another column of molecular sieves. The output gas will have an oxygen and water content each below 1 ppm in a carefully built system. The three-way stopcocks in the inlets are used to bleed the regulators and inlet lines when the cylinders are

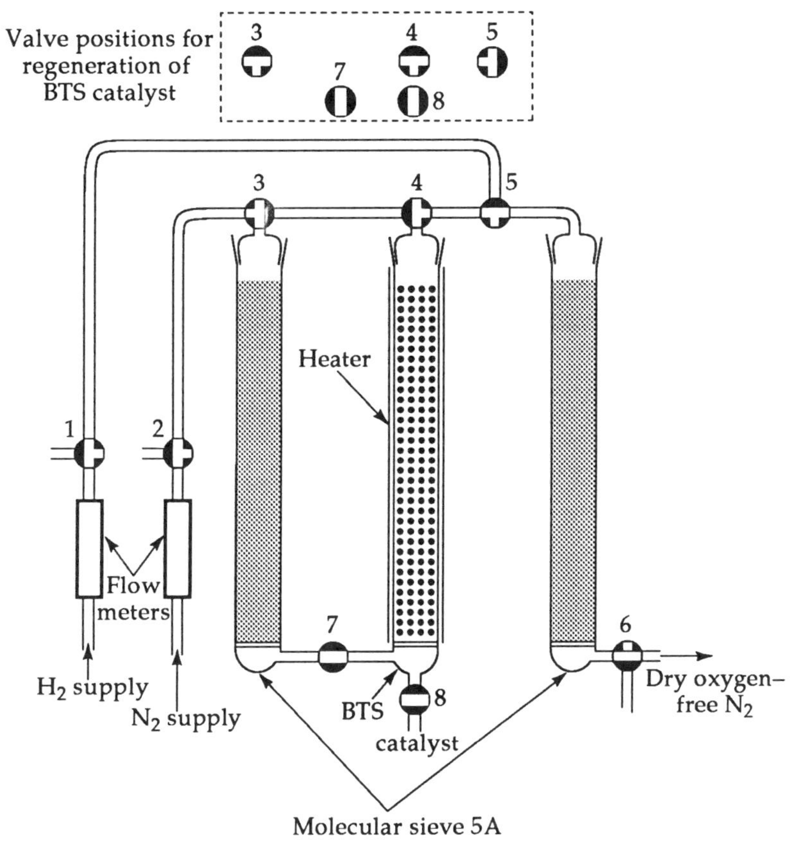

Figure 6.7 Gas purification system for removal of oxygen (BTS catalyst) and water (molecular sieves).

changed. During the regeneration cycle, the BTS catalyst is heated to about 150°C by means of a heating tape while gradually replacing the inert gas with hydrogen. During the regeneration cycle the stopcock at the bottom of the BTS column is opened to allow water produced in the regeneration step to bleed off while the stopcock between the BTS column and the left molecular sieve column is shut off. The catalyst may be regenerated many times; the extent of exhaustion of the catalyst is marked by the progress of a yellowish band as the reduced catalyst is oxidized.[22] This method appears to be superior to the use of heated copper gauze; the latter has a much lower surface area that is rapidly coated with an oxide layer.

A whole laboratory can be readily served by one purification system if the purified gas is fed into a manifold with several outlets. Copper or aluminum tubing with Swagelok fittings and needle valves are convenient for this purpose.

Where possible, gas connections should be made with metal or glass tubing. Where short flexible connections are necessary, PVC (Tygon) tubing is recommended. Most plastic materials have an appreciable permeability;[22] a 1-m length of $\frac{1}{4}$-in.-ID polyvinyl chloride tubing with $\frac{1}{16}$-in wall thickness will allow the diffusion of about 0.8 ppm of oxygen at a flow rate of 150 mL min^{-1}.

A common practice is to saturate the purified, oxygen-free gas with solvent in a gas-dispersion cylinder that is filled with a solution identical to the test solution. This avoids the evaporation of the solvent or the differential evaporation of a volatile constituent of the test solution that would otherwise accompany the continuous purging of the system. Such a practice is particularly important for volatile solvents like acetonitrile or for solutions that contain volatile constituents like ammonia.

When exclusion of both water and oxygen is necessary, gloveboxes are commercially available that can control water and oxygen levels below 1 ppm. To achieve this level requires a carefully designed system that provides for continuous circulation and purification of the inert atmosphere. Most of the commercial gloveboxes provide for positive circulation of the atmosphere through a purification train that removes water with molecular sieves and oxygen with BTS or a similar copper catalyst. Some systems incorporate a small percentage of hydrogen in the atmosphere so that the BTS catalyst will be reduced continually.

Although expensive, these units are convenient to use after a little practice. Their continued use with relatively volatile organic solvents may pose a problem, however. Organic solvents are adsorbed by the molecular-sieve columns, causing premature exhaustion of their capacity to adsorb water. Sulfur-containing compounds also poison the BTS or copper catalyst. For these reasons, solvents should be freed of oxygen before they are put in the glovebox and should be capped. Even though the atmosphere in a glovebox is maintained below 1 ppm, the water level in a solvent will not be reduced below the level present when the solvent was introduced into the glovebox. The glovebox is of little use if the solvents have not been carefully purified and protected from moisture during all transfers.

The rate of circulation in the purification train is important; a throughput not less than one box-volume per minute is recommended. The time required to achieve a specific reduction in the level of oxygen is approximated by the relation

$$t = 2.3 \frac{V}{FE} \log \frac{C_o}{C_t} \tag{6.3}$$

where C_o = initial concentration of oxygen
C_t = concentration after time t
V = box-volume
F = flow rate
E = efficiency (fraction of O_2 removed on passage through the purification train)

The introduction of 0.5 L of air into a 500-L glovebox produces a concentration of 200 ppm of oxygen. This level can be reduced to about 1 ppm in 5–10 min at a throughput of one box-volume per minute, which implies that the purification system has an efficiency of about 50%. To reduce oxygen and water below 20–50 ppm is difficult unless the circulation rate exceeds 10 box-volumes per hour.

Oxygen in the glovebox atmosphere may be monitored with a Hersch galvanic cell[24] as pictured in Figure 6.8. The sensor cathode is a fleece of graphite

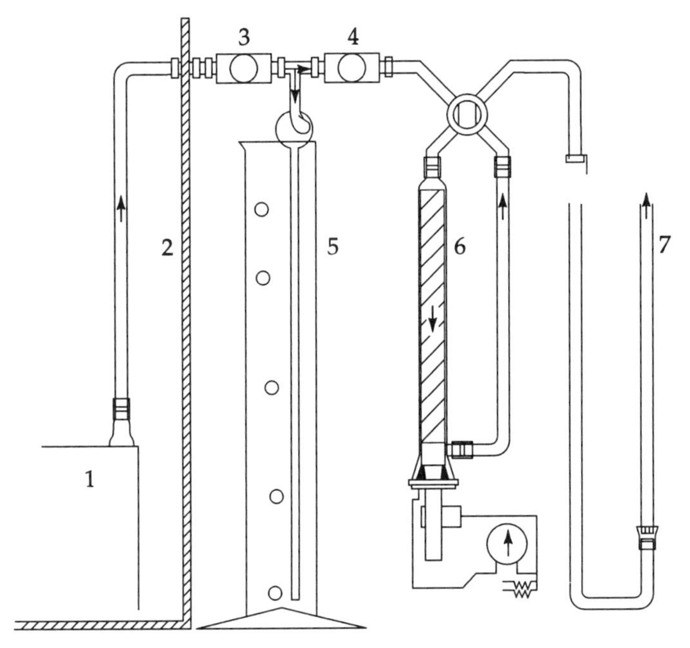

Figure 6.8 Hersch galvanic cell for the determination of oxygen at trace levels in gases.

or porous silver exposed on one side to the sample gas stream while the other side is lined by a separator soaked with sodium hydroxide. The separator in turn contacts a cadmium anode. The two electrodes are bridged by a microammeter. The maximum readout current is about 10 μA for each part per million of oxygen in the inert gas at a flow rate of 37.4 mL of inert gas per minute (with flow measured at 20°C and 1 atm). The lifetime of a tungsten filament exposed to the glovebox atmosphere may also be used as a rough indication of the oxygen content of the inert gas.[25]

Water in the glovebox atmosphere can be monitored by direct amperometric measurement.[26] The cell, shown in Figure 6.9, consists of two platinum wires closely wound in a spiral and jacketed in Teflon or glass. The approximately meter-long element is in turn wound into a helix and potted in a metal housing. The platinum wires are bridged by a high-resistance P_2O_5 electrolyte; in-use absorbed water is quantiatively electrolyzed to hydrogen and oxygen at the platinum electrodes by the application of a DC voltage greater than the decomposition potential of the water. This not only provides continuous indication of water content but also maintains the film in an absorbent condition. The current output is proportional to the water concentration in the gas flowing through the cell and can be used to measure concentrations from less than 1 ppm up to about 1000 ppm.

If a glove box is not available, Schlenk-type glassware used with medium vacuum and inert gas is a relatively inexpensive and convenient approach. The use of this glassware is described in a general reference.[27] Dissolution, recrystallization, solution transfer, and other simple operations can be carried out on the benchtop. Syringe techniques using apparatus closed with rubber septa also have been developed.[28]

The most rigorous exclusion of water and oxygen can be accomplished on a high vacuum line. There appears to be a significant psychological barrier that has to be overcome before most workers will go to the trouble of building a suitable vacuum system, so that this approach is regarded as a technique of last resort.

The capital investment in a vacuum line may be considerably less than the cost of a good glovebox, and with well-designed cells and accessories a vacuum line can be convenient to use. A simple manifold with a diffusion pump backed by a mechanical pump can serve as the core of the system.[29,30] One or more nitrogen traps are placed between the pumps and the manifold leading to the cell. In addition to a supply of liquid nitrogen for the traps, the system requires a supply of pure, dry, inert gas for backfilling and flushing the electrochemical cells after evacuation. A versatile and convenient cell is shown in Figure 6.10; its operation has been described in great detail.[31] The solid-supporting electrolyte is placed in ampoule B' and dried under vacuum; solvent is then vapor-transferred into the ampoule until the desired concentration is attained. The same is done with the reference-electrode electrolyte in ampoule A' and for the sample in ampoule C'. For polarographic work, enough mercury is added to section C to cover the platinum wire, and a small Teflon-clad stirring bar is

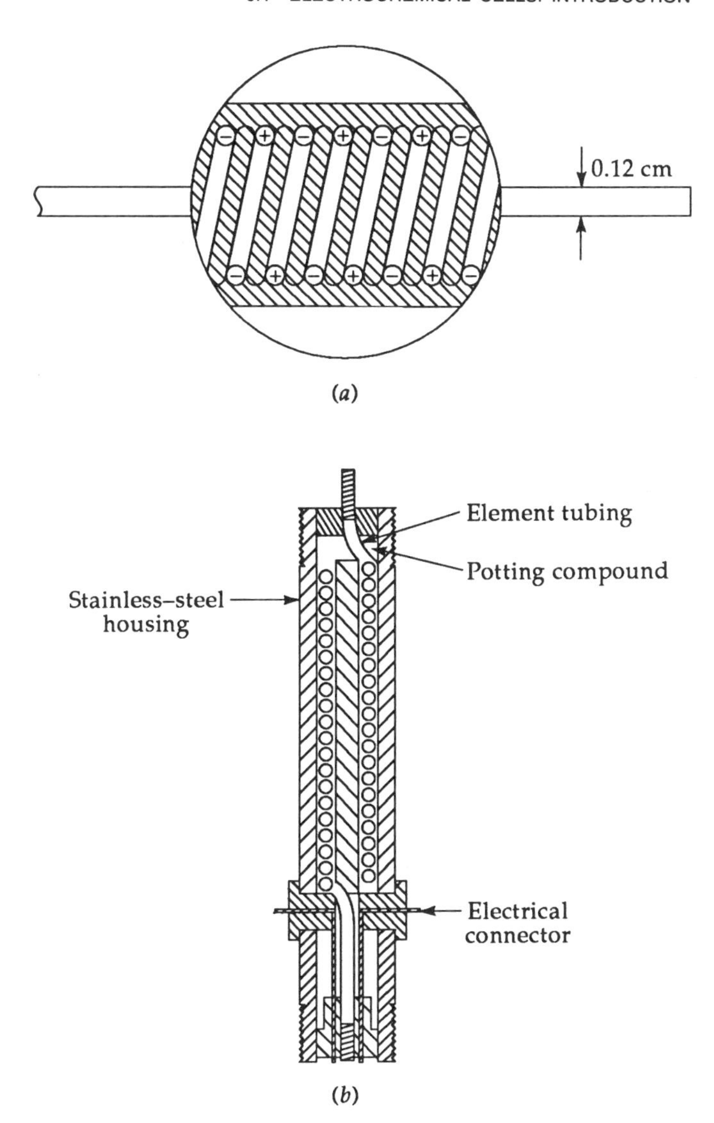

Figure 6.9 Keidel coulometric moisture analyzer for the determination of water in gases. At a flow rate of 100 mL min^{-1} a moisture content of 1 ppm gives a current of 13.2 μA.

floated on the mercury. With ampoules A', B', and C' in place, the cell is attached to the vacuum line at the joint on the top of section B. By tilting the ampoules and manipulation of the stopcocks the cell is filled and the polarographic operations carried out as described by Anderson and coworkers. The cell also may be used for cyclic voltammetry and coulometry. A similar cell has been described for use in electrosynthesis.[32]

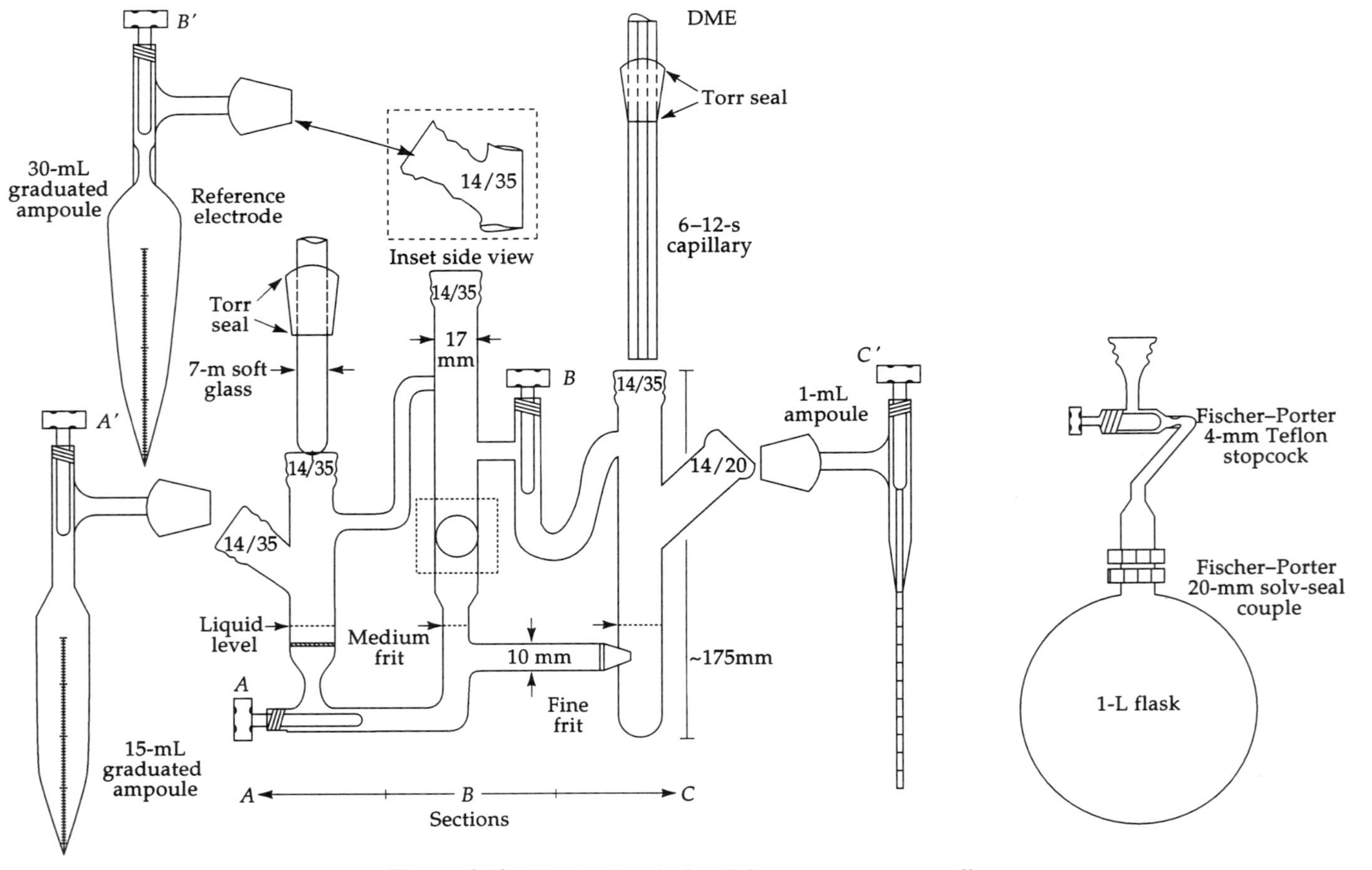

Figure 6.10 Electrochemical cell for use on a vacuum line.

6.2 DESCRIPTION OF GENERAL-PURPOSE CELLS

Cells for Voltammetry and Polarography. For ordinary voltammetric measurements in low-resistance solutions, a simple cell employing a spoutless beaker and a close-fitting rubber stopper is adequate. The stopper is drilled to accept the electrodes and the purging tubes, and to provide for addition of reagents or other necessary functions. A reference electrode of the style shown in Figure 6.11 is convenient for two-electrode voltammetry. If a three-electrode circuit is used, a higher resistance silver–silver chloride or calomel electrode of the type used in pH measurements may be used, and a platinum wire, carbon rod, or mercury pool may be used as the counter electrode. Isolation of the counter electrode usually is not necessary in a polarographic measurement because the current is so small that it produces a negligible change in solution composition.

Figure 6.11 shows one of the many variations of the popular H-cell widely used for polarography. In this design the sample compartment may be detached for easy cleaning. The sidearm of the reference-electrode compartment is closed with a glass frit, and when the cell is detached the sidearm is capped so that the reference electrode will not dry out. The stopcock allows the solution and accumulated mercury to be drained before rinsing and filling with the next sample. Strong oxidants may attack mercury, and it may be desirable to protect the accumulated mercury pool from a dropping-mercury electrode (DME) under a layer of some inert immiscible solvent that is denser than the electrolyte solution, such as chloroform. Nitrogen or another inert gas is bubbled through

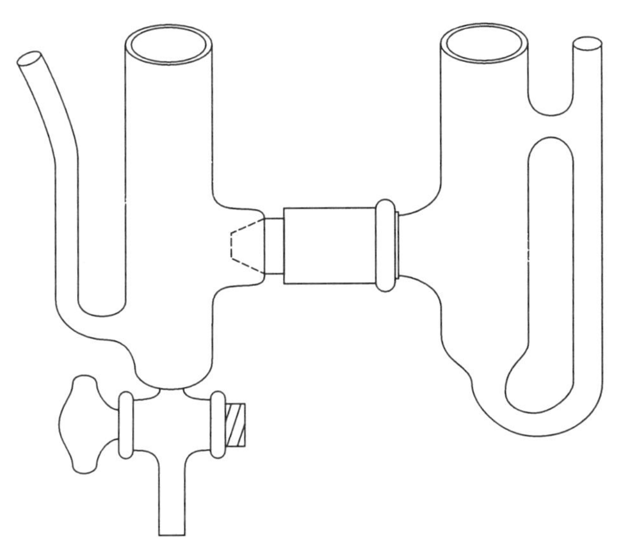

Figure 6.11 H-cell for polarography with demountable sample compartment on left side.

the solution by means of the tube sealed near the bottom of the cell. Removing 99% of the oxygen orginarily takes about 5 min; a glass-frit dispersion tube with a flow rate of 20 to 50 mL min^{-1} will cut the time to 2–3 min. A two-way stopcock (not shown) is used to pass nitrogen over the top of the solution while the polarogram is recorded. The cell top is sealed with a rubber stopper drilled to loosely accommodate the DME and purging tube used to pass inert gas over the surface of the solution; the inert gas exits around the capillary of the DME.

Another cell, pictured in Figure 6.12, is convenient for voltammetric and coulometric work except where there is the need to rigorously exclude oxygen or water. The glass cell is a standard 10-mL spoutless beaker and may be quickly changed by unsnapping the polyethylene retaining ring. The polyethylene top contains six molded holes that accommodate various electrodes, purging tubes, and salt bridges. In Figure 6.12 the cell is set up for three-electrode polarography with the counter electrode isolated by a glass frit. (For coulometric work, a second glass-fritted tube may be inserted inside the first to

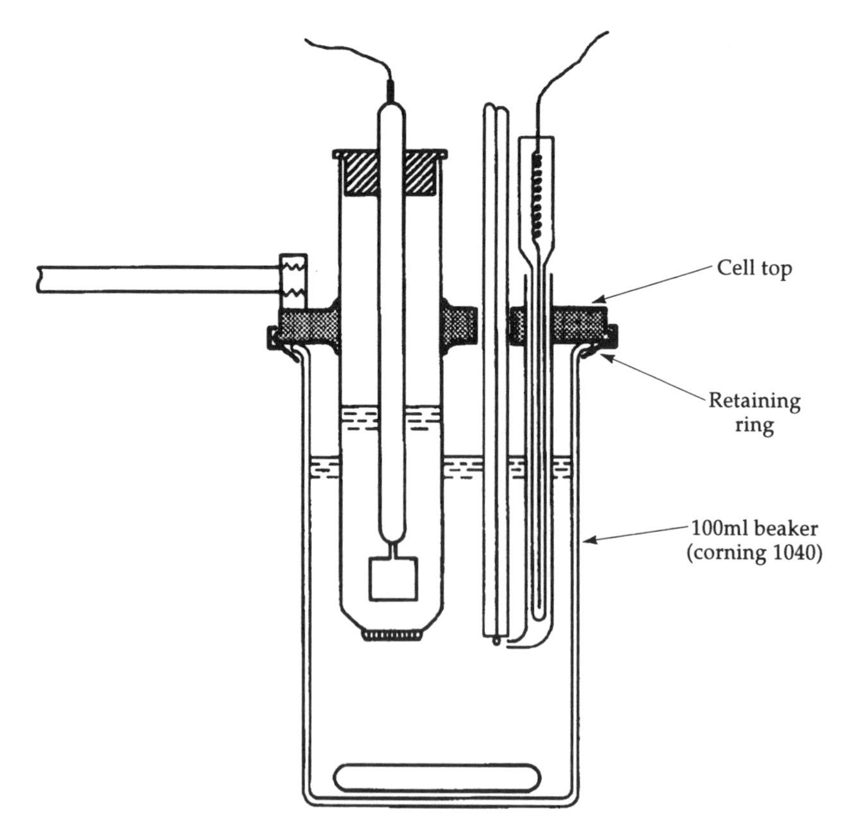

Figure 6.12 Versatile electrochemical cell with polyethylene top and retaining ring. The system includes a magnetic stirring bar, an isolated counter electrode, and a Luggin capillary reference electrode.

provide greater isolation of the counter electrode.) The reference electrode may be located in a Luggin probe as shown, or in a bridge tube like that shown in Figure 5.6*d*. This cell is well suited for routine voltammetry in organic solvents using a bridge tube filled with the organic solvent in conjunction with an aqueous or organic solvent reference electrode. The cell may be used for coulometry by inserting a platinum working electrode concentric with the glass-fritted tube holding the counter electrode; it may also be used with a mercury pool, if a platinum or tungsten contact is sealed into the bottom of the cell. A magnetic stirring bar, resting on the bottom of the cell or floating on top of the mercury pool, provides adequate stirring.

Figure 6.13 illustrates a cell that can be used for two- or three-electrode polarography. It has an integral salt bridge with provision to change the bridge solution without disrupting or disassembling the reference electrode. This allows the use of an indifferent salt-bridge electrolyte solution that prevents cross-contamination of the electrolytes in the sample and reference electrode compartments. The tapered cell can accommodate sample sizes from a few milli-

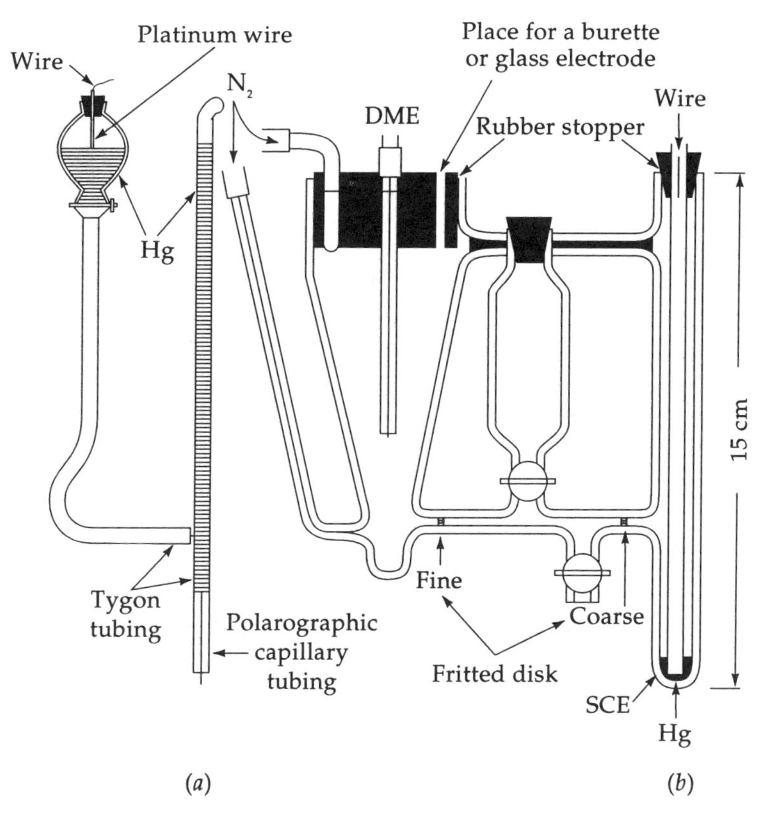

Figure 6.13 Cell for voltammetry and amperometric titrations: (*a*) dropping-mercury electrode assembly; (*b*) cell with provision to flush the salt bridge.

liters up to 50 mL; furthermore, it has a sufficiently large top that a second reference electrode may be introduced for three-electrode voltammetry, using the first reference electrode as the counter electrode. The large top also can accommodate other electrodes and a burette for specialized applications—for example, it is convenient to use with a glass electrode for studies of complexation equilibria; both the polarographic response and the solution pH can be monitored simultaneously. A slightly different cell has been described that is particularly suited to the determination of stability constants by the polarographic method.[33]

Cells for Coulometric and Preparative Electrochemistry. The controlled-current coulometric titration technique is inherently accurate and provides the basis for the precise determination of the Faraday constant to about 1 part per 100,000.[34] Two important factors in the design of cells for coulometric titrations are (1) the need to maintain the cell resistance as low as possible and (2) the necessity to provide adequate isolation of the counter electrode. Errors can arise from outflow of sample from the working-electrode compartment or by inflow of reactive products produced at the counter electrode. Some loss through glass-frit or membrane separators is inevitable because the flow of current necessarily involves transport of material. The transport may be by migration, diffusion, and hydrodynamic flow. Migration can be reduced by using a supporting electrolyte in large excess over the sample concentration. The effect of the remaining two transport mechanisms may be reduced by special cell designs. Accurate results have been obtained with a four-compartment cell (see Figure 6.14) in which the two central bridge compartments are periodically emptied into the sample compartment by applying nitrogen pressure.[34] They are refilled by applying vacuum.

A bridge compartment with sufficient hydrostatic head to ensure that fresh electrolyte flows continuously into both the working-electrode and counter-electrode compartments is another useful approach.[35] However, the volumes of the solutions in both compartments increase inconveniently during a titration. This method also suffers if the background electrolyte contains reactive impurities (e.g., oxygen).

The problem of adequate isolation has been studied theoretically and practically in the context of coulometric titrations.[36] The loss of sample by diffusion will not be greater than 0.1% if at least two fine-frit separators are used with at least 4–5 cm of electrolyte between the frit and if the time of electrolysis is short (3–15 min). A tube closed with a fine frit that contains a 12-mm layer of fine silica gel above the frit has been suggested.

Ion-exchange membranes have been used as separators,[37,38] particularly in coulometric titrations. The fixed negative charge of a cation-exchange membrane tends to exclude anions, so that current through the membrane is due primarily to cation transport with the reverse true for anion-exchange membranes.

Other devices may be useful. For example, if the counter electrode is the

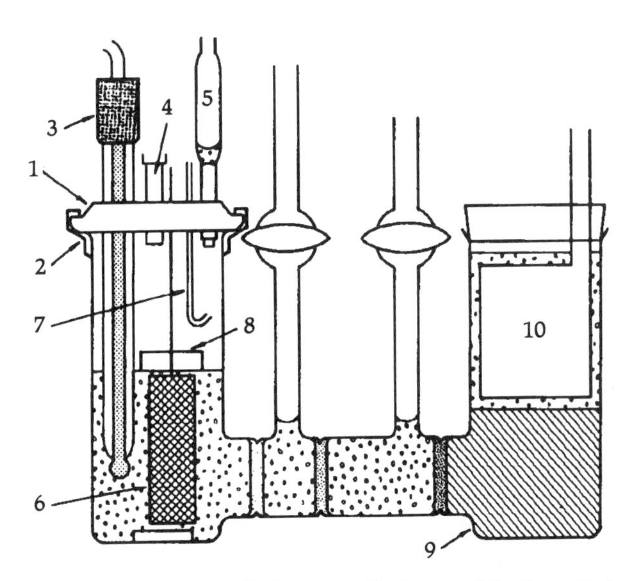

Figure 6.14 Four-compartment cell for controlled-potential electrolysis and coulometric titrations. Two central bridge compartments can be emptied to sample compartment by application of nitrogen pressure and refilled by vacuum: 1, polyethylene top; 2, lock ring; 3, combination glass–calomel electrode; 4, N_2 inlet tube; 5, N_2 outlet tube; 6, Pt gauze electrode; 7, cell rinse assembly; 8, polyethylene spray shield; 9, 0.1 M KCl in 3% agar gel; 10, Ag anode.

anode, a silver anode may be used with a halide-containing supporting electrolyte; oxidation of the silver produces a layer of silver halide on the anode that can be periodically removed. Hydrazinium chloride also has been used as an anodic depolarizer (particularly in hydrochloric acid solution to prevent formation of chlorine at a platinum anode); it is quantitively oxidized to nitrogen gas.

In both controlled potential and controlled current coulometry as low a cell resistance as possible is desirable. This obviously is affected by cell design (narrow passages between working-electrode and counter-electrode compartments should be avoided), but is is determined mainly by the specific resistance of the electrolyte. The latter typically will range from 5 Ω-cm for 0.5 M H_2SO_4 in water to 220 Ω-cm for 0.1 M tetra-*n*-butylammonium perchlorate in dimethylformamide. Typical cell resistances with these two electrolytes might range from 20 to 1000 Ω. A cell resistance as large as 1000 Ω may produce some difficulties. For example, to pass a current of 100 mA through the cell would require a potentiostat output of 100 V.

Another factor to consider is the resistance heating of the electrolyte. In the example cited, 10 W of power is dissipated by passing 100 mA through a cell with 1000 Ω of resistance. This is sufficient to raise the temperature of the electrolyte significantly unless provision for cooling is made. Finally, large potential gradients in the cell may produce electrophoretic transport of solvent through membrane or glass-frit separators, producing loss or contamination of the sample, depending on the direction of flow.

Optimization of cell design for controlled-potential coulometry. The prime contributor to instability and oscillation of the potential-control circuit is an excessive phase shift in a poorly designed cell. Ideally, a minimum-phase-shift cell will have a completely uniform potential and current distribution at the working electrode. If this cannot be realized, the reference-electrode tip should be positioned on the line of minimum separation between the working electrode and the counter-electrode separator (for reasons previously described). The reference-electrode resistance is more significant than had been realized previously, and it is important to keep this resistance as low as possible, along with the cell resistance, especially in organic solvents where resistances tend to be larger. The bridging capacitance between the counter electrode and the follower input of the reference electrode also should be minimized by use of cables in which the reference-electrode conductor is shielded or guarded separately from that of the counter electrode.

Figure 6.15 shows a mercury-pool cell that has a virtually uniform potential and current distribution because of the large parallel-spaced glass-frit separator.

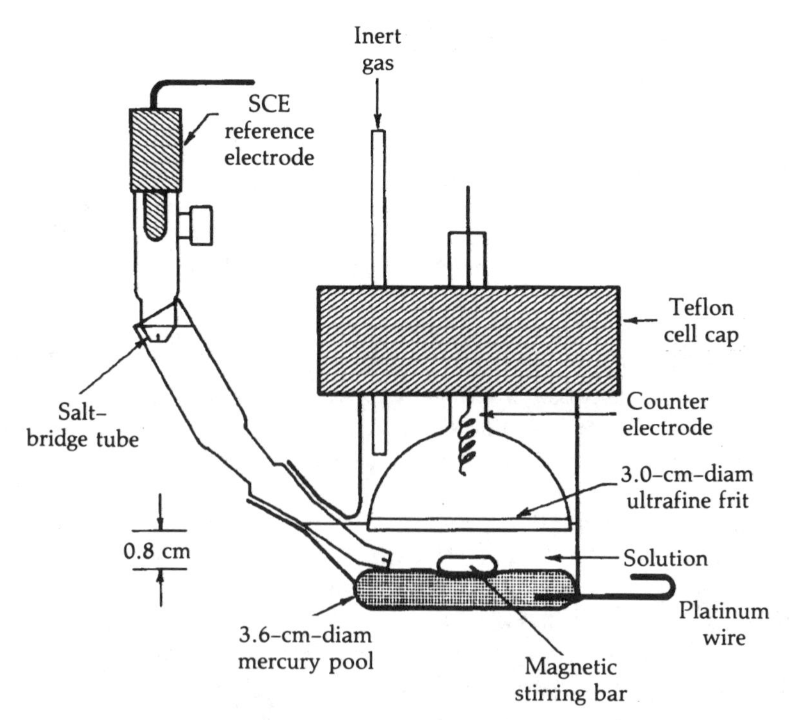

Figure 6.15 Mercury-pool cell for controlled-potential electrolysis with uniform potential and current distribution.

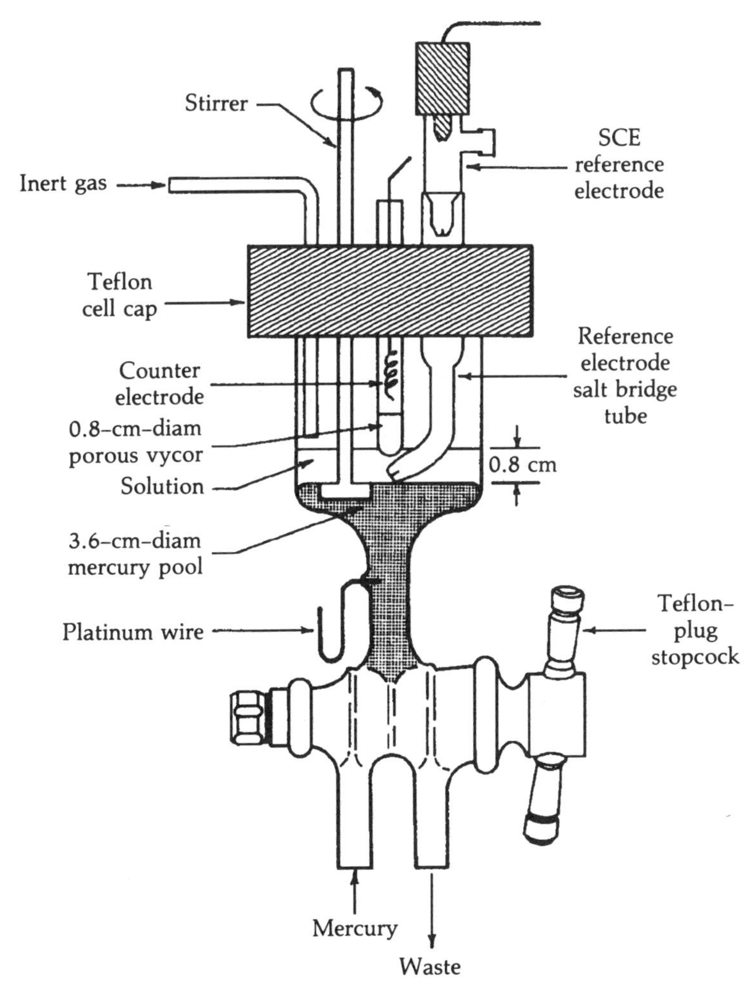

Figure 6.16 Mercury-pool cell for controlled-potential analytical coulometry.

A somewhat more convenient cell with a less ideal current distribution is shown in Figure 6.16. Note the position of the reference-electrode tip between the counter-electrode separator and the mercury pool. Both of these cells have approximately 10-cm^2 mercury-pool working electrodes with 7-mL sample solution volumes.

A platinum working electrode cell is shown in Figure 6.17. The projected area of the gauze is about 70 cm^2 with a solution volume of about 25 mL. Except for minor edge effects, this cell has a nearly uniform potential and current distribution because of the cylindrical symmetry of the platinum gauze and porus Vycor counter-electrode separator. The tip of the reference electrode is positioned close to the platinum gauze on a line representing the shortest distance between the gauze and the Vycor tube.

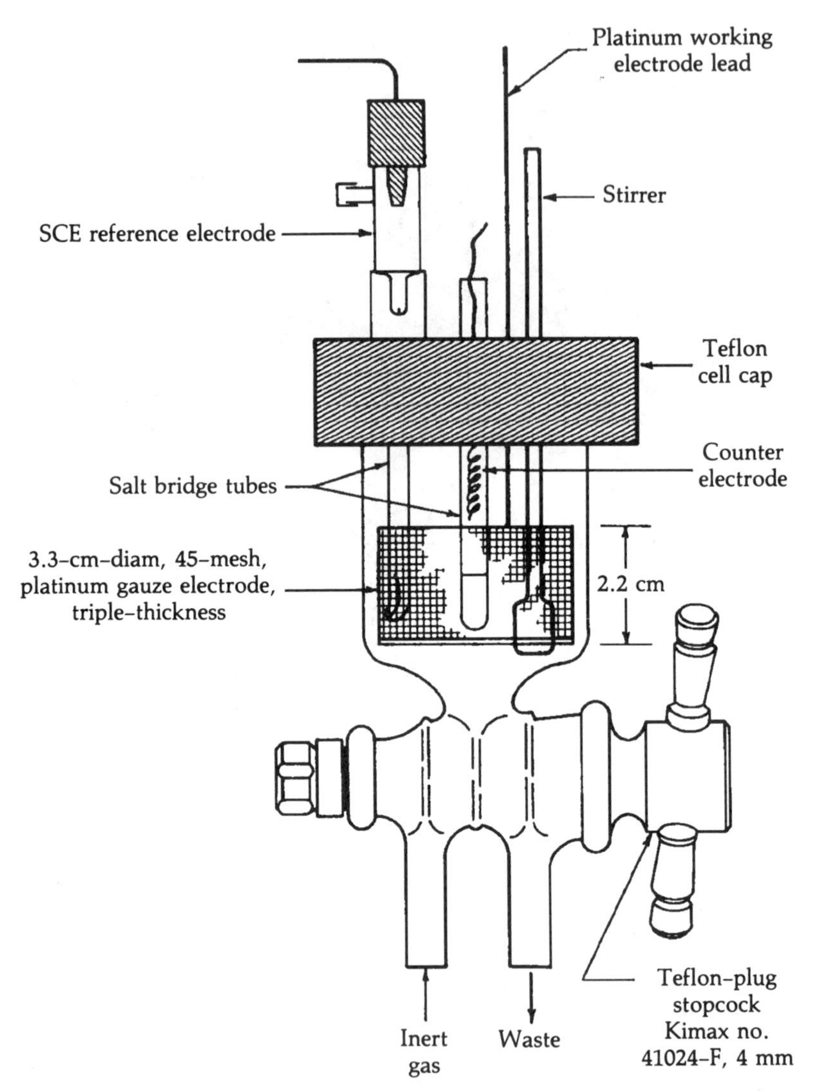

Figure 6.17 Platinum gauze cell for controlled-potential analytical coulometry.

The time required for a simple controlled-potential coulometric determination is determined by the efficiency of mass transport in the cell. The current decays exponentially according to the equation

$$i_t = i_0 \exp(-kt) \tag{6.4}$$

where i_t is the current at time t, i_0 is the initial current, and k is a function of the rate of mass transport. For a simple Nernst diffusion-layer model of mass transport k is given by the expression

$$k = \frac{DA}{\delta V} \tag{6.5}$$

where D = diffusion coefficient of the electractive species
A = electrode area
V = solution volume
δ = thickness of the diffusion layer near the electrode surface

The actual functional dependence of k on the variables is more complex than indicated by Eq. (6.5),[40] but as a rule of thumb, the ratio of solution volume to electrode area should be as small as possible and δ should be made as small as possible by efficient stirring.

Completion of electrolysis is generally taken at the time when the current has decayed to 0.1% of its initial value. For this condition

$$t = \frac{6.9}{k} \tag{6.6}$$

which means that it is desirable to have k as large as possible. Cells like those depicted in Figures 6.15–6.17 typically have k values of 0.008 s^{-1}. By increasing the efficiency of stirring and decreasing the cell volume, values of k in the range 0.01–0.02 s^{-1} can be achieved, which greatly shortens the time for a coulometric determination. Ultrasonic stirring[40] has been employed as have highly efficient rotated cells in which the solution is maintained in a thin layer over the electrode surface by centrifugal force.[41]

Cells for preparative electrochemistry. Organic and inorganic preparative electrochemistry requires that solution volumes and electrode areas be scaled up. Particular attention must be paid to optimizing the geometry of these cells in order to obtain uniform current and potential distribution over the working electrode. This is particularly important in organic electrochemistry, where products may depend critically on the potential and where the solvent–electrolyte systems used tend to have large resistances. The latter lead to large potential-control errors in a poorly designed cell. The limitations of potentiostat output and resistive heating also require that the cell be designed to obtain low resistance, and that provision for cooling be supplied. To avoid contamination of the working-electrode solution by the products of reaction at the counter electrode, some designs provide for a continuous flow of fresh electrolyte through the counter-electrode compartment.

Control of Temperature and Pressure. Diffusion coefficients in aqueous solution have a temperature coefficient of about +2% deg^{-1},[42] which means that polarographic diffusion currents or voltammetric peak currents increase about 1–2% deg^{-1}. The rates of follow-up chemical reactions of reactive species produced at the electrode surface depend even more strongly on temperature. For these reasons, temperature control to $\pm 0.1\,^{\circ}\mathrm{C}$ or better is required in all careful quantitative work. This is accomplished by immersing the cell in a temperature controlled oil or water bath, or by circulating water in a jacketed cell.[43] Temperature control in coulometry is less critical unless there is substantial cell heating by the current that passes through the cell. Coulometry usually is carried out at the ambient temperature of the laboratory.

The variable-temperature cell in which reactive intermediates can be studied at temperatures down to $-130\,^{\circ}\mathrm{C}$ (Figure 6.18) is designed with opposed, coaxial working, and Luggin-probe electrodes surrounded by a platinum-coil counter electrode. The cylindrical symmetry provides uniform current and po-

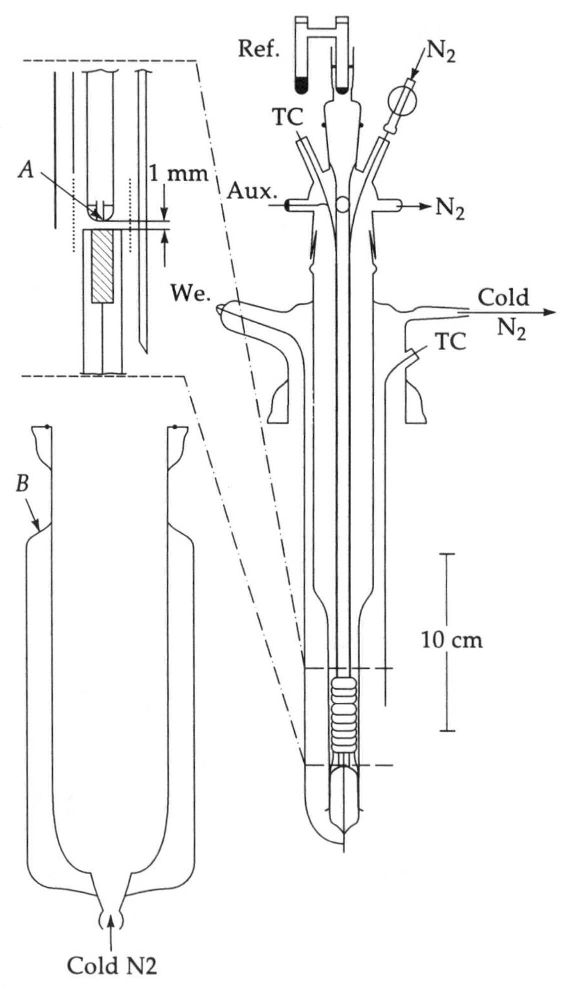

Figure 6.18 Electrochemical cell for low-temperature studies (Ref. = reference, Aux. = auxiliary, We. = working electrodes).

tential distribution across the platinum working electrode. The Luggin probe is connected to an external aqueous reference electrode maintained at room temperature. Butyronitrile/0.1 M tetra-*n*-butylammonium perchlorate proved to be the most useful low-temperature medium; the 4-mL volume of solution is deaerated by nitrogen bubbled in through a long syringe needle. Cooling of the cryostat is achieved by passing compressed dry nitrogen through a copper coil immersed in a 4-L dewar of liquid nitrogen. A vacuum-jacketed heater also is located in the inlet line. The temperature of the cooling gas, and, in turn, the cryostat is automatically regulated with a recorder–controller equipped with a sensing thermocouple located in the gas space surrounding the cell. The actual solution temperature is measured with a second thermocouple. The cell has been used to study radical-ion decay mechanisms, and for the evaluation of rate constants and activation parameters for homogeneous chemical reactions coupled to charge transfer. The stabilization of reactive intermediates by means of low-temperature electrochemistry has been reported.[45–47]

A special cell has been described[48] that employs a bimetal-thermistor electrode to measure the exotherms and endotherms associated with electrochemical reaction. This is done concurrently with cyclic voltammetric measurements.

Fused-salt systems and organic solvents with melting points above room temperature (such as dimethylsulfone) require higher temperatures. These may range from simple systems that use beakers wrapped with heating tapes to carefully designed cells that are contained in high-temperature furnaces with temperature controllers. At higher temperatures the external connectors and seals generally are contained in a water-cooled head so that conventional sealing materials can be used and the hazard of burns reduced. Several handbooks give references to a number of cell designs and describe various types of high-pressure and high-temperature apparatus.[49–52]

High-pressure cells mainly are used to contain low-boiling solvents such as liquid NH_3 or liquid SO_2 at ambient temperature or to study solvents with moderate boiling points at elevated temperature.[53] The dropping-mercury electrode (DME) can be readily operated at several thousand atmospheres, and one interesting result is that the hydrogen overpotential on mercury declines about 1 V over a range of 200°C.

Cells for Conductimetry. Reliable and precise measurements of electrolytic conductance require attention to the design of cells, electrodes, and measuring circuitry. Extraction of an ohmic resistance from AC bridge measurements is not a trivial task, particularly in solutions with high resistance (such as organic solvents) or low resistance (molten salts). Expositions of the principles are provided in monographs that emphasize aqueous solution,[54,55] and in a review of conductimetry and high-frequency oscillometry that emphasizes analytical applications.[56]

Of perhaps greater interest to electrochemists who work with organic solvents are reviews that survey measurements of the conductance of electrolytes in a number of solvents, discussions of experimental methods, cell design, and

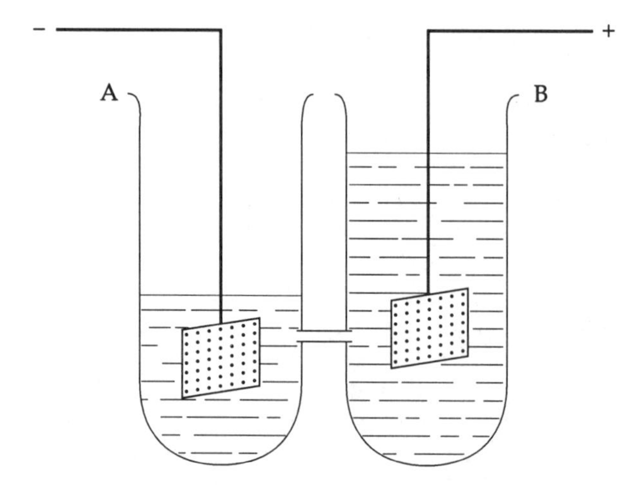

Figure 6.19 Capillary-tube DC conductance cell with platinum mesh electrodes (6 cm^2 each). Half-cell compartment volumes, 100 mL; capillary, 2 cm × 0.16 mm ID. As shown, flow from compartment B controls conductance.

a more extensive discussion of conductance data with tabulations of conductance parameters are included.

A stable and sensitive conductivity cell has been described that measures the DC current that passes through a small volume of solution confined in a small capillary tube.[60] The very simple apparatus is illustrated in Figure 6.19. When filled with 0.1 M KCl, the cell exhibits an Ohm's law behavior with an applied voltage from 5 to 100 V (approximately 5–100 mA of current). The cell also can be used for conductimetric titrations, with the titration carried out in vessel B. Because some liquid flows through the capillary during a titration, a correction factor is used to obtain the most accurate results (although corrections usually are $<0.2\%$).

Microcells. Cells that employ thin layers of solution (thin-layer electrodes) have special virtures that have been detailed in Chapter 5. In addition, the dipping thin-layer cell is especially useful for the measurement of chronopotentiometric n values.[61] It is constructed from the components of a micrometer such that the cell thickness can be varied to permit a plot of the transition time versus solution thickness. This allows the determination of n, the number of electrons involved in the electrochemical reaction. An accuracy of $\pm 4\%$ has been demonstrated for a number of compounds in different organic solvents.

In most electrochemical measurements solutions are made up to an arbitrary volume that usually is at least 1 cm^3. However a few microcells have been described for work with solution volumes that are well below 1 cm^3. The coulometric determination of silver ion in cell volumes as small as 20 μL (formed by a thin copper sheet and a cavity of beeswax) has been discussed.[62]

For work with the hanging-mercury-drop electrode (HMDE), a cell has been described[63] for use with volumes as small as 0.3 mL. An interesting cell for the microcombustion of small samples has been described.[64] After combustion, electrolyte is added and the solution volume is measured by a calibrated sidearm, then displaced by mercury into a compartment that contains a DME.

Flow and Circulation Cells. Voltammetric techniques can be used to monitor a flowing stream continuously; for example, the effluent from an ion exchange column. Of course, the material whose detection is sought must be electroactive. An effective cell design has been described (Figure 6.20) that employs a vertical-orifice dropping mercury electrode.[65] The cell possesses a low holdup time, a small volume, and a low flushout time, all of which ensure rapid

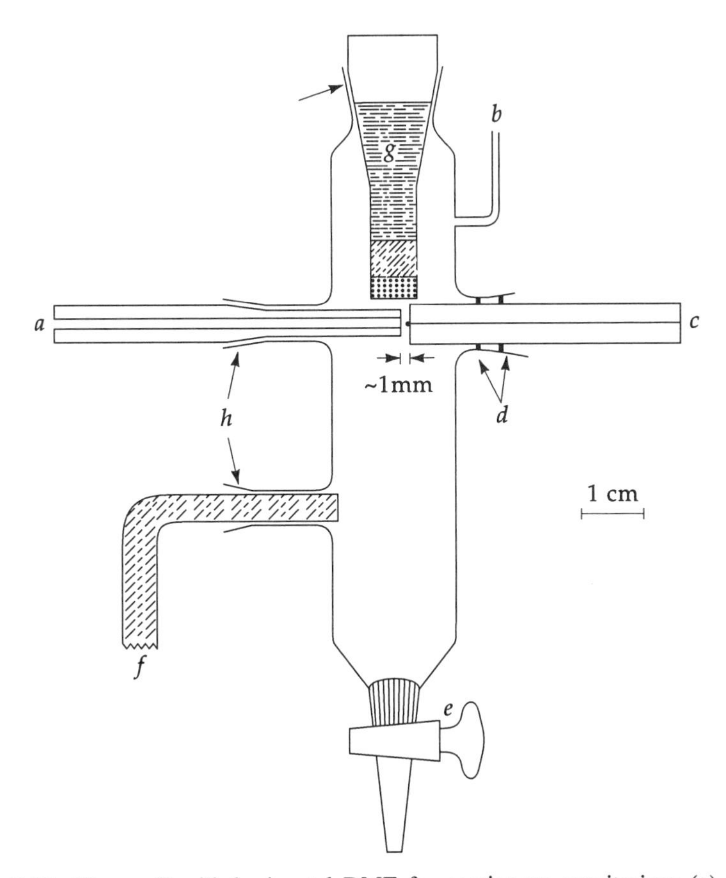

Figure 6.20 Flow cell with horizontal DME for continuous monitoring: (*a*) sample introduction tube; (*b*) solution outlet; (*c*) DME capillary; (*d*) O-ring seals; (*e*) waste mercury; (*f*) agar salt bridge to counter electrode; (*g*) SCE reference electrode with agar plug and fine glass frit; (*h*) ground-glass joints.

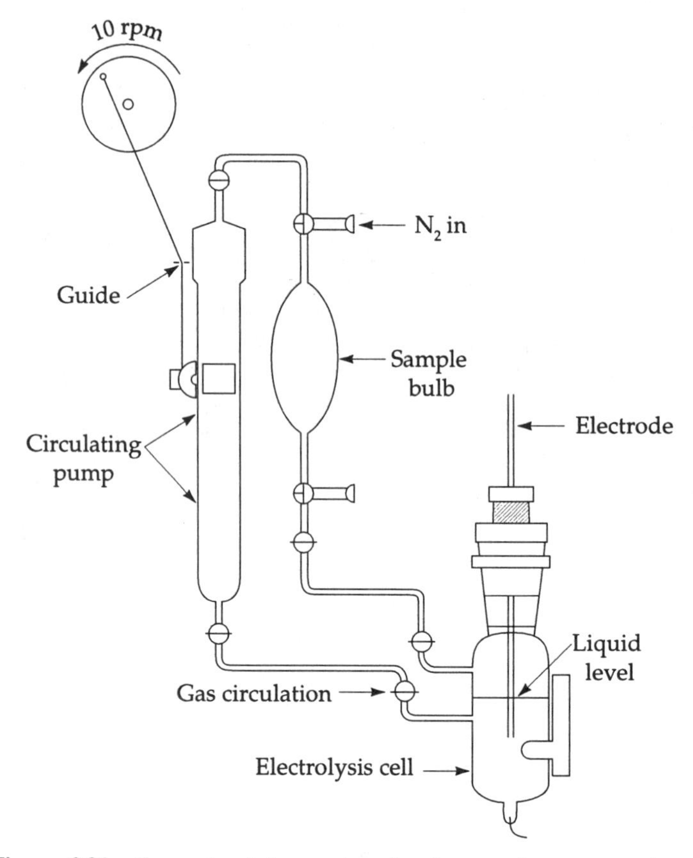

Figure 6.21 Gas-recirculating system for electroactive gas samples.

response. It also can be operated for long periods of time without dismantling or cleaning. Because of the low noise of the signal (~2 nA), high sensitivity is achieved, even in the presence of high background current. Changes of concentration as small as 3×10^{-6} M HCl in the presence of oxygen-saturated 1 M KCl can be detected with good accuracy (100-nA current signal on a 32-μA background current).

Figure 6.21 illustrates a cell that is useful for the circulation of an electroactive gas through a solution.[66] Such a system is particularly desirable when the supply of gas is so limited that it cannot be continuously bubbled through the solution. The gas is pumped by means of a loosely fitting Teflon-coated piston that is driven with a reciprocating motion by an external magnet.

Cells for Spectroelectrochemistry. Spectroscopic techniques have been used in conjunction with electrochemistry in a variety of ways, which can be grouped into three areas: the direct optical study of the electrode interface, the measurement of photon-stimulated currents, and the use of optical and ESR spectroscopy for the in situ study of reactive intermediates (particularly radical anions) that are produced electrochemically.

Direct optical studies of the electrode interface have been reviewed.[67] The principal methods that have been used include the following: ellipsometry to measure the change of condition of elliptically polarized light as it passes from one medium into another through a thin film and is reflected back into the original medium, specular reflectance or electroreflectance[68,69] to detect the formation of thin films on the metal surface and to determine the electronic structure of solid surfaces, X-ray diffraction, and laser interferometry[70] to determine concentration gradients near the electrode surface that are produced by passage of current. The references that are cited provide the reader with information about the experimental methods and the cells.

There has been a great deal of theoretical and experimental work on photon-stimulated currents. Most of this centers around the questions of whether electrons can be photoejected from an electrode surface[71] and whether photochemical excitation of the reactant or the electrode can reduce the barrier to electron transfer.[71–73] In addition, there has been some study of reactive intermediates that are produced near an electrode by flash photolysis.[74]

The third area of interest has been the observation by optical and ESR spectroscopy of intermediates that are produced electrochemically. Electron spin resonance is a useful technique for identifying species that have unpaired electrons, and reviews have documented the power of ESR for unraveling complicated reaction pathways.[75–77] A number of cells have been described for use with this technique that fall into two categories—the flow cell in which the reactive intermediate is generated externally and flows into the cavity[78] and the in situ generation system where electrodes are placed inside the resonant cavity of the spectrometer.[79]

Many optical techniques make use of transparent electrodes such as conducting tin oxide films coated on glass,[80] thin metal films coated on glass,[81] and fine gold-mesh electrodes.[82] The first two types are used mainly in the visible region of the spectrum (although thin metal grids have been used for internal reflection spectroscopy in the infrared region),[83] while the gold-mesh electrode is used in the ultraviolet, visible, and infrared. A cell that has been used with the transparent tin-oxide-coated electrode is shown in Figure 6.22. In combination with a rapid-scanning spectrophotometer, this cell provides the means to obtain spectra of intermediates that are involved in biological electron-transfer sequences.[80]

The principle of external generation also has been used for optical measurements of transient intermediates. Flow cells[84,85] and cells that provide for the circulation of the solution past an electrode and into a spectrophotometric cell in a closed loop have been described.[86,87]

Several reviews on the theory and applications of spectroelectrochemical methods have been published.[88,89]

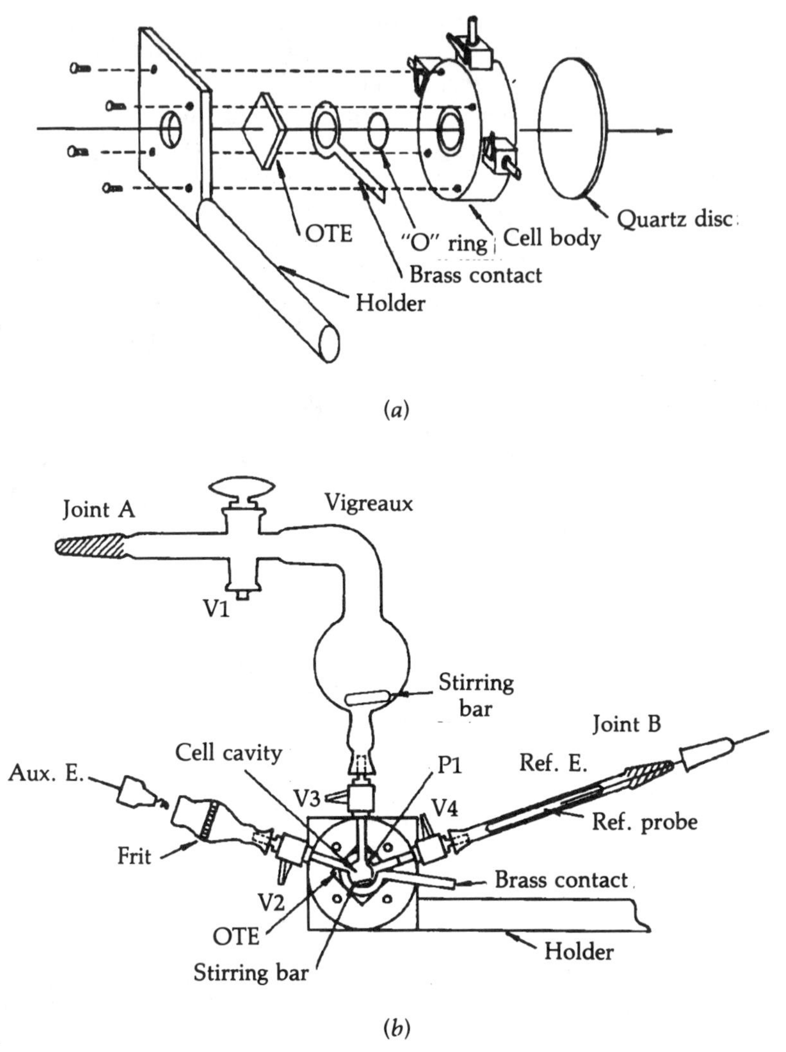

Figure 6.22 Cell system for spectroelectrochemistry by use of optically transparent electrodes (OTEs).

6.3 INSTRUMENTATION: MEASUREMENTS

Voltage. The level of precision that is required for voltage measurements in electrochemical work is summarized Table 6.5. Most electrochemical cells will have voltages in the range 0–2 V. Digital voltmeters are readily available so that precision measurements of voltage are not especially difficult unless the source impedance is very high. For the latter condition careful attention to shielding the signal source from noise pickup is necessary.

TABLE 6.5 Precision of Voltage Measurements in Electrochemistry

Type of Voltage Measurement	Level of Precision and/or Accuracy
Measurements on electrochemical cells	
Potentiometric titrations	± 5 mV
pH measurements (± 0.1 pH unit)	± 5 mV
pH measurements (± 0.01 pH unit) in measurement of blood pH and CO_2 tension	± 0.5 mV
Specific-ion electrodes ($\pm 1\%$ accuracy in activity)	$\pm \frac{0.2}{n}$ mV
Thermodynamic measurements of activity coefficients, equilibrium constants, etc.	± 0.1–0.01 mV
Voltammetric cells	
Polarographic half-wave potentials	± 1–10 mV
Voltammetric peak potentials on solid electrodes	± 5–20 mV
Miscellaneous	
Current measurements by Ohm's law, $i = E/R$	Commensurate with the precision of the resistor (generally $\pm 0.1\%$ or better)
Thermocouple measurements of temperature (± 0.1°C)	10–50 μV (dependent on the thermocouple pair)

The potentiometer is the classical instrument for the precise measurement of voltage. In a simple "student potentiometer," the precision depends on the precision of the individual resistance elements and the linearity of the slidewire. The measurement accuracy depends on the short-term stability of the battery or power-supply voltage, although slow drifts can be compensated by frequent restandardization.

One of the most satisfactory voltage standards is the Weston standard cell, shown in Figure 6.23. The cell is represented by the schematic

$$\text{Cd(10\% amalgam)}\,|\text{CdSO}_4 \cdot \tfrac{8}{3}\,\text{H}_2\text{O}|\text{CdSO}_4\text{(satd. solution)}|\,\text{Hg}_2\text{SO}_4|\text{Hg} \tag{6.7}$$

The Weston cell has a voltage of 1.018 V at 25°C, and has good voltage stability but a high temperature coefficient.[92] The Weston cell is sensitive to rough handling and should not be tipped more than 45° from the vertical and never turned upside down. The cells are best maintained at constant temperature in a thermostat that is designed to hold a group of cells whose voltages may be intercompared.

The unsaturated Weston cell (saturated at 4°C with $CdSO_4$) is more rugged and has a negligible temperature coefficient. These cells are the ones usually

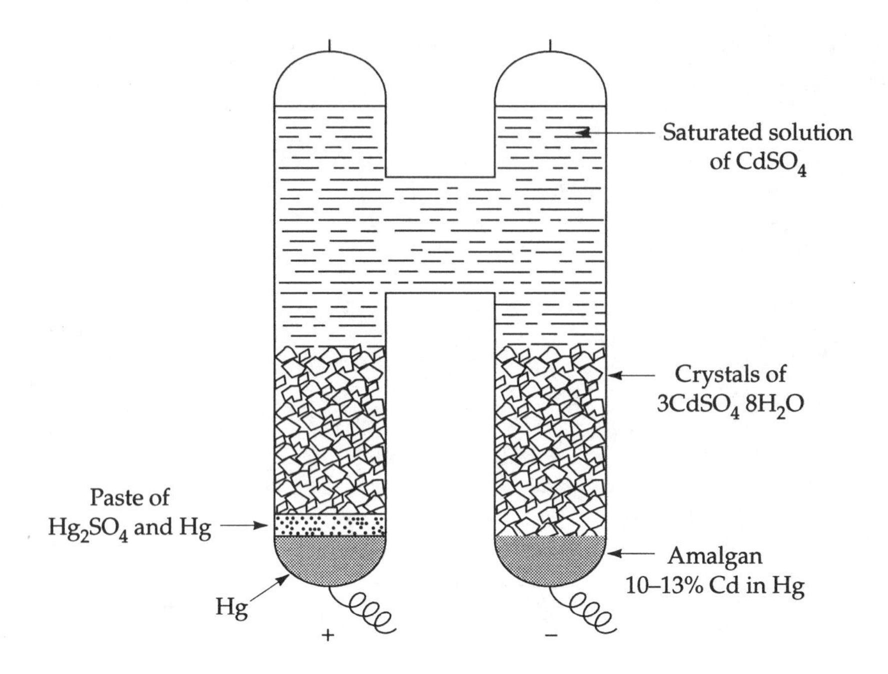

Figure 6.23 Weston standard cell.

sold commercially for incorporation into instruments as voltage standards. The voltage of the unsaturated Weston cell tends to decrease with time (the average value of a group of cells decreased 0.07 mV per year, and the average decrease of 4% of the cells was 0.28 mV per year.[93]

Solid-state devices, like the zener diode,[94] also can provide stable and precise reference voltages, but they must be calibrated against voltage standards like the Weston cell. Weston cells and zener diodes can be incorporated into voltage calibrators or voltage reference sources, which, by precise amplification, can provide a range of accurately known voltages for the calibration of voltage-measuring equipment.

Current. Digital multimeters allow the precise measurement of a wide range of currents. Currents are measured by measuring the *iR* drop across a precision resistor through which the current is passed.

The measurement of very small currents (down to 10^{-14} A) requires the use of devices that have negligibly small leakage currents at their inputs.

Bridge Measurements of Resistance, Capacitance, and Inductance. Bridge measurements provide the most direct way to compare an unknown impedance with known standard impedances. The impedance may be a pure resistance, capacitance, or inductance, or may be represented by some series or parallel combination. The DC Wheatstone bridge[95] provides a simple and direct way to measure a pure resistance. An AC supply voltage also can be used if there are not reactive components in the standard or unknown resistances and the conditions for bridge balance will be the same as in the DC bridge.

Measurement of solution conductance. Because the measurement of solution conductance (or resistance) is a common procedure that provides useful information to the electrochemist, the principles of a bridge measurement are illustrated by such an application.

The measurement of the ohmic resistance for an electrolyte is more difficult than it is for a metal wire. A variety of chemical and physical processes occur at the electrode–solution interface that must be separated from the voltage drop associated with the migration of ions through the bulk electrolyte.

The resistivity (or specific resistance) of a conductor may be thought of as the resistance of a cube of 1-cm edge, with the current assumed to be uniform and perpendicular to two opposite faces of the cube. The resistance of a conductor of uniform cross section is given by

$$R = \rho \frac{l}{A} \tag{6.8}$$

where R = resistance of the conductor in Ω,
A = uniform cross-sectional area in (cm^2)
l = length of the conductor (cm)
ρ = resistivity or specific resistance (Ω-cm)

The conductivity or specific conductance is defined as the reciprocal of the specific resistance

$$\kappa = \frac{1}{\rho} \ \Omega^{-1}\ \mathrm{cm}^{-1} \tag{6.9}$$

Because the conductance of an ionic solution depends on the number of ionic charges, it is convenient to define it in terms of the conductance per unit concentration of ionic constituent

$$\Lambda = \frac{\kappa}{C} \tag{6.10}$$

where Λ is the equivalent conductance in $cm^2/(\Omega)$ (equivalent) and C is the concentration in equivalents cm^{-3}. Because the number of equivalents cm^{-3} is equal to the normality (N) times 1000 [where N is the molarity times the number of moles of positive (or negative) charge per mole of ionic solute], the equivalent conductance is given by

$$\Lambda = \frac{1000k}{N} \ \Omega^{-1}\ \mathrm{cm}^2\ \mathrm{equivalent}^{-1} \tag{6.11}$$

Thus Λ represents the specific conductance of a hypothetical solution of the electrolyte containing 1 equivalent cm^{-3}. An experimental value of Λ may be determined by measurement of R, l/A (the cell constant), and N, and use of the relationship

$$\Lambda = \frac{1000(l/A)}{NR} \tag{6.12}$$

Most conductance cells do not have uniform cross-sectional area, which requires that the cell constant be determined by calibration with solutions of known conductivity. The cell constant, l/A, may be thought of as the ratio of *effective* length and cross-sectional area of the conducting path. Some values of the conductivity standards are tabulated in Table 6.6.

Combined with densities, molecular weights, and transference numbers (fractions of the current carried by the various ionic constituents), the conductivity yields the relative velocities of the ionic constituents under the influence of an electric field. The mobilities (velocity per unit electric field, $cm^2\ s^{-1}\ V^{-1}$) depend on the size and charge of the ion, the ionic concentration, temperature, and solvent medium. In dilute aqueous solutions of dissociated electrolytes, ionic mobilities decrease slightly as the concentration increases. The equivalent conductance extrapolated to zero electrolyte concentration may be expressed as the sum of independent equivalent conductances of the constituent ions

$$\Lambda^\circ = \lambda_+^\circ + \lambda_-^\circ \tag{6.13}$$

The concentration dependence of equivalent conductance is the principal source of our knowledge of ionic interactions (which generally lower mobilities).

The phenomena important in electrolytic conductance have been discussed[96] and are represented by the electrical equivalent circuit of a conductance cell shown in Figure 6.24*a*.

Double-layer capacitance. At each electrode–solution interface there is a substantial capacitance (represented as C_s in Figure 6-24*a*). A positively charged electrode tends to preferentially attract a layer of negative ions (and a negative

TABLE 6.6 Specific Conductivity of Standard Potassium Chloride Solution, in $\Omega^{-1}\ cm^{-1}$, for Cell Constant Determination

Temperature	KCl (g) in 1000 g of solution		
(°C)	71.1352	7.41913	0.745263
0	0.06518	0.007138	0.0007736
18	0.09784	0.011167	0.0012205
25	0.11134	0.012856	0.0014088

Source: From Jones, G.; Bradshaw, B. C., *J. Am. Chem. Soc.* **1933,** *55*, 1780.

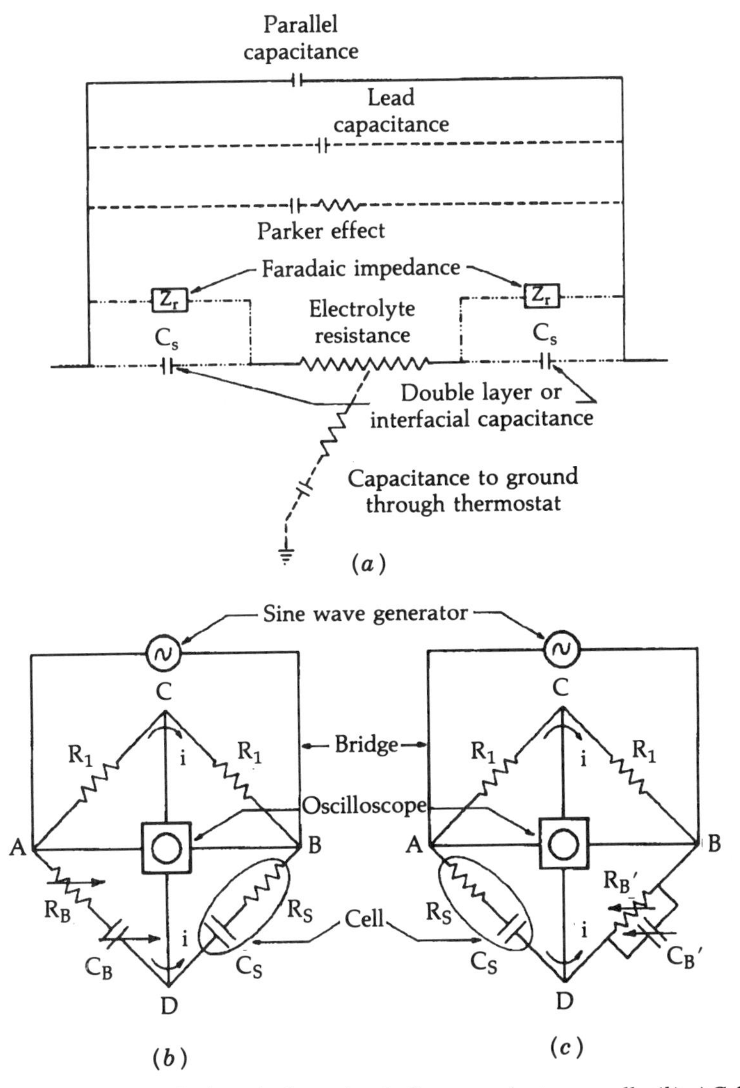

Figure 6.24 (*a*) Electrical equivalent circuit for a conductance cell; (*b*) AC bridge with the cell impedance balanced by a series *R–C* combination; (*c*) AC bridge with the cell impedance balanced by a parallel *R–C* combination (see Table 6.7).

electrode a layer of positive ions). Next to this layer of ions at the electrode surface will be a more diffuse layer of ions of opposite charge. This local ordering of ions produces a double layer that rapidly trails off in the bulk of the solution due to thermal disordering (Brownian motion). The double layer constitutes a capacitor, capable of storing charge, whose magnitude is of the order of 10 to 100 $\mu F\ cm^{-2}$ of electrode surface. Although there is usually a reasonably broad potential region where the double-layer capacitance is approximately constant, it does change with potential and can also be affected by the presence of substances adsorbed at the electrode surface.

Faradaic effects. If a small voltage is applied to a conductance cell, a charging current flows until the double-layer capacitor is charged to its equilibrium value. If the voltage is increased, the charge accumulated in the double-layer capacitor increases until a voltage is reached where electron transfer takes place across the electrode–solution interface, accompanied by oxidation at the positive electrode, and reduction at the negative electrode. This Faradaic process can be represented by a voltage-dependent resistance (nonlinear resistance) that partially short-circuits the double layer. The Faradaic process also tends to deplete the region of the double layer of the electroactive species to produce a concentration–polarization and an accompanying polarization–resistance. In the equivalent circuit, the Faradaic impedance is parallel with the double-layer capacitance.

Ohmic resistance. In the electrolytic solution, the charge carriers that constitute the flow of current are ions (rather than electrons as in a metal wire). The cations migrate toward the cathode; and the anions, toward the anode under the influence of the electric field. Velocity-dependent retarding forces oppose the accelerating force of the electric field, and result in a constant drift velocity at a given field. As the ions move through the medium, the frictional forces between ions and solvent molecules dissipate energy in the form of heat. To a close approximation the solution resistance is described by Ohm's law over a large voltage range. However, high electric fields (10^4–10^5 V cm^{-1}) can produce an additional conductance (the Wien effect). The latter results from changes in the interionic attraction that are brought about by the high field strength and the degree of dissociation of weak electrolytes.[97]

Alternating-current and frequency effects. With an AC rather than a DC voltage applied to the electrodes, the processes above reverse themselves with the period of the alternating voltage. But each process proceeds at a different rate (with a characteristic relaxation time) so that their relative contributions to energy dissipation vary with frequency. As the frequency is increased concentration–polarization can be reduced or eliminated, particularly if the electrode reaction is reversible (fast electron transfer in both directions).

Although increased frequencies (to several thousand hertz) reduce the complications of Faradaic polarization effects, other problems develop as the frequency increases. The most significant are the effects of stray capacitance[98] between connecting leads to the cell, to ground through the thermostat, or through stray currents created by a less than optimum design geometry of the cell. Just as there are no perfect gases or ideal solutions, there are no perfect (frequency-independent) electrical components—pure resistances, pure capacitances, or pure inductances. The equivalent-circuit representation of the cell shown in Figure 6-24*a* is an approximation, but one that is usually adequate.

AC conductance measurements. An AC bridge is a direct and accurate way of measuring the AC conductance of a solution to obtain the solution resistance. The method works well for resistances between 1 and 10 kΩ, and can be extended to 60 kΩ provided several experimental conditions are met. First, C_S of Figure 6.24*a* must be fairly large so that its impedance is small with respect to R_S and that the impedance of the interelectrode capacitance must be large compared to the series combination of C_S and R_S. The conditions usually are met by platinizing the conductance electrodes to obtain a large C_S (by increasing the surface area), by using a source frequency of at least 1 kHz, and by adjusting the cell constant so that the measured resistances are in the proper range.[99] Under these conditions, the cell can be represented by an ohmic resistance in series with the double-layer capacitances at the electrodes. The applied AC voltage is made small enough so that the Faradaic effects will be negligible.

AC bridge measurements provide only an absolute impedance and phase shift, so there is no way to distinguish between either a series or parallel *R–C* combination by use of a pure sinusoidal waveform. Thus the series *R–C* combination of the cell may be matched by the series combination shown in Figure 6.24*b* or the parallel combination of the Wien bridge shown in Figure 6.24*c*. Either an oscilloscope or phase-angle voltmeter[100] are recommended as null indicators because they allow both the phase shift and the imbalance potential to be nulled.

In both these bridge circuits, the upper arms will be assumed to be pure resistance arms of equal value, but the calculations can be readily extended to cover the case where they are not equal.[101] The balance conditions for the two bridge circuits of Figure 6.24*b*,*c* (summarized in Table 6.7) show that when

TABLE 6.7 Impedance Balance Conditions for the Two Bridge Circuits of Figure 6.24

Series Bridge (Figure 6.24*b*)	Parallel Bridge (Figure 6.24*c*)
$Z_S = Z_B$	$Z_S = Z'_B$
$R_S = R_B$	$R_S \neq R'_B$
$C_S = C_B$	$C_S \neq C'_B$
$Z_S = R_S + \dfrac{1}{j\omega C_S}{}^a$	$\dfrac{1}{Z'_B} = \dfrac{1}{R'_B} + j\omega C'_B{}^a$
$Z_B = R_B + \dfrac{1}{j\omega C_B}$	$R_S = \dfrac{R'_B}{1 + (\omega R'_B C'_B)^2}$
	$R'_B = R_S\left[1 + \dfrac{1}{(\omega R_S C_S)^2}\right]$
	$C'_B = \dfrac{C_S}{1 + (\omega R_S C_S)^2}$

[a] $j = \sqrt{-1}$; $\omega = 2\pi f$, where f is the frequency.

the bridge is balanced, both resistances and both capacitances in the lower arms of the series bridge will be equal. This results because equality of the two complex impedances, Z_B and Z_S, requires equality of the real parts and of the imaginary parts, independently. (The reader unfamiliar with the *j*-operator notation for representing the complex impedance is referred to Refs. 94 and 95.) A similar result would be obtained for balance of a parallel combination by another parallel combination.

In balancing a series *R–C* combination with a parallel *R–C* combination (Figure 6.24*c*), the resistances and capacitances in the two lower arms will not be equal, and the balancing resistance and capacitance will change with frequency (i.e., show frequency dispersion). The true resistance at the balance point must be calculated with the series *R–C* bridge equations of Table 6.7. If the value of $\omega C'_B R'_B$ is sufficiently small (or $\omega R_X C_S$ sufficiently large) R_S will equal R'_B and the calculation is not necessary. Fortunately this is usually true for moderately dilute aqueous solutions where typical values might be R_S = 1000 Ω, f = 1000 Hz, and C_S = 100 μF, so that R_S and R'_B will differ by only about 3 ppm. However, for measurements in molten salts (with high conductivities), or in cells with low double-layer capacitance, the calculation will be necessary.

It might seem more convenient to eliminate the calculation by employing a bridge with a series *R–C* balancing arm as in Figure 6.24*b*. However, the parallel combination is used more frequently because the parallel capacitance required to compensate a series capacitance of 100 μF at 1000 Hz with a resistance R_S = 1000 Ω is only about 300 pF. Small capacitances can be obtained with high accuracy and less frequency dependence (and less cost) than large ones.

In the most precise work, the oscillator and detector usually are coupled to the bridge by means of electrostatically shielded transformers to prevent direct coupling of the oscillator signal into the detector.

Balancing a bridge is a rather slow process, so the method cannot be used to follow a rapidly changing solution resistance (as encountered in kinetic measurements). A number of circuits reported in the literature[102–104] allow the resistance of a solution to be recorded continuously. A method that employs the successive application of constant-current pulses of equal magnitude, but of opposite sign, has been described, which works well in both low-resistance and high-resistance solutions ($\leq$ 100 kΩ) and is well suited to the fast measurement of solution resistances.[99]

6.4 SOURCES

The first edition contained the question "To build or to buy?" After 20 years the answer is clear. The progress in electrochemical instrumentation has been revolutionary. Several companies provide a broad range of simple as well as specialized instrumentation, and offer all the necessary accessories and supplies (electrodes, cells, supporting electrolytes). The commercial sources are listed in the annual Buyer Guides of *Analytical Chemistry* and *American Laboratory*.

In our view the Model CV-27 voltammograph produced by Bioanalytical Systems (BAS) is the appropriate instrument for those chemists interested in cyclic voltammetric measurements. It is relatively simple and easy to use and capable of full range of voltammetric measurements (polarography, linear-sweep and cyclic voltammetry, chronoamperometry, chronocoulometry, and controlled-potential electrolysis). When coupled with an X–Y recorder, it provides a fast and simple tool for performing most electrochemical measurements. An interface to connect a CV-27 voltammograph to a computer also is available from BAS. BAS offers a broad range of working, auxiliary, and reference electrodes. Again, in our view the BAS glassy-carbon electrode is the most inert working electrode for the majority of electrochemical experiments. Its surface can be readily renewed, and with care an electrode will remain effective for a long period of time.

A specialized version of the Model CV-27 Voltammograph (Model CV-37) is designed for use with microelectrodes and their associated low currents (nA range). For more advanced electrochemical measurements BAS provides the Model 100B/W Electrochemical Workstation, which allows one to carry out almost any modern electrochemical measurement with digital electronics and computer control.

REFERENCES

1. Schaap, W. B.; McKinney P. S., in *Polarography 1964*, Vol. 1, G. J. Hills, G. J., ed., Interscience Publishers, New York, 1966, pp. 197–214.
2. Mumby, J. E.; Perone, S. P., *Chem. Instrum.*, **1971,** *3*, 191.
3. Ilkovic, D., *Coll. Czech. Chem. Commun.*, **1936,** *8*, 13.
4. Nemec, L., *J. Electroanal. Chem.*, **1964,** *8*, 166.
5. Kolthoff, I. M.; Marshall, J. C.; Gupta, S. L., *J. Electroanal. Chem.*, **1962,** *3*, 209.
6. Devay, J., *Acta Chim. Acad. Sci. Hung.*, **1973,** *35*, 255.
7. Belew, W. L.; Fisher, D. J.; Kelley, M. T.; Dean, J. A., *Chem. Instrum.*, **1970,** *2*, 297.
8. Harrar, J. E.; Shain, I., *Anal. Chem.*, **1966,** *38*, 1148.
9. Cahan, B. D.; Nagy, A.; Genshaw, M. A., *J. Electrochem. Soc.*, **1972,** *119*, 64.
10. Booman, G. L.; Holbrook, W. B., *Anal. Chem.*, **1965,** *37*, 795.
11. Harrar, J. E.; Pomernacki, C. L., *Anal. Chem.*, **1973,** *45*, 56.
12. Booman, G. L.; Holbrook, W. B., *Anal. Chem.*, **1963,** *35*, 1793.
13. Kasper, C., *Trans. Electrochem. Soc.*, **1940,** *78*, 131.
14. Kimura, T.; Freeman, G. R., *J. Chem. Educ.*, **1973,** *50*, A85.
15. Morcos, I.; Yeager, E., *Electrochim. Acta*, **1970,** *15*, 953.

16. Eichholtz, G. G.; Nagel, A. E.; Hughes, R. B., *Anal. Chem.*, **1985,** *37*, 863.
17. Robertson, D. E., *Anal. Chem.*, **1968,** *40*, 1067.
18. Chatt, J., *Pure Appl. Chem.*, **1970,** *24*, 425.
19. Scott, F. I., Jr.; Rutherford, R. E., *Am. Lab.*, **1970,** (Aug.), 43.
20. Seitz, C. A.; Bodine, W. M.; Klingman, C. L., *J. Chromatographic Sci.*, **1971,** *9*, 29.
21. Meites, L., *Polarographic Techniques*, 2nd ed., Interscience Publishers, New York, 1965, p. 89.
22. Yates, D. J. C., *J. Chem. Educ.*, **1967,** *44*, 699.
23. Shriver, D. F., *The Manipulation of Air-Sensitive Compounds*, McGraw-Hill, New York, 1969, pp. 230–231.
24. Bahmet, W.; Hersch, P. A., *Anal. Chem.*, **1971,** *43*, 803.
25. Eubanks, I. D.; Abbott, F. J., *Anal. Chem.*, **1969,** *41*, 1708.
26. Keidel, F. A., *Anal. Chem.*, **1959,** *31*, 2043.
27. Ref. 23, p. 145.
28. Ref. 23, p. 154.
29. Bard, A. J.; Samtjama, K. S. V., *Electroanalytical Chemistry*, Vol. 4, Bard, A. J., ed., Marcel Dekker, New York, 1970, p. 304.
30. Bard, A. J., *Pure Appl. Chem.*, **1971,** *25*, 379.
31. Mills, J. L.; Nelson, R.; Shore, S. G.; Anderson, L. B., *Anal. Chem.*, **1971,** *43*, 157.
32. Schmulbach, C. D.; Omen, T. V., *Anal. Chem.*, **1973,** *45*, 820.
33. Piljac, I.; Grabaric, B.; Filipovic, I., *J. Electroanal. Chem.*, **1973,** *42*, 433.
34. Marinenko, G.; Taylor, J. K., *Anal. Chem.*, **1968,** *40*, 1645.
35. Eckfeldt, E. L.; Shaffer, E. W., Jr., *Anal. Chem.*, **1965,** *37*, 1534.
36. Lindberg, J., *J. Electroanal. Chem.*, **1972,** *40*, 265.
37. Ho, P. P. L.; Marsh, M. M., *Anal. Chem.*, **1963,** *35*, 618.
38. Feldberg, S. W.; Bricker, C. E., *Anal. Chem.*, **1959,** *31*, 1852.
39. Brown, E. R.; Smith, D. E.; Booman, G. L., *Anal. Chem.*, **1968,** *40*, 1411.
40. Bard, A. J., *Anal. Chem.*, **1963,** *35*, 1125.
41. Clem, R. G., *Anal. Chem.*, **1971,** *43*, 1853.
42. Kolthoff, I. M.; Lingane, J. J., *Polarography*, Vol. 1, 2nd ed., Interscience Publishers, New York, 1952, p. 29.
43. Buchta, R. C.; Evans, D. H., *Anal. Chem.*, **1968,** *40*, 2181.
44. Van Duyne, R. P.; Reilley, C. N., *Anal. Chem.*, **1969,** *44*, 142; 153; 158.
45. Bechgaard, K.; Parker, V. D., *J. Am. Chem. Soc.*, **1972,** *94*, 4749.
46. Miller, L. L.; Mayeda, E. A., *J. Am. Chem. Soc.*, **1970,** *92*, 5818.
47. Wilson, R. J.; Warren, L. F.; Hawthorne, M. F., *J. Am. Chem. Soc.*, **1969,** *91*, 758.
48. Graves, B. B., *Anal. Chem.*, **1972,** *44*, 993.
49. Janz, G. J., ed., *Molten Salts Handbook*, Academic Press, New York, 1967.
50. Corbett, J. D.; Duke, F. R., in *Technique of Inorganic Chemistry*, Vol. 1, Jonassen, H. B.; Weissberger, A., eds., Interscience Publishers, New York, 1963, Chap. 3.

51. Ulmer, G. C., ed., *Research Techniques for High Pressure and High Temperature*, Springer-Verlag, New York, 1971.
52. Zambonin, P. G., *Anal. Chem.*, **1969,** *41*, 868.
53. Hills, G. J.; Ovenden, P. J., in *Advances in Electrochemistry and Electrochemical Engineering*, Vol. 4, Delahay, P.; Tobias, C. W., eds., Interscience Publishers, New York, 1966, p. 185.
54. Harned, H. S.; Owen, B. B., *The Physical Chemistry of Electrolytic Solutions*, 3rd ed., Reinhold, New York, 1958, Chap. 6.
55. Robinson, R. A.; Stokes, R. H., *Electrolyte Solutions*, Butterworths Scientific Publications, London, 1955.
56. Loveland, J. W., in *Treatise on Analytical Chemistry*, Vol. 4, Interscience Publishers, New York, Part 1, 1963, Chap. 51, p. 2569.
57. Kratochvil, B.; Yeager, H. L., *Forschr. Chem. Forsch.*, **1972,** *27*, 1.
58. Daggett, H. M.; Bair, E. J.; Kraus, C. A., *J. Am. Chem. Soc.*, **1951,** *73*, 799.
59. Kay, R. L.; Hales, B. J.; Cunningham, G. P., *J. Phys. Chem.*, **1967,** *71*, 3925.
60. Hello, O., *Anal. Chem.*, **1972,** *44*, 646.
61. McClure, J. E.; Maricle, D. L., *Anal. Chem.*, **1967,** *39*, 236.
62. Lord, S. S., Jr.; O'Neill, R. C.; Rogers, L. B., *Anal. Chem.*, **1951,** *24*, 209.
63. Underkofler, W. L.; Shain, I., *Anal. Chem.*, **1961,** *33*, 1966.
64. Parkhurst, R. M., *Anal. Chem.*, **9161,** *33*, 320.
65. Scarano, E.; Bonicelli, M. G.; Forina, M., *Anal. Chem.*, **1970,** *42*, 1470.
66. Toren, P. E., *Anal. Chem.*, **1963,** *35*, 120.
67. Conway, B. E., in *Techniques of Electrochemistry*, Vol. 1, Yeager, E.; Salkind, A. J., eds., Wiley-Interscience, New York, 1972, Chap. 5.
68. Takamura, T.; Sato, Y.; Takamura, K., *J. Electroanal. Chem.*, **1973,** *41*, 31.
69. Laser, D.; Ariel, M., *J. Electroanal. Chem.*, **1972,** *35*, 405.
70. O'Brien, R. N.; Dieken, F. P., *J. Electroanal. Chem.*, **1973,** *42*, 25; 37.
71. Berg, H.; Schweiss, H.; Stutter, E.; Weller, K., *J. Electroanal. Chem.*, **1967,** *15*, 415.
72. Barker, G. C., *Electrochim. Acta,* **1968,** *13*, 1221.
73. Martinus, N.; Rayner, D. M.; Vincent, C. A., *Electrochim. Acta*, **1973,** *18*, 409.
74. Kirschner, G. L.; Perone, S. P., *Anal. Chem.*, **1972,** *44*, 443.
75. Bowers, K. W., in *Advances in Magnetic Resonance*, Vol. 1, Waugh, J. S., ed., Academic Press, New York, 1965, pp. 317–396.
76. Adams, R. N., *J. Electroanal. Chem.*, **1964,** *8*, 151.
77. Stradyn', Ya. P.; Gavar, R. A., in *Progress in Electrochemistry of Organic Compounds*, Vol. 1, Frumkin, A. N.; Ershler, A. B., eds., Plenum Press, New York, 1971, pp. 1–42.
78. Rieger, P. H.; Bernal, J.; Reinmuth, W. H.; Fraenkel, G. K., *J. Am. Chem. Soc.*, **1963,** *85*, 683.
79. Goldberg, I. B.; Bard, A. J., *J. Phys. Chem.*, **1971,** *65*, 3281.
80. Hawkridge, F. M.; Kuwana, T., *Anal. Chem.*, **1973,** *45*, 1021.
81. Von Beken, W.; Kuwana, T., *Anal. Chem.*, **1973,** *45*, 1021.
81. Von Beken, W.; Kuwana, T., *Anal. Chem.*, **1970,** *42*, 1114.

82. Petek, M.; Neal, T. E.; Murray, R. W., *Anal. Chem.*, **1971,** *43*, 1069.
83. Laser, D.; Ariel, M., *J. Electroanal. Chem.*, **1973,** *41*, 381.
84. Sioda, R. E.; Kemula, W., *J. Electroanal. Chem.*, **1971,** *31*, 113.
85. Chambers, J. Q.; Adams, R. N., *Mol. Phys.*, **1965,** *9*, 413.
86. Janata, J.; Mark, H. B., Jr., *Anal. Chem.*, **1967,** *39*, 1896.
87. Gary, A.-M.; Piemont, E.; Roynette, M.; Schwing, J. P., *Anal. Chem.*, **1972,** *44*, 198.
88. Hansen, W. N., in *Advances in Electrochemistry and Electrochemical Engineering*, Vol. 9, Miller, R. H., ed., Wiley, New York, 1973, pp. 1–60.
89. Bewick, A.; Pons, S., in *Advances in Infrared and Raman Spectroscopy*, Vol. 12, Clark, R. J. E.; Hestor, R. J. H., eds., Wiley, New York, 1985, pp. 1–63.
90. Foley, J. K.; Korzeniowski, C.; Daschbach, J. L.; Pons, S., in *Electroanalytical Chemistry*, Vol. 14, Bard, A. J., ed., Marcel Dekker, New York, 1986, pp. 309–440.
91. Gales, R. J., ed., *Spectroelectrochemistry. Theory and Practice*, Plenum Press, New York, 1988.
92. Bates, R. G., *Determination of pH—Theory and Practice*, 2nd ed., Wiley-Interscience, New York, 1973, pp. 394–399.
93. Vinal, G. W., *Primary Batteries*, Wiley, New York, 1950, Chap. 6.
94. Diefenderfer, A. J., *Principles of Electronic Instrumentation*, Saunders, Philadelphia, 1972, p. 325.
95. Malmstadt, H. V.; Enke, C. G.; Toren, E. C., Jr., *Electronics for Scientists*, Benjamin, New York, 1962, p. 273.
96. Braunstein, J.; Robbins, G. D., *J. Chem., Educ.*, **1971,** *48*, 52.
97. Kortum, G., *Treatise on Electrochemistry*, 2nd ed., Elsevier, Amsterdam, 1965, p. 202.
98. Jones, G.; Bollinger, G. M., *J. Am. Chem. Soc.*, **1931,** *53*, 411.
99. Daum, P. H.; Nelson, D. F., *Anal. Chem.*, **1973,** *45*, 473.
100. Schmidt, K., *Rev. Sci. Instrum.*, **1966,** *47*, 671.
101. Ref. 94, p. 126.
102. Colvin, D. W.; Propst, R. C., *Anal. Chem.*, **1960,** *32*, 1858.
103. Knudson, G.; Ramaley, L.; Holcombe, W. A., *Chem. Instrum.*, **1969,** *1*, 325.
104. Ford, A.; Meloan, C. E., *J. Chem. Educ.*, **1973,** *50*, 85.

CHAPTER 7

SOLVENTS AND ELECTROLYTES

7.1 INTRODUCTION

The choice of solvent in an electrochemical investigation usually is dictated by circumstances. For example, an electrochemical technique frequently is used to study a solvent–solute system that has been studied already by other techniques. The focus is on the particular system and the information that can be gleaned from the electrochemical study. However, if there is a choice of the solvent to be used, some rational criteria can be used to choose the optimum one.

The Physicochemical Properties of Solvents and Their Relevance to Electrochemistry. The solvent properties of electrochemical importance include the following: protic character (acid–base properties), anodic and cathodic voltage limits (related to redox properties and protic character), mutual solubility of the solute and solvent, and physicochemical properties of the solvent (dielectric constant and polarity, donor or solvating properties, liquid range, viscosity, and spectroscopic properties). Practical factors also enter into the choice and include the availability and cost of the solvent, ease of purification, toxicity, and general ease of handling.

Protic character. The protic character of the solvent is an important consideration because electrochemical intermediates (particularly radical anions) frequently react rapidly with protons. The classification of solvents into protic or aprotic solvents is somewhat arbitrary. A simple classification[1] is that protic solvents (such as hydrogen fluoride, water, methanol, formamide, and ammonia) are strong hydrogen-bond donors, exchange protons rapidly, and include solvents with hydrogen bound to more electronegative atoms (such as fluorine, oxygen, and nitrogen). Aprotic solvents generally have hydrogen bound only to carbon and are at best poor hydrogen-bond donors; they are weakly acidic and proton exchange occurs slowly, even with D_2O. Common aprotic solvents include aliphatic and aromatic hydrocarbons, chlorinated hydrocarbons, ethers, acetone, nitromethane, nitrobenzene, acetonitrile, benzonitrile, dimethylformamide, *N*-methyl-2-pyrrolidone, propylene carbonate, dimethyl sulfoxide, sulfolane, and hexamethylphosphoramide.

A classification of solvents can be developed on the basis of the stability of the radial anion produced by reduction of aromatic hydrocarbons, such as naphthalene and anthracene. The solvent reactions of such anions have been widely studied[2] and have generally been found to go by a sequence of reactions in either a protic solvent or in the presence of a proton donor in an aprotic solvent:[3]

$$R + e^- \rightleftharpoons R^{\bar{\cdot}} \tag{7.1}$$

$$R^{\bar{\cdot}} + \mathrm{H}X \longrightarrow R^{\cdot}\mathrm{H} + X^- \tag{7.2}$$

$$R^{\cdot}\mathrm{H} + e^- \longrightarrow R\mathrm{H}^- \tag{7.3a}$$

or

$$R^{\cdot}\mathrm{H} + R^{\bar{\cdot}} \longrightarrow R\mathrm{H}^- + R \tag{7.3b}$$

$$R\mathrm{H}^- + \mathrm{H}X \longrightarrow R\mathrm{H}_2 + X^- \tag{7.4}$$

The reversible one-electron transfer to form an anion radical ($R^{\bar{\cdot}}$) is followed by an irreversible chemical protonation to form $R^{\cdot}\mathrm{H}$, which is subsequently reduced itself (the reduction potential of the species $R^{\cdot}\mathrm{H}$, has been shown to be more positive[3] than that of the parent, R) and then undergoes another irreversible protonation reaction. In a protic solvent, the reactions proceed rapidly to the final product, $R\mathrm{H}_2$. In a rigorously purified aprotic solvent, the intermediate anion radical $R^{\bar{\cdot}}$, has an appreciable lifetime and reacts only slowly, principally with adventitious impurities in the solvent. Thus, the stability of aromatic anion radicals can be taken as a measure of the protic character of a solvent.

More recently, it has been suggested that the mechanism of reduction of aromatic hydrocarbons proceeds via the reaction of Eq. (7.3b) rather than that of Eq. (7.3a),[4] but this will have no effect on a classification that is based on the stability of the radical anion produced in the reaction of Eq. (7.1).

Both kinetic and thermodynamic factors are important in determining protic character. Although the equilibrium concentration of solvated protons in a protic solvent such as water or ethanol may be very small, there is a low activation energy for dissociation or exchange of protons, and these labile protons can rapidly react with any species having an appreciable proton affinity. Aprotic solvents usually have a lower equilibrium concentration of solvated protons (perhaps by a factor of as much as 10^{10}) and the activation energy for dissociation and exchange is substantially higher,[5,6] which results in low proton availability in the solvent.

In mechanistic studies, aprotic solvents are preferred because electrochemical intermediates (particularly anion radicals) often have sufficient stability to enable their identification by spectroscopic techniques such as ESR or UV absorption spectroscopy. Follow-up chemical reactions frequently proceed slowly enough to allow rate measurements.

Aprotic solvents may be the preferred solvents in organic polarographic analysis. Many of the compounds give ideal, one-electron reversible (diffusion controlled) waves in nonaqueous solvents, whereas the corresponding aqueous solution response frequently is irreversible or nonexistent. Similar conclusions may be justified for other dipolar aprotic solvents. The advantages of aprotic solvents over water include generally greater solubilities of organic compounds, a wider range of DC potentials for the observation of both oxidation and reduction processes, minimization of troublesome adsorption effects, and more numerous reversible electrode reactions that make the use of highly sensitive AC polarographic methods feasible. A balanced view must be taken, however. Not all the advantages cited for aprotic solvents may be realizable in any one system. Frequently an analytical sample is presented in an aqueous solution, and the removal of water and redissolution of the sample in an organic solvent may be troublesome. Where there is a choice of solvents, the behavior of the compound should be investigated in one or more dipolar aprotic solvents.

Voltage limits. The voltage limits of a solvent define the "window" of accessible electronic energy levels available for electron-transfer processes. Although the solvents may have intrinsic limits on the basis of their oxidation–reduction properties, the practical working limits depend also on the nature of the working electrode material and the composition of the supporting electrolyte. In practical terms, the voltage limits are a system property. Table 7.1 summarizes the practical working limits for a number of solvent, electrode–supporting-electrolyte systems, expressed versus the aqueous calomel or other reference electrode. Because of the uncertainty in junction potentials (see Chapter 5), there is a considerable uncertainty in the estimated values versus the aqueous calomel reference electrode. The choice of limits also is somewhat arbitrary and dependent on current density; therefore, the limits have an uncertainty of $\pm \sim 0.2$ V. These values generally are for rigorously purified solvents; the presence of proton donors will seriously diminish the negative (cathodic) limits, particularly on platinum and other solid-metal electrodes with low overvoltages for hydronium ion reduction. The anodic and cathodic limits also are affected by the composition of the supporting electrolyte and the temperature of the electrolysis–cell system. One study[7] indicates that the voltage limits are greater at low temperatures.

Several studies of chemical oxidation–reduction reactions in nonaqueous solvents have been made.[8–12] Not all dipolar aprotic solvents exhibit good stability toward oxidation or reduction. While some solvents extend the range

TABLE 7.1 Voltage Range on Mercury and Platinum in Selected Solvents and Supporting Electrolytes[a]

Supporting Electrolyte	Acetonitrile	Propylene Carbonate	Dimethyl Formamide	Dimethyl Sulfoxide
TEAP[b]				
Hg	0.6 to −2.8	0.5 to −2.5	0.5 to −3.0	0.25 to −2.8
Pt		1.7 to −1.9	1.6 to −2.1	0.7 to −1.85
TBAI[c]				
Hg	−0.6 to −2.8		−0.4 to −3.0	−0.4 to −2.85
Et_4NBF_4				
Hg	— to −2.7		— to −2.7	
Pt	+2.3 to —	+2.4 to —		
$NaClO_4$				
Hg			0.5 to −2.0	0.25 to −1.90
Pt	1.8 to −1.5		1.6 to −1.6	0.7 to −1.85

[a]Measured potentials are in volts, versus aqueous SCE.
[b]Tetraethylammonium perchlorate.
[c]Tetra-*n*-butylammonium iodide.

toward reducing conditions, others withstand oxidizing conditions. For example, dimethylformamide can be used to study reactions involving powerful reductants such as Cr(II), but is readily oxidized by mild oxidizing agents. Conversely, nitrobenzene is resistant to oxidation but quite susceptible to reduction. Although these purely chemical reactions usually parallel their electrochemical analogs, there is not always an exact correspondence between chemical and electrochemical electron-transfer reactions.

Solvent polarity: the dipole moment and dielectric constant. The net dipole moment of a molecule is given by the vector sum of the bond moments and is a function of the charge separation and geometry of the molecule. Because of the geometry factor and the possibility that bond moments may partially or exactly cancel, the dipole moment is probably a less useful measure of the ability of a solvent to promote dissociation of an ionic solute than is the dielectric constant. The orientation of solvent dipoles in an electrostatic field tends to partially cancel the electrostatic field, and in a point-charge electrostatic model, the force between positive and negative point charges is inversely proportional to the bulk dielectric constant of the medium. A large dielectric constant therefore promotes the dissociation of an ionic solute, which, in turn, lowers the solution resistance. Whenever possible, a solvent with a substantial dielectric constant should be used in electrochemical work to minimize the solution resistance. This will minimize ohmic losses and diminish the problem of potential-control errors (see Chapter 6).

The dipole moments and dielectric constants of aprotic solvents range from near-zero values for the hydrocarbons to moderate values for solvents like dimethylformamide ($\epsilon = 36.7$) and acetonitrile ($\epsilon = 36$). At dielectric constants much below 15, substantial ion association begins to take place. A value of 25 for the dielectric constant is used as the dividing line between nonpolar and dipolar solvents.[12] The choice is arbitrary because there is no sharp change of solute dissociation behavior in this region and the dielectric constants of the various solvents fall on a continuum. Others have suggested values from 15 to 30 for the dividing line.

THE GUTMANN DONOR NUMBER. Donor or ionizing solvents promote the ionization of covalent compounds for form intimate (or contact) ion pairs. This property is largely a function of the coordinating power, or Lewis base strength, of the solvent and is quantitatively expressed by the donor number. The dissociating power of a solvent, which is its ability to promote dissociation of an ionic solute or a contact ion pair, is largely a function of the dielectric constant.

Although solvents may be classified as "donor solvents" (Lewis bases) and "acceptor solvents" (Lewis acids), most of the more widely used nonaqueous solvents are donor solvents. Some acceptor solvents, such as SO_2, BrF_3, $AsCl_3$, or the liquid hydrogen halides, have proved to be useful in coordination chemistry.[13-16] Ionization is promoted in a donor solvent by solvation of cations and in an acceptor solvent by solvation of anions. For example, arsenic(III) iodide is ionized in a donor solvent *D* according to the reaction

$$2\,D + AsI_3 \longrightarrow D_2AsI_2^+I^- \tag{7.5}$$

while triphenylchloromethane is ionized in an acceptor solvent (such as liquid hydrogen chloride) according to the reaction

$$Ph_3CCl + HCl \longrightarrow Ph_3C^+HCl_2^- \tag{7.6}$$

The donor properties and protic character of a solvent are linked to a certain extent because protic solvents can solvate anions by strong hydrogen bonding as well as by ion–dipole interactions. The ease of solvation increases with decreasing size and increasing electronegativity of the anion: $SCN^- < I^-$, N_3^-, Br^-, $Cl^- \ll F^- \approx HO^-$. Anions are much less strongly solvated by dipolar aprotic solvents, because of the absence of hydrogen bonding. In this case the large polarizable anions are more easily solvated than the small, more electronegative anions, and the solvation series given above is reversed. Cations are less strongly solvated in protic solvents than anions; the reverse is true for dipolar aprotic solvents.[17]

To measure the donor power of a solvent, Gutmann[18] has proposed that one determines the enthalpy of coordination of the donor solvent toward a standard reference acceptor ($SbCl_5$) in a reference solvent of low donor power such as 1,2-dichloroethane. The quantity $-\Delta H_{D\text{-}SbCl_5}$ is easily determined at high dilution and may be considered a semiquantitative measure of the coordinating properties of a donor solvent toward an acceptor molecule. It has been termed

TABLE 7.2 Donor Number, DN_{SbCl_5}, of Some Organic Solvents

Solvent	DN_{SbCl_5}
1,2-Dichloroethane	—
Nitromethane	2.7
Nitrobenzene	4.4
Acetic anhydride	10.5
Benzonitrile	11.9
Acetonitrile	14.1
Sulfolane	14.8
Propylene carbonate	15.1
Ethylene sulfite	15.3
Propionitrile	16.1
Ethylene carbonate	16.4
n-Butyronitrile	16.6
Acetone	17.0
Ethylacetate	17.1
Water	18.0
Diethyl ether	19.2
Tetrahydrofuran	20.0
Trimethylphosphate	23.0
Tributylphosphate	23.7
Dimethylformamide	26.6
N,N-Dimethylacetamide	27.8
Tetramethylurea	29.6
Dimethylsulfoxide	29.8
N,N-Diethylformamide	30.9
N,N-Diethylacetamide	32.2
Pyridine	33.1
Hexamethylphosphoramide	38.8

the donor number, DN_{SbCl_5}. Table 7.2 summarizes the donor numbers for a number of solvents.

Although a good dissociating solvent will have a high dielectric constant, it may not necessarily behave as a good ionizing solvent. Water is both an ionizing solvent ($DN = 18$) and dissociating solvent ($\epsilon = 78$), whereas nitromethane has poor ionizing properties ($DN = 2.7$) but is a moderate dissociating solvent ($\epsilon = 36$). On the other hand, tributylphosphate is an ionizing solvent ($DN = 24$) but its dissociating properties are extremely poor ($\epsilon = 6.8$).

Solvents with a large donor number tend to associate with hydrogen donors such as water or chloroform. This may explain the finding[19] that the stability of the anthracene and naphthalene anion radicals in the presence of small amounts of added water is greater in dimethylformamide (DMF) and dimethyl sulfoxide (Me_2SO) than in acetonitrile. This stability order parallels the donor numbers of the solvents and reflects the greater tendency of DMF and Me_2SO to associate with water. The latter causes the protons on water to be less available to protonate the anion radical. Because organic solvents almost always contain traces of water in the range 10^{-3} to 10^{-2} M unless rigorously dried, solvents with large donor numbers are preferred where the maximum stability of an anion radical is desired.

LEWIS ACID–BASE CHARACTER AND STABILITY OF ANION AND CATION RADICALS. Most of the so-called dipolar aprotic solvents have appreciable Lewis base character with donor numbers greater than 10 and autoprotolysis constants smaller than 10^{-20} ($pK > 20$).[20] They solvate cations better than anions, and radical anions often have appreciable stability in the rigorously purified solvents.

Radical cations that are produced by electrochemical oxidation are not stable in solvents with appreciable base character. This results because such radicals are subject to attack by available nucleophiles, and solvents that contain donor electron pairs are good nucleophiles. Cation radicals are most stable in solvents that are good Lewis acids and show negligible basic properties. Some of the solvent systems that have been employed to stabilize electrochemically produced cation radicals include nitromethane and nitrobenzene,[21] dichloromethane,[22] trifluoroacetic acid–dichloromethane (1 : 9),[23] nitromethane–$AlCl_3$,[24] and $AlCl_3$–NaCl (1 : 1).[25] Organic chemists should be familiar with the stabilization of carbonium ions by "superacid" media.[26] These media usually contain fluorosulfuric acid, or mixtures of fluorosulfuric acid with antimony pentachloride and sulfur dioxide, and are potent solvents for the production and stabilization of organic cations.

OTHER SOLVENT POLARITY SCALES. In addition to the method proposed by Gutmann, a number of other methods have been proposed to define operationally solvent donor–acceptor power or solvent polarity. These are based on various physical phenomena and include shifts in infrared stretching frequencies of $-O-D$ and $-C=O$ bonds,[27] proton NMR shift of $CHCl_3$ in the solvent relative to that in cyclohexane,[28] retention characteristics in gas chromatography,[25] absorption frequency shifts of solvatochromic dyes,[30,31] and the solvatochromic dyes.[32] However, the Gutmann donor and acceptor numbers have achieved the most attention and are widely used.

Liquid range and vapor pressure. The liquid range and vapor pressure of a solvent may dictate special experimental requirements. Low-boiling solvents such as liquid NH_3 or SO_2 must be used at low temperature or at higher pressures at room temperature. Curiously, the earliest work was carried out with these difficult-to-handle solvents, and the techniques have been described.[32,33] Solvents such as acetonitrile or dichloromethane have substantial vapor pressures at room temperature and the practice of removing dissolved oxygen by bubbling an inert gas through the solution can lead to a loss of solvent, cooling of the solution by evaporation, and a safety hazard due to the toxic vapors. For such solvents, presaturation of the inert gas with solvent and adequate venting of the toxic vapors are necessary.

For most thermodynamic studies data must be obtained over a wide range of temperature. This requires that the solvent have a broad liquid range. Some recent work also has pointed out the advantages of operating at low temperature.[7] Hence solvents with a low melting point or glass-transition point are desirable; butyronitrile has been shown to be useful down to −130°C.

Viscosity. In many applications a low viscosity is desirable so that mass transport by diffusion or convection will extend the time range for mass transport by pure diffusional control to periods as long as 40–50 s, which can be advantageous to electroanalytical techniques such as chronopotentiometry.[34] At low temperatures the solvent may not appear to crystallize, but may form a rigid glass whose viscosity is so high that mass transport practically ceases; the experimentalist must be alert to this possibility.

The dissolution of an ionic solute sometimes greatly increases solution viscosity. A notable example is the dissolution of quaternary ammonium salts in hexamethylphosphoramide, where effective solution resistance is approximately four times that of a solution in acetonitrile or dimethylformamide of the same concentration.[35]

Solvent miscibility. An empirical miscibility scale based on the mutual miscibility of pairs of 31 standard solvents has been reported.[36] Each standard solvent is assigned a value from 1 (glycerol—low lipophilicity) to 31 (petrolatum—high lipophilicity). A liquid is assigned a miscibility number (M) by determining its miscibility with the standard solvents. The miscibility numbers of a large number of compounds are shown in Table 7.3. The rules for predicting miscibility are (1) if the M numbers of two compounds differ by 15 or less, they probably are miscibile in all proportions; (2) an M-number difference of 16 is likely to have borderline miscibility; and (3) a difference of 17 or more generally corresponds to immiscibility.

In a substantial number of cases (about 17% of those studied) immiscibility is encountered with solvents at both ends of the lipophilicity scale. This behavior is accounted for by assigning a dual M number. The first M number is always less than 16 and defines the boundary of miscibility with solvents of high lipophilicity (approximately $M + 15$). Converse statements apply to the second M number. Two liquids that have dual M numbers are usually miscible with each other. Solvents with M number 16 are likely to be miscible with nearly all the solvents in the Table 7.3; hence this central class represents the "universal" solvents.

Spectroscopic properties. The techniques of optical spectroscopy (ultraviolet, visible, and infrared spectrophotometry) are often used to examine the reactants or products of an electrode reaction. Obviously the solvent (and supporting electrolyte) must be transparent at the wavelength region of interest; all of the commonly used dipolar aprotic solvents are transparent in the visible region. However, those solvents that contain aromatic or conjugated unsatu-

TABLE 7.3 Miscibility Numbers of Organic Liquids

Miscibility Number	Liquid
23	Acetal
14	Acetic acid
12, 19	Acetic anhydride
8	Acetol
10	Acetol acetate
9, 17	Acetol formate
15, 17	Acetone
11, 17	Acetonitrile
15, 18	Acetophenone
11	*N*-Acetylmorpholine
14, 18	Acrylonitrile
8, 19	Adiponitrile
14	Allyl alcohol
22	Allyl ether
13	2-Allyloxyethanol
2	2-Aminoethanol
5	Aminoethylethanolamine
2	2-(2-Aminoethoxy)ethanol
12	1-(2-Aminoethyl)piperazine
6	1-Amino-2-propanol
25	*sec*-Amylbenzene
12	Aniline
20	Anisole
15, 19	Benzaldehyde
21	Benzene
15, 19	Benzonitrile
13	Benzyl alcohol
15, 21	Benzyl benzoate
29	Bicyclohexyl
23	Bis(2-butoxyethyl)ether
20	Bis(2-chloroethyl)ether
20	Bis(2-chloroisopropyl)ether
18	Bis(2-ethoxyethyl)ether
5	Bis(2-hydroxyethyl) thiodipropionate
6	Bis(2-hydroxypropyl)maleate
15, 17	Bis(2-methoxyethyl)ether
11, 19	Bis(2-methoxyethyl)phthalate
21	Bromobenzene
23	1-Bromobutane
25	Bromocyclohexane
27	1-Bromodecane
27	1-Bromododecane
21	Bromoethane
24	1-Bromohexane
24	1-Bromo-3-methylbutane
26	1-Bromooctane
26	2-Bromooctane
29	1-Bromotetradecane
6	1,2-Butanediol
4	1,3-Butanediol
3	1,4-Butanediol
12, 17	2,3-Butanedione
15	1-Butanol
16	2-Butanol
3	2-Butene-1,4-diol
15	2-Butene-1-ol
16	2-Butoxyethanol
15, 17	2-*iso*-Butoxyethanol
15	2-(2-Butoxyethoxy)ethanol
22	Butyl acetate
21	*iso*-Butyl acetate
22	*sec*-Butyl acetate
15	*iso*-Butyl alcohol
16	*tert*-Butyl alcohol
23	*iso*-Butyl *iso*-butyrate
26	Butyl ether
19	Butyl formate
23	Butyl methacrylate
28	Butyl oleate
26	Butyl sulfide
15	Butyraldoxime
16	Butyric acid
21	Butyric anhydride
10	Butyrolactone
14, 19	Butyronitrile
26	Carbon disulfide
24	Carbon tetrachloride
25	Castor oil
21	Chlorobenzene
23	1-Chlorobutane
27	1-Chlorodecane
11	2-Chloroethanol
19	Chloroform
22	1-Chloronaphthalene
15, 20	β-Chlorophenetole
16	*o*-Chlorophenol
23	2-Chloropropane
4	3-Chloro-1,2-propanediol
14	1-Chloro-2-propanol
20	α-Chlorotoluene
29	Coconut oil
14	*p*-Cresol
11, 18	4-Cyano-2,2-dimethylbutyraldehyde
28	Cyclohexane

TABLE 7.3 (*Continued*)

16	Cyclohexanecarboxylic acid	17	2,5-Dihydrofuran
16	Cyclohexanol	17	1,2-Dimethoxyethane
17	Cyclohexanone	13	*N,N*-Dimethylacetamide
26	Cyclohexene	10	*N,N*-Dimethylacetoacetamide
29	Cyclooctane	14	2-Dimethylaminoethanol
27	Cyclooctene	14, 19	Dimethyl carbonate
25	*p*-Cymene	12	Dimethylformamide
29	Decalin	12, 19	Dimethyl maleate
29	Decane	11, 19	Dimethyl malonate
18	1-Decanol	12, 19	Dimethyl phthalate
29	1-Decene	16	1,4-Dimethylpiperazine
14	Diacetone alcohol	16	2,5-Dimethylpyrazine
21	Diallyl adipate	22	Dimethyl sebacate
22	1,2-Dibromobutane	12, 17	2,4-Dimethylsulfolane
21	1,4-Dibromobutane	9	Dimethyl sulfoxide
20	1,2-Dibromoethane	24	Dioctyl phthalate
19	Dibromomethane	17	*p*-Dioxane
21	1,2-Dibromopropane	15, 19	*p*-Dioxene
25	1,2-Dibutoxyethane	26	Dipentene
17	*N,N*-Dibutylacetamide	23	Diphenylmethane
23	Di-*iso*-butyl ketone	25	Di-*iso*-propylbenzene
22	Dibutyl maleate	11	Dipropylene glycol
22	Dibutyl phthalate	23	Di-*iso*-propyl ketone
13	Dichloroacetic acid	12, 17	Dipropyl sulfone
21	*o*-Dichlorobenzene	29	Dodecane
20	1,4-Dichlorobutane	18	1-Dodecanol
20	1,1-Dichloroethane	29	1-Dodecene
20	1,2-Dichloroethane	14, 19	Epichlorohydrin
20	*cis*-Dichloroethylene	15, 19	Epoxyethylbenzene
21	*trans*-Dichloroethylene	5	Ethanesulfonic acid
20	Dichloromethane	14	Ethanol
20	1,2-Dichloropropane	14	2-Ethoxyethanol
20	1,3-Dichloropropane	13	2-(2-Ethoxyethoxy)ethanol
12	1,3-Dichloro-2-propanol	14, 18	2-(2-Ethoxyethoxy)ethyl acetate
26	Dicyclopentadiene	15, 19	2-Ethoxyethyl acetate
26	Didecyl phthalate	19	Ethyl acetate
1	Diethanolamine	13, 19	Ethyl acetoacetate
26	Diethoxydimethylsilane	24	Ethylbenzene
14	*N,N*-Diethylacetamide	21	Ethyl benzoate
19	Diethyl adipate	17	2-Ethylbutanol
21	Diethyl carbonate	22	Ethyl butyrate
5	Diethylene glycol	6, 17	Ethylene carbonate
12, 19	Diethylene glycol diacetate	9	Ethylenediamine
9	Diethylenetriamine	2	Ethylene glycol
18	Diethyl ketone	9, 17	Ethylene glycol bis(methoxyacetate)
14, 20	Diethyl oxalate	12, 19	Ethylene glycol diacetate
13, 20	Diethyl phthalate	8, 17	Ethylene glycol diformate
12, 21	Diethyl sulfate		

TABLE 7.3 (*Continued*)

10, 19	Ethylene monothiocarbonate
23	Ethyl ether
9	Ethylformamide
15, 19	Ethyl formate
14, 17	2-Ethyl-1,3-hexanediol
17	2-Ethylhexanol
23	Ethyl hexanoate
14	Ethyl lactate
16	*N*-Ethylmorpholine
23	Ethyl orthoformate
21	Ethyl propionate
13	2-Ethylthioethanol
21	Ethyl trichloroacetate
20	Fluorobenzene
21	1-Fluoronaphthalene
3	Formamide
5	Formic acid
10	*N*-Formylmorpholine
20	Furan
11, 17	Furfural
11	Furfuryl alcohol
1	Glycerol
3	Glycerol carbonate
13, 19	Glycidyl phenyl ether
29	Heptane
17	1-Heptanol
22	3-Heptanone
23	4-Heptanone
28	1-Heptene
26	Hexachlorobutadiene
30	Hexadecane
29	1-Hexadecene
15	Hexamethylphosphoramide
29	Hexane
5	2,5-Hexanediol
12, 17	2,5-Hexanedione
2	1,2,6-Hexanetriol
17	Hexanoic acid
17	1-Hexanol
27	1-Hexene
5	Hydroacrylonitrile
8	1-(2-Hydroxyethoxy)-2-propanol
2	2-Hydroxyethyl carbamate
1	2-Hydroxyethylformamide
12	2-Hydroxyethyl methacrylate
3	2-Hydroxypropyl carbamate
14, 17	Hydroxypropyl methacrylate
22	Iodobenzene
22	Idoethane
21	Idomethane
18	Isophorone
25	Isoprene
30	Kerosene
9	2-Mercaptoethanol
24	Mesitylene
18	Mesityl oxide
15, 19	Methacrylonitrile
4	Methanesulfonic acid
12	Methanol
8	Methoxyacetic acid
11, 19	Methoxyacetonitrile
14	3-Methoxybutanol
13	2-Methoxyethanol
12	2-(2-Methoxyethoxy) ethanol
14, 17	2-Methoxyethyl acetate
15	2-Methoxyethyl methoxyacetate
15	1-[(2-Methoxy-1-methyl)-ethoxyl]-2-propanol
5	3-Methoxy-1,2-propanediol
15	1-Methoxy-2-propanol
11, 17	3-Methoxypropionitrile
15	3-Methoxypropylamine
10	3-Methoxypropylformamide
15, 17	Methyl acetate
19	Methylal
11	2-Methylaminoethanol
19	Methyl *iso*-amyl ketone
27	2-Methyl-1-butene
26	2-Methyl-2-butene
19	Methyl-*iso*-butyl ketone
13, 19	Methyl chloroacetate
8, 17	Methyl cyanoacetate
29	Methylcyclohexane
27	1-Methylcyclohexene
28	Methylcyclopentane
17	Methyl ethyl ketone
14, 19	Methyl formate
8	2,2′-Methyliminodiethanol
20	Methyl methacrylate
13	Methyl methoxyacetate
16	*N*-Methylmorpholine
22	1-Methylnaphthalene
26	Methyl oleate
7	5-Methyloxazolidinone
29	2-Methylpentane
29	3-Methylpentane

TABLE 7.3 (*Continued*)

14	2-Methyl-2,4-pentanediol	14	PPG-400
17	4-Methyl-2-pentanol	14, 23	PPG-1000
28	4-Methyl-1-pentene	11	Propanediamine
27	*cis*-4-Methyl-1-pentene	3	1,3-Propanediol
13	1-Methyl-2-pyrrolidone	4	1,2-Propanediol
26	Methyl stearate	7, 19	Propanesulfone
23	α-Methylstyrene	15	1-Propanol
10, 17	3-Methylsulfolane	15	2-Propanol
29	Mineral spirits	19	*iso*-Propenyl acetate
14	Morpholine	15	Propionic acid
14, 20	Nitrobenzene	13, 17	Propionitrile
13, 20	Nitroethane	19	Propyl acetate
10, 19	Nitromethane	19	*iso*-Propyl acetate
15, 20	2-Nitropropane	24	*iso*-Propylbenzene
17	1-Nonanol	9, 17	Propylene carbonate
17	Nonylphenol	17	Propylene oxide
30	1-Octadecene	26	*iso*-Propyl ether
27	1,7-Octadiene	16	Pyridine
29	Octane	10	2-Pyrrolidone
26	1-Octanethiol	22	Styrene
17	1-Octanol	9, 17	Sulfolane
17	2-Octanol	13, 19	1,1,2,2-Tetrabromoethane
22	2-Octanone	19	1,1,2,2-Tetrachloroethane
28	1-Octene	25	Tetrachloroethylene
28	*trans*-2-Octene (*cis* isomer 27)	30	Tetradecane
6	3,3′-Oxydipropionitrile	29	1-Tetradecene
15, 19	Paraldehyde	7	Tetraethylene glycol
7	PEG-200	9	Tetraethylenepentamine
7	PEG-300	23	Tetraethyl orthosilicate
8	PEG-600	17	Tetrahydrofuran
25	1,3-Pentadiene	13	Tetrahydrofurfuryl alcohol
7	Pentaethylene glycol	21	Tetrahydrothiophene
9	Pentaethylenehexamine	24	Tetralin
9	Pentafluoroethanol	16	*N,N,N′,N′*-Tetramethyl-ethylenediamine
3	1,5-Pentanediol		
12, 18	2,4-Pentanedione	29	Tetramethylsilane
17	1-Pentanol	15	Tetramethylurea
23	Pentyl acetate	29	Tetrapropylene
16	*tert*-Pentyl alcohol	4	2,2′-Thiodiethanol
26	Pentyl ether	8	1,1′-Thiodi-2-propanol
31	Petrolatum (liquid)	6, 19	3,3′-thiodipropionitrile
20	Phenetole	20	Thiophene
12	2-Phenoxyethanol	23	Toluene
13, 17	1-Phenoxy-2-propanol	11, 19	Triacetin
12, 19	Phenylacetonitrile	28	Tributylamine
10	*N*-Phenylethanolamine	18	Tributyl phosphate
22	Phenyl ether	24	1,2,4-Trichlorobenzene
16	2-Picoline	22	1,1,1-Trichloroethane

TABLE 7.3 (*Continued*)

19	1,1,2-Trichloroethane	12, 17	Trimethylboroxin
20	Trichlorethylene	12	Trimethyl nitrilotripropionate
20	1,2,3-Trichloropropane	29	Tri-*iso*-butylene
27	1,1,1-Trichloro-2,2,2-trifluoroethane	27	2,4,4-Triethyl-1-pentene
		27	2,4,4-Trimethyl-2-pentene
21	Tricresyl phosphate	10	Trimethyl phosphate
2	Triethanolamine	26	Tripropylamine
26	Triethylamine	12	Tripropylene glycol
25	Triethylbenzene	20	Vinyl acetate
6	Triethylene glycol	22	Vinyl butyrate
14	Triethylene glycol monobutyl ether	26	4-Vinylcyclohexene
		26	Vinylidenenorbornene
13	Triethylene glycol monomethyl ether	29	VM&P naphtha
		23	*m*-Xylene
9	Triethylenetetramine	23	*o*-Xylene
14	Triethyl phosphate	24	*p*-Xylene
16	Trimethyl borate		

ration, or doubly bonded oxygen atoms (amides, carboxylic acids, esters, ketones), exhibit a fairly rapid cutoff in the ultraviolet region (see Table 7.4). Among the dipolar aprotic solvents only acetonitrile and ethylene carbonate (and by interference, propylene carbonate) have good transparency in the ultraviolet. For this reason acetonitrile has been used most often in spectroelectrochemistry, particularly in the study of aromatic radical anions. Acetonitrile must be purified rigorously to remove ultraviolet absorbing impurities if a UV cutoff below 200 nm is to be realized.

A number of companies supply reagent or spectroquality solvents that have been purified to remove UV-absorbing impurities. Some of them, particularly dimethyl sulfoxide, may be suitable for general electrochemical use as purchased. However, small quantities of electroactive impurities (particularly water) often are present in spectroquality solvents. Therefore, a particular batch of solvent always should be tested by measurement of the residual current with an appropriate supporting electrolyte and a platinum, gold, or carbon electrode (to test the anodic limits) and a platinum electrode (to test the cathodic limits). The voltage window or domain of electroactivity is a sensitive measure of the adequacy of the purification procedures.

Most organic solvents will show a number of absorption bands in the near and fundamental infrared regions. For IR spectroelectrochemical work the spectra of the more common solvents are useful and are available.[37]

Classification of Solvents. Solvent classification helps to identify properties useful in solvent selection for individual applications; for example, the study of acid–base reactions, oxidation–reduction reactions, inorganic coordination chemistry, organic nucleophilic displacement reactions, and electrochemistry.

TABLE 7.4 Ultraviolet Cutoff of Spectroquality Organic Solvents[a]

Solvent	UV Cutoff (nm)
Acetone	330
Acetonitrile	190
Benzene	280
Benzonitrile	299
n-Butanol	205
Isobutanol	210
Chloroform	245
1,2-Dichloroethane	225
Dichloromethane	233
N,*N*-Dimethylacetamide	270
N,*N*-Dimethylformamide	270
p-Dioxane	215
Ethyl acetate	255
Ethylene carbonate (cyclic ester)	215
Methanol	205
2-Methoxyethanol	210
1-Methyl-2-pyrrolidone	261
Dimethyl sulfoxide	262
Nitromethane	380
Pyridine	305

[a]Absorbance = 1.0 in 1-cm cell versus water.

Unfortunately, a single classification scheme suited to all areas of nonaqueous solvent study has not yet been devised; the criteria that are useful in one area often are not appropriate for another.[38]

The scheme that was originally proposed by Brønsted for acid–base systems has been discussed.[38] This divides solvents into four classes: protogenic (proton donating), protophilic (proton accepting), amphiprotic (both proton donating and accepting), and aprotic (neither proton donating or accepting). Brønsted further divided each of these classes into two subdivisions on the basis of dielectric constant, taking $\epsilon = 30$ as the dividing line.

This classification has been broadened[39,40] by replacing the Brønsted acid (proton donor) with a Lewis acid (an electron acceptor) and the Brønsted base with a Lewis base (an electron donor). (A Brønsted acid is a Lewis acid but not necessarily vice versa.) Solvent–proton interactions are therefore included as one subdivision of this classification, but many solvation reactions of cations with solvents also will be included as reactions of Lewis acid–base systems. This approach still does not solve the problem of fitting specific solvation interactions into the classification scheme. For example, acetonitrile behaves as a good Lewis base toward silver(I) ion, but a poor one toward hydronium ion. The broader scheme also does not specifically take into account hydrogen-bonding effects in hydroxylic and other solvents, which affect both the dielectric constant and the Lewis acid–base properties of a solvent. Sucrose, for example, is soluble in water ($\epsilon = 78$) because of solvent–solute hydrogen bonding; in contrast, the solubility of sugars in propylene carbonate ($\epsilon = 69$) is slight.

There also is difficulty in accommodating fused-salt systems in a classification scheme designed primarily for organic solvents, because the dielectric constants are not comparable and the measures of solvent polarity appropriate to organic solvents are not generally useful in fused salts. Therefore, fused-salt systems are not discussed, nor are they included in the general classification scheme.

In Table 7.5 is presented a solvent classification based on an extension of the basic scheme of Brønsted, along with representative solvents of each class. In this table an arbitrary value of 25 for the dielectric constant is used as the dividing line between dipolar and nonpolar solvents.[39,40] In classes 5, 6, and 8 a further subdivision into groups (based on the ability of the solvent to form a hydrogen bond by donation of a hydrogen atom) is useful. Solvents that are hydrogen-bond donors have a greater tendency to associate and to solvate anions, relative to aprotic solvents (non-hydrogen-bond donors).

7.2 ROLE OF THE SOLVENT-SUPPORTING ELECTROLYTE SYSTEM IN ELECTROCHEMISTRY

There are several ways in which the solvent-supporting electrolyte system can influence mass transfer, the electrode reaction (electron transfer), and the chemical reactions that are coupled to the electron transfer. The diffusion of an electroactive species will be affected not only by the viscosity of the medium but also by the strength of the solute–solvent interactions that determine the size of the solvation sphere. The solvent also plays a crucial role in proton mobility; water and other protic solvents produce a much higher proton mobility because of fast solvent proton exchange, a phenomenon that does not exist in aprotic organic solvents.

The medium also has important effects on the structure of the electrical double layer, a crucial factor in electrochemistry because electron transfer occurs at the double layer. In this connection three factors are important.[41] First, most of the polar organic solvents tend to be strongly oriented in the double layer at the electrode–solution interface. This strongly affects the double-layer capacitance, which is not related in a simple way to the dielectric constant of the bulk solvent. Second, adsorption phenomena, which frequently affect the course of an electrochemical reaction, are less pronounced in organic solvents than in water. This results because the energy involved in replacing a polar organic solvent molecule by a molecule of electroactive solute generally is higher in organic solvents than in water. Finally, ions of the supporting electrolyte may be specifically adsorbed in the double layer, which creates a monomolecular layer of adsorbed ions (the so-called inner Helmholtz layer) that is associated with a characteristic potential, ϕ_1. At a greater distance from the

TABLE 7.5 A Solvent Classification Scheme for Nonaqueous Solvents

Solvent Class	Dielectric Constant	Brønsted and/or Lewis Acid	Brønsted and/or Lewis Base	Examples
1. Amphiprotic	>25	+	+	Water, methanol, 1,2-ethanediol
2. Amphiprotic	<25	+	+	Ethanol, isopropanol
3. Protic	>25	+	–	Hydrogen cyanide, sulfuric acid
4. Protic	<25	+	–	Acetic acid, hydrogen halides (except HF)
5. Lewis base properties	>25	–	+	
a. Aprotic				Acetonitrile, dimethylformamide, propylene carbonate, dimethyl sulfoxide
b. Hydrogen bond donor				Formamide, 2-pyrrolidone
6. Lewis base properties	<25	–	+	
a. Aprotic				Pyridine, acetone, tetramethylurea, acetic anhydride
b. Hydrogen bond donor				Ethylenediamine
7. Negligible Lewis acid or base properties	>25	–	–	Nitromethane, nitrobenzene
8. Neglibible Lewis acid or base properties	<25	–	–	
a. Aprotic				Hexane, dichloromethane
b. Hydrogen bond donor				Chloroform

electrode surface there is presumed to be a second layer (the outer Helmholtz layer) of ions that are not specifically adsorbed, and that have their normal solvation shells with an associated potential, ϕ_2. This model of the double-layer structure has a characteristic feature—the potential does not change in a monotonic fashion in going from the bulk potential of the metal to the bulk potential of the solution. This discontinuous potential gradient can have profound effects on the rates of electron transfer.[41]

Most electron transfers that involve organic compounds have rates that tend to lie in the upper range of detection by present electrochemical techniques.[42] In the absence of adsorption or fast follow-up chemical reactions, the effect of the medium often can be isolated by measurement of the variation of half-wave potentials for one-electron, reversible systems. For a reduction reaction

$$R + e^- \longrightarrow R^{\overline{\cdot}} \tag{7.7}$$

the reduction potential (half-wave potential) relative to some reference electrode can be expressed as

$$E_{1/2} = EA + \Delta G_{\text{solv}} - K \tag{7.8}$$

where EA is the electron affinity of the neutral molecule in the gaseous state, ΔG_{solv} the difference in free energy of solvation of R and $R^{\overline{\cdot}}$ ($\Delta G_R - \Delta G_{R^{\overline{\cdot}}}$), and K a constant. Implicit in Eq. (7.8) is the assumption that there is no ion association between $R^{\overline{\cdot}}$ and the cation of the supporting electrolyte.

The half-wave potentials (corrected for changes in liquid-junction potential) for the one-electron reduction of aromatic hydrocarbons generally become more positive (the reduction is easier) as the dielectric constant of the solvent increases.[44] This is in accord with the direction of the variation in solvation energy of the radical anions that is predicted by the simple Born theory

$$\Delta G_{R^{\overline{\cdot}}} = -\frac{Ne^2}{2r}\left(1 - \frac{1}{\epsilon}\right) \tag{7.9}$$

In deriving Eq. (7.9) the ion is assumed to be a sphere of radius r in a continuous medium of dielectric constant ϵ. The changes in ΔG_{solv} that are produced by variations of the solvation energy of the neutral hydrocarbon (ΔG_R) are not expected to be large because there are no strong charge–solvent dipole interactions. Therefore, changes in ΔG_{solv} are dominated by changes in the solvation energy of the radical anion.

In the absence of secondary chemical reactions, or other complications, the transfer of a second electron is more difficult because of the repulsion; conversely, the removal of a second electron from a radical cation is more difficult because of the coulombic attraction. However, charge delocalization or structural effects can modify this behavior to the point that reversible two-electron transfers exist in which the addition of successive electrons is not distinguishable.[45]

From the point of view of the overall reaction, secondary chemical reactions play a decisive role and are strongly solvent-dependent. A radical anion that is produced in a reduction step can react by several pathways. If it is sufficiently basic, it may abstract a proton from the solvent or from the small amounts of water that are usually present; even the solute itself may act as a proton donor.[46] Neutral radicals can be produced by expulsion of a negative leaving group (such as a halide ion); these may react with one another to form dimers or with the solvent. Several examples for each type of reaction have been cited in the discussion of the role of the solvent in organic electrochemistry.[42] The use of an aprotic solvent tends to simplify the possible chemical reactions relative to those in protic solvents. This offers interesting possibilities for the study of primary anion radicals that are produced by simple electron transfer. Radical cations are more difficult to stabilize, but the addition of small amounts of finely divided alumina or trifluoroacetic anhydride to acetonitrile (to scavenge the last traces of water) prolongs their lifetime sufficiently that they may be studied by electrochemical and spectroscopic techniques.[47]

7.3 ROLE OF THE SUPPORTING ELECTROLYTE

The use of an indifferent or "inert" supporting electrolyte is indispensable in electrochemistry and affects the solvent medium in several ways: (1) it regulates cell resistance and mass transport by electrical migration; (2) it may control or "buffer" the level of hydronium ion activity in solution; (3) it may associate with the electroactive solute, as in the complexing of metal ions by certain ligands; (4) it may form ion-pair or micellar aggregates with the electroactive species; (5) it largely determines the structure of the double layer; and (6) it may impose positive or negative voltage limits because of its redox properties.

Control of Cell Resistance. All the pure organic solvents that are discussed above essentially are nonconductors. Without the addition of some electrolyte, their resistance is so great that the voltages required to pass even milliampere currents are impractically large. Therefore, the primary function of the supporting electrolyte is to provide a conducting medium. The solvent-supporting electrolyte combination should be chosen to give resistance values that are as small as possible. This will minimize uncompensated iR drop, which leads to potential-control error, and will minimize ohmic heating of the solution in preparative electrochemistry (where the currents are large).

The solubility and solution resistance of some commonly used supporting electrolytes in selected aprotic solvents are listed in Table 7.6. Table 7.7 contains conductance data for a number of other salts and solvents, while Table 7.8 indicates the extent of ion association in acetonitrile, a solvent of moderate dielectric constant. As the dielectric constant decreases, ion association in-

TABLE 7.6 Solubilities of Tetraalkylammonium Salt Electrolytes and Specific Resistances of the Solutions at 25°C

	Acetonitrile		1,2-Dimethoxyethane		Tetrahydrofuran		Dimethylformamide	
	Solubility (g 100 mL^{-1} of Solution (conc., M)	Specific Resistance (Ω-cm) (conc., M)	Solubility (g 100 mL^{-1}) of Solution (conc., M)	Specific Resistance (Ω-cm) (conc., M)	Solubility (g 100 mL^{-1}) of Solution (conc., M)	Specific Resistance (Ω-cm) (conc., M)	Solubility (g 100 mL^{-1}) of Solution (conc., M)	Specific Resistance (Ω-cm) (conc., M)
Et_4NClO_4	26(1.13)	26(0.60)	(<0.01)		(<0.01)		23(1.00)	52(0.60)
n-Bu_4NClO_4	70(2.05)	37(0.60)	31(1.10)	312(1.0)	50(1.48)	368(1.0)	79(2.29)	77(0.60)
Et_4NBF_4	37(1.69)	18(1.0)	(<0.01)		(<0.01)		27(1.24)	38(1.0)
n-Bu_4NBF_4	71(2.21)	31(1.0)	53(1.70)	228(1.0)	65(2.02)	373(1.0)	75(2.34)	69(1.0)
Et_4NBr	7.8(0.37)		(<0.01)		(<0.01)		4.1(0.19)	
n-Bu_4NBr	66(1.99)	48(0.60)	(<0.01)		4.8(0.14)		52(1.57)	106(0.60)

Source: Ref. 35.

TABLE 7.7 Limiting Ionic Conductivities of Ions λ_i° in Selected Solvents[a]

Ion	Water[b]	Acetone	Acetonitrile	Propylene Carbonate	Nitro-methane	Nitro-benezene	Dimethyl formamide	Dimethyl Sulfoxide	Sulfolane (40°C)
$H(sol)^+$	349.8	90	100		63		35	16	
$Li(sol)^+$	38.7	72.8	69.3	7.3	55		25.0	11.4	4.3
$Na(sol)^+$	50.1	78.4	76.9		58	16.3	29.9	13.5	3.6
$K(sol)^+$	73.5	80.6	83.6	12.0	60	17.8	30.8	13.9	4.0
Me_4N^+	44.9	97.7	94.5		54.5	17.1	38.9	18.6[c]	4.3
Et_4N^+	32.7	89.0	84.8	13.3	47.6	16.4	35.6	17.1	4.0
$n\text{-}Bu_4N^+$	19.5	67.3	64.1	9.3	34[c]	11.9	25.9	11.6	2.8
Cl^-	76.4	105.2	98.4	20.2	62.5	22.2	55.1	24.2	9.3
Br^-	78.1	115.9	100.7	19.4	62.9	21.6	53.6	24.3	8.9
I^-	76.8	113.0	102.4	18.8	62	20.4	52.3	24.0	7.2
NO_3^-	71.5	120.1	106.4		64	22.6	57.3	27.0	
BF_4^-			108.5						
PF_6^-			104.2						6.0
ClO_4^-	67.3	115.3	103.7	18.8	64	20.9	52.4	24.7	6.7

[a] All values at 25°C unless otherwise indicated; data from Ref. 48 unless otherwise indicated.
[b] Robinson, R. A.; Stokes, R. H. *Electrolyte Solutions*, Butterworths Scientific Publications, London, 1955, p. 452.
[c] Janz, G. J.; Tomkins, R. P. T., eds., *Nonaqueous Electrolytes Handbook*, Academic Press, New York, 1972.

TABLE 7.8 Association Constants for Some 1:1 Electrolytes in Acetonitrile[a]

	Cl^-	Br^-	NO_3^-	I^-	BF_4^-	ClO_4^-	PF_6^-
Me_4N^+	56	46		19		7	5
Et_4N^+			5	5		0	
n-Bu_4N^+		2		3			0
$Li(sol)^+$						4	
$Na(sol)^+$				0		10	
$K(sol)^+$				0		14	

Source: Ref. 48.
[a]Anions arranged in order of increasing crystallographic radius.

creases, and solution conductivity is not proportional to the concentration of electrolyte.[48] Extensive compilations of solution conductances in organic solvents are available.[48,49]

When an electric field is imposed on an electrolyte solution, the ions will tend to migrate—the cations moving toward the cathode (negative electrode) and the anions toward the anode. This migration of ions constitutes the flow of current in the cell, and each kind of ion carries a fraction of the current proportional to its mobility and concentration. For a particular value of the current, the addition of an inert electrolyte well reduce the solution resistance, which, in turn, will decrease the electric field according to Ohm's law, $E = iR$. Therefore, the mass transport of an ionic electroactive species that is caused by migration in an electric field can be reduced to a negligible level by "swamping" the solution with inert electrolyte. Then most of the current will be carried by the ions of the supporting electrolyte. In polarography and voltammetry the customary practice is to make the supporting electrolyte concentration at least 50–100 times the concentration of the electroactive species, so that electrical migration will be suppressed. This is particularly important where quantitative agreement is sought with the equations that are used to calculate limiting or peak currents. These are based on the assumption that mass transport is controlled by pure diffusion.

Control of Solution Acidity. Many inorganic and organic redox reactions involve hydronium ions, for example

$$O{=}C_6H_4{=}O + 2\,H_3O^+ + 2\,e^- \longrightarrow HO{-}C_6H_4{-}OH + 2\,H_2O \qquad (7.10)$$

The effect of proton donors on the reduction of aromatic hydrocarbons has been discussed earlier in this chapter. The importance of potential–pH relations to the understanding and utilization of the redox behavior of these systems has long been recognized; extensive potentiometric data have been obtained for only a limited number of organic compounds.[50]

With the discovery of polarography much of the activity in potentiometric studies of organic compounds tapered off. Instead, attention was focused on the measurement of the polargraphic half-wave potentials $E_{1/2}$, which provide similar free-energy information. In may cases, half-wave potentials are the only data readily obtained by electrochemical measurements. Polarographic and voltammetric techniques have provided most of the information of the involvement of hydronium ion in electrochemical processes.

The role of hydronium ion and Brønsted acids in electrode reactions has been described.[51–53] For a reversible reduction [such as that for quinone that is represented in Eq. (7.10)] of the general form

$$ox + m\,H_3O^+ + n\,e^- \longrightarrow red\text{H}_m + m\,H_2O \qquad (7.11)$$

the half-wave potential varies with pH according to

$$\frac{dE_{1/2}}{d(\text{pH})} = -\frac{2.3\,RT}{nF}\,m \qquad (7.12)$$

where m is easily evaluated once n is known.[54]

The reduction of quinone consumes ions, and if the aqueous solution is not well buffered, the hydronium ions are supplied by water to produce an excess of HO^- at the surface of the dropping mercury electrode. This will increase the pH at the electrode surface and shift the cathodic wave to more negative values. The reverse is true for the oxidation of hydroquinone, in which H_3O^+ will be liberated at the electrode surface. The result is that in an unbuffered solution the composite wave will be split into two waves, while in a well-buffered solution only a single composite wave is observed for an equimolar mixture of quinone and hydroquinone (quinhydrone). This example illustrates the fact that the buffering capacity of the system must be sufficient to react with the H_3O^+ or HO^- ions that are liberated at the electrode surface. Although this might seem simple enough to accomplish, there are several complicating factors. First, the components of some buffer systems (e.g., phosphate or citrate) may interact strongly with some species, particularly in biological systems that contain nucleotides or other metabolic intermediates. An answer to this problem is the use of one of several little-used hydronium ion buffers that are compatible with common biological media.[55] The hydronium ion activities have been assigned[56] to one of these buffers, tris(hydroxymethyl)-methylglycine, in the physiological range of pH 7.2–8.5.

The second complication is more disturbing. Calculations indicate that the pH at the electrode surface for an electrode at −1.5 V versus SCE may be as much as 2 pH units lower than in the bulk of the solution.[57,58] This results from the alteration of the distribution of charged species in the double layer due to the potential drop in the diffuse part of the double layer. The resolution of this problem must await further theoretical and experimental work.

If the electron-transfer step in an electrode reaction is preceded by a chemical reaction that involves proton transfer, the polarographic current often will be a complex function of the concentration of the electroactive species, the hydronium ion concentration, and the rate constants for proton and electron transfer. Currents controlled by the rate of a chemical reaction are called *kinetic currents* and often are observed in the reduction of electroactive acids (e.g., pyruvic acid), in which the protonated form of the acid is more easily reduced than the anion. A polarogram of pyruvic acid in unbuffered solution exhibits two waves whose relative wave heights depend on the concentration of pyruvic acid and the solution pH.[59]

The effects of solution acidity on the polarography of organic compounds have been reviewed, principally in aqueous solution. A thorough discussion of kinetic and catalytic currents that involve hydronium ion has been presented,[52] and the irreversible polarographic and voltammetric curves that involve proton transfer in unbuffered and poorly buffered solutions have been discussed.[59]

Although there is a wealth of data in the literature on acid–base behavior in aprotic solvents,[60,61] there are few examples of the use of buffers for polarography and voltammetry in aprotic solvents. This has occurred because most investigators have sought to keep all potential proton donors out of the system and thereby stabilize anion radicals. Although the picric acid–picrate system has been used as a buffer in a number of studies in aprotic solvents, its use in voltammetric work is limited because of the ease of reduction of picric acid.

Most neutral acids are much weaker acids in aprotic solvents than in water. This is due largely to their smaller dielectric constants, which increases the energy required for charge separation in the dissociation process. Table 7.9 summarizes the pK_a values for several acids in five different solvents.[62] There is a tendency in solvents with low dielectric constants, which cannot stabilize the anion of weak Brønsted acids (H*A*) by hydrogen bonding, for the anion (A^-) to hydrogen bond with the undissociated acid to yield the species AHA^-

TABLE 7.9 Comparison of pK_a Values of Several Acids in Different Solvents

Acid	Dimethyl Sulfoxide	Dimethyl Formamide	Acetonitrile	Sulfolane	Water
$HClO_4$	Strong	Strong	Strong	2.7	Strong
HSO_3CF_3	Strong	—	—	3.4	Strong
Picric	−0.3	1.2(1.4)	8.9(11.0)	—	0.38
HSO_3CH_3	1.76	—	6.1	—	—
HBr	1.0	1.8	5.5	7.4	−9
HCl	2.01	3.4	8.9	12.7	−7
H_2SO_4	1.4	3.0	7.25		−4.0
$HOOCCF_3$	3.45	—			−0.3

(called a *homoconjugate*).[63] This adds a further complication to the measurement of dissociation constants in these solvents.

Complex Formation. A number of neutral or anionic donor ligands will form complexes with metal ions. These ligands include cyanide, thicyanate, halides, amines, polyhydroxy carboxylates (such as tartrate), and polyamino carboxylates (such as EDTA). In general, the metal ion–ligand complex will be reduced at more negative potentials than the "free" metal ion (which is actually an aquo or solvent complex ion). This property can be used to eliminate interferences between overlapping polarographic waves of two metal ions by selectively complexing one of them. As an example, the determination of Tl(I) in the presence of Cu(II), Cd(II), Sn(II), Pb(II), Bi(II), and Sb(III) can be accomplished by a combination of EDTA complexation and judicious choice of pH.[64] The use of complexing agents to eliminate interferences is sometimes called "masking," and a number of applications in electroanalytical chemistry have been summarized.[65]

The use of complexing agents in organic solvents is not widespread because metal ions more often are determined in aqueous solution; this is an area where further research may prove fruitful.

Ion-Pairing and Double-Layer Effects. The reduction of metal ions and aromatic compounds in organic solvents is affected strongly by the nature of the supporting electrolyte. In particular the reductions of alkali metal ions in hexamethylphosphoramide (HMPA) are influenced significantly by the type of cation in the supporting electrolyte.[66] The addition of small amounts of tetraethylammonium ion interferes with the reduction of Na(I), Li(I), and K(I), while reduction is affected less by the presence of tetrabutylammonium ion. From conductivity studies the size of the solvated cation in HMPA appears to increase in the order: $Et_4N^+ << Bu_4N^+ \sim K(I) \sim Na(I) < Li(I)$. At extremely negative potentials the smaller solvated ions are attracted to the electrode surface such that Et_4N^+ ion can compete effectively with the considerably larger solvated Li(I) ion, while the smaller *solvated* Rb(I) and Cs(I) ions are less affected. The distance of closest approach of the supporting electrolyte cations apparently increases with increasing size of the supporting electrolyte cation (the quaternary ammonium ions). As this distance increases, the potential of ϕ_1 at the inner Helmholtz plane becomes more negative, which accelerates the desolvation and electron transfer at the inner Helmholtz plane. This leads to greater reversibility for the reduction of the alkali metal cations in the presence of tetrabutylammonium ions.

In the electroreduction of aromatic hydrocarbons, nitro compounds, and quinones in aprotic solvents, the first step is the transfer of an electron from the electrode to form a radical anion. Once the radical anion is formed, electron repulsion will decrease the facility with which a second electron transfer occurs. But solvation and ion pairing diminish the effect of electron repulsion and tend to shift the reduction potential for the addition of the second electron to more positive values. In both cases the extent of the shift will be dependent on the nature of the radical anion, the supporting electrolyte cation, and the solvent. If the radical anion is sufficiently polar, and if the electron transfer is fast, ion pairing of the radical anion with the cations of the supporting electrolyte will shift the first reduction potential to more positive values, also.

There is strong evidence[67] that at least two types of ion pairs exist. All cations in solution tend to be surrounded by solvent molecules. A small ion generates in its vicinity a more powerful electrostatic field than a large ion of the same charge; thus smaller ions tend to produce more rigid solvation shells than large ions.

The association of a cation that is surrounded by a tight solvation shell with an anion proceeds smoothly until the solvent shell comes into contact with the anion. At this stage either the structure of the ion pair, separated by solvent molecules, is preserved (Figure 7.1*a*) or the solvation shell is squeezed out in a process that leads to a contact pair. This implies that at least two types of ion pair may coexist in solution, each having its own physical and chemical properties; such two-step associations have been revealed by various relaxation experiments. However, ions that weakly interact with the solvent and do not surround themselves with tight solvation shells form contact pairs only. This situation is encountered in poorly solvated liquids and for bulky ions. Those cations that interact strongly with solvent molecules tend to form solvent-separated pairs, especially when combined with large anions.

The available solvation data have been classified according to the nature of the anion (highly polar with localized charge or low charge density with delocalized charge) and the solvating power of the solvent.[68] Nitrobenzenes and quinones give rise to polar anions with localized charges that interact sufficiently strongly with cations to cause desolvation and the formation of contact ion pairs in solvents of low or high dielectric constant. Consequently, the magnitude of the ion-pairing interaction (positive shift of the reduction potential) becomes larger as the crystallographic radius of the unsolvated cation decreases.

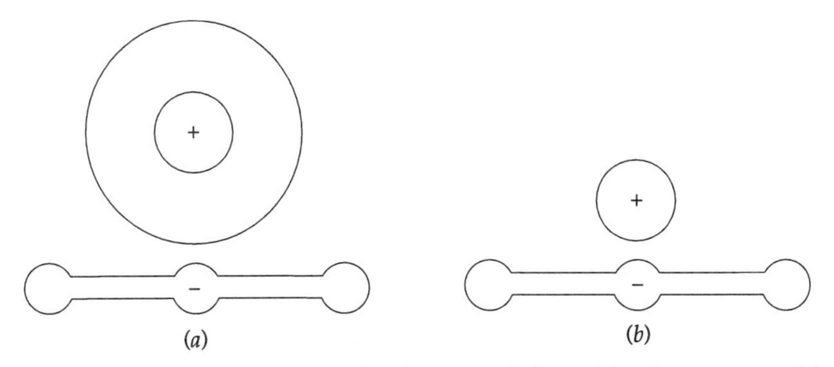

Figure 7.1 Two types of ion pairs that form in solutions: (*a*) solvent-separated ion pair; (*b*) contact ion pair.

In the case of anthracene, which forms large anions with delocalized charge, the ion-pairing effect is related to the size of the *solvated* cation; the interaction increases as the solvated cation becomes smaller.

Micellar Aggregates. Surface active agents that contain a polar head and a long hydrophobic alkyl "tail" can be used to solubilize organic compounds in water. Although these micellar aggregates may approach colloidal dimensions, electron transfer from the surface of the electrode to the organic compound in the micellar aggregate still is possible, and more or less normal polarograms can be obtained.[69]

7.4 ELECTROCHEMICAL PROPERTIES OF WATER AND SELECTED ORGANIC SOLVENTS

Water. A laboratory engaged in careful electrochemical work with aqueous solutions or in trace analysis will need facilities for the preparation and storage of highly purified water. Water commonly is contaminated with metals in both dissolved cationic form and in the form of colloidal or particulate matter that is not ionized appreciably.[70] Frequently it also is contaminated by bacteria and by organic impurities that cannot be removed by ordinary or oxidative distillation because of the steam volatility of the impurities.[71]

The purification of water has been treated exhaustively in the literature, and reviews have summarized general conclusions and recommended techniques for the ultrapurification and analysis of water.[70,72] There is a general consensus among investigators that (1) water purified in quartz or plastic apparatus contains a lower level of cationic impurities than water purified in borosilicate or metal apparatus; (2) storage of purified water for periods exceeding 30 days, even in plastic or Teflon containers, will result in an increase in cationic impurities; and (3) resistivity measurements may be used as a survey technique but cannot be relied on for an unequivocal indication of water quality because they do not indicate nonionic impurities.

Removal of cationic impurities from water. Careful analysis of water purified by various methods (see Table 7.10) indicates that the water that is obtained by passing ordinary distilled water through a small monobed deionizer (contained in polyethylene) and a submicrometer filter is equal or superior (with respect to cations) to water obtained by distillation in conventional quartz stills, and is distinctly superior to the product from systems constructed of metal.[70] From the data available in the literature, simple distillation clearly does not produce high-purity water. In practice, two effects cause contamination of the distillate. Entrainment is the major factor that prevents the perfect separation of a volatile substance from nonvolatile solids during distillation. Rising bubbles of vapor break through the surface of the liquid with considerable force and throw a fog of droplets (of colloidal dimensions) into the vapor space

TABLE 7.10 Metal Content of Water Purified by Various Methods[a]

Element	Subboiling Quartz Still, Doubly Distilled-Water Feed[b]	Doubly Distilled in Quartz, Distilled Water Feed[c]	Deionized, Monobed, Polyethylene Apparatus, Distilled-Water Feed[c]	Distilled Two-Stage Commercial Metal Still, City Water Feed,[c]	Recirculating Ultrapure System, Metal[c]	Single-Pass Deionization of Tap Water[c]	Deionization, Carbon Absorption, Deionization, Membrane Filtration[d]
Na	0.06	x	x	1.0	4		1
K	0.09	x	x	x	x	x	
Mg	0.09	0.05	0.01	8	2	2	0.5
Ca	0.08	0.07	0.03	50	10	0.2	1
Sr	0.002	x	x	x	x		
Ba	0.01	x	x	x	x		
B	x	–	–	0.01	0.5	+	3
Al	x	0.5	0.1	10	10		0.1
Tl	0.01	x	x	x	x		
Si	x	5.0	1.0	50	10	+	0.5
Sn	0.02	–	–	5	10	+	0.1
Pb	0.008	–	–	50	10	0.02	0.1
Te	0.004	x	x	x			
Ti	x	–	–	+	–		0.1
Cr	0.02	–	–	–	0.5	0.02	0.1
Mn	x	–	–	0.01	0.1	0.02	0.05
Fe	0.05	–	–	0.1	10	0.02	0.2
Ni	0.02	–	–	1.0	0.05	0.002	0.1
Cu	0.01	–	–	50	10	+	0.2
Ag	0.002	–	–	1.0	0.7	+	0.01
Zn	0.04	–	–	10	3	0.06	0.1
Cd	0.005	x	x	x	x		0.1

[a]Parts per billion by weight, ng/g; –, sought, not detected; x, not sought; +, detected, but not determined quantitatively.
[b]Kuehner et al., *Anal. Chem.* **1973,** *44*, 2050.
[c]Ref. 70.
[d]Zief and Barnard, *Chem. Tech.* **1973,** 440.

above the liquid surface. These droplets are carried into the condenser to an extent that depends on the still design and its operating conditions. A second effect is the flow of a film of water (that wets all internal surfaces of the still) toward the condenser, which is under the combined influence of capillarity and the vapor stream.

Subboiling distillation circumvents both of these problems and is considered a simple and effective technique for removing metallic or cationic impurities (Figure 7.2). In subboiling distillation, infrared heaters vaporize liquid from the surface without boiling. The vapor is condensed on a tapered cold finger, and the distillate is collected in a suitable container.

Subboiling distillation also is an effective technique for the purification of nitric, hydrochloric, hydrofluoric, perchloric, and sulfuric acids. Typical analyses indicate a 100-fold reduction in the level of metallic impurities.[73] While this method is extremely efficient for the removal of metal ions, it offers little purification from impurities with high vapor pressures, such as organic matter and many of the anions.

Removal of organic impurities. Preparation of pure water that is free from organic, surface-active contaminants is impossible by means of distillation, even from alkaline $KMnO_4$, although previously such a procedure was thought to be adequate.[71] Organic contaminants that are present in many domestic and

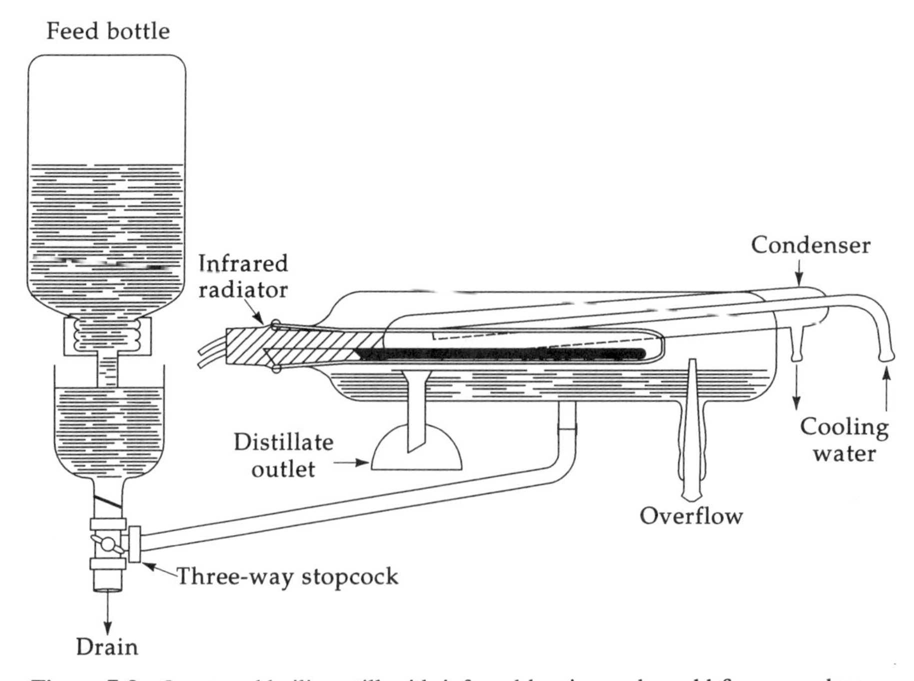

Figure 7.2 Quartz subboiling still with infrared heating and a cold-finger condenser.

industrial water supplies are steam-volatile, and hence are not removed by distillation. The criteria for detecting the presence of such contamination include (1) The inability to obtain the correct surface tension of pure water (the values generally are too low); (2) the presence of electroactivity at the platinum anode at potentials where reactions are known not to occur for pure solutions; (3) the surface blocking of electrosorption of H and OH species at platinum due to adsorption of organic contaminants; (4) the failure to obtain linear log i/E relations at the mercury electrode for oxygen-free, preelectrolyzed solutions; and (5) the indication of macromolecular contaminants by means of light scattering (usually due to the presence of bacteria).

By passing the water vapor through a column that is packed with platinum gauze and heated to 750–800°C in a stream of oxygen the organic materials can be catalytically combusted and entirely removed. The apparatus shown in Figure 7.3 is designed to recirculate the water continually through the heated zone so that the water in the reservoir is maintained free from organic impurities. The left curve of Figure 7.4 illustrates a current–potential profile for a platinum electrode in redistilled tap water; an oxidation peak that is characteristic of an organic impurity is indicated. The set of curves for the right half of Figure 7.4 indicates the decrease in height of the H adsorption and desorption peaks that results from blocking of the platinum surface by adsorbed organic impurities. The behavior on platinum in a clean aqueous solution is illustrated by Figure 5.22 of Chapter 5. The ability to reproduce every nuance of this current–potential profile can be taken as a primary criterion of the purity of aqueous solutions.

The contribution of the supporting electrolyte to the impurity burden cannot be ignored. Generally the impurities that interfere are heavy metals that may be removed by prolonged electrolysis over a mercury cathode.[74] Commercial apparatus is available for removal of heavy-metal impurities by this method. Preelectrolysis procedures that use graphite tapes also have been suggested,[75] as has the circulation of the solution through beds of charcoal.[76]

Nonaqueous Solvents. Many organic compounds are not soluble in water, and the investigator who desires to study their electrochemistry must resort to organic solvents. The solvents most often used are the so-called dipolar aprotic solvents that belong to Class 5a in the classification scheme of Table 7.5. These are solvents with moderately large dielectric constants and low proton availability. This aprotic character tends to simplify the electrochemical reactions; often the primary product is a stable radical cation or anion that is produced by removal or addition of an electron.

Although a number of solvents have been used by different workers, only a few enjoy continued favor. In Table 7.11 the physical properties of more than 50 solvents are listed (not all of them are aprotic). In the following paragraphs some of the properties and purification methods for four solvents are discussed: acetonitrile, propylene carbonate (PC), dimethylformamide (DMF), and dimethyl sulfoxide (Me_2SO). These are the most widely used solvents and prob-

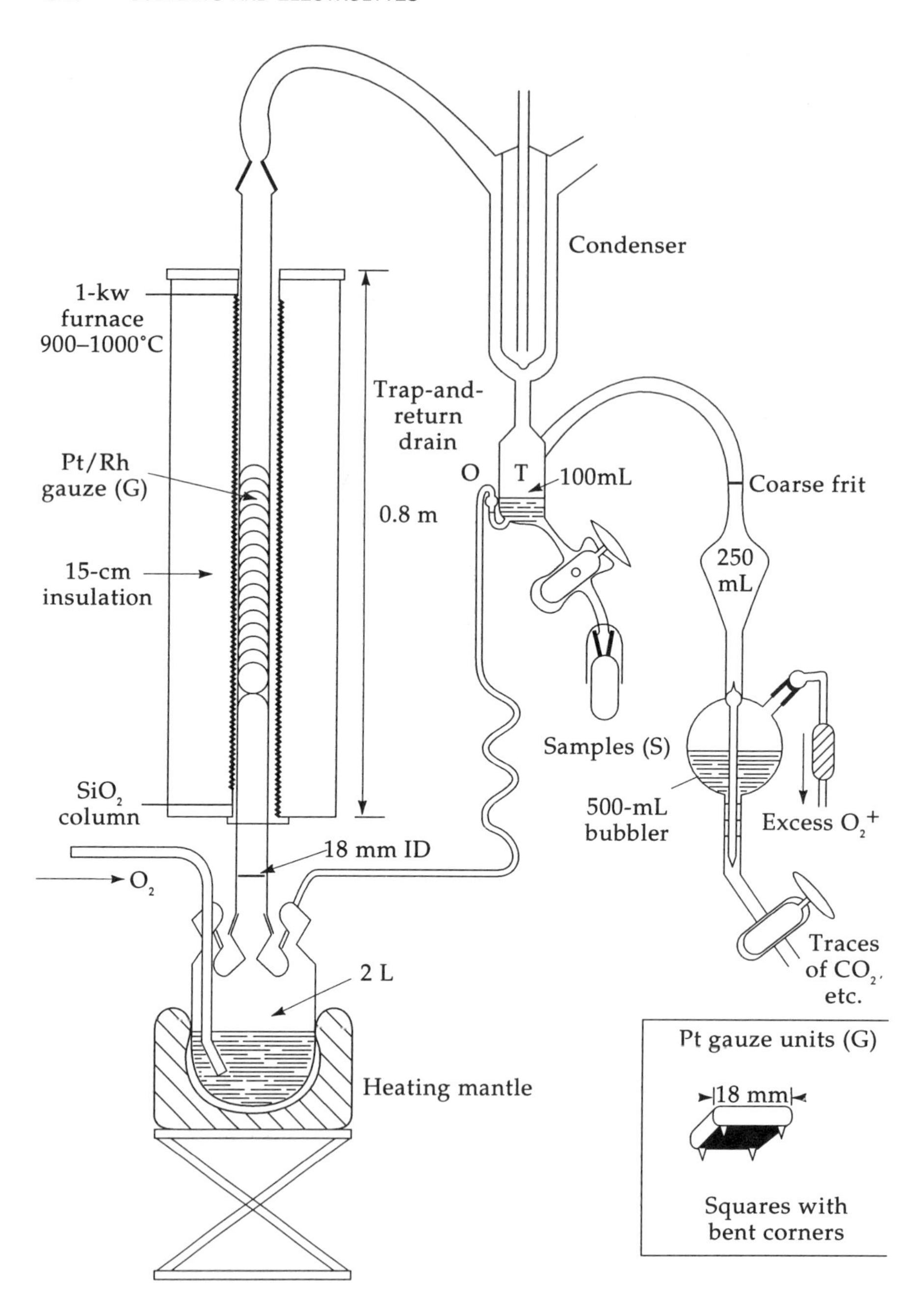

Figure 7.3 Distillation system with a catalytic combustion tube for the removal of organic impurities in water vapor.

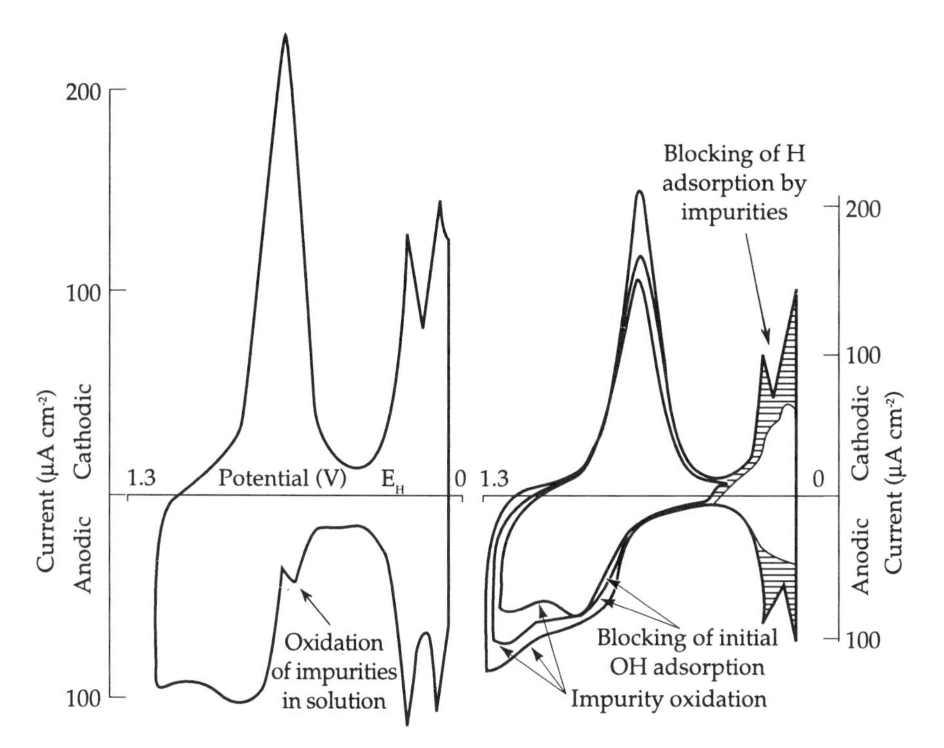

Figure 7.4 Cyclic voltammograms for a platinum electrode. Left curve for redistilled tap water with evidence of organic impurities. Right curves for distilled water that contains oxidizable organic impurities that chemisorb on the electrode.

ably fulfill the requirements of about 80% of the electrochemical uses to which organic solvents are put. More details for other solvents listed in Table 7.11 are available.[77,78]

Acetonitrile. Acetonitrile is resistant to both oxidation and reduction, is transparent in the region 200–2000 nm, and is an excellent solvent for many polar organic compounds and some inorganic salts. Its dielectric constant of 37 permits reasonably high conductivities, although there is evidence of some association (see Table 7.8). It is less basic than dimethylformamide and dimethyl sulfoxide, and therefore does not solvate alkali metal cations as strongly. However, acetonitrile forms stable complexes with Ag(I) and Cu(I) ions.

Although somewhat difficult to purify, acetonitrile is stable on storage after purification. It is toxic, with a maximum recommended limit of 40 ppm,[79] and the vapor pressure is sufficient for this to be a hazard. Its high volatility makes the removal of solvent by evaporation easy (e.g., workup of a reaction mixture for product identification). Both radical cations and anions react with traces of water in acetonitrile, and, because it does not hydrogen-bond to water as do dimethyl sulfoxide and dimethylformamide, radical anions generally have a

TABLE 7.11 Physical Properties of Selected Organic Solvents[a]

Solvent	MW	BP (°C)	FP (°C)	Vapor Pressure (mm Hg)	Density (g cm^{-3})	Dielectric Constant	Dipole Moment	Viscosity (cP)	Specific Conductance (Ω^{-1} cm^{-1})
Water	18.02	100	0	23.8	0.9970	78.4	1.87	0.89	5.49×10^{-8}
Alcohols									
Methanol	32.04	64.70	−97.68	125.03	0.7866	32.70	2.87_{20}	0.54	1.5×10^{-9}
Ethanol	46.07	78.29	114.1	59.77	0.7850	24.55	1.66_{20}	1.08	1.35×10^{-9}
2-Propanol	60.10	82.26	−88.0	45.16	0.7813	19.92	1.66_{30}	1.77_{30}	5.8×10^{-8}
1-Butanol	74.124	117.66	−88.62	6.18	0.8060	17.51	1.75	2.27_{30}	9.12×10^{-9}
1,2-Ethanediol	62.07	197.3	−13	0.12	1.1100	37.7	2.28_{20}	13.55_{30}	1.16×10^{-6}
1,2-Propanediol	76.10	187.6	−60	0.13	1.0328	32.0_{20}	2.25	56.0_{20}	
2-Methoxyethanol	76.10	124.6	−85.1	9.7	0.9602	16.93	2.04	1.60	1.09×10^{-6}
Ethers									
1,2-Dimethoxyethane (glyme)	90.12	93.0	−58	75.2	0.8621	7.20	1.71	0.46	
Bis(2-methoxyethyl) ether (diglyme)	134.18	159.76d		3.4	0.9440		1.97	0.98	
Tetrahydrofuran (THF)	72.11	66	−108.5	197	0.8892_{20}	7.58	1.75	0.46	
p-Dioxane	88.11	101.32	11.80	37.1	1.0280	2.21	0.45	1.09_{30}	5×10^{-15}
Tetrahydropyran	86.14	88	−45		0.8772	5.61	1.55	0.76	
Ketones									
Acetone	58.08	56.29	−94.7	181.72_{20}	0.7844	20.70	2.69_{20}	0.30	4.9×10^{-9}
2-Butanone (MEK)	72.11	79.64	−86.69	90.6	0.7997	18.51_{20}	2.76	0.37_{30}	3.6×10^{-9}
4-Methyl-2-pentanone (MIBK)	100.16	116.5	−84	20.0	0.7961	13.11_{20}		0.54	5.2×10^{-8}
Acids and anhydrides									
Acetic	60.05	117.90	16.66	15.43	1.0437	6.15_{20}	1.68_{30}	1.04_{30}	6×10^{-9}
Acetic anhydride	102.09	140.0	−73.1	5.1	1.0691_{30}	20.7_{19}	2.82	0.78_{30}	5×10^{-9}
Trifluoroacetic	114.02	71.78	−15.25	108	1.4785	8.55_{20}	2.28_{100}	0.86	

Esters and lactones									
Ethyl acetate	88.11	77.11	−83.93	92	0.8946	6.02	1.88	0.43	$<1 \times 10^{-9}$
γ-Butyrolactone	86.09	204	−43.52	10_{79}	1.1254	39_{20}	4.12	1.7	
Ethylene carbonate	88.06	238	36.4	$0.02_{36.4}$	1.3208	89.6_{40}	4.87		$<1 \times 10^{-7}$
Propylene carbonate[b] (PC)	102.09	241.7	−49.2		1.2	64.4		2.5	1×10^{-8}
Chlorinated alkanes									
Dichloromethane	84.93	39.75	−95.14	435.8	1.3168	8.93	1.14	0.39_{30}	4.3×10^{-11}
Chloroform	119.38	61.15	−63.55	194.8	1.4799	4.81_{20}	1.15	0.51_{30}	$<1 \times 10^{-10}$
1,2-Dichloroethane	98.96	83.48	−35.66	83.35_{20}	1.2458	10.36	1.86	0.73_{30}	4×10^{-11}
Nitro compounds									
Nitromethane	61.04	101.20	−28.55	36.66	1.1313	35.87_{30}	3.56_{20}	0.61	5×10^{-9}
Nitroethane	75.07	114.07	−89.52	20.93	1.0446	28.06_{30}	3.60_{20}	0.64	$5 \times 10^{-7}{}_{30}$
Nitrobenzene	123.11	210.80	5.76	0.28	1.1984	34.82	4.03	1.63_{30}	2.05×10^{-10}
Nitriles									
Acetonitrile (AN)	41.05	81.60	−43.84	81.81	0.7766	37.5_{20}	3.44_{20}	0.33_{30}	6×10^{-10}
Propionitrile	55.08	97.35	−92.78	44.63	0.7768	27.2_{20}	3.57_{20}	0.39_{30}	8.51×10^{-8}
Succinonitrile	80.09	267	57.88	0.008	0.9867_{60}	$56.5_{57.4}$	3.68_{30}	2.59_{60}	5.64×10^{-4}
Butyronitrile	69.11	117.94	−111.9	19.10	0.7865	20.3_{21}	3.57_{20}	0.52_{30}	
Benzonitrile	103.13	191.10	−12.75	$1_{28.2}$	1.0006	25.20	4.05	1.11_{30}	5×10^{-8}
Amines									
Butylamine	73.14	77.4	−49.1	91.75	0.7346	4.88_{20}	1.37_{20}	0.68	
Ethylenediamine	60.10	117.26	11.3	$13.1_{26.51}$	0.8860_{30}	12.9	1.90 ·	1.54	9×10^{-8}
Triethylamine	101.19	89.5	−114.7	57.07	0.7230	2.42	0.87	0.36	
Pyridine (PY)	79.10	115.26	−41.55	$20_{24.8}$	0.9782	12.4_{21}	2.37	0.88	4.0×10^{-8}
1,1,3,3-Tetramethylguanidine (TMG)[c]		160			0.9136	11.0		1.40	5×10^{-8}

TABLE 7.11 (*Continued*)

Solvent	MW	BP (°C)	FP (°C)	Vapor Pressure (mm Hg)	Density (g cm^{-3})	Dielectric Constant	Dipole Moment	Viscosity (cP)	Specific Conductance (Ω^{-1} cm^{-1})
Amides									
Formamide	45.04	210.5d	2.55	$1_{70.5}$	1.1292	111.0_{20}	3.37_{30}	3.30	$<2 \times 10^{-7}$
N-Methylformamide	59.07	~180	−3.8	0.4_{44}	0.9988	182.4	3.86	1.65	8×10^{-7}
N,N-Dimethylformamide (DMF)	73.10	153.0	−60.43	3.7	0.9440	36.71	3.86	0.80	6×10^{-8}
Acetamide	59.07	221.15	80.00	$1_{65.0}$	$0.9892_{91.1}$	59_{83}	3.44_{30}	$2.182_{91.1}$	$8.8 \times 10^{-7}{}_{83.2}$
N-Methylacetamide	73.10	206	30.55	1.5_{56}	0.9498_{30}	191.3_{32}	$4.39_{20.1}$	3.23_{35}	$2 \times 10^{-7}{}_{40}$
N,N-Dimethylacetamide (DMAC)	87.12	166.1	−20	1.3	0.9366	37.78	3.72	0.84_{30}	
1,1,3,3-Tetramethylurea (TMU)	116.16	175.2	−1.2	$10_{61.2}$	0.9687_{20}	23.06	3.47		$<6 \times 10^{-8}$
2-Pyrrolidone	85.11	245	25	10_{122}	1.107		3.55	13.3	
N-Methylpyrrolidone (1-methyl-2-pyrrolidone)	99.13	202	−24.4	4_{60}	1.0279	32.0	4.09_{30}	1.67	$1\text{–}2 \times 10^{-8}$
Sulfur compounds									
Dimethyl sulfoxide (Me_2SO)	78.13	189.0	18.54	0.600	1.0958	46.68	3.9	1.10	2×10^{-9}
Sulfolane	120.17	287.3d	28.45	5.0_{118}	1.2614_{30}	43.3_{30}	4.81	10.29_{30}	$<2 \times 10^{-8}{}_{30}$
Dimethylsulfite[d]		126	−141		1.2073_{24}	22.5		0.77_{30}	
Ethylene sulfite[e]		171			1.4375_{20}	41.8_{20}	3.68_{20}		$5 \times 10^{-8}{}_{20}$
Phosphorus compounds									
Tributylphosphate	266.32	289	< −80	0.8_{114}	0.9760	7.96_{30}	3.07	3.39	
Hexamethylphosphoramide (HMPA)	179.20	233	7.20	0.07_{30}	1.027_{20}	30_{20}	5.54	3.47_{20}	

[a]The physical properties (except BP, FP) are measured at 35°C unless the temperature is indicated by a subscript (BP—boiling point; FP—freezing point; MW—molecular weight). From Ref. 77 unless otherwise indicated.

[b]Ref. 80.

[c]Yao et al., *J. Electrochem. Soc* **1968,** *115*, 999.

[d]Gutmann and Scherhaufer, *Monatash. Chem.* **1968,** *99*, 1686.

[e]Popov et al., *J. Phys. Chem.* **1967,** *71*, 1756.

shorter half-life in acetonitrile than in Me_2SO and DMF.[19] When the last traces of water are scavenged by alumina or trifluoroacetic anhydride, stable radical cations can be produced in acetonitrile.[47] The procedures for acetonitrile purification have been reviewed.[80,81]

A procedure has been described to produce pure dry acetonitrile with good electrochemical and optical properties.[80] The method begins with practical-grade acetonitrile: step 1—reflux over anhydrous aluminum chloride (15 g L^{-1}) for 1 h prior to rapid distillation; step 2—reflux over alkaline permanganate (10 g $KMnO_4$ and 10 g Li_2CO_3 per liter) for 15 min prior to rapid distillation; step 3—reflux over potassium bisulfate (15 g L^{-1}) for 1 h prior to rapid distillation; step 4—reflux over calcium hydride (2 g L^{-1}) for 1 h prior to a careful fractionation from a packed column at high reflux ratio; the middle 80% fraction is retained. The overall yield is 60% and the product should have an UV cutoff at a wavelength below 200 nm (1-cm cell vs. water) and a 5-V voltage window (with a platinum electrode to measure the anodic limit and a DME to measure the cathodic limit).

Propylene carbonate. Propylene carbonate (4-methyldioxolone-2, PC, $C_4H_6O_3$) is a cyclic ester with a low vapor pressure, relatively low toxicity, high dielectric constant ($\epsilon = 69$), and a large liquid range (-49 to $+242$°C), although thermal decomposition begins to take place at 150°C. The solvent can be supercooled by about 25°C. Hydrolysis in the presence of acids or bases is the principal chemical decomposition reaction. It is relatively nonhygroscopic and noncorrosive.

Propylene carbonate is produced via a reaction of CO_2 and propylene. It is a clear, colorless (when pure) liquid capable of dissolving a variety of organic and inorganic compounds. It has been recommended as especially suitable for electrochemistry because of its low background currents and resistance to oxidation.[82,83] Both radical cations and anions are stable in the solvent. The solvent has been used widely in battery research.[84]

The impurities of propylene carbonate that have been identified by gas chromatography[85] include CO_2, water, propylene oxide, allyl alcohol, 1,2-propylene glycol, 1,3-propylene glycol, and lesser amounts of unidentified materials. The solvent may be purified readily by vacuum fractional distillation, with a reflux ration of 10:1 at 0.5–1 torr. At this pressure the column head temperature is 72–75°C. Another suggestion is to heat propylene carbonate with sodium carbonate and potassium permanganate (both 10 g L^{-1}) for 2 h prior to vacuum distillation.

Dimethylformamide. Dimethylformamide (DMF) is a colorless, mobile liquid that is miscible with water. Its slight amine odor results form hydrolysis of the DMF by absorbed water to give dimethylamine and formic acid; the purified solvent cannot be stored for more than a day or so at room temperature without some decomposition. DMF, with a dielectric constant of 37, is a good solvent for a wide range of organic compounds and it also will dissolve many inorganic perchlorates, iodides, and lithium chloride. Nitrates are soluble, but they decompose.[86] The vapor is toxic and individuals should not be exposed for long periods to a concentration greater than 10 ppm by volume.[87]

DMF has been widely used as an electrochemical solvent, especially for the reduction of aromatic hydrocarbons.[88] The polarography of a number of metal ions in DMF also has been reviewed.[89] In general, the voltage range attained in reductions is comparable to acetonitrile and dimethyl sulfoxide, but DMF is less suitable for the study of oxidations. It has been suggested that the cyclic amide, *N*-methylpyrrolidone, may have most of the favorable properties of DMF, but with less tendency to hydrolyze.[90,91] However, it is less available and more expensive.

Although reagent-grade DMF may be used directly in some work, particularly if it has been kept free of water, it is commonly purified just before use by drying with molecular sieves and vacuum distillation. One purification procedure[92] is to dry reagent-grade DMF with Linde AW-500 molecular sieves (which are superior to the more commonly used Linde 4A) for at least 24 h. The DMF is slurried with about 60 g of P_2O_5 per 500 mL and distilled under nitrogen at 2.5–8 torr, which corresponds to a distillation temperature of 33–49°C. The first fraction of 100 mL is discarded, and the second fraction of about 300 mL is collected and stored under nitrogen in a freezer at −20°C. The purified solvent, which should be used within 48 h after preparation, contains less than 10 ppm of water (as measured by Karl Fischer titration) and less than 4×10^{-6} M acidic or basic impurities. The specific conductance should be $3 \times 10^{-7}\ \Omega^{-1}\ \mathrm{cm}^{-1}$.

Dimethyl sulfoxide (Me_2SO). The applications of Me_2SO in electrochemistry have been thoroughly reviewed.[93] It is a particularly useful solvent because it has a high dielectric constant and is sufficiently resistant to both oxidation and reduction to provide a fairly wide potential range. It is, however, not as resistant as acetonitrile or propylene carbonate to oxidation and these latter two solvents are preferred over Me_2SO for this purpose.

Dimethyl sulfoxide is rather polar, with a high donor number, but the claims that it is a strongly associated liquid have been refuted.[94] The measurements of the Kirkwood correlation factor for dimethylformamide, dimethyl sulfoxide, and nitrobenzene gave a value near 1.0 over a wide temperature range for these three solvents. This indicates that their properties are due primarily to large, but nonspecific, molecular dipole–dipole interactions. There is evidence that Me_2SO strongly associates with water,[95] which may account for the fact that radical anions generally are more stable in Me_2SO than in acetonitrile when both solvents contain traces of water.[19]

The reagent-grade solvent generally contains few impurities beyond traces of water and can be used without further purification for many electrochemical purposes. For some applications the presence of traces of water can significantly affect the properties of the solvent. For example, dimethyl sulfoxide and dimethylformamide that contain 100 ppm of water dissolve NaCN and NaN_3 to a much greater extent than do rigorously purified solvents.[96] Dimethyl sulfoxide can be purified by drying over molecular sieves, distilling under vacuum, and then flash-vacuum distilling from a small quantity of freshly prepared KNH_2 (under vacuum in a rotary evaporator) and holding the product under vacuum for several hours to remove traces of NH_3. The product should contain less than 10 ppm of water by Karl Fischer titration and less than 5×10^{-6} M acidic or basic impurities.

7.5 PREPARATION AND PURIFICATION OF SUPPORTING ELECTROLYTES

In both polarographic and preparative electrochemistry in aprotic solvents the custom is to use tetraalkylammonium salts as supporting electrolytes. In such solvent-supporting electrolyte systems electrochemical reductions at a mercury cathode can be performed at -2.5 to -2.9 V versus SCE. The reduction potential ultimately is limited by the reduction of the quaternary ammonium cation to form an amalgam, $(R_4N\cdot)Hg_n$, $n = 12$–13. The tetra-n-butyl salts are more difficult to reduce than are the tetraethylammonium salts and are preferred when the maximum cathodic range is needed. On the anodic side the oxidation of mercury occurs at about $+0.4$ V versus SCE in a supporting electrolyte that does not complex or form a precipitate with the Hg(I) or Hg(II) ions that are formed.

By use of platinum, gold, or carbon electrodes, a much more positive potential can be attained than on mercury; the anodic limit depends on the negative counterion. Halide counterions do not provide a good anodic range because they are readily oxidized; perchlorate or tetrafluoroborate counterions generally are preferred. The tetrafluoroborate ion gives a slightly greater anodic range than does perchlorate ion, which can be oxidized in the presence of tetrafluoroborate ion.[97] The exact nature of the oxidation product is somewhat obscure, but recent evidence indicates that the eventual product is ClO_2.[98] Hexafluorophosphates are slightly more resistant toward anodic oxidation than tetrafluoroborates.[99]

A number of quaternary ammonium perchlorates and tetrafluoroborates have been prepared, and their solubilities and specific resistance values have been measured in several solvents (see Table 7.6).[35] The resistance values are about 10-fold higher in ethers like tetrahydrofuran and 1,2-dimethyoxyethane relative to solutions in acetonitrile and dimethylformamide. This is due to the greater ion association in the ethers because of their smaller dielectric constants.

Concentrated solutions of supporting electrolytes in hexamethylphosphoramide (HMPA) become very viscous, which causes a twofold increase in resistance over solutions in acetonitrile and dimethylformamide. The increased viscosity may be due to the greater solvation of the cations by the more basic HMPA.

For routine work, the tetrafluoroborate salts are recommended because they are somewhat easier to purify and dry. They also may be safer than the perchlorates, which under conditions of high temperature and high acidity can decompose violently. However, tetraethylammonium perchlorate, which is particularly convenient to prepare because it can be recrystallized from water, has been prepared in many laboratories in batches as large as a kilogram. Obviously care must be taken to see that the initial reaction mixture that contains perchloric acid is never allowed to become overheated or evaporated to dryness.

The following procedures have been used[35] for the preparation of quaternary ammonium perchlorates and tetrafluoroborates. They can be scaled up to prepare larger quantities when needed.

1. *Tetra-n-butylammonium tetrafluoroborate.* A solution of 8.4 g (25 mmol) of $(n\text{-Bu})_4NBr$ in a minimum volume of water (~18 mL) is treated with 3.6 mL (~26 mmol) of aqueous 48–50% HBF_4. The resulting mixture is stirred at 25°C for 1 min and the crystalline salt is collected on a filter, washed with water until the washings are neutral, and dried. The crude salt (6.3 g or 79%, MP 155–175°C) is recrystallized three times from ethyl acetate–pentane mixtures to separate 6.0 g (75%) of $(n\text{-Bu})_4NBF_4$ as white needles; MP (after drying) 162–162.5°C.

2. *Tetraethylammonium tetrafluoroborate.* A solution of 5.3 g (25 mmol) of Et_4NBr in about 8 mL of water is reacted with HBF_4 and concentrated. Next, it is diluted with ethyl ether and filtered to yield 4.6 g (85%) of the crude salt; MP 375–378°C with decomposition. Two recrystallizations from a methanol–petroleum ether (BP 30–60°C) mixture yields 3.7 g (69%) of pure Et_4NBF_4 as white needles; MP (after drying) 377–378°C with decomposition.

3. *Tetra-n-butylammonium perchlorate.* A saturated aqueous solution of 8.4 g (25 mmol) of $(n\text{-Bu})_4NBr$ in 18 mL of water is treated with 2.1 mL (~26 mmol) of aqueous 70–72% $HClO_4$. After the resulting insoluble perchlorate salt has been collected, washed with cold water, and dried, the yield is 8.0 g (94%); MP 197–199°C. Two recrystallizations from an ethyl acetate–pentane mixture yield 7.6 g (90%) of pure $(n\text{-Bu})_4NClO_4$ as white needles that are dried at 100°C under reduced pressure; MP 212.5–213.5°C.

4. *Tetraethylammonium perchlorate.* The same procedure is used to convert 5.3 g (25 mmol) of Et_4NBr (in about 8 mL of water) to 4.7 g (81%) of the crude perchlorate salt, which is cooled before filtration. The salt is recrystallized from water to yield 4.3 g (75%) of the pure Et_4NClO_4 as white needles that are dried at 100°C under reduced pressure; MP 351–352.5°C with decomposition.

5. *Tetra-n-butylammonium hexafluorophosphate.*[100] 100 g (270 mmol) of $(n\text{-Bu})_4NI$ is dissolved in a minimum quantity of acetone. To this is added, with stirring, an acetone solution that contains 50 g (310 mmol) of $(NH_4)PF_6$. It is important to maintain at least a 5% excess of PF_6^-. The resultant mixture is filtered to remove the precipitate of NH_4I. Then water is added to precipitate $(n\text{-Bu})_4NPF_6$, which is collected by filtration and washed several times with water to remove all ammonium salts. The crude product is again precipitated

TABLE 7.12 Water Contents and Dielectric Constants of Some Burdich & Jackson Solvents

Solvent	Maximum Water Content (%)	Dielectric Constant
Acetonitrile	0.01	37.5 (20°C)
Dimethylacetamide	0.03	37.38 (25°C)
Dimethylformamide	0.03	36.71 (25°C)
Dimethyl sulfoxide	0.04	46.68 (20°C)
Dioxane	0.05	0.45 (20°C)
Propylene carbonate	0.04	69.0 (23°C)
Pyridine	0.01	12.4 (20°C)
Tetrahydrofuran	0.03	7.38 (20°C)

from a 5% solution of $(NH_4)PF_6$ to ensure complete removal of iodide. The white solid is recrystallized from ethanol–water and dried in vacuo at 100°C for 10 h. The yield is about 80%.

7.6 SOURCES

As we pointed out in previous chapters, the quality and purity of the solvent and supporting electrolyte used is important in electrochemical measurements. For most measurements in aprotic solvents it is necessary to keep water levels as low as possible. Earlier in this chapter procedures were described for purifying solvents and supporting electrolytes. However, it is tedious work, which requires time and energy. Moreover, it is not possible to obtain as low water levels as those available from specialized companies. From our own experience the solvents purified by Burdich & Jackson (''distilled in glass grade''), a division of Baxter, can be used in electrochemical measurements without further purification (most attempts to improve their materials result in higher H_2O levels). Table 7.12 lists maximum water contents and dielectric constants for several Burdick & Jackson solvents that are frequently used by electrochemists. However, the actual water level in most cases is much lower.

Several fine chemicals companies sell salts that can be used as supporting electrolytes without further purification.

REFERENCES

1. Parker, A. J., *Chem. Rev.* **1969,** *69*, 1.
2. Peover, M. E., in *Electroanalytical Chemistry*, Bard, A. J., ed., Marcel Dekker, New York, 1967, pp. 1–51.
3. Hoijink, G. J.; van Schooten, J.; de Boer, E.; Aalbersberg, W. Y., *Rec. Trav. Chim.* **1954,** *73*, 355.

4. Fujihira, M.; Suzuki, H.; Hayano, S., *J. Electroanal. Chem.* **1971,** *33*, 393.
5. Caldin, E. F., *Fast Reactions in Solution*, Wiley, New York, 1964, pp. 239–251.
6. Ritchie, C. D., in *Solute-Solvent Interactions*, Coetzee, J. F.; Ritchie, C. D., eds., Marcel Dekker, New York, 1969, pp. 246–271.
7. Van Duyne, R. P.; Reilley, C. N., *Anal. Chem.* **1972,** *44*, 142.
8. Meites, L.; Zuman, P.; et al., eds., *CRC Handbook Series in Inorganic Electrochemistry*, Vols. 1–7, CRC Press, Boca Raton, Fla., 1980–1986.
9. Meites, L.; Zuman, P.; et al., eds., *Electrochemical Data*, Wiley: New York, 1974.
10. Meites, L.; Zuman, P.; et al., eds., *CRC Handbook Series in Organic Electrochemistry*, Vols. 1–4, CRC Press, Boca Raton, Fla., 1980–1986.
11. Gritzner, G., *Pure Appl. Chem.* **1990,** *62*, 1839.
12. Kratochvil, B., *Crit. Rev. Anal. Chem.* **1971,** *1*, 415; *Rec. Chem. Prog.* **1966,** *27*, 253.
13. Gutmann, V., *Coordination Chemistry in Non-Aqueous Solutions*, Springer-Verlag, New York, 1968, Chap. 5.
14. Gutmann, V., *Fortschr. Chem. Fortsch.* **1972,** *27*, 61.
15. Covington, A. K.; Dickinson, T., *Physical Chemistry of Organic Solvent Systems*, Plenum Press, New York, 1973.
16. Gutmann, V., *The Donor-Acceptor Approach to Molecular Interactions*, Plenum Press, New York, 1978.
17. Martin, D.; Weise, A.; Niclas, H. J., *Angew. Chem. Int. Ed.* **1967,** *6*, 318.
18. Gutmann, V.; Wychera, E., *Inorg. Nucl. Chem. Lett.* **1966,** *2*, 257.
19. Jezorek, J. R.; Mark, H. B., Jr., *J. Phys. Chem.* **1970,** *74*, 1627.
20. Kolthoff, I. M., in *Polarography 1964*, Hills, G. J., ed., Interscience Publishers, New York, 1966, p. 2.
21. Marcoux, L.; Malachesky, P.; Adams, R. N., *J. Am. Chem. Soc.* **1967,** *89*, 5766.
22. Phelps, J.; Santhanam, K. S. V.; Bard, A. J., *J. Am. Chem. Soc.* **1967,** *89*, 1752.
23. Bechgaard, K.; Parker, V. D., *J. Am. Chem. Soc.* **1972,** *94*, 4749.
24. Bauer, D.; Foucault, A., *J. Electroanal. Chem.* **1972,** *39*, 385.
25. Jones, H. L.; Boxall, L. G.; Osteryoung, R. A., *J. Electroanal. Chem.* **1972,** *38*, 476.
26. Gillespie, R. J., *Accts. Chem. Res.* **1968,** *1*, 202.
27. Kagiya, T.; Sumida, Y.; Inoue, T., *Bull. Chem. Soc. Japan* **1968,** *41*, 767, 773, 779.
28. Taft, R. W.; Gurka, D.; Joris, L.; von Schleyer, P. R.; Rakshys, J. W., *J. Am. Chem. Soc.* **1969,** *91*, 4801.
29. Harris, H. G.; Prausnitz, J. M., *J. Chromatographic Sci.* **1969,** *7*, 685.
30. Reichardt, C., Angew. *Chem. Int. Ed.* **1966,** *4*, 29.
31. Reichardt, C.; Dimroth, K., *Fortschr. Chem. Forsch.* **1968,** *11*, 1.
32. Popov, A. I., in *Technique of Inorganic Chemistry*, Jonassen, H. B.; Weissberger, A., eds., Interscience Publishers, New York, 1972, p. 37.

33. Lagowski, J. J., ed., *The Chemistry of Non-Aqueous Solvents*, Vol. 1, Academic Press, New York, 1966, Chap. 7; Vol. 2, 1967, Chap. 6; Vol. 3, 1970, Chap. 2.
34. Adams, R. N., *Electrochemistry at Solid Electrodes*, Marcel Dekker, New York, 1969, p. 33.
35. House, H. O.; Feng, E.; Peet, N. P., *J. Org. Chem.* **1971,** *36*, 2371.
36. Godfrey, N. B., *Chem. Tech.* **1972,** 359.
37. Ponchert, C. J., *The Aldrich Library of Infrared Spectra*, 3rd ed., Aldrich Chemical, Milwaukee, 1981.
38. Bates, R. G., in *Solute–Solvent Interactions*, Coetzee, J. F.; Ritchie, C. D., eds., Marcel Dekker, New York, 1969, pp. 50–53.
39. Kratochvil, B., *Crit. Rev. Anal. Chem.* **1971,** *1*, 415.
40. Kratochvil, B., *Rec. Chem. Prog.* **1966,** *27*, 253.
41. Cauquis, G., in *Organic Electrochemistry*, Baizer, M. M., ed., Marcel Dekker, New York, 1973, Chap. 1.
42. Savéant, J. M., *J. Electroanal. Chem.* **1971,** *29*, 87.
43. Ref. 2, p. 40.
44. Ref. 2, pp. 11–15.
45. Bard, A. J.; Phelps, J., *J. Electroanal. Chem.* **1973,** *25*, App. 2.
46. Mark, H. B., Jr., *Rec. Chem. Prog.* **1968,** *29*, 217.
47. Hammerich, O.; Parker, V. D., *Electrochim. Acta*, **1973,** *18*, 537.
48. Kratochvil, B.; Yeager, H. L., *Fortsch. Chem. Forsch.* **1972,** *27*, 1.
49. Janz, G. J.; Tomkins, R. P. T., eds., *Nonaqueous Electrolytes Handbook*, Academic Press, New York, 1972.
50. Clark, W. M., *Oxidation–Reduction Potentials of Organic Systems*, Williams & Wilkins, Baltimore, 1960.
51. Zuman, P., in *Progress in Polarography*, Zuman, P.; Meites, L., eds., Wiley-Interscience, New York, 1972, pp. 73–156.
52. Mairanovskii, S. G., *Catalytic and Kinetic Waves in Polarography*, Plenum Press, New York, 1968.
53. Elving, P. J., *Pure Appl. Chem.* **1963,** *7*, 423.
54. Meites, L., *Polarographic Techniques*, 2nd ed., Interscience Publishers, New York, 1965, p. 282.
55. Good, N. E.; Winget, G. D.; Winter, W.; Connolly, T. N.; Izawa, S.; Singh, R. M., *Biochemistry* **1966,** *5*, 467.
56. Bates, R. G.; Roy, R. N.; Robinson, R. A., *Anal. Chem.* **1973**, *45,* 1663.
57. Ref. 52, p. 164.
58. Mairanovskii, S. G., *J. Electroanal. Chem.* **1962,** *4*, 161.
59. Guidelli, R.; Pezzatini, G.; Foresti, M. L., *J. Electroanal. Chem.* **1973,** *43*, 83, 95.
60. Davis, M. M., *Acid–Base Behavior in Aprotic Organic Solvents*, National Bureau of Standards Monograph 105, U.S. Government Printing Office, Washington, D.C., 1968.
61. Ritchie, C. D., in *Solute–Solvent Interactions*, Coetzee, J. F.; Ritchie, C. D., eds., Marcel Dekker, New York, 1969, Chap. 4.

62. Benoit, R. L.; Buisson, C., *Electrochim. Acta* **1973,** *18*, 105.
63. Ref. 60, p. 88.
64. Temmerman, E.; Verbeek, F., *J. Electroanal. Chem.* **1968,** *19*, 423.
65. Perrin, D. D., in *Chemical Analysis,* Elving, P. J.; Kolthoff, I. M., eds., Wiley-Interscience, 1970, Chap. 10.
66. Izutsu, K.; Sukura, S.; Kuroki, K.; Fujinaga, T., *J. Electroanal. Chem.* **1971,** *32*, App. 11.
67. Szwarc, M., *Accts. Chem. Res.* **1969,** *2*, 87.
68. Avaca, L. A.; Bewick, A., *J. Electroanal. Chem.* **1973,** *41*, 405.
69. Westmoreland, P. G.; Day, R. A., Jr.; Underwood, A. I., *Anal. Chem.* **1972,** *44*, 737.
70. Hughes, R. C.; Murau, P. C.; Gundersen, G., *Anal. Chem.* **1971,** *43*, 691.
71. Conway, B. E.; Angerstein-Kozlowska, H.; Sharp, W. B. A.; Criddle, E. E., *Anal. Chem.* **1973,** *45*, 1331.
72. Mitchell, J. W., *Anal. Chem.* **1973,** *45*, 492A.
73. Kuehner, E. C.; Alvarez, R.; Paulsen, P. J.; Murphy, T. J., *Anal. Chem.* **1973,** *44*, 2050.
74. Azzam, A. M.; Bockris, J. O'M.; Conway, B. E.; Rosenberg, H., *Trans. Faraday Soc.* **1950,** *46*, 918.
75. Rosen, M.; Bauer, H. H.; Elving, P. J., *J. Electrochem. Soc.* **1970,** *117*, 878.
76. Jenkins, D. A.; Weedon, C. J., *J. Electroanal. Chem.* **1971,** *31*, App. 13.
77. Riddick, J. A.; Bunger, W. B., in *Techniques of Chemistry*, Weissberger, A., ed., Wiley-Interscience, New York, 1986.
78. Mann, C. K., in *Electroanalytical Chemistry*, Vol. 3, Bard, A. J., ed., Marcel Dekker, New York, 1969, pp. 57–134.
79. Ref. 77, p. 805.
80. Walter, M.; Ramaley, L., *Anal. Chem.* **1973,** *45*, 165.
81. Coetzee, J. F., *Pure Appl. Chem.* **1966,** *13*, 429.
82. Nelson, R. F.; Adams, R. N., *J. Electroanal. Chem.* **1967,** *13*, 184.
83. Clark, D. B.; Fleischmann, M.; Pletcher, D., *J. Electroanal. Chem.* **1973,** *42*, 133.
84. Jasinski, R., in *Advances in Electrochemistry and Electrochemical Engineering*, Vol. 8, Tobias, C. W., ed., Wiley-Interscience, New York, 1971, p. 253–335.
85. Jasinski, R. J.; Kirkland, S., *Anal. Chem.* **1967,** *39*, 1663.
86. Ref. 78, p. 76.
87. Ref. 77, pp. 839–840.
88. Dietz, R.; Peover, M. E., *Discuss. Faraday Soc.* **1968,** *45*, 154.
89. Headridge, J. B.; Ashraf, M.; Dodds, H. L. H., *J. Electroanal. Chem.* **1968,** *16*, 116.
90. Lund, H.; Iversen, P., in *Organic Electrochemistry*, Baizer, M. M., ed., Marcel Dekker, New York, 1973, p. 221.
91. Breant, M.; Sue, J. L., *J. Electroanal. Chem.* **1972,** *40*, 89.
92. Ritchie, C. D.; Megerle, G. H., *J. Am. Chem. Soc.* **1967,** *89*, 1447.
93. Butler, J. N., *J. Electroanal. Chem.* **1967,** *14*, 89.

94. Amey, R. L., *J. Phys. Chem.* **1968,** *72*, 3358.
95. Rasmussen, D. H.; McKenzie, A. P., *Nature* **1968,** *220*, 1315.
96. Ritchie, C. D.; Skinner, G. A.; Badding, V. G., *J. Am. Chem. Soc.* **1967,** *89*, 2063.
97. Fleischmann, M.; Pletcher, D., *Tetrahedron Lett.* **1968,** 6255.
98. Bontempelli, G.; Magno, F.; Mazzochin, G. A., *J. Electroanal. Chem.* **1973,** *42*, 57.
99. Fleischmann, M.; Pletcher, D., *Tetrahedron Lett.* **1968,** 6255.
100. Fry, A. J.; Britton, W. E., in *Laboratory Techniques in Electroanalytical Chemistry*, Kissinger, P. T.; Heineman, W. R., eds., Marcel Dekker, New York, 1984, pp. 367–382.

CHAPTER 8

HYDRONIUM IONS (H_3O^+), BRØNSTED ACIDS (HA), AND MOLECULAR HYDROGEN (H_2)

8.1 INTRODUCTION

The most fundamental redox process in electrochemistry is the reductive transformation of hydronium ion [$H_3O^+_{(aq)}$] at a platinum electrode to molecular hydrogen [$H_2(g)$][1]

$$2\ H_3O^+_{(aq)} + 2\ e^- \overset{\text{Pt}}{\rightleftharpoons} H_2(g) \qquad E°,\ 0.0000\ \text{V vs. NHE} \tag{8.1}$$

When properly engineered and with [$H_3O^+_{(aq)}$] at unity activity and P_{H_2} at unit fugacity, this electrode system is the thermodynamic reference standard for measurements of electrochemical potentials, and is referred to as the *normal hydrogen electrode* (NHE), which is alternatively called *standard hydrogen electrode* (SHE):

$$E = E^{\circ}_{\text{NHE}} + \frac{2.30RT}{2F} \log \frac{[H_3O^+_{(aq)}]^2}{P_{H_2}} \tag{8.2}$$

($E^{\circ}_{\text{NHE}} \equiv 0.0000$ V at all temperatures, 0–100°C)
[$2.30RT/2F = 0.059/2$ at 25°C]

The latter equation for the NHE also is the defining basis for the potentiometric measurement of hydronium ion activity [H_3O^+] and molecular hydrogen fugacity (P_{H_2}):

$$pH_a \equiv -\log [H_3O^+] = \frac{(E^\circ_{NHE} - E_{ind})}{0.059} - \log (P_{H_2})$$

$$= \frac{-E_{ind}}{0.059} \quad (8.3)$$

$$(E^\circ_{NHE} = 0.000 \text{ V}; P_{H_2} = 1.00 \text{ atm})$$

$$\log (P_{H_2}) = (E^\circ_{NHE} - E_{ind}) \frac{2}{0.059} + \log [H_3O^+]^2$$

$$= -E_{ind}\left(\frac{2}{0.059}\right) \quad (8.4)$$

$$\{E^\circ_{NHE} = 0.000 \text{ V}; [H_3O^+] = 1.000\}$$

Electrochemical measurement of pH_a via Eq. (8.3) senses hydronium ion activity rather than its concentration. Hence, electrochemical evaluations of dissociation constants (K_{HA}) yield thermodynamic quantities

$$HA + H_2O \rightleftharpoons H_3O^+ + A^- \quad K_{HA} \quad (8.5)$$

$$K_{HA} = [H_3O^+][A^-]/[HA]$$

In contrast, spectrophotometric measurements via indicator dyes sense concentrations and yield nonthermodynamic dissociation constants (concentrations rather than activities).

8.2 HYDRONIUM ION (H_3O^+) REDUCTION

Although the NHE is fundamental to electrochemistry, it does not represent the primary electron-transfer step for hydronium ion reduction at an inert (glassy-carbon) electrode:[2]

$$H_3O^+_{(aq)} + e^- \overset{GCE}{\rightleftharpoons} H\cdot_{(aq)} \qquad E^\circ, -2.10 \text{ V vs. NHE} \quad (8.6)$$

where GCE means a glassy-carbon electrode.

The −2.10-V difference in standard potential (E°) between the latter and that for the NHE [Eq. (8.1)] is due to the platinum electrode, which stabilizes the hydrogen atom (H·) via formation of a Pt—H covalent bond

$$H_3O^+_{(aq)} + Pt(s) + e^- \longrightarrow Pt\text{—}H(s) \qquad E^\circ, 0.000 \text{ V} \quad (8.7)$$

$$\Delta(-\Delta G_{BF}) = [0.00 - (-2.10)]96.48 \text{ kJ mol}^{-1} (\text{eV})^{-1}$$

$$= 203 \text{ kJ mol}^{-1}$$

where $-\Delta G_{BF}$ is the free energy of bond formation.

In this system the platinum-electrode surface does not consist of free platinum atoms, but must undergo homolytic Pt—Pt bond breakage [ΔH_{DBE}, 100 kJ mol^{-1} per Pt· (est)] before a Pt—H bond is formed. Thus, the Pt—H bond-formation energy ($-\Delta G_{BF}$) is estimated to be 301 kJ mol^{-1} on the basis of the electrochemical data [for the gas-phase Pt—H molecule, $\Delta H_{DBE} \leq 335$ kJ mol^{-1} or ($-\Delta G_{BF} \approx 301$ kJ mol^{-1})].[3] Determination of the reduction potential for $H_3O^+_{(aq)}$ at other metal electrodes (M) provides a convenient means for estimating M—H bond energies [$\Delta(-\Delta G_{BF})$].

Table 8.1 presents estimates of the differential formation energies for M—H(s) bonds [$\Delta(-\Delta G_{BF})$] at various metal electrodes.[3] On the basis of these and Eq. (8.7), approximate *formal reduction potentials* [$E^{\circ\prime}_{M,\, H_3O^+/M-H(s)}$] have been estimated for the reduction of H_3O^+ (unit activity) at the several metal electrodes (Table 8.1). In our view direct evaluations of $\Delta(-\Delta G_{BF})_{M-H(s)}$ via electrochemistry will provide useful insights to metal–hydrogen bonds and metal-catalyzed hydrogenations.

Likewise, when the elementary process of Eq. (8.6) [$H_3O^+_{(aq)}$ reduction] is done in the presence of molecules (X) that stabilize hydrogen atoms (H·) via covalent bond formation, the shift in the reduction potential is a measure of the X—H bond energy. For example,[4]

$$H_3O^+_{(aq)} + \cdot O_2\cdot + e^- \longrightarrow H{-}OO\cdot + H_2O \qquad E^\circ,\ +0.12\ \text{V} \qquad (8.8)$$

$$-\Delta G_{BF} = [0.12\text{-}(-2.10)]96.5$$
$$= 215\ \text{kJ mol}^{-1}$$

$$H_3O^+_{(aq)} + HOO\cdot + e^- \longrightarrow HOO{-}H + H_2O \qquad E^\circ,\ +1.44\ \text{V} \qquad (8.9)$$

$$-\Delta G_{BF} = [1.44 - (-2.10)]96.5$$
$$= 342\ \text{kJ mol}^{-1}$$

$$H_3O^+_{(aq)} + HO\cdot + e^- \longrightarrow HO{-}H + H_2O \qquad E^\circ,\ +2.72\ \text{V} \qquad (8.10)$$

$$-\Delta G_{BF} = [2.72 - (-2.10)]96.5$$
$$= 466\ \text{kJ mol}^{-1}$$

$$H_3O^+_{(aq)} + O{=}\dot{N}{=}O + e^- \longrightarrow H{-}ONO \qquad E^\circ,\ +1.07\ \text{V} \qquad (8.11)$$

$$-\Delta G_{BF} = [1.07 - (-2.10)]96.5$$
$$= 306\ \text{kJ mol}^{-1}$$

$$H_3O^+_{(aq)} + \cdot N{=}O + e^- \longrightarrow \tfrac{1}{2}\,(NO{-}H)_2 \qquad E^\circ,\ +0.71 \qquad (8.12)$$

$$-\Delta G_{BF} = [0.71 - (-2.10)]96.5$$
$$= 272\ \text{kJ mol}^{-1}$$

Another factor that affects the reduction of hydronium ion is its solvent sheath (solvation energy). By convention the pK_a of $H_3O^+_{(aq)}$ in water is defined

TABLE 8.1 Estimates of the Differential Formation Energies for M—H(s) Bonds [$\Delta(-\Delta G_{BF})$] at Metal–Electrode Surfaces [M(s)], and Approximate Formal Reduction Potentials for H_3O^+ at Metal Electrodes ($E^{\circ\prime}_{M,H_3O^+/M-H}$)

Electrode	$(\Delta H_{DBE})_{M-H(g)}$ [a] (kJ mol^{-1})	$(\Delta H_{DBE})_{M-M(g)}$ [a] (kJ mol^{-1})	$(\Delta G_{DBE})_{M-M(s)}$ (kJ mol^{-1}) (est)	$\Delta(\Delta G_{BF})_{M-H(s)}$ [b] (kJ mol^{-1})	$(E^{\circ\prime}_{M,H_3O^+/M-H})$ V vs NHE
Pt	335	358	100	203	0.00
Pd	234	71	<10	201	−0.03
Ni	252	203	59	159	−0.46
Li	238	110	42	163	−0.42
Fe	180	100	25	121	−0.85
Pb	176	89	21	121	−0.85
Hg	40	8	<10	25	−1.84

[a] Ref. 3.

[b] $$\Delta(-\Delta G_{BF})_{m-H(s)} = [(\Delta H_{DBE})_{M-H(g)} - T\Delta S] - (\Delta G_{DBE})_{M-M(s)}$$
$$= [(\Delta H_{DBE})_{M-M(g)} - 32.6] - (\Delta G_{DBE})_{M-M(s)} = (E^{\circ\prime} + 2.10)\ 96.5\ \text{kJ mol}^{-1}\ (\text{eV})^{-1}$$

to have a value of 0.000

$$(H_2O)_nH_3O^+_{(aq)} \rightleftharpoons H_3O^+_{(aq)} + nH_2O \qquad (K_a)_{H_2O},\ 1.000 \tag{8.13}$$

In other solvents (sol) the dissociation is shifted according to the relation

$$(\text{sol})_nH_3O^+_{(aq)} \rightleftharpoons H_3O^+_{(aq)} + n\ \text{sol} \qquad (K_a)_{\text{sol}} \tag{8.14}$$

and

$$[H_3O^+_{(aq)}] = (K_a)_{\text{sol}} \frac{[(\text{sol})_nH_3O^+_{(aq)}]}{[\text{sol}]^n} \tag{8.15}$$

Note: In our experience even the most rigorously dried solvents contain 20–100 ppm of H_2O (1–5 mM). If H_2O is a stronger Brønsted base than the solvent (e.g., MeCN), $H_3O^+_{(aq)}$ will be formed; otherwise Hsol^+ is formed.

From Eqs. (8.3) and (8.14)

$$E_{\text{Pt}} = 0.059 \log (K_a)_{\text{sol}} + 0.059 \log \frac{[(\text{sol})_nH_3O^+_{(aq)}]}{[\text{sol}]^n P_{H_2}^{1/2}} \tag{8.16}$$

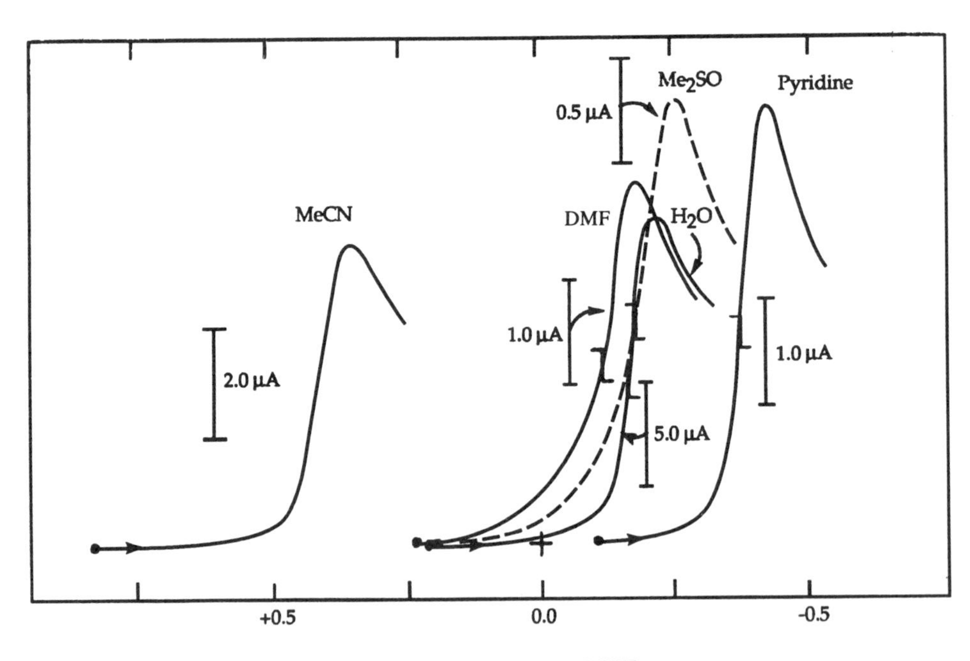

Figure 8.1 Cyclic voltammograms for 5 mM solutions of $(H_3O)ClO_4$ in MeCN, DMF, H_2O, Me_2SO, and py at a freshly resurfaced platinum electrode (area 0.46 mm^2). Scan rate 0.1 V s^{-1}; the solutions contained 0.5 M tetraethylammonium perchlorate (TEAP).

where

$$0.059 \log (K_a)_{sol} = E^{\circ\prime}{}_{H_3O^+_{(sol)}} = -0.059 \, pK_{a(sol)} \tag{8.17}$$

When voltammetric measurements are utilized, the half-wave potential ($E_{1/2}$) approximates the condition where $[(sol)_nH_3O^+_{(aq)}] = P^{1/2}_{H_2}$ at the platinum-electrode surface. With the further approximation that the activity of the solvent, $[sol]^n$, is unity, Eq. (8.15) is simplified to

$$pK_{a(sol)} = \frac{-E^{\circ\prime}_{H_3O^+_{(sol)}}}{0.059} \approx \frac{E_{1/2}}{0.059} \tag{8.18}$$

Figure 8.1 illustrates cyclic voltammograms for the reduction of $(sol)_nH_3O^+_{(aq)}$ in several solvents.[5] Evaluations of $pK_{a(sol)}$ via Eq. (8.17) and the half-wave potentials for the voltammograms of Figure 8.1 are summarized in Table 8.2.

8.3 BRØNSTED ACID (H*A*) REDUCTION AND EVALUATION OF $pK_{a(sol)}$

Brønsted acids (H*A*) undergo dissociation in any solvent to yield the solvated aquahydronium ion $[(sol)_nH_3O^+_{(aq)}]$ {from residual H_2O; or $[(sol)_nHsol^+]$ for basic solvents}, which further dissociates to $H_3O^+_{(aq)}$ [Eq. (8.13)], and the solvated conjugate base $[A^-(sol)_n]$

$$HA(sol) + H_2O \rightleftharpoons H_3O^+_{(aq)} + A^-(sol) \; (K_a)_{sol} \tag{8.19}$$

and

$$[H_3O^+_{(aq)}] = (K_a)_{sol} \frac{[HA(sol)]}{[A^-(sol)]} \tag{8.20}$$

Substitution of Eq. (8.19) into Eq. (8.2) gives

$$E_{Pt} = 0.059 \log (K_a)_{sol} + 0.059 \log \frac{[HA(sol)]}{[A^-(sol)]} - \frac{0.059}{2} \log P_{H_2} \tag{8.21}$$

This equation, in conjunction with voltammetric measurements of half-wave potentials ($E_{1/2}$) for the reduction of Brønsted acids at a platinum electrode in any solvent, permits the evaluation of $(K_a)_{sol}$ [pK_a(sol)]:[5]

$$\begin{aligned} E_{1/2} &= 0.059 \log (K_a)_{sol} + 0.059 \log \frac{\kappa_{A^-}\kappa_{H_2}}{\kappa_{HA}} \\ &= -0.059 \, pK_a + \epsilon \end{aligned} \tag{8.22}$$

TABLE 8.2 Effective Acidities (pK_a) of Various Brønsted Acids in Four Aprotic Solvents and Water[a]

Brønsted Acid (HA)	pK_a [$E_{1/2}$ (V) vs. NHE][b]					
	MeCN	DMF	py	Me_2SO	H_2O	pK, H_2O (lit.)[c]
$(H_3O)ClO_4$	−8.8 (+0.452)	0.7 (−0.108)	4.6 (−0.338)	2.6 (−0.218)	0.00 (−0.172)	0.000
$(pyH)ClO_4$	1.8 (−0.108)	3.1 (−0.183)	5.7 (−0.338)	4.4 (−0.258)	4.9 (−0.308)	5.2
2,4-$(NO_2)_2$PhOH	4.3 (−0.253)	4.5 (−0.268)	5.5 (−0.323)	4.9 (−0.288)		4.0
$(NH_4)ClO_4$		9.6 (−0.568)	7.3 (−0.433)	11.7 (−0.693)	8.7 (−0.531)	9.2
$(Et_3NH)Cl$	10.0 (−0.593)	9.9 (−0.583)	7.6 (−0.448)	12.7 (−0.748)	10.1 (−0.613)	10.7
PhC(O)OH	7.9 (−0.468)	11.5 (−0.678)	11.6 (−0.683)	13.6 (−0.803)	3.2 (−0.208)	4.2
PhOH	16.0 (−0.943)	19.4 (−1.148)	20.1 (−1.188)	20.8 (−1.232)	9.2 (−0.563)	10.0
p-EtOPhOH	19.3 (−1.138)	21.5 (−1.268)	21.8 (−1.288)	23.8 (−1.403)	9.6 (−0.583)	10.2
blank [H_2O(sol)]	30.4 (−1.776)	34.7 (−2.030)	30.5 (−1.782)	36.7 (−2.148)	14.8[d] (−0.738)	14.0[d]

[a]The pK_a values are accompanied by the half-wave potentials ($E_{1/2}$) for 5 mM substrate [0.5 M TEAP (Me_2SO, DMF, MeCN), 0.5 M TBAP (py), and 0.5 M KCl (H_2O)] at a platinum electrode (area 0.4 mm^2). For these conditions $pK_a(sol) = -E^{\circ\prime}_{HA(sol)}/0.059 = -(E_{1/2} + \epsilon)/0.059$ with $\epsilon = 0.07$ V for MeCN, DMF, py, and Me_2SO, and $\epsilon = -0.17$ V for H_2O; Ref. 5.

[b]Precision, ±0.5 pK_a units (±0.03 V).

[c]Refs. 6–9.

[d]Standard state for bulk H_2O, unit activity.

where κ_{A^-}, κ_{HA} are parameters that relate to diffusion coefficients, activity coefficients, and P_{H_2} for a given experimental system. Figure 8.2 illustrates cyclic voltammograms for several Brønsted acids (H*A*) in dimethylformamide. The half-wave potentials ($E_{1/2}$) for these voltammograms, as well as those for other Brønsted acids and solvents, provide a measure via Eq. (8.21) of the thermodynamic acidities [pK_a(sol)].

Tables 8.2 and 8.3 summarize such evaluations for several Brønsted acids in four organic solvents (ϵ, 0.07 V) and water (ϵ, -0.17 V).[5] In the case of water, comparisons with literature values[6-9] confirm that the methodology is sound. Most literature p*K* values for Brønsted acids in organic solvents are in terms of concentrations (from spectrophotometric measurements) rather than $H_3O^+_{(aq)}$ activity (pK_a). The platinum electrode directly senses the activity of $H_3O^+_{(aq)}$ via Eq. (8.12). Because solvent has such a major and selective effect on the activity of $H_3O^+_{(aq)}$ (see first line of Table 8.2), the differences between concentration-based dissociation constants (p*K*) and activity-based constants (pK_a, Tables 8.2 and 8.3) are dramatic. For example, phenol in MeCN (p*K*, 26.6;[10] pK_a, 16.0) and in Me_2SO (p*K*, 16.4;[10] pK_a, 20.8) exhibits a reversal; it dissociates more in Me_2SO than in MeCN, but is more acidic [greater $H_3O^+_{(aq)}$ activity] in MeCN. Other examples include $(Et_3NH)Cl$ in MeCN (p*K*, 18.5;[11] pK_a, 10.0) and in Me_2SO (p*K*, 10.5;[11] pK_a, 12.7) (again, greater dissociation in Me_2SO and greater acidity in MeCN); PhC(O)OH in MeCN (p*K*,

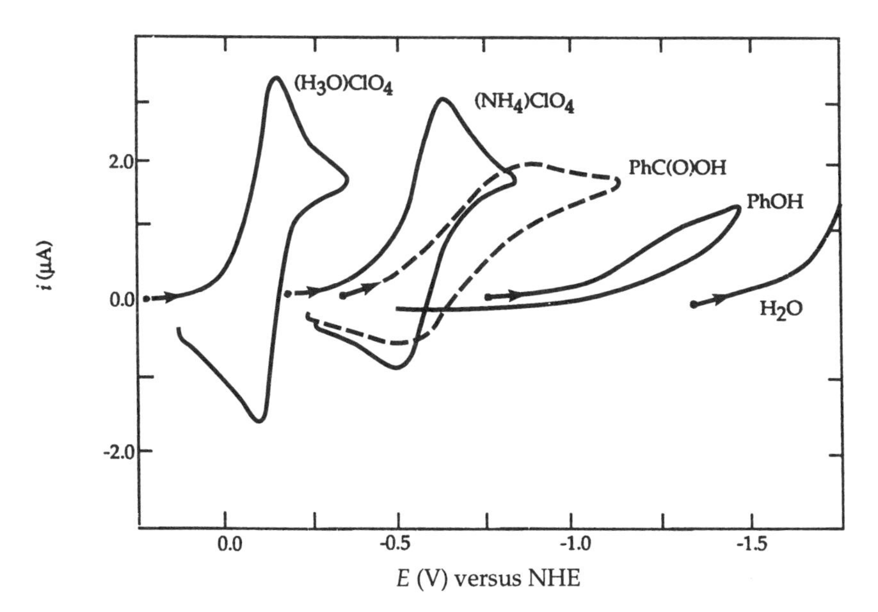

Figure 8.2 Cyclic voltammograms for 5 mM solutions of Brønsted acids (H*A*) in dimethylformamide (DMF) at a freshly surfaced platinum electrode (area 0.46 mm^2). Scan rate 0.1 V s^{-1}; solutions contained 0.5 M tetraethylammonium perchlorate (TEAP).

TABLE 8.3 Effective Proton-Activity Dissociation Constants (K_a) for Brønsted Acids in Dimethylformamide (0.5 M TEAP).

Brønsted Acids	pK_a
$(H_3O)ClO_4$	0.7
$(pyH)ClO_4$	3.1
2,4-Dinitrophenol	4.5
Ascorbic acid	9.6
$(NH_4)ClO_4$	9.6
Triethylamonium chloride $[(Et_3NH)Cl]$	9.9
3-Hydroxytyramine hydrobromide (dopamineHBr)	10.2
tert-Butylammonium perchlorate $[(t\text{-}BuNH_3)ClO_4]$	10.5
tert-Butylammonium chloride $[(t\text{-}BuNH_3)Cl]$	10.7
Diisopropylammonium chloride $[(i\text{-}Pr)_2NH_2)Cl]$	10.7
Benzoic acid	11.5
Acetic acid	12.4
Pyrocatechol	14.9
3-Hydroxytyramine (dopamine)	17.2
3,5-Di-*tert*-butylcatechol	18.1
2,4-Pentanedione	18.2
Phenol	19.4
8-Quinolinol	20.0
p-Ethoxyphenol	21.5
2,4-Dimethylphenol	22.3
2-Methoxy-4-methylphenol	22.3
Pyrocatechol monoanion	23.3
3-Hydroxytyramine monoanion ($dopamine^-$)	24.5
3,5-di-*tert*-Butylcatechol monoanion ($3,5\text{-}DTBCH^-$)	27.3
Methanol	31.1
$[H_2O(DMF)]$	34.7

20.7;[12] pK_a, 7.9), in Me_2SO (pK, 11.1;[11] pK_a 13.6), in DMF (pK, 11.6;[11] pK_a, 11.5), and in H_2O (pK, 4.2;[6] pK_a, 3.2); and H_2O in MeCN (pK_a, 30.4) and in Me_2SO (pK, 31.4;[13] pK_a, 36.7).

8.4 OXIDATION OF DISSOLVED DIHYDROGEN (H_2)

Molecular hydrogen (H_2; ΔH_{DBE}, 435 kJ mol^{-1}) is resistant to electrochemical oxidation at inert electrodes (glassy carbon or passivated metals; Ni, Au, Hg, Cu). At passivated Pt and Pd dissolved H_2 exhibits only broad, diffuse, anodic voltammetric peaks with irreproducible peak currents that are not proportional to the partial pressure of dissolved H_2 (P_{H_2}). However, with freshly preanodized Pt and Pd electrodes well-defined oxidation peaks for H_2 are obtained, which have peak currents that are proportional to P_{H_2} (Figure 8.3).[14] The surface

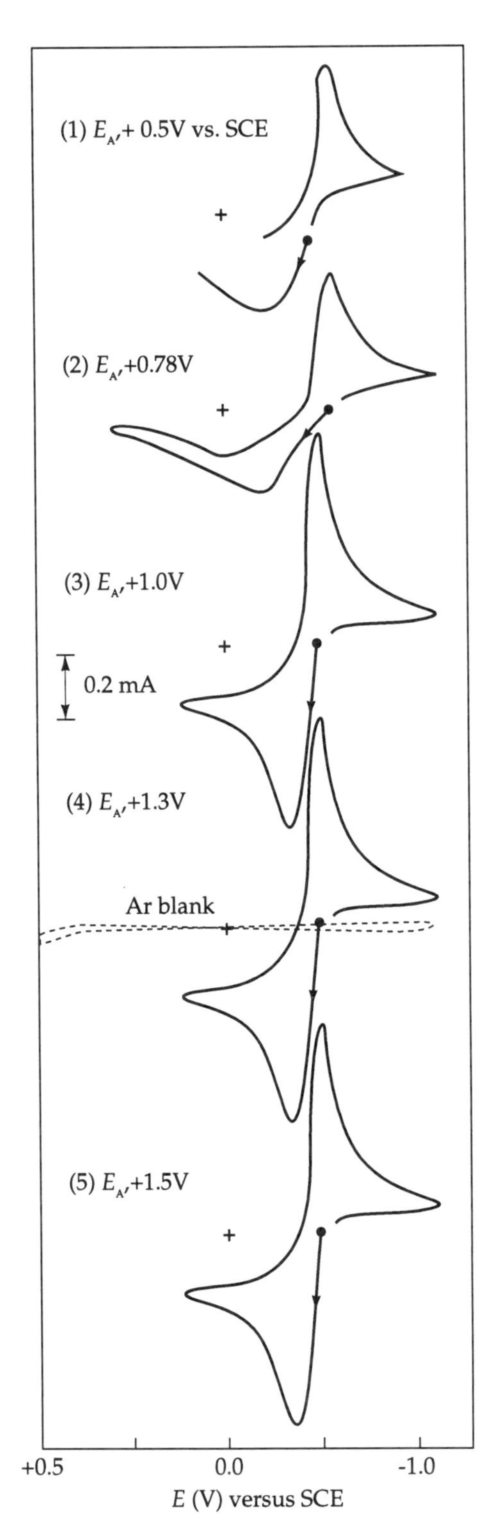

Figure 8.3 Voltammograms for dissolved H_2 (1 atm) in Me_2SO (0.5 M TEAP) at a Pt electrode (area 0.458 cm^2). Cyclic voltammograms initiated at the rest potential (-0.5 V vs. SCE) after the electrode was preanodized for 2 min at the indicated activation potential E_A; scan rate 0.1 V s^{-1}. The dashed (--) voltammogram represents the response of activated electrodes in the presence of an argon atmosphere. Oxidation current is in a downward direction.

conditioning produces a fresh reactive metal-oxide surface [$Pt^{II}(OH)_2(s)$], which on exposure to H_2 becomes an oxide-free metal surface (Pt*):

$$Pt^{II}(OH)_2(s) + H_2 \longrightarrow Pt^*(s) + 2\ H_2O \quad (8.23)$$

In turn, the clean surface reacts with a second H_2 to form two Pt—H bonds:

$$2\ Pt^*(s) + H_2(P_{H_2}) \overset{\text{fast}}{\rightleftharpoons} 2\ Pt{-}H(s) \qquad K_{eq},\ \sim 1\ \text{atm}^{-1} \quad (8.24)$$

The value of K_{eq} is estimated on the basis of the Pt—H bond-formation energy from metallic platinum [$\Delta(-\Delta G_{BF})$, 203 kJ mol^{-1}; Eq. (8.7)] and the dissociative bond energy for H_2(ΔG_{DBE}, ~402 kJ mol^{-1}).

This activated platinum-electrode system is the equivalent of an NHE, and therefore conforms to the Nernst expression of Eq. (8.2). Rearrangement gives

$$\log P_{H_2} = -E_{Pt}\frac{2}{0.059} + \log [H_3O^+_{(aq)}]^2 \quad (8.25)$$

which responds to the overall reaction

$$H_2(P_{H_2}) + 2\ Pt^*(s) \overset{K_{eq}}{\rightleftharpoons} 2\ Pt{-}H(s) \xrightarrow[H_2O]{-2e^-} 2\ H_3O^+_{(aq)} + 2\ Pt(s) \quad (8.26)$$

Although there is a long-standing tradition of writing an electron-transfer sequence with electron removal (ionization) from the hydrogen atom to produce a proton (Pt—H(s) $\xrightarrow{-e^-}$ Pt + H^+) followed by hydration to give the observed hydronium ion [$H^+ + H_2O \longrightarrow H_3O^+_{(aq)}$], the ionization potential for a free hydrogen atom (H·) is 13.6 eV[15] and even greater for bound hydrogen [Pt—H(s)]. Electron removal from an aqueous solution at pH 0, with a glassy-carbon electrode (GCE) (free of H_2 and Pt—H) occurs via the lowest energy path, which is oxidation of the solvent

$$H_2O \xrightarrow{-e^-} [H_2^+O\cdot] \xrightarrow{H_2O} H_3O^+ + HO\cdot \qquad E^\circ,\ +2.72\ \text{V vs. NHE} \quad (8.27)$$

In the presence of Pt—H(s) (and H_2), this process is facilitated via direct formation of an H—OH bond

$$H_2O + Pt{-}H(s) \xrightarrow[H_2O]{-e^-} Pt(s) + H_3O^+ + HO{-}H \qquad E^\circ,\ 0.00\ \text{V} \quad (8.28)$$

$$\Delta(-\Delta G_{BF}) = (2.72 - 0.00)96.5 = 263\ \text{kJ mol}^{-1}$$

TABLE 8.4 Activation Potentials and Current Sensitivities for the Voltammetric Oxidation of Dissolved Hydrogen at Activated Platinum Electrodes in Five Solvents (Scan Rate 0.1 V s^{-1}); Solubilities and Diffusion Coefficients for H_2 in Several Solvents

Solvent	Electrode Activation $(E_A)_{opt}$ (V) versus SCE	$E_{1/2}$[a] (V) versus NHE	Sensitivity (mA cm^{-2} atm^{-1})	$[H_2]$ (1 atm H_2)[b] (mM)	D_{H_2}[c] (cm^2 s^{-1})	D_{H_2}[d,e] (cm^{-2} s^{-1})
H_2O	+1.3	−0.18	0.78 ± 0.01	0.78	2.5×10^{-5}	2.4×10^{-5}
Me_2SO	+1.3	−0.17	0.84 ± 0.01	1.12	1.0×10^{-5}	3.0×10^{-5}
DMF	+1.8	−0.12	2.24 ± 0.02	1.96	2.3×10^{-5}	5.2×10^{-5}
py	+1.0	−0.28	1.45 ± 0.02	2.00	0.9×10^{-5}	1.0×10^{-5}
MeCN	+3.0	+0.43				

[a] $E_{1/2} = (E_{p,a} + E_{p,c})/2$; $E_{NHE} = E_{SCE} + 0.242$ V.
[b] Ref. 16.
[c] Determined by linear-sweep voltammetry.
[d] Determined by rotated-disk voltammetry; for potential step to −0.1 V versus SCE, $i_L = 0.62\, n\, FAD^{1/2}\omega^{1/2}\nu^{-1/6}$; Ref. 14.
[e] $\pm 0.5 \times 10^{-5}$ cm^2 s^{-1}.

Addition of the Pt—H(s) bond energy [$\Delta(-\Delta G_{BF}) = 203$ kJ mol^{-1}, Eq. (8.7)] to the differential bond-formation energy for H—OH gives an electrochemical measure of the HO—H bond energy [$-\Delta G_{BF} = 465.7$ kJ mol^{-1} (literature value, 465.3 kJ mol^{-1})[3]].

Figure 8.3 illustrates cyclic voltammograms at a platinum (Pt) electrode for a dimethyl sulfoxide (Me_2SO) solution that is saturated with H_2 (1 atm). For each voltammogram the Pt electrode has been activated via preanodization at the indicated potential for 2 min prior to initiation of the voltammetric scan.

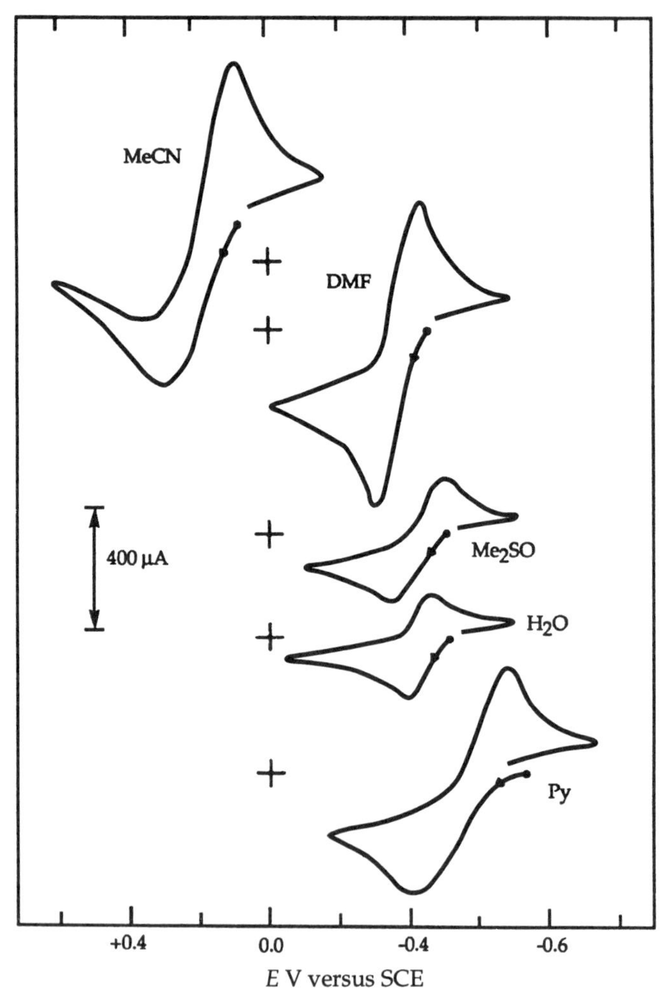

Figure 8.4 Cyclic voltammograms for dissolved H_2 (1 atm) in MeCN, DMF, Me_2SO, H_2O, and py at a freshly preanodized Pt electrode (area 0.23 cm^2). The activation was for 2 min at the $(E_A)_{opt}$ values of Table 8.4 for the various solvents. Scan rate 0.1 V s^{-1}; each solution contained 0.5 M supporting electrolyte.

The optimum preanodization potentials $(E_A)_{opt}$ for platinum in four aprotic solvents and water are summarized in Table 8.4.

Figure 8.4 illustrates the voltammograms with activated platinum electrodes for dissolved H_2 in five solvents. The peak potential shifts with the basicity of the solvent and the separation between the potentials for the anodic and cathodic peaks reflects the unbuffered solvent matrix and the system's conformity to Eq. (8.2). The voltammogram for dissolved H_2 in MeCN indicates severe adsorption effects, which precludes this solvent system for quantitative determinations via peak-current measurements. The other solvents yield anodic peak currents that are proportional to the concentration of dissolved H_2 and its diffusion coefficient (D_{H_2}). The H_2 concentration, in turn, is dependent on the partial pressure of dissolved H_2 (P_{H_2}) and its solubility in a particular solvent. Figure 8.5 summarizes the voltammetric peak currents ($i_{p,a}$) as a function of P_{H_2} in H_2O, Me_2SO, DMF, and py. The slopes of the linear curves are proportional

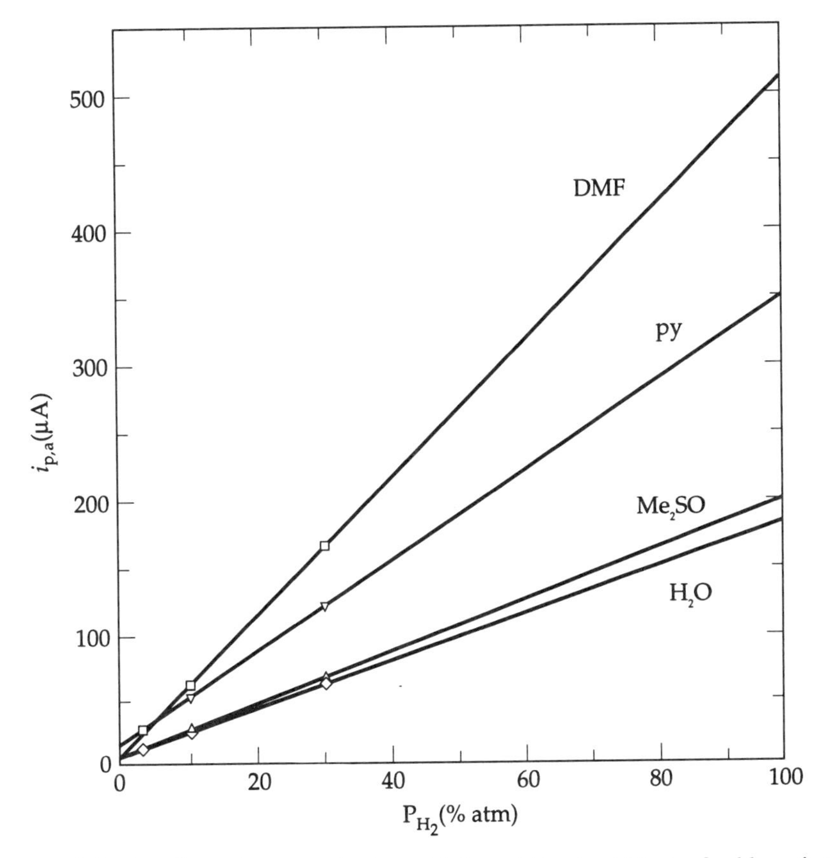

Figure 8.5 Anodic peak current for the oxidation of dissolved H_2 at a freshly activated Pt electrode (area 0.23 cm^2) as a function of H_2 partial pressures in DMF, py, Me_2SO, and H_2O solutions (0.5 M supporting electrolyte). Scan rate 0.1 V s^{-1}.

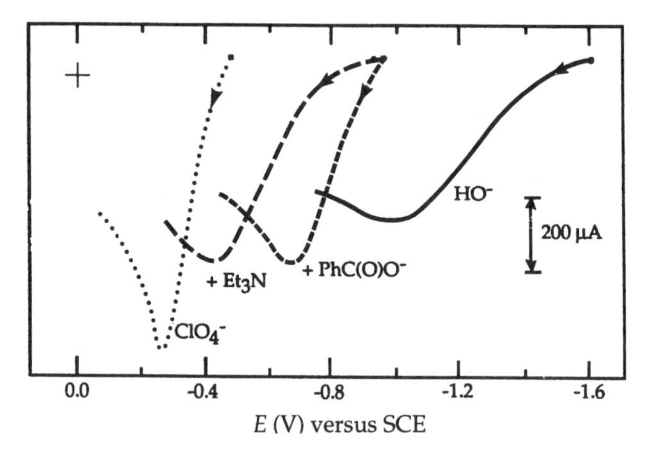

Figure 8.6 Voltammograms for the oxidation of dissolved H_2 (1 atm) in DMF [(0.5 M TEAP) (Et_4NClO_4)] and with 20 mM additions of triethylamine (Et_3N), tetraethylammonium benzoate ($PhC(O)O^-$), and tetraethylammonium hydroxide (HO^-). The Pt electrode (area 0.23 cm^2) was preanodized at +1.8 V versus SCE for 2 min prior to each voltammogram; scan rate 0.1 V s^{-1}.

to the solubility of H_2 and to the square root of the H_2 diffusion coefficient.

$$i_{p,a} = 2.67 \times 10^5 n^{3/2} A D_{H_2}^{1/2} (C_{H_2})_{1\ \text{atm}}\ \nu^{1/2} \tag{8.29}$$

On the basis of the slopes and the solubilities of H_2 in the respective solvents,[16] the diffusion coefficients for H_2 (D_{H_2}) have been evaluated via Eq. (8.29) and are summarized in Table 8.4. Rotated platinum-disk electrodes also provide a measure of D_{H_2}; values for such measurements are included in Table 8.4.[14]

Figure 8.6 illustrates the effect of basic substrates on the oxidation of H_2 in DMF. As the basicity of the substrate increases, the anodic peak potential for H_2 shifts from −0.3 V versus SCE (ClO_4^-) to −1.0 V (HO^-) and the peak current decreases. The potential shifts are consistent with Eq. (8.28) and the H-atom-facilitated oxidation of water [bases (B) facilitate formation of [$H_2{}^+O\cdot$] via proton transfer; ($B + H_2{}^+O\cdot \longrightarrow BH^+ + HO\cdot$). In the limit hydroxide ion in H_2O at unit activity becomes the primary substrate with its oxidation product ($HO\cdot$) coupling with H_2 (1 atm) via Pt—H(s)

$$HO^- + Pt{-}H(s) \xrightarrow{-e^-} Pt + HO{-}H \qquad E°, -0.83\ \text{V vs. NHE} \tag{8-30}$$

$$-\Delta G_{BF},\ 466\ \text{kJ mol}^{-1}$$

This combination [HO^-/Pt—H(s), H_2] is a strong one-electron reductant that will react spontaneously with most oxidants [O_2, HOOH, high-valence metal oxides ($Mn^{IV}O_2$, $Fe^{III}{}_2O_3$, $Ag^I{}_2O$, $Cu^{II}O$, $Mn^{VII}O_4^-$, $Cr^{VI}O_4^{2-}$)]. As a result, these represent interferences to the voltammetric determination of dissolved H_2 concentrations.

REFERENCES

1. Ross, P. N., "Hydrogen," in *Standard Potentials in Aqueous Solution*, Bard, A. J.; Parsons, R.; Jordan, J., eds., Marcel Dekker, New York, 1985, Chap. 3.
2. Parsons, R., *Handbook of Electrochemical Constants*, Butterworths, London, 1959.
3. Lide, D. R., ed., *CRC Handbook of Chemistry and Physics*, 71st ed., CRC, Boca Raton, Fla., 1990, pp. **9**-86–98.
4. Sawyer, D. T., *Oxygen Chemistry*, Oxford University Press, New York, 1991, Chap. 2.
5. Barrette, W. C., Jr.; Johnson, H. W., Jr.; Sawyer, D. T., *Anal. Chem.* **1984,** *56,* 1890.
6. Albert, A.; Sergeant, E. P., *Ionization Constants of Acids and Bases*, Wiley, New York, 1962.
7. King, E. J., *Acid–Base Equilibria*, Macmillan, New York, 1965.
8. Cookson, R. F., *Chem. Rev.* **1974,** *74,* 5.
9. Lowry, T. H.; Richardson, K. S., *Mechanism and Theory in Organic Chemistry*, Harper & Row, New York, 1981, pp. 263ff. and references cited therein.
10. Ritchie, C. D.; Uschold, R. E., *J. Am. Chem. Soc.* **1968,** *90,* 2821.
11. Kolthoff, I. M.; Chantooni, M K., Jr.; Bhowmik, *S. J., Am. Chem. Soc.* **1968,** *90,* 23.
12. Coetzee, J. F., *Prog. Phys. Org. Chem.* **1967,** *4,* 45.
13. Olmstead, W. N.; Margolin, Z.; Bordwell, F. G., *J. Org. Chem.* **1980,** *45,* 3295.
14. Barrette, W. C., Jr.; Sawyer, D. T., *Anal. Chem.* **1984,** *56,* 6553.
15. Lide, D. R., ed., *CRC Handbook of Chemistry and Physics*, 71st ed., CRC, Boca Raton, Fla., 1990, pp. **10**-180–181, **10**-210–211.
16. Young, C. L., ed., *I. U.P.A.C. Solubility Data Series*, Pergamon Press, New York, 1981, Vols. 5/6 (*Hydrogen* and *Deuterium*).

CHAPTER 9

DIOXYGEN SPECIES (O_2, HOO·, $O_2^{\overline{\cdot}}$, HOOH, HOO^-), OZONE (O_3), AND ATOMIC OXYGEN

9.1 INTRODUCTION; REDOX THERMODYNAMICS

Although the electrochemistry of hydronium ion and molecular hydrogen is fundamental, the electrochemical characterization of oxygen species (O_2, HOO·, $O_2^{\overline{\cdot}}$, HOOH, HOO^-, O, HO·, $O^{\overline{\cdot}}$, H_2O, HO^-, O_3, $O_3^{\overline{\cdot}}$) and of the oxygen component of molecules (e.g., M_xO_y, oxy anions and radicals, quinones) is its most important and unique application. In general, electrochemical measurements provide the only direct means for the evaluation of the electron-transfer thermodynamics of oxygen species and oxygen-containing molecules. Also, amperometric sensors for O_2 are the most common analytical methodology for its assay in blood, gas streams, biotreaters, and process streams.

Molecular Oxygen. The reduction of molecular oxygen is influenced by the solution matrix and its acidity. Thus, the redox thermodynamics of O_2 are directly dependent upon hydronium ion activity:

$$O_2 + 4\,H_3O^+ + 4\,e^- \longrightarrow 6\,H_2O \qquad (E°)_{H_2O},\ +1.23\text{ V vs. NHE} \qquad (9.1)$$

which, in turn, depends on the reaction matrix. Table 9.1 summarizes the pK_a values for a series of Brønsted acids in several aprotic solvents and water.[1] In acetonitrile the activity values for pK_a range from -8.8 for $(H_3O)ClO_4$ to 30.4 for H_2O. This means that the formal potential ($E°'$) for the reaction of Eq. (9.1) in acetonitrile (MeCN) is +1.79 V versus NHE in the presence of 1 M $(H_3O)ClO_4$ and -0.53 V in the presence of 1 M $[(n\text{-Bu})_4N]OH$. Another limiting factor with respect to the chemical energy flux for oxidative metabolism

TABLE 9.1 Effective pK_a Values for Brønsted Acids in Aprotic Solvents and Water

	Solvent [0.5 M $(Et_4N)ClO_4$][a]				
Brønsted Acid	MeCN	DMF	Me_2SO	py	H_2O(lit)
$(H_3O)ClO_4$	−8.8	0.7	2.6	4.6	0.0 (0.0)
p-MePhS(O)$_2$OH	−3.8	—	—	—	—
(pyH)ClO_4	1.8	3.1	4.4	5.7	4.9 (5.2)
2,4-$(NO_2)_2$PhOH	4.3	4.5	4.9	5.5	— (4.0)
$(NH_4)ClO_4$	5.9	9.6	11.7	7.3	8.7 (9.2)
$(Et_3NH)Cl$	10.0	9.9	12.7	7.6	10.1 (11.0)
PhC(O)OH	7.9	11.5	13.6	11.6	3.2 (4.2)
2-Py-C(O)OH (PAH)	8.6	—	—	—	— (5.5)
PhOH	16.0	19.4	20.8	20.1	9.2 (9.9)
p-EtOPhOH	19.3	21.5	23.8	21.8	9.6
H_2O	30.4	34.7	36.7	30.5	14.8 (14.0)

[a]MeCN, acetonitrile; DMF, dimethylformamide; Me_2SO, dimethyl sulfoxide; py, pyridine.
Source: Ref. 1.

and respiration is the solubility of O_2. Because of its nonpolar character, dioxygen is much more soluble in organic solvents than in H_2O (Table 9.2).[2]

The reduction potentials for O_2 and various intermediate species in H_2O at pH 0, 7, and 14 are summarized in Table 9.3;[3–6] similar data for O_2 in MeCN at pH −8.8, 10.0, and 30.4 are presented in Table 9.4.[2,7,8] Potentials for the reduction of ozone (O_3) in aqueous and acetonitrile solutions are presented in Table 9.5.[6,7]

The reduction manifolds for O_2 (Tables 9.3 and 9.4) indicate that the limiting step (in terms of reduction potential) is the first electron transfer to O_2, and that an electron source adequate for the reduction of O_2 will produce all the other reduced forms of dioxygen ($O_2^{\bar{\cdot}}$, HOO·, HOOH, HOO^-, HO·, H_2O, HO^-) via reduction, hydrolysis, and disproportionation steps (Scheme 9.1).[9,10]

TABLE 9.2 Solubilities of O_2 (1 atm) in Various Solvents

Solvent	$[O_2]_{1\,atm}$ (mM)
H_2O	1.0
Me_2SO	2.1
DMF	4.8
py	4.9
MeCN	8.1
Hydrocarbons	~10
Fluorocarbons	~25

Source: Ref. 1.

TABLE 9.3 Standard Reduction Potentials for Dioxygen Species in Water [O_2, 1 atm (1.0 mM)] (Formal Potentials for O_2 at Unit Activity)

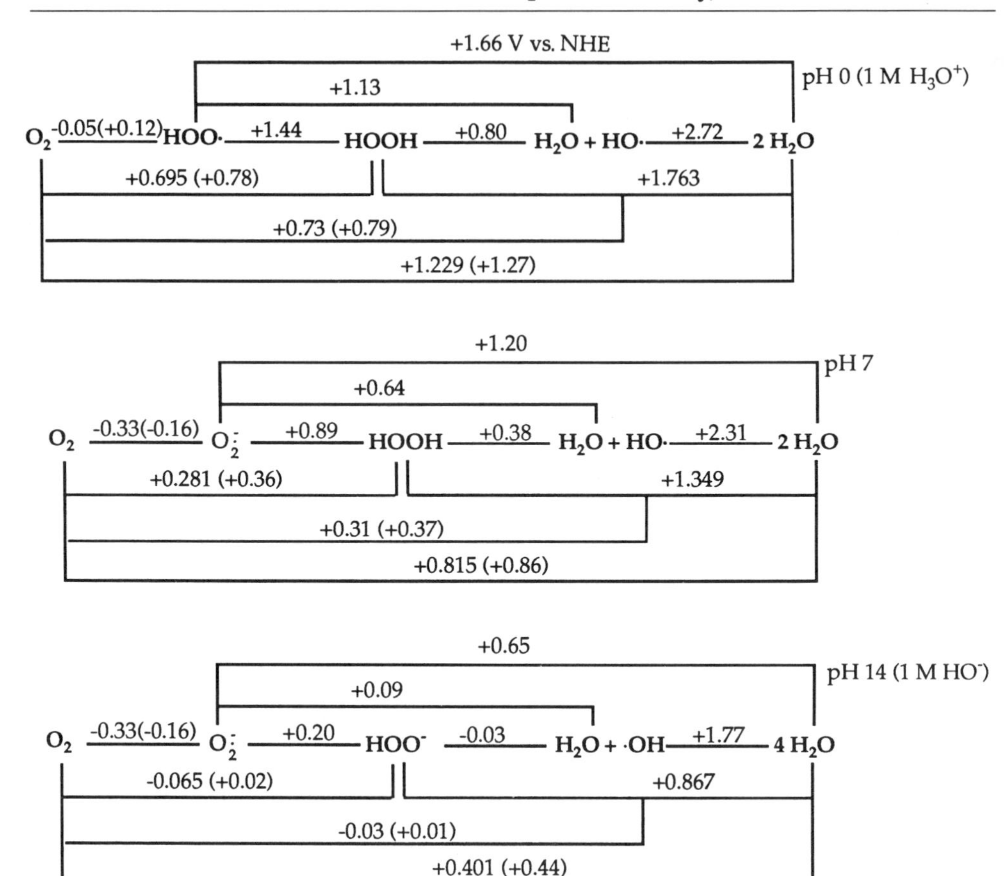

$$O_2 + e^- \underset{}{\overset{-0.33\ \mathrm{V}}{\rightleftharpoons}} O_2^{\cdot-} \xrightarrow{\mathrm{HA}} \mathrm{HOO}\cdot + A^-$$

$$\mathrm{HOO}\cdot + \mathrm{HOO}\cdot \rightarrow \mathrm{HOOH} + O_2 \quad k, \sim 10^6\ \mathrm{M}^{-1}\ \mathrm{s}^{-1}$$

$$\mathrm{HOO}\cdot + O_2^{\cdot-} \rightarrow \mathrm{HOO}^- + O_2 \quad k, \sim 10^8\ \mathrm{M}^{-1}\ \mathrm{s}^{-1}$$

$$\mathrm{HOO}^- \xrightarrow{\mathrm{HA}} \mathrm{HOOH} + A^-$$

$$\mathrm{HOO}^- \xrightarrow{\mathrm{HOOH}} O_2 + H_2O + HO^-$$

Scheme 9.1 Formation and degradation of superoxide ion ($O_2^{\cdot-}$) in aqueous solutions.

TABLE 9.4 Formal Reduction Potentials for Dioxygen Species in Acetonitrile [O_2 at 1 atm (8.1 mM)]

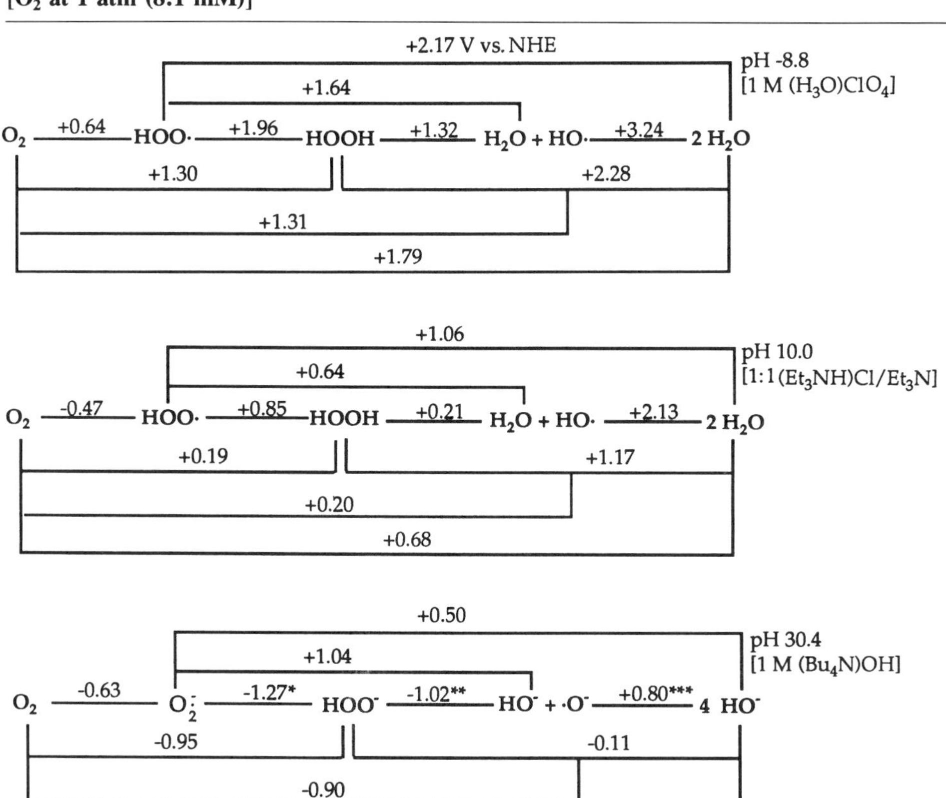

*($O_2^{\cdot -}$ $\xrightarrow{-1.51\ V}$ HOOH) **(HOOH $\xrightarrow{-0.90\ V}$ $HO^- + HO\cdot$) ***($HO\cdot$ $\xrightarrow{+0.92\ V}$ HO^-)

Thus, the most effective means to activate O_2 is the addition of an electron (or hydrogen atom; $H_3O^+ + e^- \longrightarrow H\cdot$), which results in significant fluxes of several reactive oxygen species.

Superoxide Ion. The dominant characteristic of $O_2^{\cdot -}$ in any medium is its ability to act as a strong Brønsted base via formation of $HOO\cdot$,[11,12] which reacts with itself or a second $O_2^{\cdot -}$ (Scheme 9.1). Within water, superoxide ion is rapidly converted to dioxygen and hydroperoxide ion:

$$2\ O_2^{\cdot -} + H_2O \rightleftharpoons O_2 + HOO^- + HO^- \qquad K \sim 2.5 \times 10^8 \text{ atm} \qquad (9.2)$$

Such a proton-driven disproportionation process means that $O_2^{\cdot -}$ can deprotonate acids much weaker than water ($\leq pK_a \approx 23$).[13]

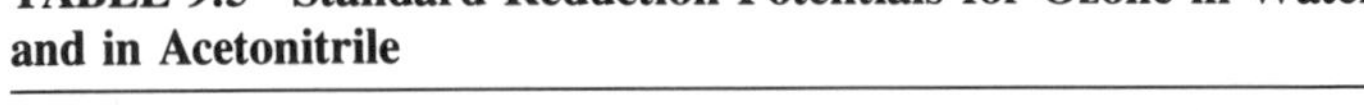

TABLE 9.5 Standard Reduction Potentials for Ozone in Water and in Acetonitrile

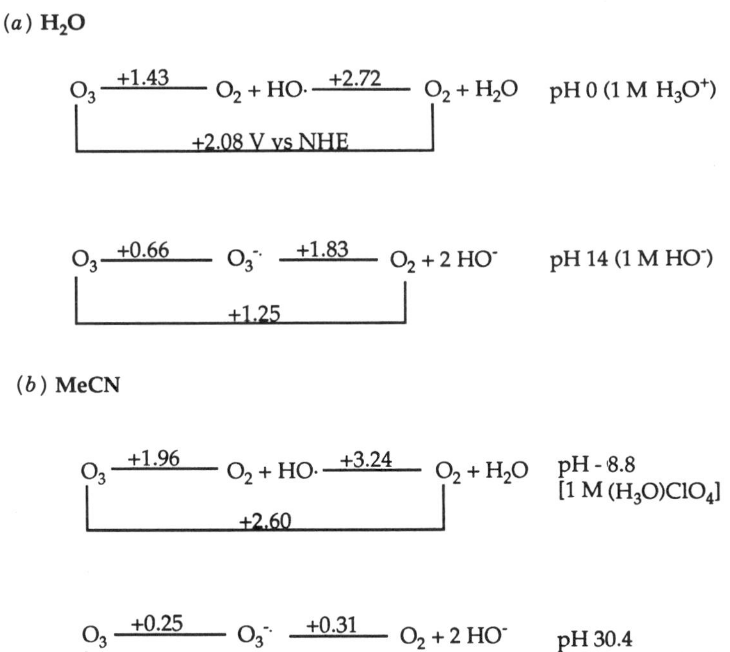

The data of Tables 9.3 and 9.4 indicate that $O_2^{\bar{\cdot}}$ is a moderate one-electron reducing agent [cytochrome-*c* (Fe^{III}) is reduced in H_2O[14,15] and iron(III) porphyrins in dimethylformamide]:

$$(DMF)^+Fe^{III}(TPP) + O_2^{\bar{\cdot}} \longrightarrow (DMF)Fe^{II}(TPP) + O_2 \qquad \Delta E,\ +0.7\ V \tag{9.3}$$

Although $O_2^{\bar{\cdot}}$ reacts with proton sources to form $HOO\cdot$ (which disproportionates via a second $O_2^{\bar{\cdot}}$), with limiting fluxes of protons to control the rate of $HOO\cdot$ formation from $O_2^{\bar{\cdot}}$, the rate of decay of $HOO\cdot$ is enhanced by reaction with the allylic hydrogens of excess 1,4-cyclohexadiene (1,4-CHD).[25] Because $HOO\cdot$ disproportionation is a second-order process, low concentrations favor hydrogen–atom abstraction from 1,4-CHD. This is especially so for Me_2SO, in which the rate of disproportionation for $HOO\cdot$ is the slowest (PhCl > MeCN > H_2O > DMF > Me_2SO).[16]

Atomic Oxygen. The electron-transfer potentials for the reduction of ground-state atomic oxygen in water and in acetonitrile are summarized in Tables 9.6

TABLE 9.6 Redox Potentials in Aqueous Media for Species That Contain Atomic Oxygen

	$E°$ (V) versus NHE
pH 0	
$O(g) + 2\ H_3O^+ + 2\ e^- \longrightarrow 3\ H_2O$	+2.43
$O(g) + H_3O^+ + e^- \longrightarrow HO\cdot + H_2O$	+2.14
$HO\cdot + H_3O^+ + e^- \longrightarrow 2\ H_2O$	+2.72
$O_3(g) + 2\ H_3O^+ + 2\ e^- \longrightarrow O_2(g) + 3\ H_2O$	+2.08
$O_3(g) + H_3O^+ + e^- \longrightarrow HO\cdot + O_2 + H_2O$	+1.43
$HOOH + 2\ H_3O^+ + 2\ e^- \longrightarrow 4\ H_2O$	+1.76
$HOIO_3 + H_3O^+ + 2\ e^- \longrightarrow IO_3^- + 2\ H_2O$	+1.6
$HOCl + 2\ H_3O^+ + 2e^- \longrightarrow HCl + 3\ H_2O$	+1.49
pH 7	
$O(g) + 2\ H_3O^+ + 2\ e^- \longrightarrow 3\ H_2O$	+2.01
$O(g) + H_3O^+ + e^- \longrightarrow HO\cdot + H_2O$	+1.71
$HO\cdot + H_3O^+ + e^- \longrightarrow 2\ H_2O$	+2.31
$O_3(g) + 2\ H_3O^+ + 2\ e^- \longrightarrow O_2(g) + 3\ H_2O$	+1.66
$HOOH + 2\ H_3O^+ + 2\ e^- \longrightarrow 4\ H_2O$	+1.35
$IO_4^- + 2\ H_3O^+ + 2\ e^- \longrightarrow IO_3^- + 3\ H_2O$	+1.2
pH 14	
$O(g) + H_2O + 2\ e^- \longrightarrow 2\ HO^-$	+1.60
$O(g) + H_2O + e^- \longrightarrow HO\cdot + HO^-$	+1.31
$O(g) + e^- \longrightarrow O^{\cdot-}$	+1.43
$HO\cdot + e^- \longrightarrow HO^-$	+1.89
$HOH + e^- \longrightarrow H\cdot + HO^-$	−2.93
$O^{\cdot-} + H_2O + e^- \longrightarrow 2\ HO^-$	+1.77
$O_3(g) + H_2O + 2\ e^- \longrightarrow O_2(g) + 2\ HO^-$	+1.25
$O_3(g) + e^- \longrightarrow O_3^{\cdot-}$	+0.66
$O_3^{\cdot-} + H_2O + e^- \longrightarrow O_2(g) + 2\ HO^-$	+1.83
$HOO^- + H_2O + 2\ e^- \longrightarrow 3\ HO^-$	+0.87
$IO_4^- + H_2O + 2\ e^- \longrightarrow IO_3^- + 2\ HO^-$	+0.8
$ClO^- + H_2O + 2\ e^- \longrightarrow Cl^- + 2\ HO^-$	+0.89
$ClO^- + e^- \longrightarrow Cl^- + O^{\cdot-}$	+0.02

and 9.7.[3,4,17] Selected values for the reduction of the oxygen atom in O_3, $HOIO_3$, HOOH, and HOCl also are included.[3] Because the oxygen atom in a water molecule essentially is without charge, these reduction processes represent the facilitated reduction of hydronium ions via stabilization of the product $H\cdot$ atom through strong covalent-bond formation with atomic oxygen. Thus, the standard-state reduction of hydronium ions to hydrogen atoms

$$2\ H_3O^+ + 2\ e^- \longrightarrow 2\ H\cdot + 2\ H_2O \qquad E°,\ -2.10\ \text{V vs. NHE} \tag{9.4}$$

is shifted to a much more favored process in the presence of atomic oxygen:

$$O(g) + 2\ H_3O^+ + 2\ e^- \longrightarrow 3\ H_2O \qquad E°,\ +2.43\ \text{V} \tag{9.5}$$

TABLE 9.7 Redox Potentials in MeCN [$H_3O^+ \equiv 1$ M $(H_3O)ClO_4$; $HO^- \equiv 1$ M $[(n\text{-}Bu)_4N]OH(MeOH)$] for Species That Contain Atomic Oxygen

	$E^{\circ\prime}$ (V) versus NHE
Acid, pH (−8.8)	
$O(g) + 2\ H_3O^+ + 2\ e^- \longrightarrow 3\ H_2O$	+2.95
$O(g) + H_3O^+ + e^- \longrightarrow HO\cdot + H_2O$	+2.66
$HO\cdot + H_3O^+ + e^- \longrightarrow 2\ H_2O$	+3.24
$O_3(g) + 2\ H_3O^+ + 2\ e^- \longrightarrow O_2(g) + 3\ H_2O$	+2.60
$HOOH + 2\ H_3O^+ + 2\ e^- \longrightarrow 4\ H_2O$	+2.28
$HOCl + 2\ H_3O^+ + 2\ e^- \longrightarrow HCl + 3\ H_2O$	+2.0
Base, pH 30.4	
$O(g) + H_2O + 2\ e^- \longrightarrow 2\ HO^-$	+0.63
$O(g) + H_2O + e^- \longrightarrow HO\cdot + HO^- \longrightarrow O^{\cdot-}(H_2O)$	+0.34
$HO\cdot + e^- \longrightarrow HO^-$	+0.92
$O^{\cdot-} + H_2O + e^- \longrightarrow 2\ HO^-$	+0.80
$HOH + e^- \longrightarrow H\cdot + HO^-$	−3.90
$O_3 + H_2O + 2\ e^- \longrightarrow O_2 + 2\ HO^-$	+0.28
$ClO^- + H_2O + 2\ e^- \longrightarrow Cl^- + 2\ HO^-$	−0.08
$HOO^- + H_2O + 2\ e^- \longrightarrow 3\ HO^-$	−0.10

Likewise, the electron-transfer reduction of H_2O (with uncharged hydrogen and oxygen atoms), which must overcome the stabilization of the strong O—H bonds, results in a −1 charge for oxygen rather than hydronium ion reduction:

$$H_2O + e^- \longrightarrow H\cdot + HO^- \qquad E^\circ,\ -2.93 \text{ V vs. NHE} \tag{9.6}$$

In contrast, an equivalent charge-density change for the neutral oxygen in $HO\cdot$ is strongly favored:[9]

$$HO\cdot + e^- \longrightarrow HO^- \qquad E^\circ,\ +1.89 \text{ V} \tag{9.7}$$

and reduction of the combination of the zero-charge oxygen of water with −1-charge $O^{\cdot-}$ yields two −1-charge hydroxide ions:[4]

$$O^{\cdot-} + H_2O + e^- \longrightarrow 2\ HO^- \qquad E^\circ,\ +1.77 \text{ V} \tag{9.8}$$

9.2 ELECTROCHEMISTRY OF DIOXYGEN

The ground state of molecular oxygen ($\cdot O_2\cdot$, $^3\Sigma_g^-$) has two unpaired electrons in degenerate $2\pi_g$ orbitals (Table 9.8), and is referred to as *dioxygen* by most contemporary biologists and biochemists. When dioxygen is reduced by electron transfer, a series of intermediate basic dioxygen and monooxygen species are produced that may take up one or two hydronium ions (H_3O^+) from the media ($O_2^{\cdot-}$, $HOO\cdot$, HOO^-, HOOH, $\cdot O^-$, $HO\cdot$, HO^-, and H_2O). The ther-

TABLE 9.8 Molecular-Orbital Diagrams for O_2 and $O_2^{\overline{\cdot}}$.

	O_2			
Orbital	$^3\Sigma_g^-$	$^1\Delta_g$	$^1\Sigma_g^+$	$O_2^{\overline{\cdot}}$
$2p\sigma_u$	—	—	—	—
$2p\pi_g$	↑ ↑	↑↓	↑ ↓	↑↓ ↑
$2p\pi_u$	↑↓ ↑↓	↑↓ ↑↓	↑↓ ↑↓	↑↓ ↑↓
$2p\sigma_g$	↑↓	↑↓	↑↓	↑↓
$2s\sigma_u$	↑↓	↑↓	↑↓	↑↓
$2s\sigma_g$	↑↓	↑↓	↑↓	↑↓

modynamics for the various reduction steps in aqueous solutions have been evaluated by numerous techniques, but all are fundamentally based on the calorimetry associated with the reaction

$$O_2(g) + 2\ H_2(g) \longrightarrow 2\ H_2O(l) \qquad -\Delta G^\circ = nE^\circ_{cell}F \tag{9.9}$$

If redox potentials are relative to the normal hydrogen electrode

$$2\ H_3O^+ + 2\ e^- \longrightarrow H_2(g)\ E^\circ_{H_3O^+/H_2} \equiv 0.000 \tag{9.10}$$

then the standard redox potential for the four-electron reduction of dioxygen ($E^\circ_{O_2/H_2O}$) can be calculated from the calorimetric data for the reaction of Eq. (9.9) under standard-state conditions:

$$O_2(g) + 4\ H_3O^+_{(aq)} + 4\ e^- \longrightarrow 2\ H_2O(l) \tag{9.11}$$

$$E^\circ_{O_2/H_2O} \equiv E^\circ_{cell} - E^\circ_{H_3O^+/H_2} = +1.229\ \text{V vs. NHE}$$

Table 9.3 summarizes the redox potentials for the reduction of various dioxygen species in aqueous media at pH 0, 7, and 14. For those couples that involve dioxygen itself, formal potentials are given in parentheses for O_2 at unit activity ($\sim 10^3$ atm; $[O_2] \approx 1$ mM at 1 atm partial pressure).

Although an early voltammetric investigation demonstrated that dioxygen in aqueous solutions is reduced to hydrogen peroxide at passivated platinum electrodes,[18]

$$O_2 + 2\ H_3O^+ + 2\ e^- \longrightarrow \text{HOOH} + 2\ H_2O \qquad E_{p,c},\ +0.05\ \text{V vs. NHE} \tag{9.12}$$

the pH independence of the reduction potential and the irreversibility of the process confirm that this represents the overall reduction process and that the mechanism is *not* a single, concerted, two-electron reduction. Subsequent voltammetric studies have established that the reduction process depends on the surface of solid electrodes in aqueous media.[19,20]

The pH-dependent reduction potential for O_2 that is observed with activated metal surfaces is due to the reduction of the freshly formed metal hydroxide film; for example

$$Pt^{II}(OH)_2(s) + 2\ H_3O^+_{(aq)} + 2\ e^- \longrightarrow Pt(s) + 4\ H_2O$$

$$E^\circ,\ +0.67\ \text{V vs. NHE} \tag{9.13}$$

After the metal oxide is reduced electrochemically, it is reformed by chemical reaction with O_2:

$$2\ Pt(s) + O_2 + 2\ H_2O \longrightarrow 2\ Pt^{II}(OH)_2(s) \tag{9.14}$$

Hence, the metal oxide is a catalyst for the reduction of O_2 to H_2O. Such a mechanism accounts for the dependence of the reduction potentials and the rates of reaction on the electrode material and pH.

The older literature on the electrochemistry of dioxygen in acidic media attributes the difference between the thermodynamic potential for the four-electron reduction (O_2/H_2O, +1.23 V vs. NHE; Table 9.3) and the observed value at a freshly activated platinum electrode [+0.67 V vs. NHE, Eq. (9.13)] to *overvoltage* (or kinetic inhibition). Likewise, the difference between the thermodynamic potential for the two-electron reduction (O_2/HOOH, +0.70 V vs. NHE, Table 9.3) and the observed value at passivated electrodes [+0.05 V vs. NHE, Eq. (6.12)] was believed to be due to the kinetic inhibition of the two-electron process.

Electron-Transfer Reduction of O_2. Within aqueous solutions the most direct means to the electron-transfer reduction of dioxygen is by pulse radiolysis. Irradiation of an aqueous solution by an electron beam yields (almost instantly; 10^{-12} s) solvated electrons [e^-(aq)], hydrogen atoms (H·), and hydroxyl radicals (HO·). If the solution contains a large excess of sodium formate [$Na^{+-}O(O)CH$] and is saturated with O_2, then the radiolytic electron flux efficiently and cleanly reduces O_2 to superoxide ion ($O_2^{\bar{\cdot}}$):[21–25]

$$\cdot O_2 \cdot + e^-(aq) \longrightarrow O_2^{\bar{\cdot}} \tag{9.15}$$

$$HC(O)O^- + HO\cdot \longrightarrow \cdot C(O)O^- + H_2O \tag{9.16}$$

$$\cdot O_2 \cdot + \cdot C(O)O^- \longrightarrow O_2^{\bar{\cdot}} + CO_2 \tag{9.17}$$

$$\cdot O_2 \cdot + H\cdot \longrightarrow HOO\cdot \xrightarrow{H_2O} H_3O^+ + O_2^{\bar{\cdot}} \quad pK_a, 4.9 \tag{9.18}$$

This represents electrodeless electrochemistry, whereby the only process is electron transfer and the only product is the one-electron adduct of O_2, the superoxide ion ($O_2^{\bar{\cdot}}$). Through the use of redox mediator dyes and spectrophotometry, the electron-transfer thermodynamic reduction potential for O_2 has been evaluated:

$$O_2(g, 1\ atm) + e^- \rightleftharpoons O_2^{\bar{\cdot}} \quad (E^\circ)_{pH\,5-14}, \ -0.33\ V\ vs.\ NHE; \ -0.16\ V\ vs.\ NHE\ for\ O_2\ at\ unit\ activity \tag{9.19}$$

This relation in combination with the O_2/HOOH couple at pH 7.0 of Table 9.3 yields the one-electron reduction potential for $O_2^{\bar{\cdot}}$:

$$O_2^{\bar{\cdot}} + H_3O^+ + e^- \rightleftharpoons HOOH \quad (E^{\circ\prime})_{pH7}, +0.89\ V\ vs.\ NHE \tag{9.20}$$

A direct electrochemical measurement of the reduction potential for the $O_2/O_2^{\bar{\cdot}}$ couple in aqueous solutions is complicated by the rapid proton-induced disproportionation reactions for $O_2^{\overline{\cdot}}$. The *hydroperoxyl radical* (HOO·) is a moderately weak acid (roughly equivalent to acetic acid)[21]

$$HOO\cdot + H_2O \rightleftharpoons H_3O^+ + O_2^{\overline{\cdot}} \qquad pK_a,\ 4.89 \tag{9.21}$$

which undergoes a rapid homolytic disproportionation

$$HOO\cdot + HOO\cdot \xrightarrow{k_{bi},\ 8.6 \times 10^5\ M^{-1}s^{-1}} HOOH + O_2 \qquad K,\ 10^{25}\ atm \tag{9.22}$$

as well as an even faster electron-transfer process with $O_2^{\overline{\cdot}}$ at pH 4.89:

$$HOO\cdot + O_2^{\overline{\cdot}} \xrightarrow[H_3O^+]{k_{bi},\ 1.0 \times 10^8\ M^{-1}s^{-1}} HOOH + O_2 + H_2O \tag{9.23}$$

Figure 9.1 illustrates the electrochemical reduction of O_2 at platinum electrodes in aqueous media (1.0 M $NaClO_4$). The top curve represents the cyclic voltammogram (0.1 V s^{-1}) for O_2 at 1 atm (~1 mM), and the lower curve is the voltammogram with a rotated-disk electrode (900 rpm, 0.5 V min^{-1}). Both processes are totally irreversible with two-electron stoichiometries and half-wave potentials ($E_{1/2}$) that are independent of pH. The mean of the $E_{1/2}$ values for the forward and reverse scans of the rotated-disk voltammograms for O_2 is 0.0 V versus NHE. If the experiment is repeated in media at pH 12, the mean $E_{1/2}$ value also occurs at 0.0 V.

The aqueous electrochemical reduction of O_2 at inert electrodes occurs via an initial reversible electron-transfer process [analogous to the homogeneous process of pulse radiolysis, Eq. (9.19)]

$$O_2 + e^- \rightleftharpoons O_2^{\overline{\cdot}} \qquad (E_{1/2})_{pH\,12},\ 0.0\ V\ vs.\ NHE \tag{9.24}$$

which is followed by two rapid chemical steps:

$$O_2^{\overline{\cdot}} + H_2O \rightleftharpoons HOO\cdot + HO^- \qquad K = 10^{-14}/10^{-4.9} = 8 \times 10^{-10} \tag{9.25}$$

$$HOO\cdot + HOO\cdot \rightleftharpoons HOOH + O_2 \qquad K,\ 10^{25}\ atm \tag{9.26}$$

The rate and extent of the hydrolysis reaction [Eq. (9.25)] is suppressed under alkaline conditions, and the lifetime of $O_2^{\overline{\cdot}}$ is increased such that the $E_{1/2}$ potential may be shifted to more negative potentials. This is the result of thermodynamic effects via the Nernst equation rather than kinetic (overvoltage) effects. The overall post-electron-transfer chemistry is given by the relation

$$2\ O_2^{\overline{\cdot}} + 2\ H_2O \rightleftharpoons HOOH + O_2 + 2\ HO^- \qquad K,\ 10^9\ M\ atm \tag{9.27}$$

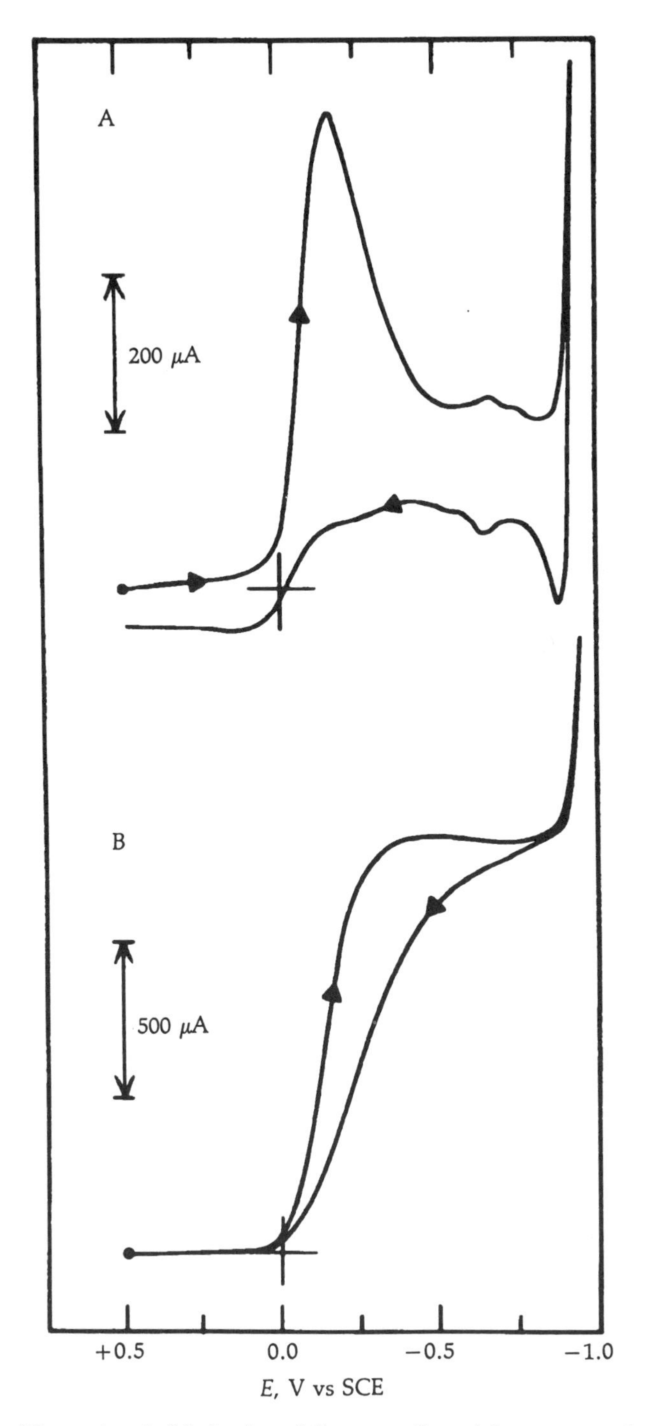

Figure 9.1 Electrochemical behavior of dioxygen (1 atm) in aqueous solutions (1 M $NaClO_4$) at a platinum electrode (area 0.458 cm^2): (*A*) the cyclic voltammogram was initiated at the rest potential with a scan rate of 0.1 V s^{-1}; (*B*) the rotated-disk (400-rpm) voltammogram was obtained with a scan rate of 0.5 V min^{-1}.

The net electrochemical process is the sum of Eqs. (9.24) and (9.27):

$$O_2 + 2\ H_2O + 2\ e^- \longrightarrow HOOH + 2\ HO^- \qquad (9.28)$$

and is observed with inert electrodes in aqueous media. With metal electrodes there is a tendency for the intermediate species ($O_2^{\cdot -}$ and $HOO\cdot$) to react and form electroactive metal oxides.

Aprotic Media. In the absence of proton sources dioxygen is reversibly reduced to superoxide ion, as illustrated by Figure 9.2 for a pyridine solution (0.1 M TEAP):[26,27]

$$O_2 + e^- \rightleftharpoons O_2^{\cdot -} \qquad E^{\circ\prime},\ -0.64\ \text{V vs. NHE} \qquad (9.29)$$

The second reduction is an irreversible one-electron process:

$$O_2^{\cdot -} + e^- \xrightarrow{H_2O} HOO^- + HO^- \qquad E_{p,c},\ -1.8\ \text{V} \qquad (9.30)$$

$$HO^- + HOO^- + O_2 \rightleftharpoons 2\ O_2^{\cdot -} + H_2O \qquad K,\ 10^{-9}\ \text{M}^{-1}\ \text{atm}^{-1} \qquad (9.31)$$

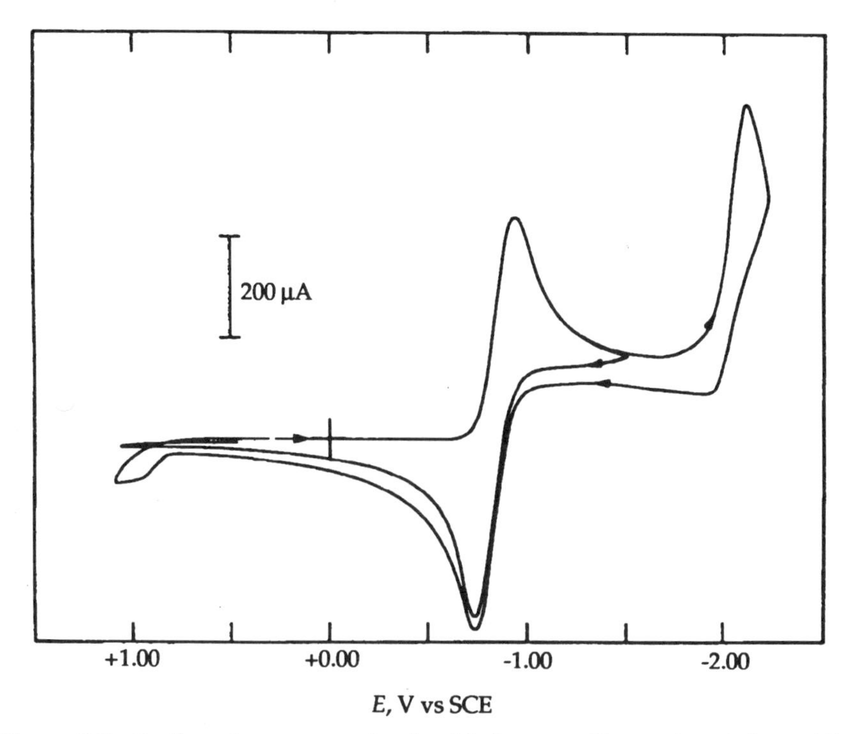

Figure 9.2 Cyclic voltammogram for 5 mM dioxygen (O_2, at 1 atm) in pyridine (0.1 M tetrapropylammonium perchlorate) at a platinum electrode (area 0.23 cm^2). Scan rate, 0.1 V s^{-1}. Saturated calomel electrode (SCE) versus NHE, +0.244 V.

In alkaline pyridine solutions HOOH also decomposes to give $O_2^{\bar{\cdot}}$:[28]

$$HOO^- + HOOH \xrightarrow[k,\ 10^4 M^{-1} s^{-1}]{py} O_2^{\bar{\cdot}} + H_2O + \frac{1}{n}[\dot{p}y(OH)]_n \qquad (9.32)$$

There is a general view that the electrochemical reduction of O_2 in aprotic media is independent of media and electrode materials, but the cyclic voltammograms of Figure 9.3 and the electrochemical data of Table 9.9 provide clear evidence that both have a significant effect on the reversibility. Although the

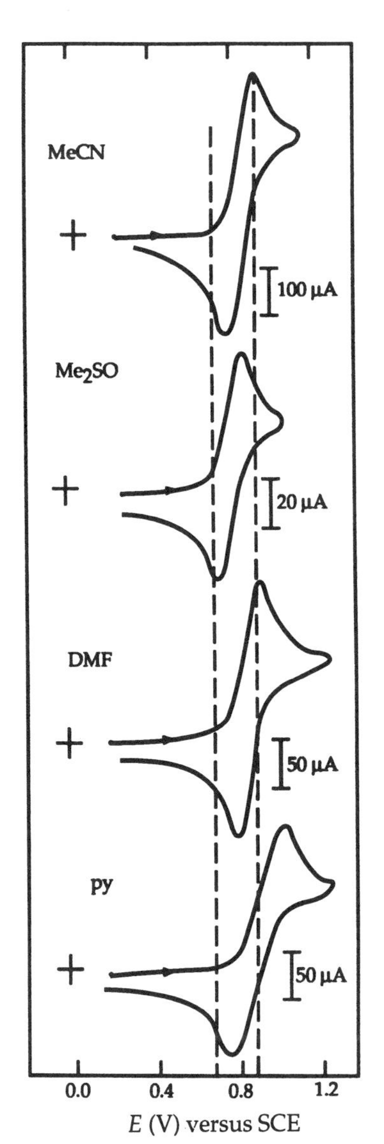

Figure 9.3 Cyclic voltammograms for O_2 (at 1 atm) in acetonitrile (MeCN) (8.1 mM), dimethyl sulfoxide (Me_2SO) (2.1 mM), dimethylformamide (DMF) (4.8 mM), and pyridine (py) (4.9 mM) at a vitreous carbon inlay electrode (area 0.20 cm^2). All solutions contained 0.1 M TEAP as supporting electrolyte, and the scan rate was 0.1 V s^{-1}.

TABLE 9.9 Redox Potentials for the Reduction of Dioxygen (1 atm) at Cyclic Voltammetry with Four Different Electrodes in Four Aprotic Solvents (0.1 M Tetraethylammonium Perchlorate)

A. Electrochemical Peak Potentials; Scan Rate 0.1 V s^{-1}; V versus NHE

	C			Pt			Au			Hg		
Solvent	$E_{p,c}$	$E_{p,a}$	ΔE_p	$E_{p,c}$	$E_{p,a}$	ΔE_p	$E_{p,c}$	$E_{p,a}$	ΔE_p	$E_{p,c}$	$E_{p,a}$	ΔE_p
Me_2SO	−0.62	−0.46	0.16	−0.87	−0.70	0.17	−0.66	−0.44	0.22			
DMF	−0.71	−0.53	0.18	−0.72	−0.52	0.20	−0.74	−0.54	0.20			
py	−0.76	−0.52	0.24	−0.79	−0.52	0.27	−0.76	−0.51	0.25			
MeCN	−0.71	−0.54	0.71	−0.94	−0.36	0.56	−0.88	−0.41	0.47	−0.88	−0.37	0.51

B. Formal Reduction Potentials for the $O_2/O_2^{\cdot-}$ couple (molar concentrations for O_2 and $O_2^{\cdot-}$)

$E^{\circ\prime}$ (V) versus NHE			
H_2O	−0.16	py	−0.64
Me_2SO	−0.54	MeCN	−0.63
DMF	−0.62		

peak separation (ΔE_p) varies with solvent and with electrode material, the median potentials $(E_{p,c} + E_{p,a})/2$ are essentially independent of electrode material and provide a reasonable measure for the formal reduction potential ($E^{\circ\prime}$) for the $O_2/O_2^{\cdot-}$ couple. The average values for $E^{\circ\prime}$ are summarized in Table 9.10 together with the value for an aqueous solvent at pH 7.

Reference to Table 9.9 confirms that the reduction potential for the $O_2/O_2^{\cdot-}$ couple shifts to more negative values as the solvating properties of the media decrease. The heat of hydration ($-\Delta H_{aq}$) for gaseous $O_2^{\cdot-}$ is 418 kJ mol^{-1} (100 kcal mol^{-1}), which is consistent with the unique strong solvation of anions by water. Hence, if the $E^{\circ\prime}_{O_2/O_2^{\cdot-}}$ values for the $O_2/O_2^{\cdot-}$ couple are affected primarily by the degree of solvation of $O_2^{\cdot-}$ (i.e., the solvation energy for O_2 is assumed to be small and about the same for different solvents), the relative solvation energies for $O_2^{\cdot-}$ will be H_2O >> Me_2SO > DMF > py ~ MeCN.

The variation in the peak-separation values (ΔE_p) for the cyclic voltammetric data of Table 9.9 may be interpreted in terms of heterogeneous electron-transfer kinetics, but the most reasonable explanation is uncompensated resistance (especially for py and MeCN) and surface reactions (especially for the metal electrodes).

Protic and Electrophilic Substrates. When dioxygen is reduced in the presence of an equimolar concentration of a strong acid ($HClO_4$) in dimethylformamide (DMF) (Figure 9.4), a new irreversible peak occurs at −0.13 V versus SCE (+0.11 V vs. NHE) in addition to the regular quasireversible couple at −0.86 V versus SCE.[2] [In the absence of O_2, strong acids ($HClO_4$) in DMF exhibit a reversible one-electron couple at −0.37 V.] The peak height at −0.13 V increases linearly as the concentration of $HClO_4$ is increased up to a mole ratio of 4 : 1 relative to O_2 (equivalent to a two-electron reduction). This increase in peak height is at the expense of the reversible couple at −0.86 V. The addition of excess phenol to a dioxygen/DMF solution causes the one-electron process to become a two-electron reduction to yield HOOH as the major product (Figure 9.4); the reverse scan indicates that phenoxide ion is formed during the reduction of O_2. Similar effects are observed when other moderately weak acids are present in excess, but the addition of a fivefold excess (relative to O_2 concentration) of H_2O or *n*-butanol does not affect the electrochemistry of O_2 in DMF.

The cyclic voltammogram of Figure 9.4 indicates that a new process occurs when O_2 is reduced in the presence of a strong acid. For example, in DMF an electron is first transferred to H_3O^+ to give $H\cdot$, which couples with O_2:

$$O_2 + (DMF)H_3O^+ + e^- \longrightarrow HOO\cdot + DMF + H_2O$$

$$E_{p,c}, \ -0.11 \text{ V vs. NHE} \qquad (9.33)$$

The $HOO\cdot$ species rapidly disproportionates to HOOH, which is not electroactive in DMF at potentials less negative than that for the $O_2/O_2^{\cdot-}$ couple (−0.64

TABLE 9.10 Examples of Organic Substrates that Enhance the Voltammetric Peak Height for the Reduction of O_2 via Post-Electron-Transfer Reactions by $O_2^{\bar{\cdot}}$ in DMF (Source of $O_2^{\bar{\cdot}}$; $O_2 + e^- \longrightarrow O_2^{\bar{\cdot}}$)

	MeCl
Rate-controlling step	$MeCl + O_2^{\bar{\cdot}} \longrightarrow MeOO\cdot + Cl^-$; k, 80 $M^{-1}\ s^{-1}$
Overall stoichiometry	$2\ MeCl + 2\ O_2^{\bar{\cdot}} \longrightarrow MeOOMe + 2\ Cl^- + O_2$
Electron stoichiometry (in the limit as $\nu \longrightarrow 0$)[a]	$2\ MeCl + O_2 + 2\ e^- \longrightarrow MeOOMe + 2\ Cl^-$; $2\ e^-/O_2$
	CCl_4
Rate-controlling step	$CCl_4 + O_2^{\bar{\cdot}} \longrightarrow Cl_3COO\cdot + Cl^-$; k, 1300 $M^{-1}\ s^{-1}$
Overall stoichiometry	$CCl_4 + 6\ O_2^{\bar{\cdot}} \longrightarrow CO_4^{2-} + 4\ Cl^- + 4\ O_2$
Electron stoichiometry (in the limit as $\nu \longrightarrow 0$)	$CCl_4 + 2\ O_2 + 6\ e^- \longrightarrow CO_4^{2-} + 4\ Cl^-$; $3\ e^-/O_2$
	MeC(O)OPh
Rate-controlling step	$MeC(O)OPh + O_2^{\bar{\cdot}} \longrightarrow MeC(O)OO\cdot + PhO^-$; k, 160 $M^{-1}\ s^{-1}$
Overall stoichiometry	$2\ MeC(O)OPh + 4\ O_2^{\bar{\cdot}} \longrightarrow 2\ MeC(O)O^- + 2\ PhO^- + 3\ O_2$
Electron stoichiometry (in the limit as $\nu \longrightarrow 0$)	$2\ MeC(O)OPh + O_2 + 4\ e^- \longrightarrow 2\ MeC(O)O^- + 2\ PhO^-$; $4\ e^-/O_2$

[a] ν represents the voltammetric scan rate at a platinum electrode (0.23 cm^2) for the reduction of O_2 in the presence of excess organic substrate (Ref. 2).

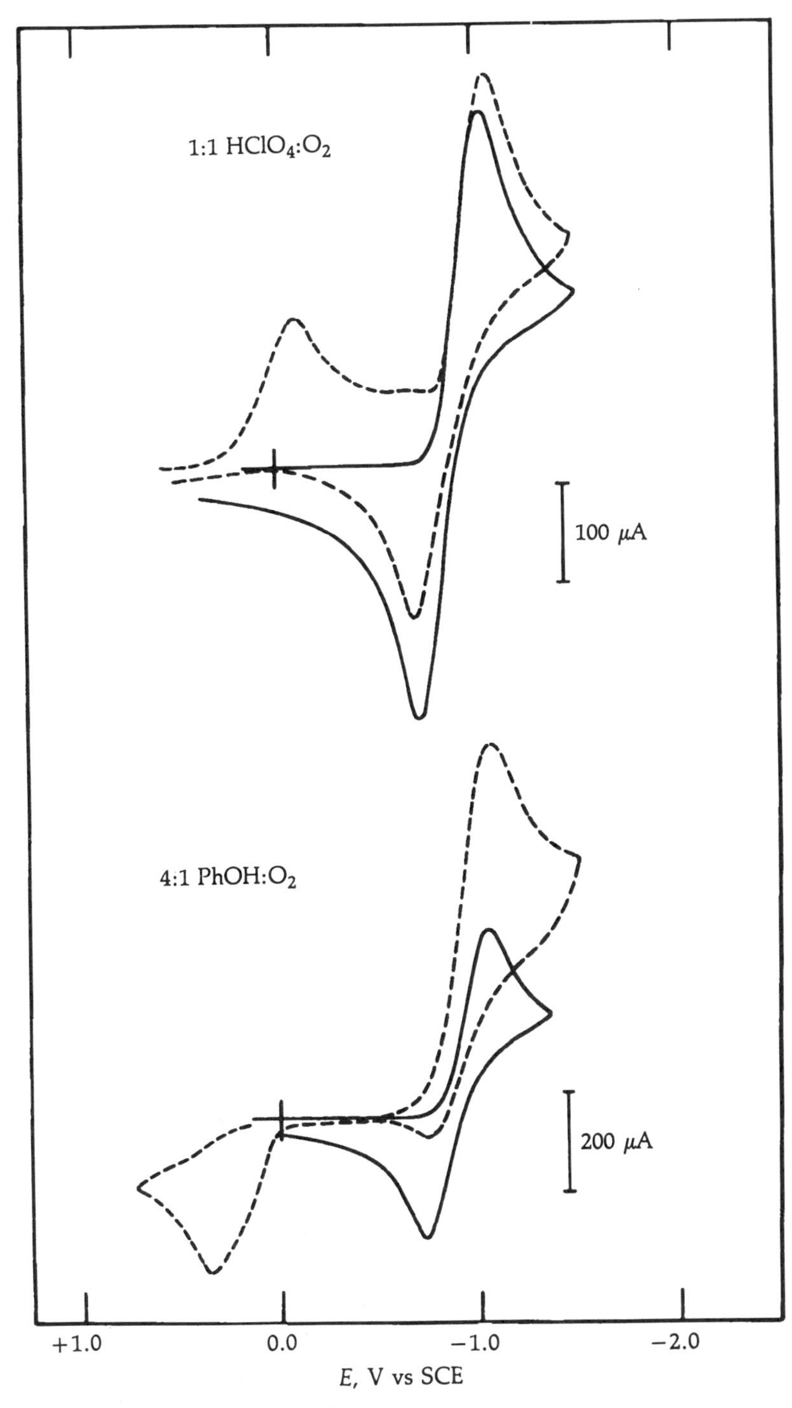

Figure 9.4 Cyclic voltammograms for 4.8 mM O_2 (at 1 atm) in DMF (0.1 M TEAP) (solid lines) and in the presence of an equimolar concentration of $HClO_4$ and in the presence of a 4 : 1 mole ratio of phenol (PhOH) (dashed lines); platinum electrode, area 0.23 cm^{-2}; scan rate 0.1 V s^{-1}.

V vs. NHE). A reasonable pathway for the facile disproportionation is the formation of a dimer with subsequent dissociation to dioxygen and peroxide:[16]

$$HO_2\cdot + HO_2\cdot \longrightarrow [\text{cyclic dimer}] \longrightarrow HOOH + O_2,$$
$$k > 10^4\ M^{-1}\ s^{-1} \tag{9.34}$$

With adequate hydronium-ion fluxes, $HOO\cdot$ is transformed to HOOH via a second electron–proton transfer ($HOO\cdot + H_3O^+ + e^- \longrightarrow HOOH$ $E^{\circ\prime}$, +0.8 V vs. NHE). Achievement of a full two-electron peak height for the process of Eq. (9.33) requires a ratio of at least four $(DMF)H_3O^+$ ions per O_2 molecule in DMF. This results because $(H_3O)ClO_4$ forms $(DMF)H_3O^+$, and the diffusion coefficient for the latter is much smaller than that for O_2. Furthermore, the flux of $(DMF)H_3O^+$ to the electrode must be twice that for O_2 to achieve the second cycle of Eq. (9.33) with the products of Eq. (9.34).

Because superoxide ion is an effective nucleophile, the presence of electrophilic substrates with effective leaving groups (alkyl halides, acyl halides, esters, and anhydrides) in a dioxygen solution causes an apparent increase in the electron stoichiometry for the cathodic voltammogram of O_2 and a decrease in the reverse peak height (these substrates are not electroactive within the voltage range for O_2 reduction). The extent of this effect is dependent on (1) the concentration of the substrate, (2) the stoichiometry for the substrate/$O_2^{\cdot-}$ reaction, (3) the rate constant for the latter, and (4) the voltammetric scan rate. Illustrative examples are summarized in Table 9.10. When an excess of a reactive substrate such as CCl_4 is present, the peak height for O_2 reduction is more than doubled at a scan rate of 0.1 V s^{-1} (the process is totally irreversible); the cyclic voltammogram is similar in appearance to that for the 4:1 PhOH/O_2 system (Figure 9.4).

The electrochemical reduction of O_2 in aprotic media is dramatically changed by the presence of electroinactive metal cations.[2] Figure 9.5 illustrates the effect of a fivefold excess of $[Zn^{II}(OH_2)_6](ClO_4)_2$, $[Zn^{II}(\text{dimethylurea})_6](ClO_4)_2$, and $[Zn^{II}(bpy)_3](ClO_4)_2$ on the cyclic voltammetry of O_2 in DMF at a platinum electrode. Prior to each reductive scan the electrode has been repolished; a second scan yields a much reduced peak current. In the presence of an excess concentration of Zn(II) cations the reduction of O_2 is a totally irreversible process, and the electrodes (Pt, Au, and C) are passivated after the initial negative scan.

For slow scan rates in the presence of 10- to 30-fold excess concentrations (relative to the O_2 concentration) of several divalent metal cations, the potentials for O_2 reduction are shifted to more positive values and the peak currents are increased by a factor of 2 or 4. Table 9.11 summarizes the results for such experiments with platinum electrodes in Me_2SO (0.1 M TEAP). Apparently these cations act as Brønsted acids relative to the O_2 reduction process and thereby cause it to be shifted to more positive potentials in the following order:

$$Fe^{II}(OH_2)_6^{2+} > Mn^{II}(OH_2)_6^{2+} > Co^{II}(OH_2)_6^{2+} \sim Zn^{II}(OH_2)_6^{2+}$$
$$\sim Cd^{II}(OH_2)_6^{2+} \sim Li^{I}(OH_2)_4^{+} \sim TEA^{+}$$

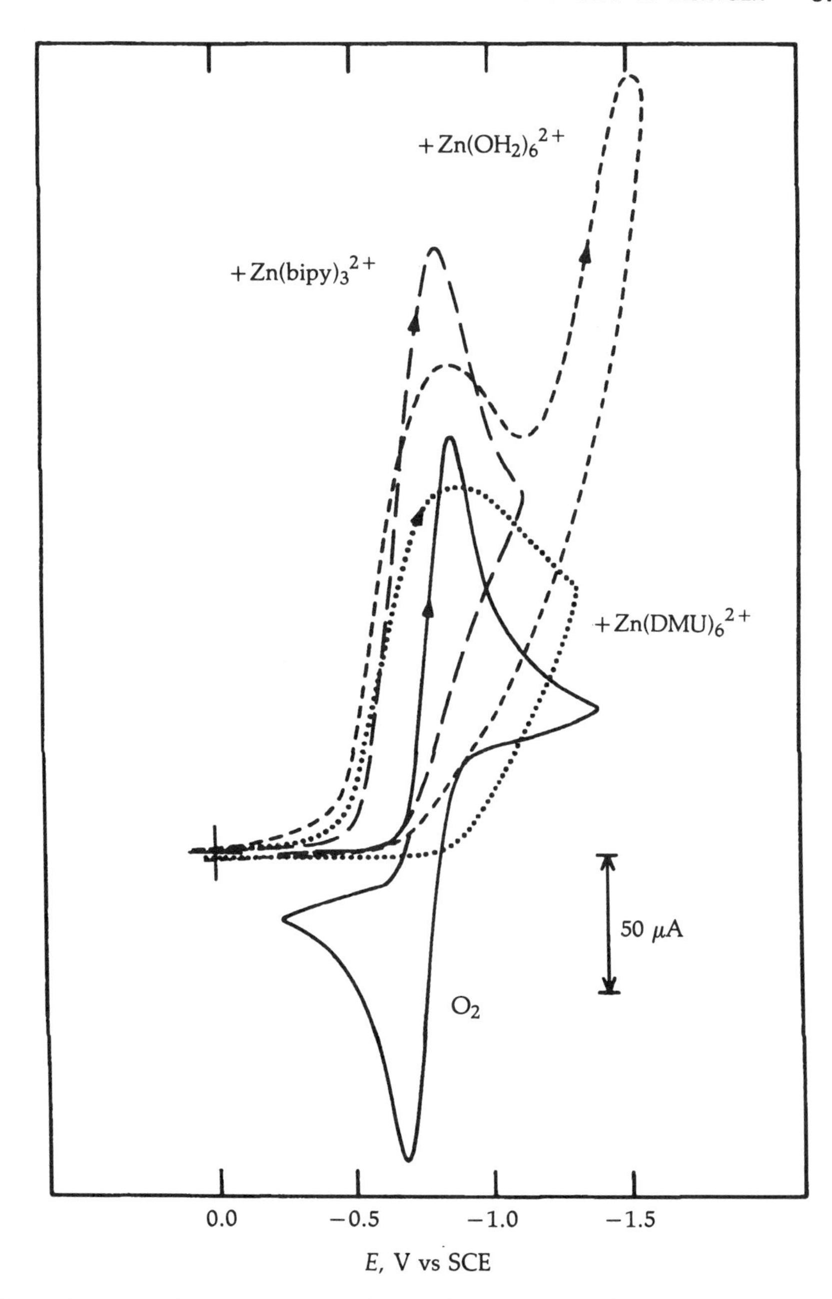

Figure 9.5 Cyclic voltammograms in Me_2SO (0.1 M TEAP) of 2.1 mM O_2 (at 1 atm) (solid line) and in the presence of 50 mM $Zn^{II}(DMU)_6(ClO_4)_2$ (DMU, dimethylurea) (dotted line); 50 mM $Zn^{II}(OH_2)_6(ClO_4)_2$ (short dashed line); and 50 mM $Zn^{II}(bpy)_3(ClO_4)_2$ (bipy, 2,2′-bipyridine) (long dashed line); platinum electrode, area 0.23 cm^2; scan rate 0.1 V s^{-1}.

TABLE 9.11 Voltammetric Reduction of 2.1 mM O_2 in the Presence of a 10- to 30-Fold Excess (Relative to O_2 Concentration) of Metal Cations in Me_2SO (0.1 M TEAP)[a] at a Pt Electrode (0.02 V s^{-1})

Metal[b]	$E_{p,c}$ (V) versus NHE	$n(e^-/O_2)$
—	−0.56	1
$Li^I(OH_2)_4^+$	−0.55	1
$Zn^{II}(OH_2)_6^{2+}$	−0.43	2
$Cd^{II}(OH_2)_6^{2+}$	−0.43	2
$Fe^{II}(OH_2)_6^{2+}$	−0.28	2
$Mn^{II}(OH_2)_6^{2+}$	−0.37	4
$Co^{II}(OH_2)_6^{2+}$	−0.42	4

[a]Tetraethylammonium perchlorate.
[b]Added as the hydrated perchlorate salts (Ref. 2).

where TEA^+ = tetraethylammonium cation from the supporting electrolyte, TEA-(ClO_4). On the basis of the electron stoichiometries of Table 9.11, the HOO· product disproportionates to HOOH and O_2:

$$Zn^{II}(OH_2)_6^{2+} + O_2 + e^- \longrightarrow HOO\cdot + (H_2O)_5^+Zn^{II}(OH) \tag{9.35}$$
$$HOO\cdot \xrightarrow{HOO\cdot} HOOH + O_2$$

The $Mn^{II}(OH_2)_6^{2+}$ and the $Co^{II}(OH_2)_6^{2+}$ complexes catalyze the rapid decomposition of HOOH to give the metal oxides and O_2, and an overall stoichiometry of four electrons per O_2.

Most of these metal–oxygen intermediates are expected to be reactive toward organic substrates and electrode surfaces. Hence, the presence of metal cations enhances the electron stoichiometry for the reduction of O_2, but frequently passivates the electrode surface. Thus, the formation of $(H_2O)_4Zn^{II}(O_2)$ on the surface of a platinum electrode probably initiates a metathesis reaction:

$$(H_2O)_4Zn^{II}(O_2)(s) + Pt^{II}O(s) \longrightarrow Pt^{II}(O_2)(s) + Zn^{II}O(s) + 4\ H_2O \tag{9.36}$$

A similar process may occur between the electrode surface and the HOOH that is produced via the reaction of eq. (9.35):

$$Pt^{II}O + HOOH \longrightarrow Pt^{II}(O_2) + H_2O \tag{9.37}$$

In the case of Mn(II) ions, the presence of HOOH from the proton-induced disproportionation of $O_2^{\overline{\cdot}}$ [Eq. (9.34)] can lead to the production of HO· radicals:

$$(H_2O)_4^{2+}Mn^{II} + HOOH \longrightarrow (H_2O)_4^{2+}Mn^{III}(OH) + HO\cdot \quad (9.38)$$

$$Pt(s) + HO\cdot \longrightarrow [\cdot Pt^{I}(OH)] \xrightarrow{HO\cdot} Pt^{II}O(s) + H_2O \quad (9.39)$$

Because the weakest solvation of $O_2^{\overline{\cdot}}$ occurs in MeCN (of the solvents that have been considered in Table 9.9), superoxide ion should exhibit its maximum reactivity in this solvent. The peak separations for the $O_2/O_2^{\overline{\cdot}}$ cyclic voltammograms with metal electrodes strongly support this proposition (Figure 9.6 and Table 9.9). To account for the extreme peak separation for the O_2 couple at platinum in MeCN, a reaction sequence that involves the formation of platinum peroxides is proposed:

$$Pt^{II}O(Pt)(s) + 2\ O_2^{\overline{\cdot}} \underset{2\,HO^-}{\xrightarrow{H_2O}} 2\ Pt^{II}(O_2)(s) \underset{H_2O}{\xrightarrow{2\,Pt^{II}O}} 2\ Pt^{II}(OH)OOPt^{II}(OH)(s) \quad (9.40)$$

$$Pt^{II}(OH)OOPt^{II}(OH)(s) + 2\ HO^- \longrightarrow 2\ Pt^{II}O(s) + O_2 + 2\ H_2O + 2\ e^- \quad (9.41)$$

The latter process would be expected to occur at a potential more positive than that for the oxidation of $O_2^{\overline{\cdot}}$. Processes similar to the reactions of Eqs. (9.40) and (9.41) have been observed with the electrochemical reduction of HOOH at mercury electrodes in DMF and MeCN. The almost reversible cyclic voltammogram for O_2 at a glassy-carbon electrode in MeCN indicates that specific reactions occur between metal electrodes and $O_2^{\overline{\cdot}}$ in a poorly solvating medium for anions, such as MeCN.

Electrode Material Effects. Figure 9.6 illustrates the cyclic voltammograms for the reduction of O_2 in MeCN (0.1 *M* TEAP) at glassy carbon, platinum, gold, and mercury electrodes.[3] The peak potentials and peak separations for the reduction of O_2 and reoxidation of $O_2^{\overline{\cdot}}$ with these electrodes in MeCN, Me_2SO, DMF, and py are summarized in Table 9.9. Clearly, the larger peak separations (ΔE_p) for the metal electrodes, especially in MeCN, indicate that the apparent irreversibility for the $O_2/O_2^{\overline{\cdot}}$ couple is due to the reaction of $O_2^{\overline{\cdot}}$, or its disproportionation product (HOO^-), with the metal surface. This effect is largest in MeCN because it is the poorest solvating agent for $O_2^{\overline{\cdot}}$, which is equivalent to minimizing its deactivation.

The significant effects of media and electrode material on the reduction potentials and voltammetric peak currents for dissolved dioxygen require substantial knowledge of the media and appropriate calibration of the electrochem-

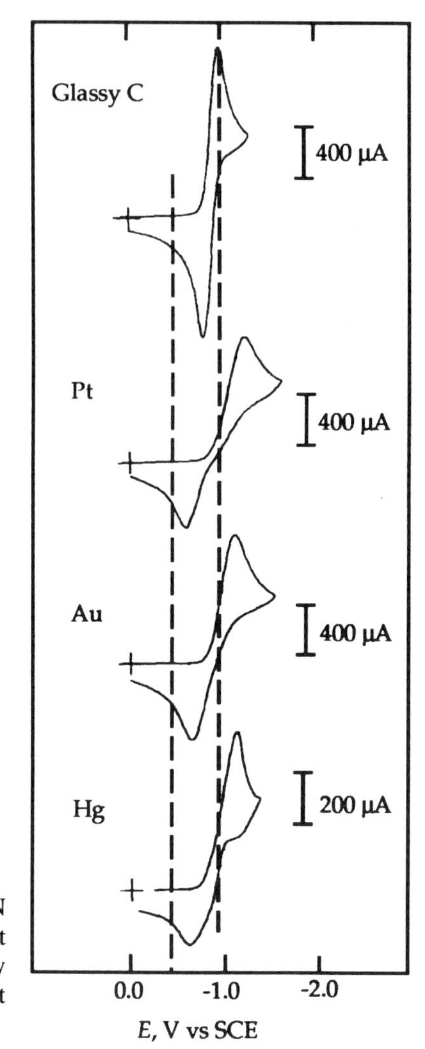

Figure 9.6 Cyclic voltammograms in MeCN (0.1 M TEAP) of 8.1 mM O_2 (at 1 atm) at vitreous carbon, platinum, gold, and mercury electrodes; electrode areas, 0.74 cm^2 (except mercury, 0.3 cm^2); scan rate 0.1 V s^{-1}.

ical electrode system before identification and quantitative determinations of O_2 are possible. Some of the interferences (e.g., CH_3Cl, $HCCl_3$, and CCl_4) may have substantial permeability through the membranes of the polarographic oxygen-membrane sensors (Clark electrode), which will cause substantial positive errors. However, these problems can be turned to an advantage. The enhanced reduction currents for the reduction of O_2 (at 1 atm) in the presence of alkyl halides, esters, and acyl halides can be used for their specific determination (with appropriate calibration).

Transition-Metal Complexes (ML_x); O_2 Reduction Catalysts. Figure 9.7 illustrates the cyclic voltammograms for the reduction and oxidation of $Fe^{III}Cl_3$,

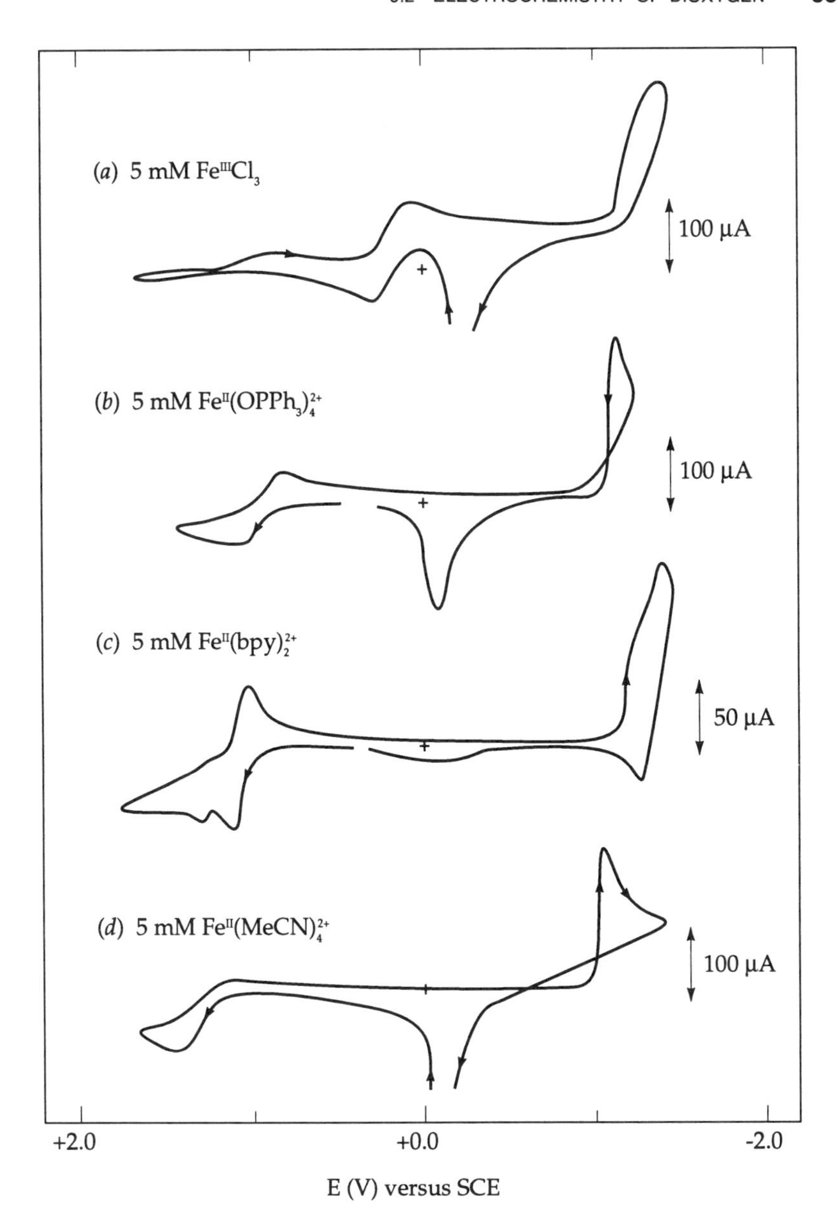

Figure 9.7 Cyclic voltammograms in MeCN [0.1 M $(Et_4N)ClO_4$] for (*a*) 5 mM $Fe^{III}Cl_3$, (*b*) 5 mM $Fe^{II}(OPPh_3)_4^{2+}$, (*c*) 5 mM $Fe^{II}(bpy)_2^{2+}$, and (*d*) 5 mM $Fe^{II}(MeCN)_4^{2+}$. Conditions: glassy-carbon electrode (area 0.09 cm^2); scan rate 0.1 V s^{-1}; SCE versus NHE, +0.244 V.

$Fe^{II}(OPPh_3)_4^{2+}$, $Fe^{II}(bpy)_2^{2+}$, and $Fe^{II}(MeCN)_4^{2+}$ in acetonitrile (MeCN) under an argon atmosphere. When dioxygen (O_2) (1 atm, 8.1 mM; $E_{1/2}$, −0.87 V vs. SCE in MeCN) is present, these iron complexes each exhibit a new irreversible reduction peak at potentials that are significantly less negative than those for the iron(II/I) couples and the $O_2/O_2^{\overline{\cdot}}$ couple (Figure 9.8).[29]

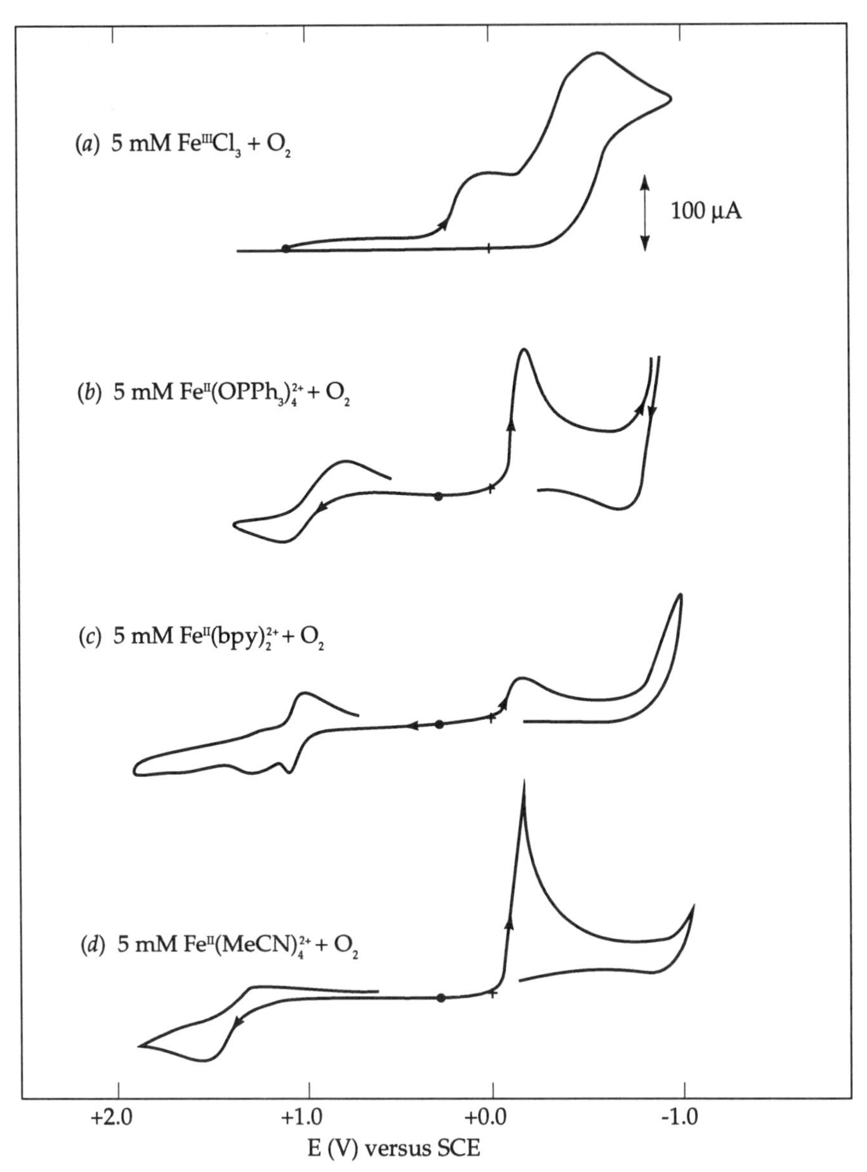

Figure 9.8 Cyclic voltammograms in MeCN [0.1 M $(Et_4N)ClO_4$] with 1 atm O_2 (8.1 mM) for (*a*) 5 mM $Fe^{III}Cl_3$, (*b*) 5 mM $Fe^{II}(OPPh_3)_4^{2+}$, (*c*) 5 mM $Fe^{II}(bpy)_2^{2+}$, and (*d*) 5 mM $Fe^{II}(MeCN)_4^{2+}$.

Within a 2:1 pyridine/acetic acid [$(py)_2HOAc$] solution matrix the $Fe^{II}(PA)_2$ complex (PAH; picolinic acid, 2-carboxylpyridine) is reversibly oxidized at +0.2 V versus SCE and is not reduced before the solvent (−0.8 V).[30] In the presence of O_2 [1 atm, 3.4 mM; $E_{p,c}$, −0.75 V in $(py)_2HOAc$] the $Fe^{II}(PA)_2$ complex exhibits a new two-electron reduction at −0.2 V that produces HOOH. The latter, in turn, reacts with excess $Fe^{II}(PA)_2$ to produce two $(PA)_2Fe^{III}OH$ ($E_{1/2}$, +0.2 V) via Fenton chemistry (k, 2×10^3 M^{-1} s^{-1}).[30,31] The 1:1 combination of $Fe^{II}(PA)_2$ and *t*-BuOOH undergoes a similar autoxidation to give $(PA)_2Fe^{III}OH$ ($E_{1/2}$, +0.2 V, Figure 9.9). The presence of *c*-C_6H_{12} causes some of the *t*-BuOOH to be consumed (Figure 9.9). In the presence of O_2 the 1:1 $Fe^{II}(PA)_2$/*t*-BuOOH combination undergoes an irreversible two-electron reduction ($E_{p,a}$, −0.1 V) that is an electrocatalytic cycle (O_2 + 2HOAc $\xrightarrow{2e^-}$ HOOH + $2AcO^-$) with up to 10 electrons per $Fe^{II}(PA)_2$ at slow scan rates (Figure 9.9*c*,*d*).

Figures 9.10 and 9.11 illustrate the cyclic voltammograms in MeCN for $Cu^I(bpy)_2^+$ and $Co^{II}(bpy)_2^{2+}$ in the absence and the presence of O_2. Again, new reduction peaks occur for the ML_x/O_2 combinations at less negative potentials. With $Co^{II}(bpy)_2^{2+}$ the presence of added base (H_2O, Et_3N, or py) influences the extent of the potential shift and whether the reduction is a one-electron/Co(II) or two-electron/Co(II) process (Figure 9.11).

Table 9.12 summarizes the reduction potentials for O_2, the metal complexes (ML_x), and their dioxygen adducts [$ML_x(O_2)$]. For each system the reduction potential is a measure of its electrophilicity, which is enhanced via formation of O_2 adducts. Reasonable reaction pathways are presented for each system that are in accord with the observed products and electron stoichiometry. Some would suggest that electron transfer to free O_2 occurs rather than to an adduct, with the shift to less negative potentials due to ligand ($O_2^{\cdot-}$) stabilization by the metal centers. However, the reduction peak potential for O_2 is unaffected by the presence of $Zn^{II}(bpy)_3^{2+}$ or $Zn^{II}(TPPCl_8)$,[32,33] which contrasts sharply with the effect for the comparable iron and copper complexes. The shift in the reduction potential for ML_x and $ML_x(O_2)$ (ΔE) is a measure of the ML_x—(O_2) bond-formation free energy {$-\Delta G_{BF} = \Delta E \times 96.5$ *k*J mol^{-1}; for Cl_2-$Fe^{III}(O_2)$, [−0.34–(−1.21)] × 96.5 = 84 kJ mol^{-1}.[34] Table 9.13 summarizes the estimated $-\Delta G_{BF}$ values for the several reduced $ML_x(O_2)$ adducts.

The addition of *t*-BuOOH to the $Cu^I(bpy)_2^+/O_2$ system induces the formation of a reactive intermediate that oxygenates saturated hydrocarbons:[35]

$$Cu^I(bpy)_2^+ + t\text{-BuOOH} + O_2 \xrightarrow[(MeCN)_4py]{}$$

$$[L_2Cu^{III}(OOBu)(O_2) + pyH^+] \xrightarrow[-0.4\ V]{2e^-} [L_2Cu^I(OOH)(OOBu)^-$$

$$\Big\downarrow 2\ c\text{-}C_6H_{12}$$

$$c\text{-}C_6H_{10}(O) + c\text{-}C_6H_{11}OH + Cu^IL_2^+ + H_2O \qquad (9.42)$$

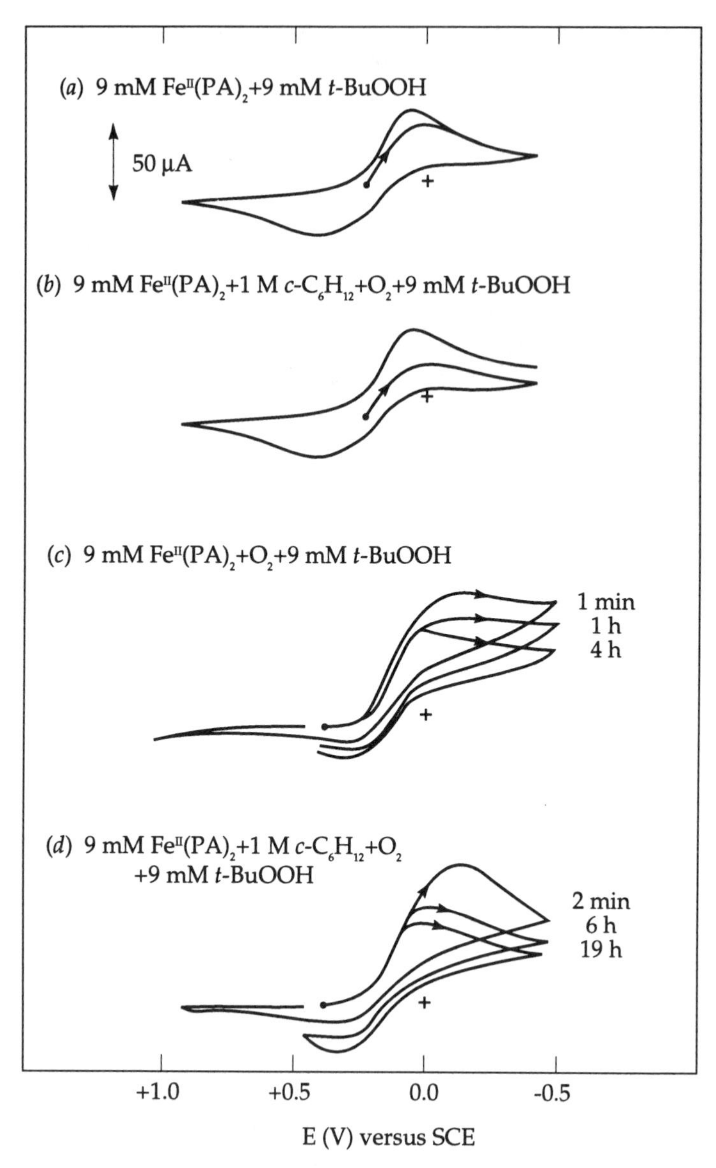

Figure 9.9 Cyclic voltammograms in $(py)_2HOAc$ [0.1 M $(Et_4N)ClO_4$] for the combination of (*a*) 9 mM $Fe^{II}(PA)_2$ and 9 mM *t*-BuOOH; (*b*) 9 mM $Fe^{II}(PA)_2$, 9 mM *t*-BuOOH, and 1 M *c*-C_6H_{12}; (*c*) 9 mM $Fe^{II}(PA)_2$ and 9 mM *t*-BuOOH in the presence of O_2 (1 atm, 3.4 mM); and (*d*) solution *c* plus 1 M *c*-C_6H_{12}. The duration of the reactions for solutions *c* and *d* is indicated on the voltammograms.

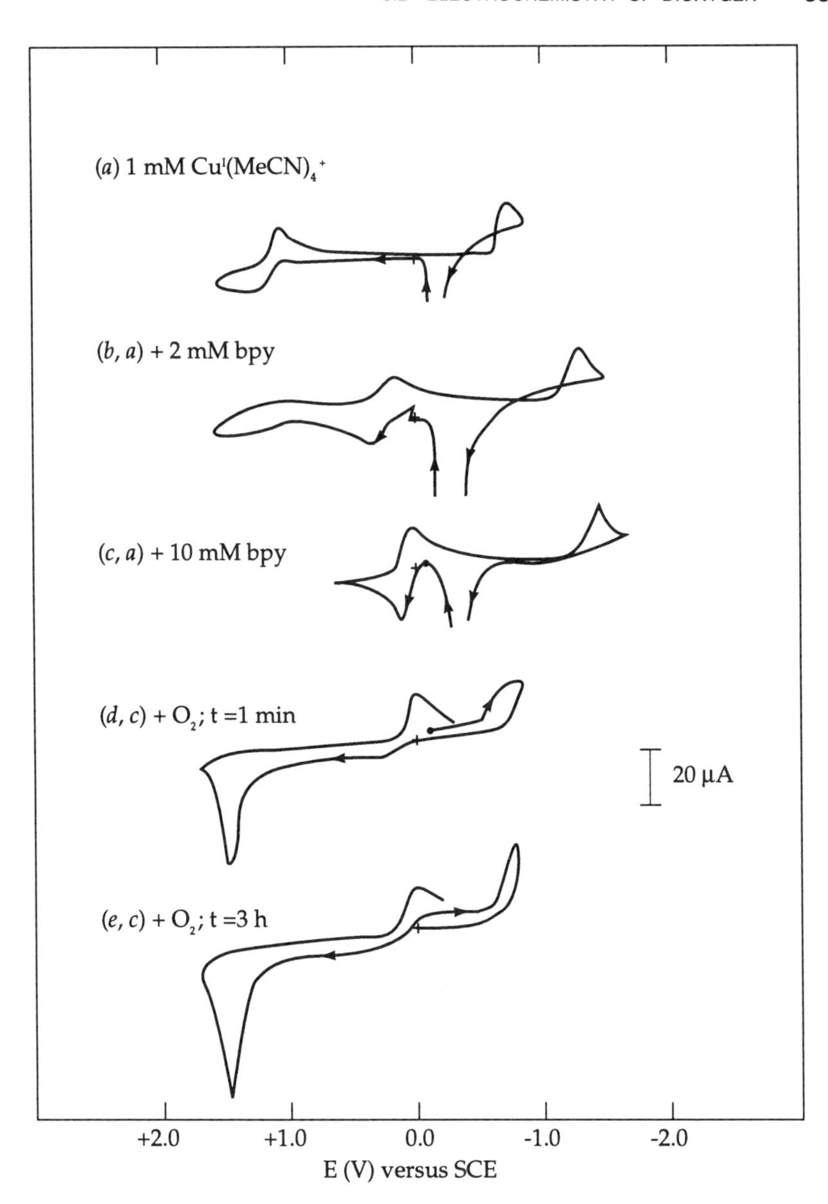

Figure 9.10 Cyclic voltammograms for (*a*) 1 mM $Cu^{I}(MeCN)_4^+$; (*b*, *a*) plus 2 mM bpy; (*c*, *a*) plus 10 mM bpy; (*d*, *c*) plus O_2, after 1 min; and (*e*, *d*) after 3 h in MeCN [0.1 M $(Et_4N)ClO_4$].

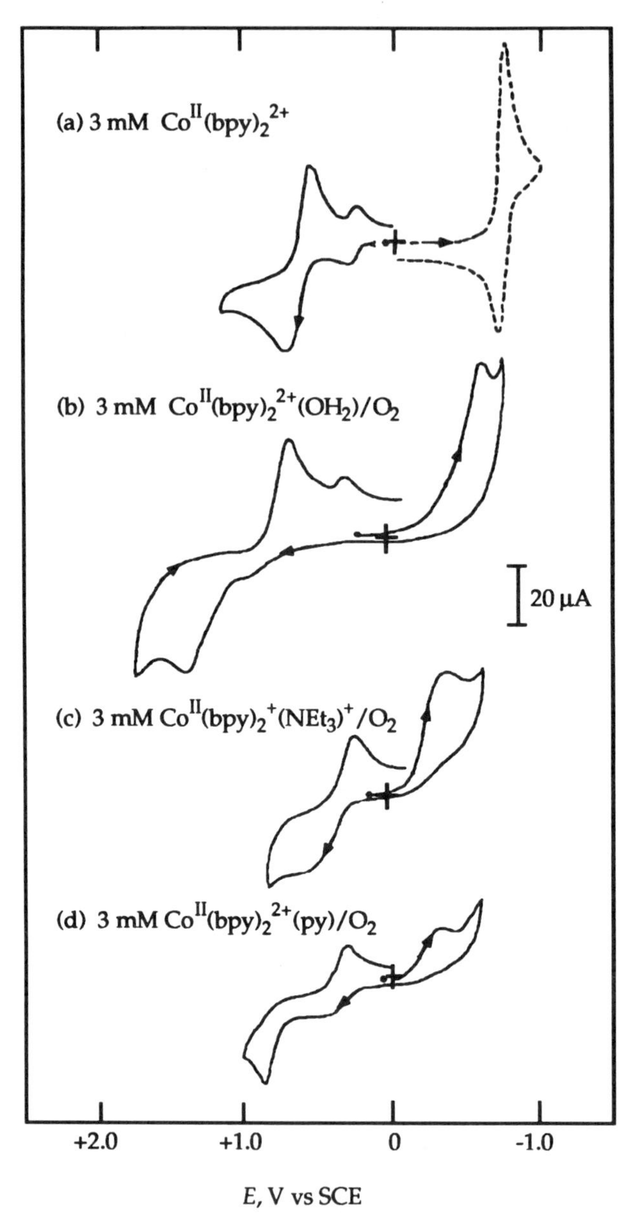

Figure 9.11 Cyclic voltammograms of (*a*) 3 mM $Co^{II}(bpy)_2^{2+}$ and (*b–d*) 3 mM $Co^{II}(bpy)_2^{2+}$ (B) complexes (B = H_2O, Et_3N, py) in the presence of 8.1 mM O_2 (1 atm) in MeCN (0.1 M TEAP). Scan rate 0.1 V s^{-1}; glassy-carbon working electrode (0.09 cm^2); SCE versus NHE, +0.244; (––) initial negative scan.

TABLE 9.12 Electrochemical Reduction Potentials for O_2, Metal Complexes [FeL_x, $Cu^{I}(bpy)_2$, $Co^{II}(bpy)_2(py)^{2+}$], and Their Combination in Aprotic Media

Reaction	$E_{1/2}$ (or $E_{p,c}$) (V) versus SCE
O_2, HOO·, and HOOH	
(1) $O_2 + H_3O^+ \xrightarrow[MeCN]{e^-} HOO\cdot + H_2O$	+0.40
(2) $O_2 \underset{MeCN}{\overset{e^-}{\rightleftarrows}} O_2^{\cdot-}$	−0.87
(3) $O_2 \xrightarrow[(py)_2HOAc]{e^-} HOO\cdot + AcO^-$	−0.75
(4) $HOO\cdot + H_3O^+ \xrightarrow[MeCN]{e^-} HOOH + H_2O$	+1.72
(5) $HOO\cdot + HOO\cdot \xrightarrow[MeCN]{e^-} HOOH + O_2^{\cdot-}$	+0.40
(6) $HOOH \xrightarrow[MeCN]{e^-} HOO^- + H\cdot$	−3.0
ML_x and ML_x/O_2 in MeCN	
(7) $Fe^{II}(Cl_2) \xrightarrow{e^-} Fe^{I}Cl_2^- \xrightarrow{e^-} Fe(s) + 2\,Cl^-$	−1.21
$\quad \updownarrow O_2$	
$[Cl_2Fe^{IV}(O_2)] \xrightarrow{e^-} Cl_2^-Fe^{III}(O_2) \xrightarrow[2\,H_2O]{e^-} Fe^{II}(OH)_2 + 2\,Cl^- + HOOH$	−0.34
(8) $Fe^{II}(OPPh_3)_4^{2+} \xrightarrow{e^-} Fe^{I}L_4^+ \xrightarrow{e^-} Fe(s) + 4\,L$	−1.17
$\quad \updownarrow O_2$	
$[L_2^{2+}Fe^{IV}(O_2)] \xrightarrow{e^-} L_4^+Fe^{III}(O_2) \xrightarrow[2\,H_2O]{e^-} Fe^{II}(OH)_2(s) + 4\,L + HOOH$	−0.15

TABLE 9.12 (*Continued*)

Reaction	$E_{1/2}$ (or $E_{p,c}$) (V) versus SCE
ML_x and ML_x/O_2 in MeCN	
(9) $Fe^{II}(bpy)_2^{2+} \underset{}{\overset{e^-}{\rightleftharpoons}} Fe^{I}L_2^+ \overset{e^-}{\rightleftharpoons} FeL_2$	−1.19
$\downarrow\uparrow O_2$	
$[L_2^{2+}Fe^{IV}(O_2)] \xrightarrow{e^-} L_2^+Fe^{III}(O_2) \xrightarrow[2\,H_2O]{Fe^{II}L_2^{2+},\,e^-} 2\,L_2^+Fe^{II}OH + HOOH$	−0.14
(10) $Cu^{I}(bpy)_2^+ \xrightarrow{e^-} Cu(s) + 2\,L$	−1.17
$\downarrow\uparrow O_2$	
$[L_2^+Cu^{III}(O_2)] \xrightarrow{e^-} L_2Cu^{I}OO\cdot \xrightarrow{Cu^{I}L_2^+,\,e^-} 2\,Cu^{II}O(s) + 4\,L$	−0.15
(11) $Cu^{I}(bpy)_2^+ + t\text{-BuOOH} \overset{py}{\rightleftharpoons} \underset{\mathbf{1}}{[L_2Cu^{I}OOBu + pyH^+]} \overset{O_2}{\rightleftharpoons} \mathbf{1}(O_2) \xrightarrow{2\,e^-} L_2Cu^{I}(OOBu)(OOH)^- + py$	−0.4
(12) $Co^{II}(bpy)_2^{2+}(py) \overset{e^-}{\rightleftharpoons} L_2^+Co^{I}(py)$	−0.94
$\uparrow\downarrow O_2$	
$[L_2^{2+}(py)Co^{III}OO\cdot] \xrightarrow{e^-} L_2^+Co^{III}(O_2) \xrightarrow[py]{H_2O} L_2^+Co^{III}(OH)(OOH)$	−0.57
$\downarrow$ py, $-2\,e^-$	+1.25
$L_2^{2+}Co^{III}(py)^+ + H_2O + O_2$	

	Reaction	E
(16)	$Co^{II}(TPPCl_8) \underset{(DMF)}{\overset{e^-}{\rightleftharpoons}} (Cl_8TPP)Co^-$	−0.75
	$\downarrow\uparrow O_2$	
	$[(TPP)Co^{III}OO\cdot] \xrightarrow[(DMF)]{e^-} (TPP)Co^{III}OO^-$	−0.62
(17)	$Fe^{II}(TPPCl_8) \underset{(DMF)}{\rightleftharpoons} (Cl_8TPP)Fe^-$	−1.04
	$\downarrow\uparrow O_2$	
	$[(Cl_8TPP)Fe^{IV}(O_2)] \xrightarrow[(DMF)]{e^-} (Cl_8TPP)Fe^{III}OO^-$	−0.65
	$Fe^{II}(PA)_2$, $Fe^{II}(PA)_2/t$-BuOOH, and $Fe^{II}(PA)_2/t$-BuOOH/O_2 in $(py)_2HOAc$	
(18)	$Fe^{II}(PA)_2 \xrightarrow{e^-} NR$	< −1.1
	$\downarrow\uparrow O_2$	
	$[L_2Fe^{IV}(O_2)] \xrightarrow[HOAc]{e^-} L_2Fe^{III}OOH + AcO^-$	−0.2
	$L_2Fe^{III}OOH \xrightarrow[HOAc]{e^-} HOOH + Fe^{II}L_2 + 2\,AcO^-$	
(19)	$Fe^{II}(PA)_2 + t\text{-BuOOH} \rightleftharpoons [L_2^-Fe^{II}OOBu + pyH^+]$ (**1**)	
	$[L_2^-Fe^{II}OOBu + pyH^+] \xrightarrow{Fe^{II}L_x} 2\,L_2Fe^{III}OH \overset{2e^-}{\rightleftharpoons} 2\,Fe^{II}L_2 + 2\,H_2O + 2\,AcO^-$	−0.2
	1 $\downarrow\uparrow O_2$	
	$\mathbf{1}(O_2) \xrightarrow[2\,HOAc]{2e^-} Fe^{II}L_2 + HOOH + t\text{-BuOOH}$	−0.1

TABLE 9.13 Apparent $L_xM(O_2)$ Bond Energies ($-\Delta G_{BF}$)

$L_xM(O_2)$	$-\Delta G_{BF}$ (kJ mol^{-1})[a]
$Cl_2^-Fe^{III}(O_2)$ (side-on, η^2-O–O)	84
$(Ph_3PO)_4^+Fe^{III}(O_2)$ (side-on, η^2-O–O)	100
$(bpy)_2^+Fe^{III}(O_2)$ (side-on, η^2-O–O)	100
$(bpy)_2Cu^I—OO\cdot$	63
$(bpy)_2^+Co^{III}(O_2)$ (side-on, η^2-O–O)	38
$(Cl_8TPP)Co^{III}—OO^-$	13
$(Cl_8TPP)Fe^{III}—OO^-$	38

[a] $\Delta G_{BF} = [(E_{1/2})_{ML_x(O_2)} - (E_{1/2})_{ML_x}] \times 96.5$ kJ mol^{-1}; Ref. 34.

Similar activation occurs with the $Fe^{II}(PA)_2$ complex:[36]

$$Fe^{II}(PA)_2 + t\text{-BuOOH} + O_2 \xrightarrow[(py)_2HOAc]{} [L_2^-Fe^{IV}(OOBu)(O_2) + pyH^+] \xrightarrow{2\ c\text{-}C_6H_{12}} c\text{-}C_6H_{10}(O) + c\text{-}C_6H_{11}OH + Fe^{II}L_2 + H_2O \quad (9.43)$$

The stimulated electroreduction of O_2 by $Fe^{II}(PA)_2$ [Table 9.12, reactions (18) and (19)] can be used to induce the ketonization of methylenic carbons:[31,36]

$$O_2 + 2\ HOAc + 2\ e^- \xrightarrow[-0.2\,V]{Fe^{II}(PA)_2} HOOH + 2\ AcO^- \quad (9.44)$$

$$Fe^{II}(PA)_2 + HOOH + O_2 \longrightarrow$$

$$[L_2^{-}Fe^{III}OOH(O_2) + pyH^{+}] \xrightarrow[-0.1\ V]{2e^-} 2\ HOOH + Fe^{II}L_2 + 2\ AcO^-$$

$$\downarrow Fe^{II}L_2 \mid c\text{-}C_6H_{12} \qquad \xrightarrow[HOAc]{2e^-,\ +0.2\ V} 2\ Fe^{II}L_2 \qquad (9.45)$$

$$c\text{-}C_6H_{10}(O) + 2\ L_2Fe^{III}OH$$

$$\text{Overall: } c\text{-}C_6H_{12} + 2\ O_2 + 4\ HOAc + 4\ e^- \xrightarrow[-0.3\ V,\ 80\text{–}90\%]{Fe^{II}(PA)_2} c\text{-}C_6H_{10}(O) + 4\ AcO^- + 3\ H_2O \qquad (9.46)$$

The same chemistry can be accomplished with the electrons and acid replaced by reductases (PhNHNHPh or $PhCH_2SH$):[37]

$$c\text{-}C_6H_{12} + 2\ O_2 + 2\ PhNHNHPh\ (\text{or } 4\ PhCH_2SH) \xrightarrow[(py)_2HOAc]{Fe^{II}(DPAH)_2} c\text{-}C_6H_{10}(O) + 3\ H_2O + 2\ PhN{=}NPh\ (40\%\ \text{efficient})\ (\text{or } 2\ PhCH_2SSCH_2Ph,\ 53\%)$$

$$(DPAH_2 = 2,6\text{-dicarboxylpyridine}) \qquad (9.47)$$

The results in Table 9.12 confirm that transition-metal complexes can facilitate multielectron reduction pathways for O_2. Within the group of complexes the iron(II) systems are the most effective. Thus, $Fe^{II}(bpy)_2^{2+}$ in MeCN catalyzes a two-electron process at a potential that is 0.73 V less negative than the uncatalyzed one-electron process (−0.87 V):

$$O_2 + 2\ H_2O + 2\ e^- \xrightarrow[MeCN,\ -0.14\ V]{Fe^{II}(bpy)_2^{2+}} HOOH + 2\ L_2^{+}Fe^{II}OH \qquad (9.48)$$

$$\downarrow$$

$$Fe^{II}(bpy)_2^{2+} + Fe^{II}(OH)_2(s)$$

In practice, this system needs a buffer (e.g., py/pyH$^+$) to serve as a proton source and prevent precipitation of the catalyst. Although the $Fe^{II}(PA)_2$/$(py)_2HOAc$ system is similar to this,

$$O_2 + 2\,HOAc + 2\,e^- \xrightarrow[\substack{(py)_2HOAc \\ -0.2\ V}]{Fe^{II}(PA)_2} HOOH + 2\,AcO^- \tag{9.49}$$

$$HOOH + 2\,AcO^- \xrightarrow[k,\ 2\times 10^3\ M^{-1}\,s^{-1}]{2\,Fe^{II}(PA)_2} 2\,L_2Fe^{III}OH$$

$$2\,L_2Fe^{III}OH \xrightarrow[\substack{2\,e^- \\ +0.2\ V}]{HOAc} 2\,Fe^{II}L_2 + 2\,H_2O + 2\,AcO^-$$

the HOOH product rapidly autoxidizes $Fe^{II}(PA)_2$ to $(PA)_2Fe^{III}OH$, which, in turn, is reduced. Hence, the overall four-electron electrocatalytic reduction of O_2 at -0.2 V is 0.55 V less negative than the uncatalyzed overall two-electron process in this medium:

$$O_2 + 4\,HOAc + 4\,e^- \xrightarrow[(py)_2HOAc]{Fe^{II}(PA)_2} 2\,H_2O + 4\,AcO^- \tag{9.50}$$

$$-0.2\ V\ vs.\ SCE$$

However, Nature remains the best catalyst designer for the four-electron reduction of O_2 via *cytochrome-c oxidase* (+0.3 V vs. SCE). Even with this four-metal-centered protein the reduction potential is 0.3 V less positive than the thermodynamic limit (Table 9.3).

Chemical Reduction. In addition to its reduction by solvated electrons from pulse radiolysis, O_2 also can be reduced via electron transfer by several chemical agents. Because there are no apparent kinetic barriers to single-electron transfer to O_2, any reductant with a sufficiently negative one-electron reduction potential $[(E^{\circ\prime}_R)$ more negative than $E^{\circ\prime}_{O_2/O_2^-}]$ should reduce O_2 to superoxide ion. This is believed to be the basis for the in situ generation of superoxide ion in biological matrices by reduced ferridoxin $[(Fe_4S_4)^{2-}_{red}]$ and the semiquinone-like state of flavins ($Fl^-\cdot$) in the xanthine–xanthine oxidase system:[14,38]

$$(Fe_4S_4)^{2-}_{red} + O_2 \longrightarrow (Fe_4S_4)^-_{ox} + O_2^{\cdot-} \tag{9.51}$$

$$Fl^-\cdot + O_2 \longrightarrow Fl_{ox} + O_2^{\cdot-} \tag{9.52}$$

Also, the introduction of methyl viologen (MV^{2+}, paraquat) in biological systems results in the formation of its one-electron reduction product, $MV^+\cdot$, which mediates the production of $O_2^{\cdot-}$:

$$MV^+\cdot + O_2 \longrightarrow MV^{2+} + O_2^{\cdot-} \tag{9.53}$$

In turn, the latter species is believed to be the cytotoxic agent that results from the application of methyl viologen on plants engaged in photosynthesis (photon-driven electron transfer).[39]

Reduction of O_2 by Atom Transfer. In contrast to the preceding electron-transfer mechanism with inert electrodes, the reduction of dioxygen at freshly preoxidized metal electrodes is pH-dependent, yields negligible amounts of hydrogen peroxide, and occurs at the same potential as that for the metal oxide of the electrode material. A plausible explanation for these observations is represented by the reaction sequence

$$M^{II}(OH)_2 + 2\ H_3O^+ + 2\ e^- \longrightarrow M + 4\ H_2O \qquad (9.54a)$$

$$M^{II}(OH)_2 + 2\ e^- \longrightarrow M + 2\ HO^- \qquad (9.54b)$$

$$2\ M + O_2 + 2\ H_2O \longrightarrow 2\ M^{II}(OH)_2 \qquad (9.55)$$

In this mechanism, fresh metal oxide, $M^{II}(OH)_2$, is electrochemically reduced and reformed by a chemical oxidation of the surface by the dioxygen in the solution. The scheme satisfies the requirement that the reduction potential and rate of reaction be dependent on the electrode material, M, and pH. The net reaction is

$$O_2 + 2\ H_2O + 4\ e^- \longrightarrow 4\ HO^- \qquad (9.56)$$

or

$$O_2 + 4\ H_3O^+ + 4\ e^- \longrightarrow 6\ H_2O \qquad (9.57)$$

and does not involve formation of peroxide. Representative experimental reduction potentials at platinum are +0.05 V versus NHE at pH 2 and −0.65 V at pH 13, respectively (Figure 9.12). Although Eq. (9.54) clearly represents the rate-determining step in the reduction of dioxygen at preoxidized electrodes, some dioxygen may react directly at freshly reduced electrode sites to form trace amounts of peroxide via an electron-transfer mechanism.

Because the electron-transfer reduction of O_2 is a reversible one-electron process, the hope for an electron-transfer catalyst is futile. Atom-transfer catalysts [Eqs. (9.54) and (9.55)] can promote more extensive reduction and higher potentials (those that correspond to the reduction of the metal oxide).

The proposition of a one-electron mechanism for the electron-transfer reduction of dioxygen and the associated conclusions present significant ramifications relative to the development of improved fuel cells and metal–air batteries. To date the practical forms of such systems have used strongly acidic or basic electrolytes. Such solution conditions normally cause atom transfer to be the dominant reduction process for molecular oxygen at metal electrodes. Hence, the search for effective catalytic materials should be in this context rather than in terms of a one-electron-transfer process.

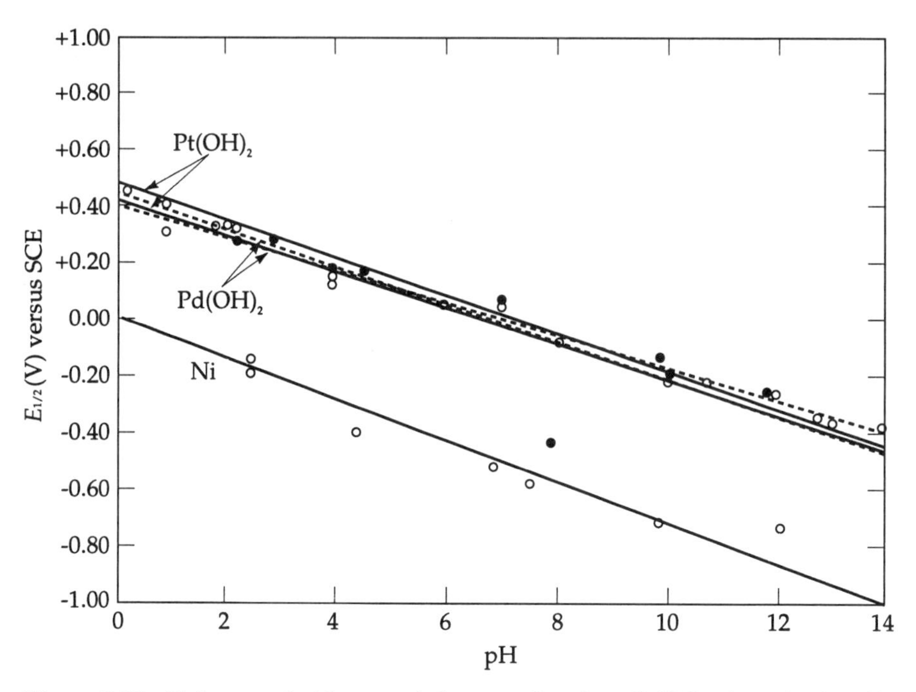

Figure 9.12 Voltammetric (slow-scan) data as a function of pH for the reduction (*a*) of freshly formed metal oxides [$Pt^{II}(OH)_2$ and $Pd^{II}(OH)_2$] (closed circles) and (*b*) of O_2 (1 atm) with preoxidized electrodes [$Pt^{II}(OH)_2$, $Pd^{II}(OH)_2$, $Ni^{II}(OH)_2$] (open circles). All solutions contained 0.1 M K_2SO_4; H_2SO_4 and NaOH were used to adjust the pH.

The third step of the scheme, whereby the metal is spontaneously oxidized to metal oxide [Eq. (9.55)], is the general equation for the corrosion of metals. This appears to involve the splitting of the oxygen–oxygen bond as oxygen atoms are transferred to the metal. Although the mechanism of this reaction is unclear, extensive thermodynamic data are available for the *oxygenation* of metals by molecular oxygen.

Concerted One-Electron Reductions. Reduction of O_2 in the presence of excess zinc cation [$Zn^{II}(bpy)_2^{2+}$], (tetraphenylporphinato)iron(III) ion [$(H_2O)^+Fe^{III}TPP$], and cuprous ion [$Cu^I(MeCN)_4^+$] results in formation of metal–dioxygen adducts. Figures 9.13 and 9.14 illustrate the cyclic voltammograms for ($Fe^{III}TPP$)Cl and [$Cu^I(MeCN)_4$]ClO_4, respectively, in the absence and presence of O_2. Reaction schemes for the three metal–O_2 systems are outlined:

(a) $$Zn^{II}(bpy)_2^{2+} + O_2 + e^- \xrightarrow{-0.7\,V\,vs.\,SCE} [Zn^{II}(bpy)_2OO\cdot]^+ \quad (9.58)$$

$$[Zn^{II}(bpy)_2OO\cdot]^+ + e^- \longrightarrow Zn^{II}(bpy)_2(O_2) \quad (9.59)$$

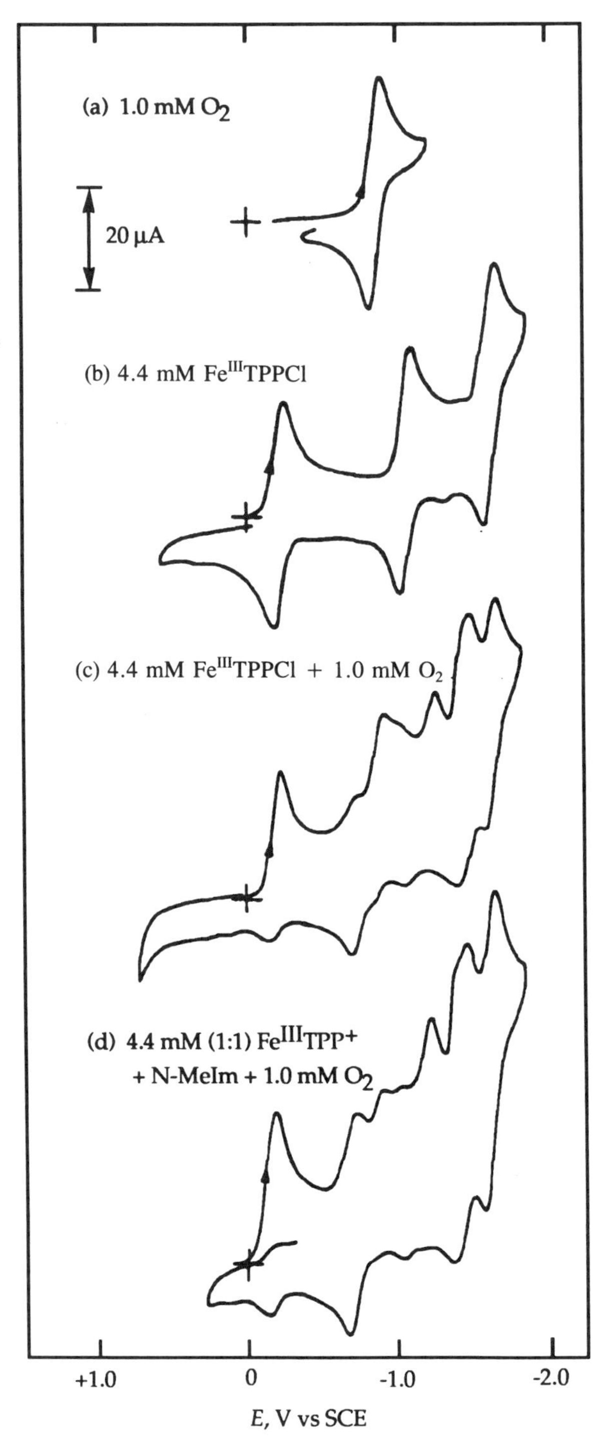

Figure 9.13 Cyclic voltammograms in DMF (0.1 M TEAP) of (*a*) O_2, (*b*) $Fe^{III}TPPCl$, (*c*) O_2 + $Fe^{III}TPPCl$, and (*d*) O_2 + $Fe^{III}TPP(N\text{-}MeIm)^+$. Measurements were made with a glassy-carbon electrode (area 0.11 cm^2) at a scan rate of 0.1 V s^{-1}; temperature 25°C.

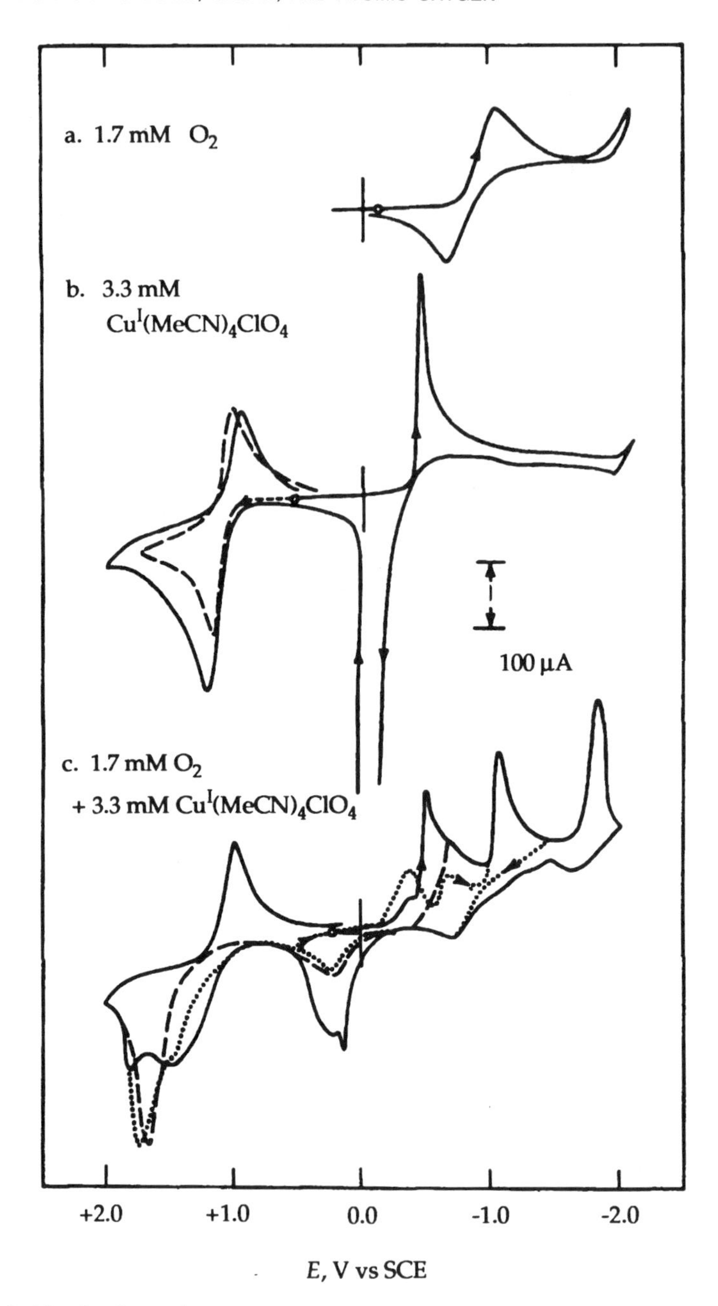

Figure 9.14 Cyclic voltammograms in MeCN (0.1 M TEAP) of (*a*) O_2, (*b*) $[Cu^I(MeCN)_4]ClO_4$, (*c*) O_2 + $[Cu^I(MeCN)_4]ClO_4$. Measurements were made with a platinum electrode (area 0.23 cm^2) at a scan rate of 0.1 V s^{-1}; temperature 25°C.

(b) $$Fe^{II}TPP + O_2 + e^- \xrightarrow[Me_2SO]{-0.7\,V\,vs.\,SCE} TPPFe^{III}OO^- \quad (9.60)$$

$$Fe^{III}TPP^+ + O_2 + 2\,e^- \xrightarrow[DMF,\,Me-Im]{-0.8\,V\,vs.\,SCE} TPPFe^{III}OO^- \xrightarrow[2x]{H_2O}$$

$$(TPP)Fe^{III}\text{-}O\text{-}Fe^{III}(TPP) + HOOH + O_2 + 2\,HO^- \quad (9.61)$$

(c) $$[Cu^I(MeCN)_4]ClO_4 + O_2 + e^- \xrightarrow{-0.5\,V} Cu^{II}(O_2)(s) \quad (9.62)$$

$$Cu^{II}(O_2)(s) + e^- \xrightarrow{-1.1\,V} Cu^I(O_2)^- \xrightarrow[H_2O]{e^-} Cu^{II}O(s) + 2\,HO^- \quad (9.63)$$

$$\cdot Cu^{II}O(s) + e^- \xrightarrow[H_2O]{-1.8\,V} Cu^IOH(s) + HO^- \quad (9.64)$$

$$Cu^{II}(O_2)(s) \xrightarrow{+1.9\,V} \cdot Cu^{II}(MeCN)_4^{2+} + O_2 + 2\,e^- \quad (9.65)$$

$$2\,Cu^I(MeCN)_4^+ + O_2 + 2\,e^- \longrightarrow Cu^IOOCu^I(s) \longrightarrow 2\,\cdot Cu^{II}O(s) \quad (9.66)$$

$$Cu^IOOCu^I(s) + 2\,HO^- \underset{-0.3\,V}{\overset{+0.2\,V}{\rightleftharpoons}} HOCu^{II}OOCu^{II}OH(s) + 2\,e^- \quad (9.67)$$

When the metal ion cofactor is electroactive at potentials equal or more positive than that for the one-electron reduction of O_2, then an effective two-electron reduction is accomplished by two sequential one-electron reductions [see Eq. (9.61)]. Such a process is similar to the favored O_2-activation mechanism for cytochrome *P*-450.[40]

Applications. At one time the main route for the production of hydrogen peroxide was the electrolytic reduction of O_2 in acidic media:

$$O_2 + 2\,H_3O^+ + 2\,e^- \longrightarrow HOOH + 2\,H_2O \quad (9.68)$$

Although current industrial production is via chemical reduction, the electrolysis process is convenient for small quantities of HOOH.

The main application of dioxygen electrochemistry has been and continues to be in the realm of analytical chemistry. The preceding sections of this chapter provide ample evidence that for carefully controlled conditions the current (i_{O_2}) for the reduction of O_2 is directly proportional to its concentration [which is directly proportional to its partial pressure (P_{O_2})]:

$$i_{O_2} = k[O_2] = k'P_{O_2} \quad (9.69)$$

Although this has been known since the very beginning of electrochemistry in the nineteenth century, commercial electroanalytical instrumentation has been available only for the past 35 years. The direct reduction of O_2 in aqueous solutions (especially biological and water treatment and sewage samples) is subject to positive interference by electroactive components of the sample (transition metals, oxygenated organic molecules, peroxides) and negative interference from organic matter, proteins, lipids, and carbohydrates in the sample, which film the electrode. All these problems were alleviated by separating the sample solution (or gas sample) and the electrode surface (and the electrolysis cell) with a gas-permeable membrane (0.025-mm polyethylene or Teflon). This ingenious advance was discovered in 1954 by Leland C. Clark, Jr., and is the basis for dioxygen sensors and analytical instrumentation for biological fluids, sewage and water-treatment plants, and gas mixtures.[41] Because the polymer membrane constitutes the limit for O_2 diffusion to the electrode, the Clark electrode gives a current response that is proportional to the partial pressure of O_2 (P_{O_2}). Thus, it will give an identical signal for a pure O_2 gas sample (1 atm, 44 mM) and an aqueous solution saturated with O_2 at 1 atm (O_2 concentration, 1 mM).

9.3 REDUCTION OF HOOH

The electrochemical reduction of HOOH in pyridine yields superoxide ion.[42] That a reduction process generates a species with a higher oxidation state than HOOH is surprising. The cyclic voltammogram for the reduction of HOOH in pyridine at a platinum working electrode exhibits a broad cathodic peak at −0.95 V versus NHE and anodic peaks for the reverse scan at −0.50 V and −0.15 V versus NHE. The latter is characteristic of electrolytically generated H_2. Controlled-potential reduction of HOOH in pyridine at −1.0 V versus NHE (with argon degassing) results in a solution that exhibits an anodic cyclic voltammogram that is characteristic of $O_2^{\bar{\cdot}}$. ESR studies of the reduced solution confirm the presence of superoxide ion.

The products and the observed electron stoichiometries for the electrochemical reduction of HOOH are consistent with a mechanism in which the primary step is a one-electron transfer

$$\mathrm{HOOH} + e^- \longrightarrow \mathrm{HOO}^- + \tfrac{1}{2}\,\mathrm{H_2} \qquad E^{\circ\prime},\ -1.0\ \mathrm{V\ vs.\ NHE} \tag{9.70}$$

followed by a chemical reaction with another hydrogen peroxide molecule:

$$\mathrm{HOOH} + \mathrm{HOO}^- \rightleftharpoons [\mathrm{HOOOH}] + \mathrm{HO}^- \xrightarrow[k,\,3\times10^{9}\,\mathrm{M^{-1}\,s^{-1}}]{\mathrm{py}} \frac{1}{n}\,[\mathrm{p\dot{y}(OH)}]_n + \mathrm{H_2O} + \mathrm{O_2^{\bar{\cdot}}} \tag{9.71}$$

The resultant of HO· is trapped by the pyridine solvent to yield a stable solution of $O_2^{\bar{\cdot}}$ [Eq. (9.71)].[39]

In acetonitrile there is no evidence of superoxide ion, either by ESR or by cyclic voltammetry, from the electrochemical reduction of HOOH. This can be explained by the slow rate of reaction of HO· with CH_3CN (k, 4×10^6 M^{-1} s^{-1}), which favors direct disproportionation:

$$[HOOOH] + HO^- \longrightarrow H_2O + [HOOO^-] \qquad (9.72)$$
$$[HOOO^-] \longrightarrow HO^- + O_2$$

The reduction of hydrogen peroxide in aqueous solution appears to be analogous to that in CH_3CN, with the mechanism represented by the reaction of Eq. (9.70) followed by the reactions of Eqs. (9.71) and (9.72). Thus, the reduction of HOOH yields H_2 and HOO^- initially, in a one-electron step. The final products are the result of the reaction of HOO^- and HOOH, and are analogous to those for the base-induced decomposition of HOOH.[43]

9.4 OXIDATION OF HOOH AND HOO^-

In acetonitrile HOOH is oxidized to O_2 via an electron-transfer/chemical/electron-transfer (ECE) mechanism:

$$HOOH \xrightarrow{-e^-} [HOOH]^{+\cdot} \xrightarrow{H_2O} HOO\cdot + H_3O^+ \qquad E°', +2.0 \text{ V vs. NHE} \qquad (9.73)$$

$$HOO\cdot \xrightarrow[H_2O]{-e^-} O_2 + H_3O^+ \qquad E°', +0.6 \text{ V vs. NHE} \qquad (9.74)$$

Although HOO^- reacts rapidly with most organic solvents, it persists long enough in pyridine to permit its electrochemical oxidation via a similar ECE mechanism:[44]

$$HOO^- \xrightarrow{-e^-} [HOO\cdot] \xrightarrow{HO^-} H_2O + O_2^{\cdot-} \qquad E°, -0.37 \text{ V vs. NHE} \qquad (9.75)$$

$$O_2^{\cdot-} \xrightarrow{-e^-} O_2 \qquad E°, -0.66 \text{ V vs. NHE} \qquad (9.76)$$

9.5 SUMMARY

The redox chemistry of dioxygen and its reduction products is heavily dependent on mechanistic pathway, substrate, and solution acidity. For those circumstances that are limited by direct electron transfer, the redox mechanisms for dioxygen species involve one-electron steps. To achieve the full oxidizing potential and energetics for O_2 requires catalysts that facilitate atom-transfer mechanisms and/or stabilize dioxygen intermediates. The dominant characteristic of $O_2^{\bar{\cdot}}$ (the primary product from the electron-transfer reduction of O_2) is its ability to act as a Brønsted base. Under aprotic conditions $O_2^{\bar{\cdot}}$ is a strong nucleophile, a moderate one-electron reducing agent, and a hydrogen-atom-transfer oxidant of basic substrates such as hydroxylamines and reduced flavins. The base-induced disproportionation of HOOH leads to the formation of $O_2^{\bar{\cdot}}$ and HO· in aprotic media. The variety of oxidation–reduction, hydrolytic, atom-transfer, and disproportionation reaction pathways for dioxygen species is ample evidence of the unique chemical versatility of O_2 and HOOH.

REFERENCES

1. Barrette, W. C., Jr.; Johnson, H. W., Jr.; Sawyer, D. T., *Anal. Chem.* **1984,** *56,* 1890.
2. Sawyer, D. T.; Chiericato, G., Jr.; Angelis, C. T.; Nanni, E. J., Jr.; Tsuchiya, T., *Anal. Chem.* **1982,** *54,* 1720.
3. Parsons, R., *Handbook of Electrochemical Constants*, Butterworths, London, 1959.
4. Schwarz, H. A.; Dodson, R. W., *J. Phys. Chem.* **1984,** *88,* 3643.
5. Wilshire, J.; Sawyer, D. T., *Accts. Chem. Res.* **1979,** *12,* 105.
6. Bard, A. J.; Parsons, R.; Jordan, J., eds., *Standard Potentials in Aqueous Solution*, Marcel Dekker, New York, 1985.
7. Cofré, P.; Sawyer, D. T., *Anal. Chem.* **1986,** *58,* 1057.
8. Cofré, P.; Sawyer, D. T., *Inorg. Chem.* **1986,** *25,* 2089.
9. Sawyer, D. T.; Roberts, J. L.; Jr.; Tsuchiya, T.; Srivatsa, G. S., "Generation of Activated Oxygen Species by Electron-Transfer Reduction of Dioxygen in the Presence of Protons, Chlorinated Hydrocarbons, Methyl Viologen, and Transition Metal Ions," in *Oxygen Radicals in Chemistry and Biology*, Bors, W.; Saran, M.; Tait, D., eds., Walter de Gruyter, Berlin, 1984, pp. 25–34.
10. Roberts, J. L., Jr.; Morrison, M. M.; Sawyer, D. T., *J. Am. Chem. Soc.* **1978,** *100,* 329.
11. Roberts, J. L., Jr.; Sawyer, D. T., *Israel J. Chem.* **1983,** *23,* 430.
12. Chin, D.-H.; Chiericato, G. Jr.; Nanni, E. J., Jr.; Sawyer, D. T., *J. Am. Chem. Soc.* **1982,** *104,* 1296.
13. Gibian, M. J.; Sawyer, D. T.; Ungerman, T.; Tangpoonpholvivat, R.; Morrison, M. M., *J. Am. Chem. Soc.* **1979,** *101,* 640.
14. Fridovich, I., "The Biology of Superoxide and Superoxide Dismutase—In Brief," in *Oxygen and Oxy Radical in Chemistry and Biology*, Rodgers, M. A. J.; Powers, E. L., eds., Academic Press, New York, 1981, pp. 197–204.
15. Fee, J. A., "Is Superoxide Toxic and Are Superoxide Dismutases Essential for Aerobic Life?," in *Oxygen and Oxy Radicals in Chemistry and Biology*, Rodgers, M. A. J.; Powers, E. L., eds., Academic Press, New York, 1981, pp. 205–239.

16. Sawyer, D. T.; McDowell, M. S.; Yamaguchi, K. S., *Chem. Res. Toxicol.* **1988,** *1,* 97.
17. Tsang, P. K. S.; Cofré, P.; Sawyer, D. T., *Inorg. Chem.* **1987,** *26,* 3604.
18. Laitinen, H. A.; Kolthoff, I. M., *J. Phys. Chem.* **1941,** *45,* 1061.
19. Sawyer, D. T.; Interrante, L. V., *J. Electroanal. Chem.* **1961,** *2,* 310.
20. Sawyer, D. T.; Day, R. J., *Electrochim. Acta* **1963,** *8,* 589.
21. Bielski, B. H. J., *Photochem. Photobiol.* **1978,** *28,* 645.
22. Bielski, B. H. J.; Allen, A. O., *J. Phys. Chem.* **1977,** *81,* 1048.
23. Ilan, Y. A.; Meisel, D.; Czapski, G., *Israel J. Chem.* **1974,** *12,* 891.
24. Behar, D.; Czapski, G.; Rabini, J.; Dorfman, L. M.; Schwartz, H. A., *J. Phys. Chem.* **1970,** *74,* 3209.
25. Rabini, J.; Nielsen, S. O., *J. Phys. Chem.* **1969,** *73,* 3736.
26. Sawyer, D. T.; Roberts, J. L., Jr., *J. Electroanal. Chem.* **1966,** *12,* 90.
27. Sawyer, D. T.; Seo, E. T., *Inorg. Chem.* **1977,** *16,* 499.
28. Roberts, J. L., Jr.; Morrison, M. M.; Sawyer, D. T., *J. Am. Chem. Soc.* **1978,** *100,* 329.
29. Sawyer, D. T.; Kang, C.; Qiu, A.; Sobkowiak, A., "Redox Mechanisms for the Activation of Dioxygen (O_2) in the Presence of Iron, Copper, and Cobalt Complexess," in *Fifth International Symposium on Redox Mechanisms and Interfacial Properties of Molecules of Biological Importance*, Schultz, F. A.; Taniguchi, I., eds., The Electrochemical Society: Pennington, N.J., 1993, pp. 216–231.
30. Cofré, P.; Richert, S. A.; Sobkowiak, A.; Sawyer, D. T., *Inorg. Chem.* **1990,** *29,* 2645.
31. Sheu, C.; Richert, S. A.; Cofré, P.; Ross, B., Jr.; Sobkowiak, A.; Sawyer, D. T.; Kanofsky, J. R., *J. Am. Chem. Soc.* **1990,** *112,* 1936.
32. Sawyer, D. T., *Oxygen Chemistry*, Oxford University Press, New York, 1991, Chap. 2.
33. Tsang, P. K. S.; Sawyer, D. T., *Inorg. Chem.* **1990,** *29,* 2848.
34. Tung, H.-C.; Chooto, P.; Sawyer, D. T., *Langmuir,* **1991,** *7,* 1635.
35. Sobkowiak, A.; Qiu, A.; Liu, X.; Llobet, A.; Sawyer, D. T., *J. Am. Chem. Soc.* **1993,** *115,* 609.
36. Kang, C.; Redman, C.; Cepak, V.; Sawyer, D. T., *Bioorg. Med. Chem.,* **1993,** *1,* 125.
37. Sheu, C.; Sobkowiak, A.; Jeon, S.; Sawyer, D. T., *J. Am. Chem. Soc.* **1993,** *112,* 879.
38. McCord, J. M.; Fridovich, I., *J. Biol. Chem.* **1968,** *243,* 5733.
39. Nanni, E. J., Jr.; Angelis, C. T.; Dickson, J.; Sawyer, D. T., *J. Am. Chem. Soc.* **1981,** *103,* 4268.
40. Ortiz de Montellano, P. R., ed., *Cytochrome P-450: Structure, Mechanism, and Biochemistry*, Plenum Press, New York, 1986.
41. Clark, L. C., Jr., *Trans. Am. Soc. Artif. Intern. Organs* **1956,** *2,* 41.
42. Morrison, M. M.; Roberts, J. L., Jr.; Sawyer, D. T., *Inorg. Chem.* **1979,** *18,* 1971.

43. Roberts, J. L., Jr.; Morrison, M. M.; Sawyer, D. T., *J. Am. Chem. Soc.* **1978,** *100,* 329.

44. Sawyer, D. T.; Tsang, P. K. S.; Jeon, S.; Nicholson, M., ''Nucleophilic and Reductive Remediation Strategies for Hazardous Halogenated Hydrocarbons (PCBs, HCB, PCP, TCE, CCl_4, DCBP, EDB, DDT, and DDE),'' in *Industrial Environmental Chemistry*, Sawyer, D. T.; Martell, A. E.; eds., Plenum Press, New York, 1992, pp. 181–212.

CHAPTER 10

METALS AND METAL COMPLEXES

10.1 OXIDATION OF METALS

The transformation of metal-electrode surfaces via electrooxidation to their metallooxides, solvated ions, and metal complexes is fundamental to most anodic electrochemical processes (batteries, electrorefining, anodic-stripping analysis, and reference electrodes). Although this is traditionally represented as the removal of one (or more) electrons from a metal atom at the electrode surface to give a metal ion [e.g., Ag(s) $\xrightarrow{-e^-}$ Ag^+; $E°$, +0.80 V vs. NHE], the gas-phase ionization potential [e.g., Ag·(g) $\xrightarrow{-e^-}$ Ag^+(g); IP, 7.6 eV] is far greater than the observed oxidation potential.[1] The difference is attributed to the solvation energy for the metal ion [e.g., $Ag^+ + n\ H_2O \rightarrow Ag^+(aq)$; $-\Delta G(aq) \approx$ 293–418 kJ mol^{-1}]. However, such a sequential path would not obviate the 7.6-V energy barrier for the initial step and is in conflict with the observed thermodynamic reversibility for many metal/solvated-metal-ion redox couples.

All reactions, and particularly redox processes, occur via the easiest and lowest-energy pathway that is available (mechanistically feasible) to the system. In the case of the metal-electrode/electrolyte interface undergoing anodic transformations, the electrons can come from (1) surface metal atoms (energy limit; first ionization potential), (2) solvent molecules (energy limit; oxidation potential of solvent), (3) electrolyte anions (energy limit; oxidation potential of anions), and (4) base ligands (energy limit; oxidation potential of ligand). Without exception, all metal electrodes are electrochemically transformed via path 2, 3, or 4, and never via path 1. This general conclusion will be illustrated for silver and copper electrodes in aqueous and acetronitrile (MeCN) solutions that contain inert electrolyte, chloride ion (Cl^-), or bipyridine (bpy).

Metal–X Bond Energies. In aqueous solutions at pH 0 the silver electrode is oxidized reversibly:

$$Ag(s) + 2\ H_2O \xrightarrow{-e^-} Ag^I(OH_2)_2^+ \qquad E°,\ +0.80\ V\ vs.\ NHE \qquad (10.1)$$

The gas-phase ionization potential for a silver atom is 7.6 eV. In contrast, water is oxidized (gives up an electron) at much lower potentials:

$$2\ H_2O \xrightarrow{-e^-} [(H_2O)H_2O^+\cdot] \longrightarrow H_3O^+ + HO\cdot \qquad E°,\ 2.72\ V \quad (E°)_{pH5},\ +2.42\ V \qquad (10.2)$$

At a silver electrode the latter process is facilitated via formation of a Ag^I—OH_2^+ bond; the shift in oxidation potential from +2.42 to +0.80 V is a measure of the bond-formation energy ($-\Delta G_{BF}$):

$$-\Delta G_{BF} = (+2.42 - 0.80)96.5 = 156\ kJ\ mol^{-1} \qquad (10.3)$$

At pH 14 the anodic process is the oxidation of HO^-:

$$HO^- \xrightarrow{e^-} HO\cdot \qquad E°,\ +1.89 \quad (E°)_{pH7},\ +2.30\ V \qquad (10.4)$$

which is facilitated via formation of a Ag^I—OH bond [$-\Delta G_{BF} = (1.89 - 0.34)96.5 = 150\ kJ\ mol^{-1}$]:

$$Ag(s) + HO^- \xrightarrow{-e^-} Ag^I\text{—}OH(s) \qquad E°,\ +0.34\ V \quad (E°)_{pH7},\ +0.75\ V \qquad (10.5)$$

The data of Eqs. (10.1) and (10.5) can be combined to give a value for the solubility product (K_{sp}) for AgOH(s):

$$Ag^I(OH_2)_2^+ + HO^- \longrightarrow Ag^I\text{—}OH(s) + 2\ H_2O \qquad (10.6)$$

$$[Ag^I(OH_2)_2^+][HO^-] = K_{sp};\ \log K_{sp} = (0.34 - 0.80)/0.059 = -7.8$$

In the presence of chloride ion, metal electrodes facilitate its oxidation

$$Cl^- \xrightarrow{-e^-} Cl\cdot \qquad E°,\ +2.41\ V\ vs.\ NHE \quad (E°')_{MeCN},\ +2.24\ V \qquad (10.7)$$

via formation of metal–chlorine covalent bonds, for instance

$$Ag(s) + Cl^- \xrightarrow{-e^-} Ag{-}Cl(s) \qquad E°, +0.22 \text{ V} \tag{10.8}$$

Hence, the differential bond-formation energy $[\Delta(-\Delta G_{BF})]$ (Ag—Cl bond energy, minus the energy required to break Ag—Ag bond at the Ag(s) surface) is given by the difference in oxidation potentials [Eqs. (10.7) and (10.8)]:

$$\Delta(-\Delta G_{BF}) = (2.41 - 0.22)96.5 = 212 \text{ kJ mol}^{-1} \tag{10.9}$$

Because the escape energy for a Ag· atom from Ag(s) is 285 kJ mol^{-1}, a reasonable approximation for the breakage of a single bond is 95 kJ mol^{-1} [(1/3)285 kJ].[2] When combined with Eq. (10.9), this gives a reasonable value for $-\Delta G_{BF}$:

$$Ag\cdot + Cl\cdot \longrightarrow Ag{-}Cl \qquad -\Delta G_{BF} = 212 + 95 = 307 \text{ kJ mol}^{-1} \tag{10.10}$$

The literature value for the dissociative bond energy (ΔH_{DBE}) of Ag—Cl is 341 kJ mol^{-1}, which is equivalent to an estimated $-\Delta G_{BF}$ value of 307 kJ mol^{-1} [$-\Delta G_{BF} = \Delta H_{DBE} - T\Delta S$ = 341–33 (est) = 308 kJ mol^{-1}].[2] Thus, the proposition that metal-electrode oxidations are solvent or ligand-centered with potentials that reflect the metal-solvent/ligand bond-formation free energies ($-\Delta G_{BF}$) is supported by independent bond-energy data. The data of Eqs. (10.1) and (10.8) provide a measure of the solubility product for AgCl(s):

$$Ag^I(OH_2)_2^+ + Cl^- \longrightarrow Ag^I{-}Cl(s) + 2\ H_2O \tag{10.11}$$

$$[Ag^I(OH_2)_2^+][Cl^-] = K_{sp};\ \log K_{sp} = (0.22 - 0.80)/0.059 = -9.8$$

Similar results are observed for a silver electrode in the presence of Br^-:

$$Br^- \xrightarrow{-e^-} Br\cdot \qquad E°, +1.51 \text{ V} \tag{10.12}$$

$$Ag(s) + Br^- \xrightarrow{-e^-} Ag^I{-}Br(s) \qquad E°, +0.07 \text{ V} \tag{10.13}$$

which gives a measure of the Ag^I—Br bond energy:

$$\Delta(-\Delta G_{BF}) = (1.51 - 0.07)96.5 = 141 \text{ kJ mol}^{-1} \tag{10.14}$$

The latter, when combined with the bond-dissociation energy of Ag(s) ($\Delta H_{DBE}/3$ = 95 kJ mol^{-1}), gives a value for $-\Delta G_{BF}$ of 235 kJ mol^{-1} [literature, $-\Delta G_{BF} = \Delta H_{DBE} - T\Delta S = (293 \pm 29) - 33$ (est) $= 259 \pm 29$ kJ mol^{-1}].[2]

Another important example is the oxidation of Cl^- at a mercury [$Hg_2(l)$] electrode to form calomel [mercurous chloride, $Hg_2Cl_2(s)$, Cl—Hg^{II}—Hg^{II}—Cl(s)]:

$$Hg_2(l) + Cl^- \xrightarrow{-e^-} [Cl{-}Hg{-}Hg\cdot] \xrightarrow[Cl^-]{-e^-} Cl{-}Hg{-}Hg{-}Cl(s)$$

$$E°, +0.27 \text{ V} \tag{10.15}$$

The potential shift for the $Cl^-/Cl\cdot$ couple from +2.41 V [Eq. (10.7)] to +0.27 V in the presence of $Hg_2(l)$ is a measure of the [Cl—HgHg] bond energy $[-\Delta G_{BF} = (2.41 - 0.27)96.5 = 207 \text{ kJ mol}^{-1}]$.

Similar metal-facilitated oxidations of H_2O and of Cl^- occur for all metal electrodes. The respective potentials for the oxidation of each at a copper electrode are

$$Cu_{(s)} + 2\,H_2O \underset{}{\overset{-e^-}{\rightleftharpoons}} (H_2O)Cu^I(OH_2)^+ \qquad (E°)_{pH0}, +0.52 \text{ V} \tag{10.16}$$

$$[-\Delta G_{BF}, 192 \text{ kJ mol}^{-1}]$$

$$Cu_{(s)} + Cl^- \overset{-e^-}{\rightleftharpoons} Cu^I{-}Cl \qquad E°, +0.14 \text{ V} \tag{10.17}$$

$$[-\Delta G_{BF}, 218 \text{ kJ mol}^{-1}]$$

Additional redox data for oxidations of H_2O/HO^- at Cu, Ag, and Au electrodes in aqueous and acetronitrile (MeCN) solutions are summarized in Table 10.1.[3,4] At pH 0 with an iron electrode the water oxidation of Eq. (10.2) is shifted by −3.12 V:

$$Fe_{(s)} + 4\,H_2O \overset{-2e^-}{\rightleftharpoons} [(H_2O)_2Fe^{II}(OH_2)_2]^{2+} \qquad (E°)_{pH0}, -0.40 \text{ V} \tag{10.18}$$

which indicates that the $[H_2O(H_2O\cdot)^+]$ species is stabilized by a strong $(H_2O)_2(H_2O)^+Fe^{II}{-}OH_2^+$ covalent bond $[-\Delta G_{BF}, \sim 297 \text{ kJ mol}^{-1}]$.

TABLE 10.1 Redox Potentials ($E°$) for the $M^I(OH_2)_2/M$ and M^IOH/M, HO^- Couples of Cu, Ag, and Au in H_2O and in MeCN (0.1 M Tetraethylammonium Perchalorate)

		$E°$ (V) versus NHE[a]	
	M	$M^I(solv.)_n^+/M$	M^IOH/M, HO^-
H_2O[b]	Cu	+0.52	−0.36
	Ag	+0.80	+0.34
	Au	+1.7	
	$H_2O^{\cdot+}/H_2O$; $HO\cdot/HO^-$(GC)	+2.72	+1.89
MeCN[c]	Cu	+0.19	−0.79
	Ag	+0.54	−0.30
	Au	+1.58	−0.19
	$H_2O^{\cdot+}$(MeCN)/$H_2OHO\cdot/HO^-$(GC)	+3.2	+0.92

[a]SCE = +0.24 V versus NHE.
[b]Ref. 3.
[c]Ref. 4.

10.2 OXIDATION–REDUCTION OF TRANSITION-METAL COMPLEXES

Metal–Ligand Bond Energies. In an analogous fashion, the removal of an electron (oxidation) from water via Eq. (10.2) is aided by the presence of transition-metal ions [e.g., $Cu^{I}(OH_2)_6^{+}$, $Fe^{II}(OH_2)_6^{2+}$, and $Ce^{III}(OH_2)_6^{3+}$, each with one, two, and three $M{-}OH_2^{+}$ covalent bonds, respectively]:

$$Cu^{I}(OH_2)_6^{+} + H_2O \xrightarrow[(E^\circ)_{pH0},\ +0.16]{-e^-} [(H_2O)_5^{+}Cu^{II}{-}OH_2^{+}(OH_2]) \longrightarrow (H_2O)_5^{+}Cu^{II}{-}OH + H_3O^{+} \quad [-\Delta G_{BF},\ 247\ kJ\ mol^{-1}] \tag{10.19}$$

$$Fe^{II}(OH_2)_6^{2+} + 2\,H_2O \xrightarrow[(E^\circ)_{pH0},\ +0.77\,V]{-e^-} [(H_2O)_5^{2+}Fe^{II}{-}OH_2^{+}(OH_2)] \longrightarrow (H_2O)_5^{2+}Fe^{III}{-}OH + H_3O^{+} \quad [-\Delta G_{BF},\ 188\ kJ\ mol^{-1}] \tag{10.20}$$

$$Ce^{III}(OH_2)_6^{3+} + H_2O \underset{(E^\circ)_{pH0},\ +1.60\,V}{\overset{-e^-}{\rightleftarrows}} [(H_2O)_5^{3+}Ce^{IV}{-}OH_2^{+}(OH_2)] \longrightarrow (H_2O)_5^{3+}Ce^{IV}{-}OH + H_3O^{+} \quad [-\Delta G_{BF},\ 105\ kJ\ mol^{-1}] \tag{10.21}$$

In none of these examples has the potential for removal of an electron approached the ionization potentials of the metals. Although traditional treatments attribute the potentials of Eqs. (10.1), (10.15), and (10-17)–(10-21) to the removal of electrons from the metals, coupled with large ionic solvation energies, this requires a pathway with the ionization potential as a kinetic barrier. Furthermore, the spontaneous reaction of iron with acidified water is driven by the formation of $Fe{-}OH_2^{+}$ and $H{-}H$ covalent bonds that facilitate hydrogen-atom transfer from water (rather than electron transfer from iron):

$$Fe_{(s)} + 2\ H_3O^{+} \xrightarrow{H_2O} (H_2O)_4Fe^{II}(OH_2)_2^{2+} + H_2 \tag{10.22}$$

{*Note:* To ionize a gas-phase iron atom ($Fe \xrightarrow{-3e^-} Fe^{3+}$) requires 54.8 eV (5290 kJ mol^{-1});[1] in turn this species reacts on dissolution into liquid water [$Fe^{3+}(g) + 7\ H_2O_{(l)} \rightarrow (H_2O)_5^{2+}Fe^{III}{-}OH + H_3O^{+}$, $-\Delta H \approx 4180$ kJ mol^{-1} (5290 − 1110)]; the net energy change often is ascribed as the solvation energy for $Fe^{3+}(g)$ (heat of hydration)}.

Within an aprotic solvent (e.g., MeCN) oxidation of metals and metal complexes also is ligand-centered with the potential determined by the oxidation potential of the ligand and the metal–ligand covalent-bond formation free energy ($-\Delta G_{BF}$). For example, the free bipyridine (bpy) ligand in acetronitrile is oxidized near the solvent limit at a glassy carbon electrode (GC) (Figure 10.1):[5]

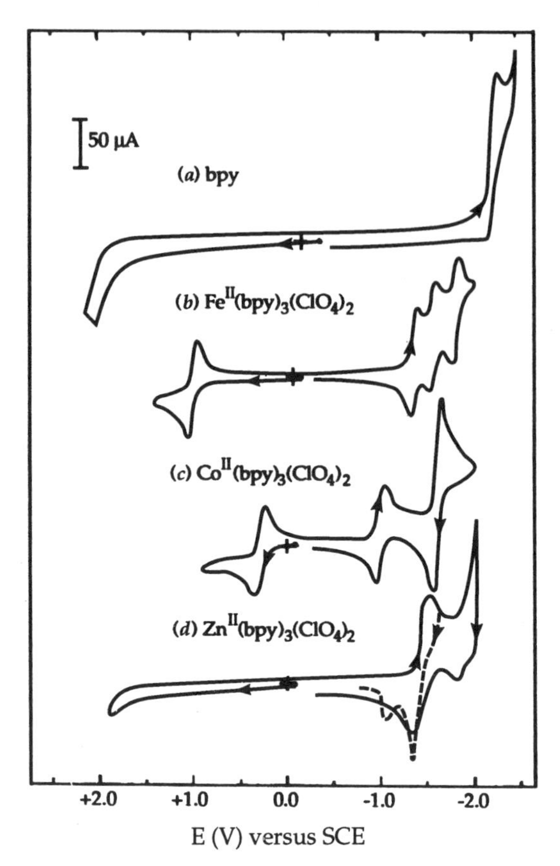

Figure 10.1 Cyclic voltammograms of 3 mM solutions in MeCN (0.1 M tetraethylammonium perchlorate): (*a*) bpy; (*b*) $Fe(bpy)_3^{2+}$; (c) $Co(bpy)_3^{2+}$; (*d*) $Zn(bpy)_3^{2+}$. Conditions scan rate 0.1 V s^{-1}; 25°C; glassy-carbon working electrode (0.09 cm^2); SCE versus NHE, +0.244 V.

$$(\text{bpy}) \xrightarrow[\text{MeCN}]{-e^-} \text{bpy}^{+\cdot} \qquad E^{\circ\prime}, +2.32 \text{ V vs. NHE} \tag{10.23}$$

but at a copper electrode the oxidation occurs at a negative potential:[6]

$$\text{Cu(s)} + 2\ \text{bpy} \xrightarrow{-e^-} \text{Cu}^{\text{I}}(\text{bpy})_2^+ \qquad E^{\circ\prime}, -0.16 \text{ V} \tag{10.24}$$

Even more striking is the reduction of $(bpy)_2Cu^I(OH_2)^+$ at a glassy-carbon electrode, which occurs at −1.04 V versus NHE. The difference (−0.88 V) is due to the Cu—Cu bond energy (85 kJ mol^{-1}) that must be overcome in the metal-oxidation process. Reduction of $(bpy)_2Cu^ICl$ occurs at essentially the same potential as that for $Cu^I(MeCN)_4Cl$ (−1.01 V vs. NHE):[7]

$$(\text{bpy})_2\text{Cu}^{\text{I}}\text{Cl} + e^- \longrightarrow \text{Cu} + 2\ \text{bpy} + \text{Cl}^- \qquad E^{\circ\prime}, -1.06 \text{ V} \tag{10.25}$$

The difference between this value and that for the $Cl\cdot/Cl^-$ couple [Eq. (10.7), +2.24 V vs. NHE] is a measure of the $(bpy)_2Cu^I{-}Cl$ bond energy $\{-\Delta G_{BF} = [2.24 - (-1.06)]96.5 = 319\ kJ\ mol^{-1}$ [the value for gas-phase Cu—Cl(g) is $351 \pm 4\ kJ\ mol^{-1}$][2]}.

Figure 10.1 illustrates that the oxidation of the $Fe^{II}(bpy)_3^{2+}$ complex is reversible and ligand-centered:[5]

$$Fe^{II}(bpy)_3^{2+} \xrightarrow[MeCN]{-e^-} Fe^{III}(bpy)_3^{3+} \qquad E°',\ +1.30\ V \tag{10.26}$$

(Noteworthy are the three reversible one-electron reductions for this complex). The electron that is removed from the $Fe^{II}(bpy)_2^{2+}$ complex comes from the ligands to give $bpy^{+\bullet}$, which couples with one of the four unpaired electrons of the iron(II) center (d^6sp) to give a third covalent bond [$Fe^{III}(bpy)_3^{3+}$, d^5sp^2; $S = \frac{5}{2}$]. The difference in oxidation potentials for $Fe^{II}(bpy)_3^{2+}$ and free bpy [Eq. (10.23)] is a measure of the $Fe^{III}{-}bpy^+$ bond energy [$-\Delta G_{BF} = (2.32 - 1.30)96.5 = 99\ kJ\ mol^{-1}$]. The potential that would be required to remove an electron from the d^6sp manifold of the iron(II) center of $Fe^{II}(OH_2)_6^{2+}$ or $Fe^{II}(bpy)_3^{2+}$ would be greater than the first ionization potential of atomic iron (7.9 eV).[1]

Table 10.2 summarizes the oxidation potentials for ligands (L) and their $M^{II}L_3$ complexes with zinc(II), manganese(II), iron(II), and cobalt(II). The difference in the potentials for the free and complexed ligands is a measure of the metal(III)–ligand bond-formation energies ($-\Delta G_{BF}$); these are summarized in Table 10.3.[5] For this group of complexes the order of metal(III)–nitrogen bond energies, $Co^{III}(bpy)_3^{3+} > Fe^{III}(bpy)_3^{3+} > Mn^{III}(byp)_3^{3+}$; and of metal(III)–oxygen bond energies, $Fe^{III}(acac)_3 > Co^{III}(acac)_3 > Mn^{III}(acac)_3$; is consistent with their relative stability constants. With the picolinate (PA^-) ligand there is a combination metal–oxygen covalent bonding and nitrogen-base donor in-

TABLE 10.2 Oxidation Potentials for Ligands (*L*) and Their ML_3 Complexes with Zn(II), Mn(II), Fe(II), and Co(II) in MeCN (0.1 M Tetraethylammonium Perchlorate)

	$E_{1/2}$[b] (V) versus NHE[c]				
Ligand (L or L^-)[a]	$L/L^{+\bullet}(L^-/L\cdot)$	$Zn^{II}L_3$	$Mn^{II}L_3$	$Fe^{II}L_3$	$Co^{II}L_3$
H_2O	+2.8	+2.8	+2.8	+1.84	+2.8
bpy	+2.32	> +2.5	+1.55	+1.30	+0.58
8 Q^-	+0.21	+0.22	−0.06	−0.41	−0.57
$acac^-$	+0.55	+0.58	+0.18	−0.42	−0.35
PA^-	+1.50	+1.54	+0.60	+0.20	+0.04

[a]Key: bpy, 2,2′bipyridine; 8 Q^-, 8-quinolinate; $acac^-$, acetylacetonate; PA^-, picolinate (2-carboxylate pyridine).

[b]$E_{1/2}$ taken as $(E_{p,a} + E_{p,c})/2$ for reversible couples of $Mn^{II}L_3$ and $Fe^{II}L_3$ complexes; as $E_{p,a/2} + 0.03$ V for L (or L^-) and $Zn^{II}L_3$; and as $E_{p,c/2} - 0.03$ V for $Co^{II}L_3$ complexes that exhibit separated redox couples.

[c]SCE versus NHE, +0.244 V.

TABLE 10.3 Apparent Metal–Ligand Covalent-Bond-Formation Free Energies ($-\Delta G_{BF}$) for Several Manganese, Iron, and Cobalt Complexes

Complex	$-\Delta G_{BF}$ (kJ mol^{-1})[a]
Manganese	
$(8Q)_2Mn^{III}—8Q$	25
$(acac)_2Mn^{III}—acac$	38
$(PA)_2Mn^{III}—PA$	92
$[(bpy)_2Mn^{III}—bpy]^{3+}$	>96[b]
Iron	
$(8Q)_2Fe^{III}—8Q$	63
$(acac)_2Fe^{III}—acac$	96
$(PA)_2Fe^{III}—PA$	130
$[(bpy)_2Fe^{III}—bpy]^{3+}$	>121[b]
$[(Ph_3PO)_3Fe^{III}—OPPh_3]^{3+}$	>126[b]
$[(MeCN)_4Fe^{III}—OH_2]^{3+}$	96
Cobalt	
$(8Q)_2Co^{III}—8Q$	70
$(acac)_2Co^{III}—acac$	88
$(PA)_2Co^{III}—PA$	146
$[(bpy)_2Co^{III}—bpy]^{3+}$	>192[b]

[a] $(-\Delta G_{BF}) = [E_{1/2[ZnL_3^-/ZnL_2(L\cdot)]} - E_{1/2(ML_3^-/M(\cdot L)L_2)}] \times 96.5$ kJ mol^{-1}.
[b] $(-\Delta G_{BF}) = [E_{p,a(ZnL/ZnL^{+\cdot})} - E_{p,a(ML/M-L^+)}] \times 96.5$ kJ mol^{-1}; L = $(bpy)_3$ or $(Ph_3PO)_4$.

teraction, which shifts the bond-energy order, $Co^{III}(PA)_3 > Fe^{III}(PA)_3 > Mn^{III}(PA)_3$. All the data are consistent with ligand-centered redox processes.

Table 10.4 summarizes the oxidation potentials for a variety of ligands (L) in acetronitrile (MeCN).[7] Their relative Lewis basicity (nucleophilicity) increases as their oxidation potential becomes less positive. However, the potential at which L is oxidized (and $L\cdot$ is reduced) within an ML_x complex is shifted by the $M—L$ covalent-bond energy ($-\Delta G_{BF}$).

Figures 10.2 and 10.3 illustrate the electrochemistry for several copper(II) and copper(I) complexes in MeCN. The redox potentials for these copper complexes and their ligands are summarized in Table 10.5 (related data for aqueous media are given in Table 10.6).[3,8–13] In addition, the shift in redox potential (ΔE) for the free ligand (L) and when bonded in a complex (CuL_x) is tabulated. This quantity is a measure of the apparent copper–ligand covalent-bond-formation free energy ($-\Delta G_{BF}$):

$$-\Delta G_{BF} = (\Delta E)96.5 \text{ kJ mol}^{-1} \tag{10.27}$$

Table 10.7 summarizes the copper–ligand bond energies for the various complexes.

TABLE 10.4 Redox Potentials for Ligands in Acetonitrile [0.1 M $(Et_4N)ClO_4$]

Ligand (L)[a]	$E_{p,a}$ (V) versus SCE[b]	$E_{p,c}$ (V) versus SCE[b]
H_2O	2.80	
py	2.30	−2.75
bpy	2.15	−2.25
tpy	2.00	−1.5, −2.5
Cl^-	2.00	
$PhC(O)O^-$	1.45	
PA^-	1.34	
AcO^-	1.30	
$DPAH^-$	1.20	
$HOC(O)O^-$	1.15, 1.55	
HO^-	0.68	
$PhCH_2O^-$	0.50	
DPA^{2-}	0.25, 1.25	
$TDTH^-$	−0.05	

[a]*Key*: bpy, 2,2′-bipyridine; tpy, 2,2′:6′,2″-terpyridine; PA^-, picolinate anion; $DPAH^-$, 2,6-carboxyl, carboxylatopyridine anion; DPA^{2-}, 2,6-dicarboxylatopyridine dianion; $HOC(O)O^-$, bicarbonate anion; $TDTH^-$, toluene-3,4-dithiol anion.

[b]$E_{p,a}$, anodic peak potential; $E_{p,c}$, cathodic peak potential. Glassy-carbon electrode (GCE); scan rate 0.1 V s^{-1}.

[c]Saturated calomel electrode (SCE); E_{SCE}, +0.244 V versus NHE.

The dianion of toluene-3,4-dithiol (TDT^{2-}) forms unique complexes [$M^{II}(TDT)_2^{2-}$] with transition metals that are readily oxidized via a ligand-centered process to $M^{III}(TDT)_2^-$.[14] Figure 10.4 illustrates the cyclic voltammetry for the latter complexes of Cu, Ni, Co, and Fe. Not only does each $M^{III}(TDT)_2^-$ complex undergo a reversible one-electron reduction, but the Ni(III), Co(III), and Fe(III) complexes also exhibit a somewhat reversible oxidation to the M(IV) valence state. For example

$$Fe^{II}(TDT)_2^{2-} \xrightarrow[-0.83\,V]{-e^-} Fe^{III}(TDT)_2^- \xrightarrow[+0.10\,V]{-e^-} Fe^{IV}(TDT)_2 \qquad (10.28)$$

Table 10.8 summarizes the redox potentials for this group of complexes and the estimated $M-S$ bond energies ($-\Delta G_{BF}$) in the $M^{III}(TDT)_2^-$ and $M^{IV}(TDT)_2$ complexes. These are based on the oxidation-potential difference (ΔE) between the $M^{II}(TDT)_2^{2-}$ complex and $Zn^{II}(TDT)_2^{2-}$ (not able to form a third covalent bond) and the ΔE between $M^{III}(TDT)_2^-$ and $Cu^{III}(TDT)_2^-$ (with filled valence-electron shell and unable to form a fourth covalent bond), respectively [$-\Delta G_{BF} = (\Delta E)96.5$ kJ mol^{-1}].

Although most iron(II) complexes are oxidized by hydrogen peroxide (HOOH) via Fenton chemistry,

$$2\ Fe^{II}L_x + HOOH \longrightarrow 2\ L_xFe^{III}OH \qquad (10.29)$$

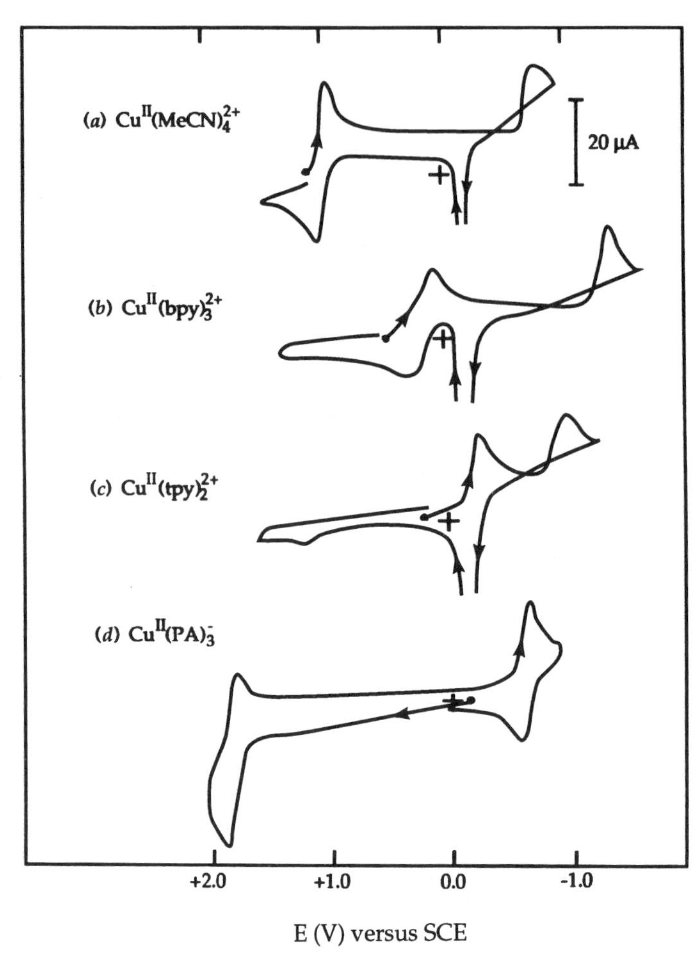

Figure 10.2 Cyclic voltammograms for (*a*) 1 mM $[Cu^{II}(MeCN)_4](ClO_4)_2$; (*b*, *a*) plus 3 equiv of bpy; (*c*, *a*) plus 2 equiv of typ; and (*d*, *a*) plus 3 equiv of PA^- in MeCN [0.1 M $(Et_4N)ClO_4$]. Scan rate 0.1 V s^{-1}; glassy-carbon electrode (GCE) (0.9 cm^2); SCE versus NHE, +0.244 V.

within MeCN the combination of $Fe^{II}(OPPh_3)_4^{2+}$ and HOOH (1 : 10) yields a unique purple complex [λ_{max}, 576 nm (ϵ 1770 M^{-1} s^{-1})], $(Ph_3PO)_4^{2+}$-$Fe^{III}OOH$.[15] The reversible one-electron, ligand-centered oxidation of $Fe^{II}(OPPh_3)_4^{2+}$ at +1.2 V versus SCE is replaced by an irreversible two-electron oxidation at +1.9 V (Figure 10.5):

$$(Ph_3PO)_4^{2+}Fe^{III}OOH + 3\ H_2O \xrightarrow{-2e^-} (Ph_3PO)_4^{2+}Fe^{III}OH + O_2 + 2\ H_3O^+ \quad (10.30)$$

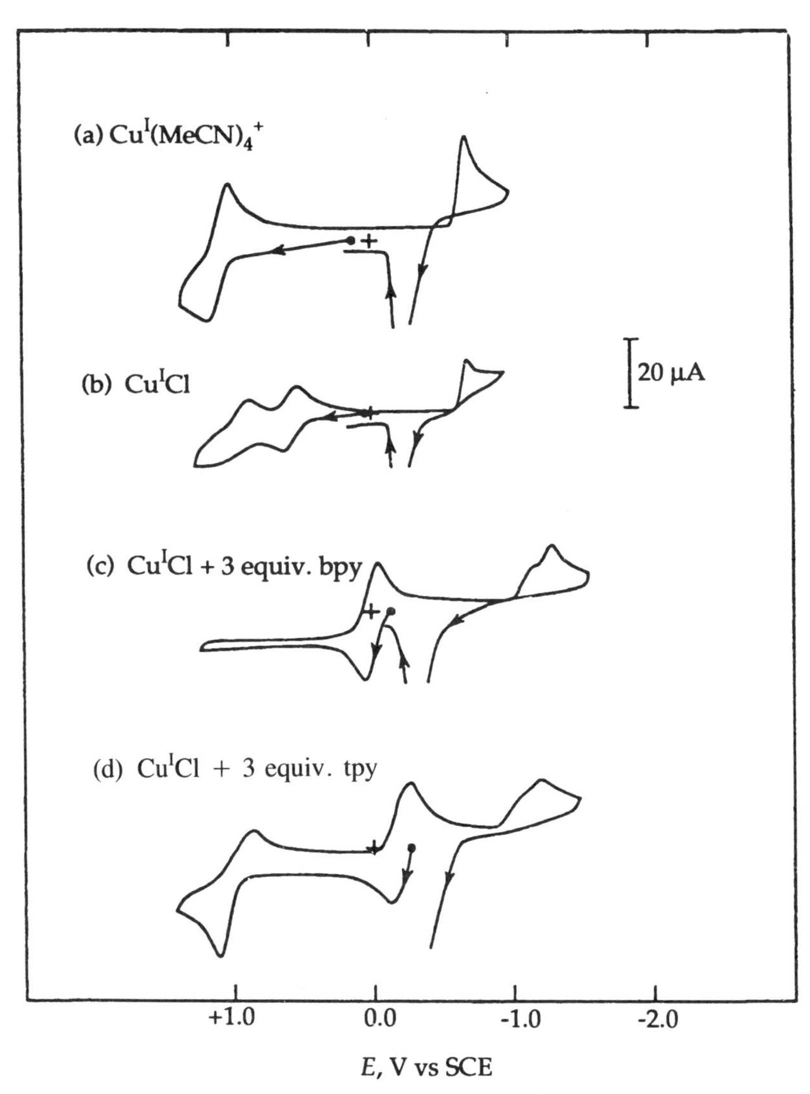

Figure 10.3 Cyclic voltammograms for (*a*) 1 mM $Cu^I(MeCN)_4^+$; (*b*) 1 mM Cu^ICl; (*c*, *b*) plus 3 mM bpy; and (*d*, *b*) plus 3 mM tpy in MeCN [0.1 M $(Et_4N)ClO_4$]. Scan rate 0.1 V s^{-1}; GCE (0.09 cm^2); SCE versus NHE, + 0.244 V.

Whereas $Fe^{II}(OPPh_3)_4^{2+}$ is reduced by two electrons at −1.1 V to give metallic iron, the $(Ph_3PO)_4^{2+}Fe^{III}OOH$ complex is reduced in several steps to give an iron oxide:

$$(Ph_3PO)_4^{2+}Fe^{III}OOH + e^- \xrightarrow[+0.3\,V]{} (Ph_3PO)_4^{+}Fe^{II}OOH \xrightarrow[-0.1\,V]{e^-}$$

$$(Ph_3PO)_4Fe^{I}OOH \xrightarrow[-1.8\,V]{e^-} Fe^{II}O + HO^- + 4\,L \qquad (10.31)$$

TABLE 10.5 Redox Potentials for Copper Complexes and Their Ligands in MeCN

Electrode Reaction	$E_{1/2}$[a] (V) versus SCE[b]	ΔE[c] (V)
$bpy - e^- \longrightarrow bpy^{+\cdot}$	2.1	
$Cu^{II}(bpy)_2^{2+} + e^- \longrightarrow Cu^I(bpy)_2^+$	0.1	2.0
$Cu^{II}(OH)(bpy)_2^+ + e^- \longrightarrow Cu^I(OH)(bpy)_2$	−0.1	2.2
$Cu^{II}(OAc)(bpy)_2^+ + e^- \longrightarrow Cu^I(OAc)(bpy)_2$	−0.1	2.2
$tpy - e^- \longrightarrow tpy^{+\cdot}$	1.9	
$Cu^{II}(tpy)_2^{2+} + e^- \longrightarrow Cu^I(tpy)_2^+$	−0.2	2.1
$Cu^I(tpy)_2^+ + e^- \longrightarrow Cu + 2\ tpy$	−0.9	2.8
$PA^- - e^- \longrightarrow PA\cdot$	1.3	
$Cu^{II}(PA)_3^- + e^- \longrightarrow Cu^I(PA) + 2\ PA^-$	−0.6	1.9
$Cu^I(PA) + e^- \longrightarrow Cu + PA^-$	−1.6	2.9
$AcO^- - e^- \longrightarrow AcO\cdot$	1.2	
$Cu^I(OAc)(MeCN)_4 + e^- \longrightarrow Cu + 4\ MeCN + AcO^-$	−1.2	2.4
$Cu^I(OAc)(bpy)_2 + e^- \longrightarrow Cu + 2\ bpy + AcO^-$	−1.3	2.5
$PhC(O)O^- - e^- \longrightarrow PhC(O)O\cdot$	1.4	
$Cu^{II}[OC(O)Ph]_2 + e^- \longrightarrow Cu^I[OC(O)Ph] + PhC(O)O^-$	−0.25	1.65
$Cu^I[OC(O)Ph] + e^- \longrightarrow Cu + PhC(O)O^-$	−1.3	2.7
$PhCH_2O^- - e^- \longrightarrow PhCH_2O\cdot$	0.4	
$Cu^{II}(OCH_2Ph)_2(bpy)_2 - e^- \longrightarrow Cu^I(OCH_2Ph)_2^-(bpy)_2$	−0.4	0.8
$DPAH^- - e^- \longrightarrow DPAH\cdot$	1.2	
$Cu^{II}(DPAH)(DPA)^- + e^- \longrightarrow Cu^I(DPA)^- + DPAH^-$	−0.5	1.7
$DPA^{2-} - e^- \longrightarrow DPA^{\bar{\cdot}}$	0.2	
$Cu^I(DPA)^- + e^- \longrightarrow Cu + DPA^{2-}$	−1.8	2.0
$Cl^- - e^- \longrightarrow Cl\cdot$	2.0	
$Cu^{II}Cl_2(MeCN)_4 + e^- \longrightarrow Cu^ICl(MeCN)_4 + Cl^-$	0.56	1.44
$Cu^{II}Cl_2(bpy)_2 + e^- \longrightarrow Cu^ICl(bpy)_2 + Cl^-$	0.02	1.98
$Cu^{II}Cl_2(tpy) + e^- \longrightarrow Cu^ICl(tpy) + Cl^-$	−0.1	2.1
$Cu^ICl(MeCN)_4 + e^- \longrightarrow Cu + 4\ MeCN + Cl^-$	−1.2	3.2
$Cu^ICl(bpy)_2 + e^- \longrightarrow Cu + 2\ bpy + Cl^-$	−1.25	3.25
$Cu^ICl(tpy) + e^- \longrightarrow Cu + tpy + Cl^-$	−1.15	3.15
$HO^- - e^- \longrightarrow HO\cdot$ (at pH 7 in H_2O)	2.1	
$Cu^I(OH)(H_2O)_3 + e^- \longrightarrow Cu + 3\ H_2O + HO^-$	−0.3	2.4
$Cu^I(OH)(bpy)_2 + e^- \longrightarrow Cu + 2\ bpy + HO^-$	−1.3	3.4

[a] $E_{1/2}$ taken as $E_{p,a/2} + 0.03$ V for the irreversible reductions and $E_{p,c/2} - 0.03$ V for the irreversible oxidations.

[b] Saturated calomel electrode (SCE) versus NHE, +0.244 V.

[c] $\Delta E = E_{1/2(L^+/L)} - E_{1/2(Cu^{II}/Cu^I)}$ or $\Delta E = E_{1/2(L^+/L)} - E_{1/2(Cu^I/Cu\cdot)}$.

TABLE 10.6 Standard Reduction Potentials of Copper Complexes in H_2O

	$E°$ (V) versus SCE[a]	
Complex	Cu(II/I)	Cu(I/0)
$Cu^{II}(CN)_2$	0.88	−0.68
$Cu^{II}I^+$	0.62	−0.42
$Cu^{II}(py)_4^{2+}$	0.45	
$Cu^{II}L_2^{2+}$	0.35	
(2,9-dimethyl-1,10-phenanthroline)$_2$		
$Cu^{II}Cl^+$	0.32	−0.12
$Cu^{II}Cl_2$	0.22	
$Cu^{II}L_4^{2+}$ (imidazole)	0.11	
$Cu^{II}Cl_3^-$	0.10	
Cu^{II}(1,10-phenanthroline)$_2^{2+}$	−0.07	
Cu^{II}(aq)	−0.08	0.28
$Cu^{II}(bpy)_2^{2+}$	−0.12	
$Cu^{II}(NH_3)_4^{2+}$	−0.14	−0.34
Cu^{II}(alanine)$_2$	−0.37	
Cu^{II}(glycine)$_2$	−0.40	
Cu^{II}(oxalate)$_2$	<-0.44	
$Cu^{II}(en)_2^{2+}$	−0.60	

[a]Ref. 3, 8–13; SCE versus NHE, +0.244 V.

The electrochemical reduction of permanganic acid [$HOMn^{VII}(O)_3$], which is traditionally represented as a metal-centered electron transfer to change Mn^{7+} to Mn^{6+}, is another example of a ligand-centered process:

$$HOMn^{VII}(O)_3 + H_3O^+ + e^- \longrightarrow Mn^{VI}(O)_3 + 2\,H_2O$$

$$E°,\ +1.51\text{ V vs. NHE} \qquad (10.32)$$

Comparison of this with the reduction of free hydroxyl radical (HO·)

$$HO\cdot + H_3O^+ + e^- \longrightarrow 2\,H_2O \qquad E°,\ +2.72\text{ V} \qquad (10.33)$$

provides a measure of the $HO{-}Mn^{VII}(O)_3$ bond energy [$-\Delta G_{BF} = (2.72 - 1.51)96.5 = 117$ kJ mol^{-1}]. The other strong oxidants [$(HO)_2Cr_2^{VI}(O)_5$ and $HOCe^{IV}(OH_2)_5^{3+}$] that are used for aqueous redox titrations are reduced by a similar path:

$$\underset{-\Delta G_{BF},\ 130\text{ kJ mol}^{-1}}{HO{-}Cr^{VI}(O)_2OCr^{VI}(O)_2OH} + H_3O^+ + e^- \xrightarrow[+1.36\text{ V}]{} (O)_2Cr^{V}OCr^{VI}(O)_2OH + 2H_2O \qquad (10.34)$$

TABLE 10.7 Apparent Metal–Ligand Covalent-Bond-Formation Free Energies ($-\Delta G_{BF}$) for Several Copper Complexes

Metal–Ligand Bond	$-\Delta G_{BF}$ (kJ mol^{-1})
$(bpy)^+{-}Cu^{II}(bpy)_2^+$	192
$(bpy)^+{-}Cu^{II}(OH)(bpy)$	213
$(bpy)^+{-}Cu^{II}(OAc)(bpy)$	213
$(tpy)^+{-}Cu^{II}(tpy)^+$	201
$(typ)^+{-}Cu^{I}(tpy)$	268
$PA{-}Cu^{II}(PA)_{2^-}$	180
$PA{-}Cu^{I}$	280
$AcO{-}Cu^{I}(MeCN)_4$	230
$AcO{-}Cu^{I}(bpy)_2$	238
$PhC(O)O{-}Cu^{II}[OC(O)Ph]$	155
$PhC(O)O{-}Cu^{I}$	259
$PhCH_2O{-}Cu^{II}(OCH_2Ph)(bpy)_2$	75
$DPAH{-}Cu^{II}(DPA)^-$	163
$DPA^-{-}Cu^{I}$	192
$Cl{-}Cu^{II}Cl(MeCN)_4$	138
$Cl{-}Cu^{II}Cl(bpy)_2$	192
$Cl{-}Cu^{II}Cl(tpy)$	201
$Cl{-}Cu^{I}(MeCN)_4$	310
$Cl{-}Cu^{I}(bpy)_2$	314
$Cl{-}Cu^{I}(tpy)$	305
$HO{-}Cu^{I}(H_2O)_3$	230
$HO{-}Cu^{I}(bpy)_2$	326

$$\underset{-\Delta G_{BF},\ 109\ \mathrm{kJ\ mol^{-1}}}{HO{-}Ce^{IV}(OH_2)_5^{3+}} + H_3O^+ + e^- \xrightarrow[+1.61\,V]{} Ce^{III}(OH_2)_6^{3+} + H_2O \qquad (10.35)$$

An important point in these electron-transfer reductions is that the primary electron acceptor is the hydronium ion (H_3O^+), which is transformed to a hydrogen atom (H·) that reacts with HO· (either free or bound via a covalent bond to the metal center).

Under alkaline conditions $Mn^{VII}O_4^-$ is reduced via direct electron addition to one of the bound oxygen atoms:

$$^-OMn^{VII}(O)_3 + e^- \longrightarrow {}^-OMn^{VI}(O)_2O^- \qquad E°,\ +0.55\ \text{vs. NHE} \qquad (10.36)$$

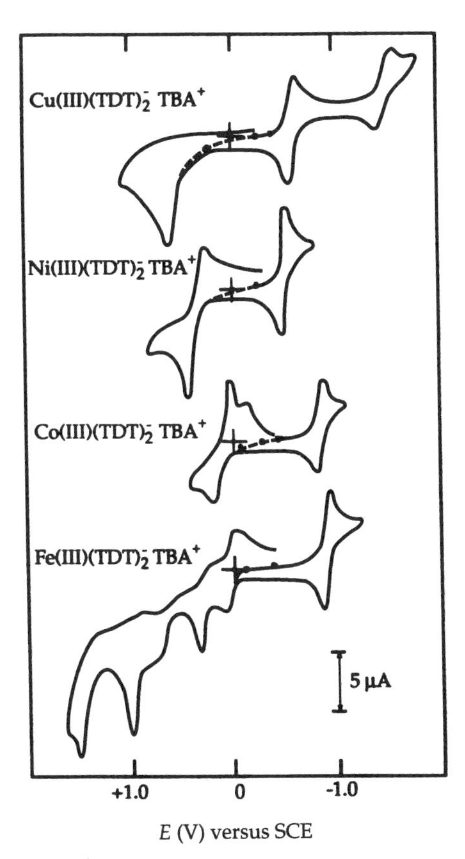

Figure 10.4 Cyclic voltammograms in MeCN (0.1 M TEAP) of $[M(III)(TDT)_2][Bu_4N]$ complexes (M = Cu, Ni, Co, and Fe). Scan rate 0.1 V s^{-1}; Pt electrode area 0.11 cm^2.

TABLE 10.8 Electrochemical Oxidation Potentials for $M^{II}(TDT)_2^{2-}$ Complexes in MeCN (0.1 M TEAP)

Metal	$E_{p,a}$ (V) versus SCE		$-\Delta G^{BF}$ (kJ mol^{-1})	
(M)	1st oxidation	2nd oxidation	M^{III}—S	M^{IV}—S
($TDTH^-$)	−0.05 (irreversible)			
Zn	+0.18 (irreversible)			
Cu	−0.53	+0.62	69	
Ni	−0.47	+0.44	63	18
Co	−0.73	+0.20	88	41
Fe	−0.83	+0.10, +0.32	97	50
Mn	−0.63	+0.22 (irreversible)	78	38

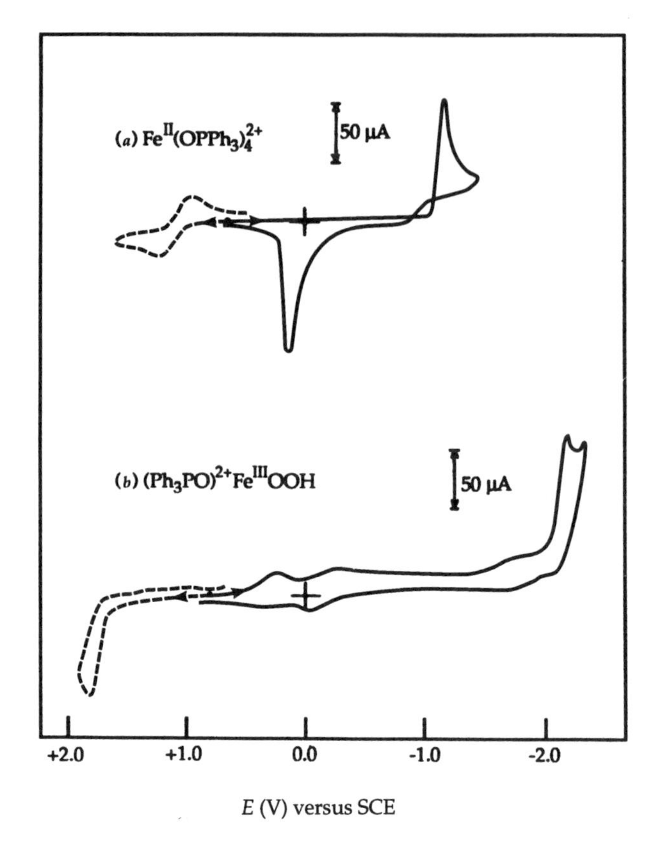

Figure 10.5 Cyclic voltammograms in MeCN for (*a*) 3 mM $Fe^{II}(OPPh_3)_4(ClO_4)_2$ and (*b*) 3 mM $(Ph_3PO)_4^{2+}Fe^{III}OOH$. Scan rate 0.1 V s^{-1}; glassy-carbon working electrode (area 0.11 cm^2).

The extent of the stabilization of the oxygen atom in $Mn^{VII}O_4^-$ is indicated by the reduction potential for a free $\cdot O \cdot$:

$$\cdot O \cdot + e^- \longrightarrow \cdot O^- \qquad (E^\circ)_{pH\,14},\ +1.43\ V \tag{10.37}$$

Although metalloporphyrins often are classified as coordination complexes, they are much closer to organometallic compounds with their strong metal–nitrogen covalent bonds. Therefore, the electrochemistry of these important redox-active molecules is discussed within the context of organometallic compounds (Chapter 13). Given that the redox chemistry of coordination complexes is ligand-centered, they represent a special class of organometallic compounds with metal–oxygen and metal–nitrogen covalent bonds.

REFERENCES

1. Lide, D. R., ed., *CRC Handbook of Chemistry and Physics*, 71st ed., CRC, Boca Raton, Fla., 1990, pp. **10**-180–181, **10**-210–211.
2. Lide, D. R., ed., *CRC Handbook of Chemistry and Physics*, 71st ed., CRC, Boca Raton, Fla., 1990, pp. **9**-86–98.
3. Bard, A. J.; Parsons, R.; Jordan, J., eds., *Standard Potentials in Aqueous Solution*, Marcel Dekker, New York, 1985.
4. Sawyer, D. T.; Chooto, P.; Tsang, P. K. S., *Langmuir* **1989,** *5*, 84.
5. Richert, S. A.; Tsang, P. K. S.; Sawyer, D. T., *Inorg. Chem.* **1989,** *28*, 2471.
6. Sobkowiak, A.; Qiu, A.; Llobet, A.; Sawyer, D. T., *J. Am. Chem. Soc.* **1993,** *115*, 609.
7. Qiu, A., Ph.D. dissertation, Texas A&M University: College Station, Tex., 1992, p. 53.
8. Philips, C. S. G.; Williams, R. J. P., *Inorganic Chemistry*, Oxford University Press, Oxford, 1966, Vol. 2, p. 316.
9. Fee, J. A., *Structure and Bonding* **1975,** *23*, 1.
10. Moore, G. R.; Williams, R. J. P., *Coord. Chem. Rev.* **1976,** *18*, 125.
11. Ochiai, E.-I., *Bioinorganic Chemistry*, Allyn & Bacon, Boston, 1978, p. 236.
12. Wilkinson, G.; Gillard, R. D.; McCleverty, J. A., *Comprehensive Coordination Chemistry*, Pergamon Press, Oxford, 1987, Vol. 5, p. 594.
13. James, B. R.; Williams, R. J. P., *J. Chem. Soc.* **1961,** 2007.
14. Sawyer, D. T.; Srivatsa, G. S.; Bodini, M. E.; Schaefer, W. P.; Wing, R. M., *J. Am. Chem. Soc.* **1986,** *108*, 936.
15. Sawyer, D. T.; McDowell, M. S.; Spencer, L.; Tsang, P. K. S., *Inorg. Chem.* **1989,** *28*, 1166.

CHAPTER 11

NONMETALS (SULFUR, NITROGEN, AND CARBON COMPOUNDS)

11.1 INTRODUCTION

Although hydronium ion (H_3O^+) (Chapter 8) and dioxygen (O_2) (Chapter 9) are the most studied of the molecules and ions without metal atoms, several of the molecules that contain sulfur, nitrogen, or carbon also are electroactive. The results for representative examples are presented to illustrate the utility of electrochemical measurements for the evaluation of the redox thermodynamics and bond energies for non-metal-containing molecules. In particular, the electrochemistry for several sulfur compounds [S_8, SO_2, $HS(CH_2)_3SH$], nitrogen compounds [$\cdot NO$, $HON{=}O$, N_2O, H_2NOH, hydrazines (RNHNHR'), amines, phenazine], and carbon compounds (CO_2, CO, NC^-) is summarized and interpreted.

Some of the considerations for electron-transfer processes that have been discussed in previous chapters are fundamental to the electrochemistry of these examples. Thus, reductive processes always involve the most electrophilic (acidic, positive-charge density) center (substrate or substrate–matrix combination) that produces the least basic (nucleophilic) product. Under acidic conditions the primary reactant often is the hydronium ion (H_3O^+) to give a hydrogen atom that couples with the substrate via covalent bond formation; for instance

$$\text{PhN{=}NPh} + 2\,H_3O^+ + 2\,e^- \xrightarrow[\text{Me}_2\text{SO}]{} \text{PhNHNHPh}$$

$$E_{p,c},\ -0.5\ \text{V vs. SCE} \qquad (11.1)$$

In contrast, oxidations always involve the most nucleophilic (basic, negative-charge density) center (substrate or substrate–base combination) that produces the least acidic (electrophilic) product. Under basic conditions (or neutral aqueous solutions) the primary reactant often is the base [B^-; HO^-, $HOC(O)O^-$, AcO^-, PhO^-] to give a hydroxyl radical ($HO\cdot$) that adds to the substrate or, more often, abstracts a hydrogen atom, for example

$$\mathrm{PhNHNHPh} + 2\,\mathrm{HO^-} \xrightarrow[\mathrm{MeCN}]{-2e^-} \mathrm{PhN{=}NPh} + 2\,\mathrm{H_2O}$$

$$E_{p,a},\ -1.1\ \mathrm{V\ vs.\ SCE} \qquad (11.2)$$

11.2 SULFUR

Elemental Sulfur (S_8). Figure 11.1 illustrates the electrochemical reduction of S_8 in dimethyl sulfoxide (Me_2SO) at a gold electrode.[1] Although the cyclic voltammogram appears to be a simple sequence of two two-electron-per-S_8, pseudoreversible reductions; controlled-potential coulometry at -0.7 V versus SCE indicates that 2.7 electrons per S_8 are consumed to give S_6^{2-}. Controlled-potential coulometry at -1.5 V versus SCE consumes 4.0 electrons per S_8 to produce two S_4^{2-} ions. On the basis of these results and associated spectroscopic studies,[1] the reduction sequence in Me_2SO invovles two ECE steps with ring-

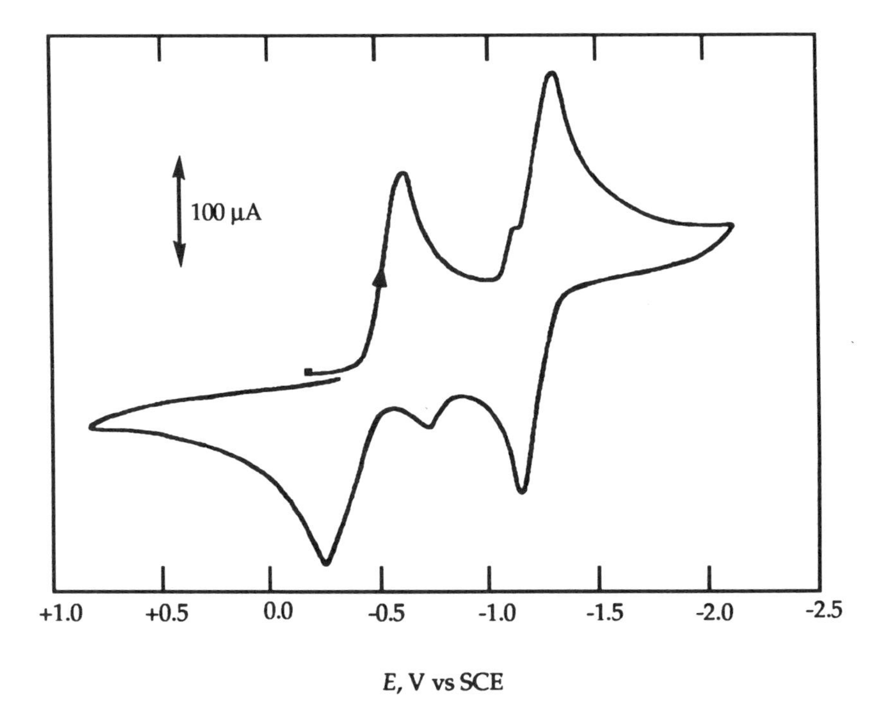

Figure 11.1 Cyclic voltammogram of 2.9 mM S_8 in Me_2SO (0.1 M TEAP) at a scan rate of 0.1 V s^{-1} (Au electrode).

opening and chain-breaking the two chemical steps:

$$c\text{-}S_8 + e^- \xrightarrow[E_{p,c}, -0.60 \text{ V vs. SCE}]{} \cdot S_8^{\cdot -} \xrightarrow{e^-} {}^-S_8^- \xrightarrow[E_{p,c}, -1.29 \text{ V}]{e^-} S_4^{2-} + S_4^{\cdot -} \xrightarrow{e^-} S_4^{2-} \quad (11.3)$$

$$4\, S_8^{2-} \rightleftharpoons 4\, S_6^{2-} + S_8 \qquad K, 8 \times 10^{-8} \text{ M} \quad (11.4)$$

$$S_8^{2-} + S_4^{2-} \rightleftharpoons 2\, S_6^{2-} \qquad K, 2.5 \times 10^1 \quad (11.5)$$

$$S_6^{2-} \rightleftharpoons 2\, S_3^{\cdot -} \qquad K_{diss}, 9 \times 10^{-3} \text{ M} \quad (11.6)$$

$$\lambda_{max}, 618 \text{ nm } (\varepsilon\ 4.5 \times 10^3 \text{ M}^{-1} \text{ cm}^{-1})$$

The appearance of the bright blue solution ($S_3^{\cdot -}$) during the course of electrolysis is dramatic:

$$3\, S_8 + 8\, e^- \xrightarrow{-0.6 \text{ V vs. SCE}} 8\, S_3^{\cdot -} \quad (11.7)$$

$$8\, S_3^{\cdot -} \rightleftharpoons 4\, S_6^{2-} \qquad K_D, 1.1 \times 10^2 \text{ M}^{-1}$$

In aprotic solvents hydroxide ion (HO^-) is an equally effective reductant:[2]

$$3\, S_8 + 8\, HO^- \longrightarrow 3\, S_3^{\cdot -} + 4\, HOOH \quad (11.8)$$

The resultant HOOH reacts with Me_2SO when it is the solvent

$$HOOH + Me_2SO \xrightarrow{HO^-} Me_2S(O)_2 + H_2O \quad (11.9)$$

and with $S_3^{\cdot -}$ in MeCN solutions

$$S_3^{\cdot -} + 2\, HOOH \longrightarrow \tfrac{3}{8} S_8 + O_2^{\cdot -} + 2\, H_2O \quad (11.10)$$

Because $S_3^{\cdot -}$ is the chromophore in the mineral lapis lazuli,[3] the chemistry of Eq. (11.8) may be the basis of its geologic formation.

Sulfur Dioxide (SO_2). In aprotic solvents SO_2 undergoes a reversible one-electron reduction (Figure 11.2):[4]

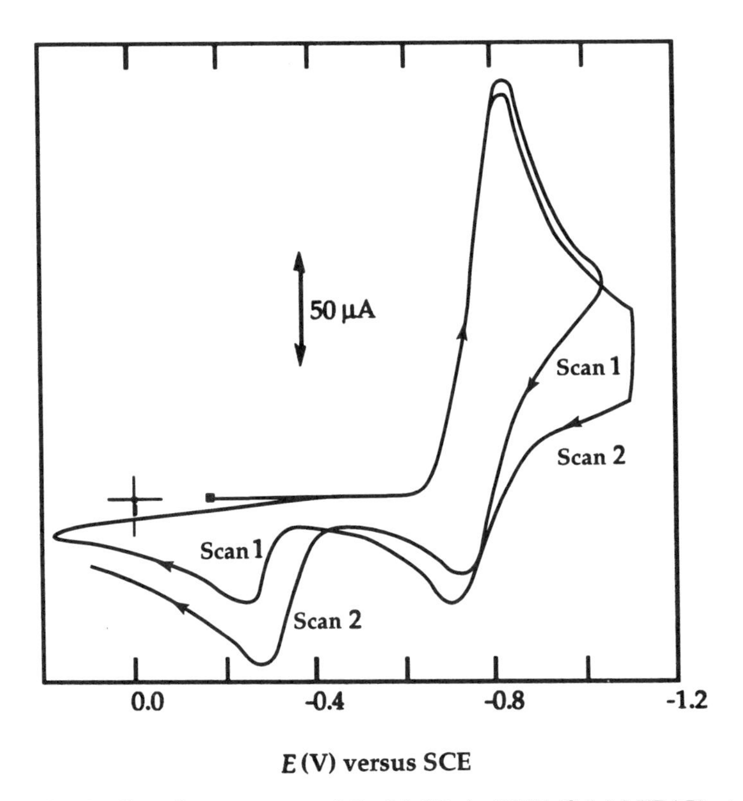

Figure 11.2 Cyclic voltammograms of 5 mM SO_2 in DMF (0.1 M TEAP) at a scan rate of 0.1 V s^{-1} (platinum electrode). Scan 2 held at −1.1 V for 30 s.

$$SO_2 + e^- \underset{DMF}{\rightleftarrows} SO_2^{\bar{\cdot}} \qquad E_{1/2}, -0.75 \text{ V vs. SCE} \tag{11.11}$$

However, the product species interacts with excess SO_2 to give a blue complex that is significantly more difficult to oxidize:

$$SO_2^{\bar{\cdot}} + 2\,SO_2 \longrightarrow (SO_2)_2\dot{S}(O)O^- \tag{11.12}$$

$$(SO_2)_2\dot{S}(O)O^- \xrightarrow{e^-} 3\,SO_2 \qquad E_{p,a}, -0.25 \text{ V vs. SCE}$$

Complete electrolysis yields $SO_2^{\bar{\cdot}}$ which dimerizes to dithionite ion (colorless):

$$2\,SO_2^{\bar{\cdot}} \underset{DMF}{\rightleftarrows} {}^-O_2SSO_2^- \qquad K_D, 2.4 \times 10^1 \text{ M}^{-1} \tag{11.13}$$

In aqueous solutions SO_2 is a moderately strong Brønsted acid:

$$SO_2 + H_2O \rightleftharpoons H_3O^+ + HOS(O)O^- \qquad pK_a, 1.9 \qquad (11.14)$$

Hence, for solutions at pH 2 or greater the reductive electrochemistry is limited to H_3O^+. However, at pH 1 the dominant species is SO_2, and with a mercury electrode, the reduction of H_3O^+ to H_2 is shifted to negative potentials (−2.0 V vs. SCE). For such conditions SO_2 exhibits a well-defined, diffusion-controlled reduction (Figure 11.3):[5]

$$SO_2 + H_3O^+ + e^- \underset{}{\overset{H_2O}{\rightleftharpoons}} HO\dot{S}{=}O \underset{}{\overset{H_3O^+,\, e^- \;\; H_2O}{\rightleftharpoons}} HOSOH$$

$$E_{1/2}, -0.37 \text{ V vs. SCE} \qquad (11.15)$$

The electroactive species (H_3O^+) is reduced to a hydrogen atom ($H\cdot$) that is stabilized via bond formation with the oxygens of SO_2. The hydrogenated product, HOSOH, is unstable and decomposes to elemental sulfur (S_8)

$$2\,HOSOH \longrightarrow \tfrac{1}{8}S_8 + SO_2 + 2\,H_2O \qquad E_{p,c}, -0.28 \text{ V vs. SCE}$$

$$\tfrac{1}{8}S_8 \xrightarrow{Hg(l)} Hg^{II}S(s) \xrightarrow[2\,H_3O^+]{2\,e^-} Hg(l) + H_2S \qquad (11.16)$$

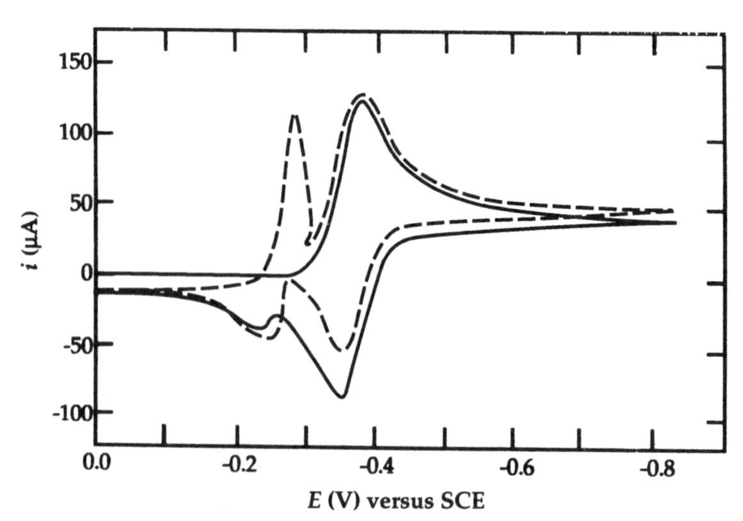

Figure 11.3 Cyclic voltammograms of 1 mM Na_2SO_3 in 0.1 M $HClO_4$. Dashed line represents second potential sweep. Scan rate 0.05 V s^{-1}.

which reacts with the electrode to give a surface deposit of $Hg^{II}S$ (reduced at a prewave in a second reductive scan; Figure 11.3).

Because SO_2 is electrophilic (acidic), it is extremely resistant to direct electron-transfer oxidation. However, in aqueous solutions at pH 1, SO_2 facilitates the oxidation of H_2O [$2\ H_2O \xrightarrow{-e^-} HO\cdot + H_3O^+$; $E^{\circ\prime}$, +2.42 V vs. SCE (pH 1)] at gold electrodes via covalent-bond formation to give sulfuric acid [$(HO)S(O)_2$] in an ECE process (Figure 11.4):[6]

$$SO_2 + 2\ H_2O \xrightarrow{-e^-} HO\text{—}\dot{S}(O)_2 + H_3O^+ \qquad E_{p,a}, +0.40 \text{ V vs. SCE}$$

$$-\Delta G_{BF} = (2.4 - 0.4)\ 96.5 = 192 \text{ kJ mol}^{-1}$$

$$\xrightarrow[2\ H_2O]{-e^-} (HO)_2S(O)_2 + H_3O^+ \qquad (11.17)$$

The peak current is proportional to the SO_2 concentration and its diffusion coefficient, which makes this anodic process suitable as a voltammetric monitor for dissolved SO_2 or gas-phase SO_2 via a gas-permeable membrane.[7]

Propane-1,3-Dithiol [$HS(CH_2)_3SH$]. In acetonitrile the electrochemical oxidation of thiols (*R*SH) is facilitated via the presence of bases (H_2O/HO^-). Ease of oxidation is further facilitated with dithiols that can form cyclic disulfides. Figure 11.5 illustrates cyclic voltammograms at a gold electrode for

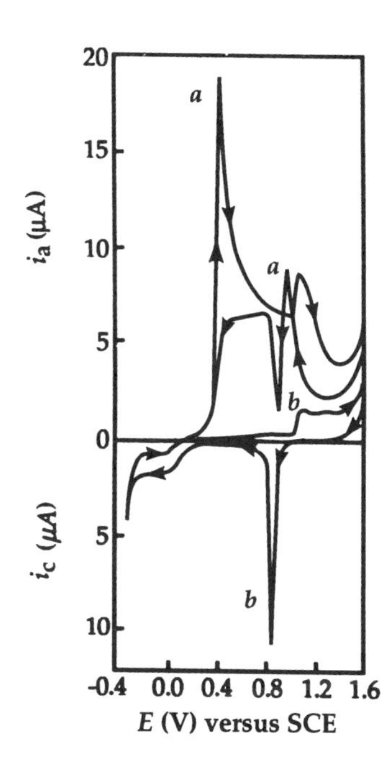

Figure 11.4 Voltammetric oxidation of dissolved sulfur dioxide at a gold electrode in 0.1 M H_2SO_4: cyclic scans.

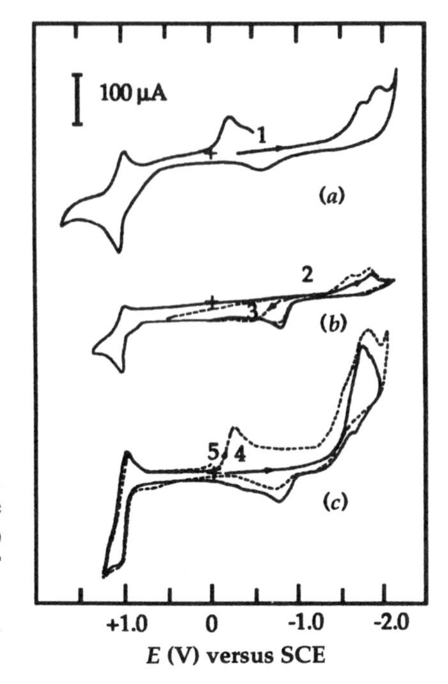

Figure 11.5 Cyclic voltammograms of 1 mM solutions in 0.1 M TEAP in acetronitrile at a gold electrode, scan rate 0.1 V s^{-1}: (*a*) propane-1,2-dithiol (H_2PDT); (*b*) H_2PDT plus 2 equiv of HO^-; (*c*) 4, TMDS; 5, TMDS plus 2 equiv of H_3O^+; 1, 2, 4, and 5, initial negative scans; 3, initial positive scan.

$HS(CH_2)_3SH$ and $\overline{S(CH_2)_3S}$ (in pure solvent, as well as the in presence of H_3O^+ or HO^-).[8] Again, an ECE mechanism is indicated:

$$HS(CH_2)_3SH + H_2O \xrightarrow{-e^-} HS(CH_2)_3S\cdot + H_3O^+ \qquad E_{1/2}, +1.0 \text{ V vs. SCE}$$

$$\xrightarrow[H_2O]{-e^-} \overline{S(CH_2)_3S} + H_3O^+ \tag{11.18}$$

With excess base the dianion is formed, which is easily oxidized by electron transfer or reaction with O_2:

$$^-S(CH_2)_3S^- \xrightarrow{-2e^-} \overline{S(CH_2)_3S} \qquad E_{p,a}, +0.73 \text{ V vs. SCE} \tag{11.19}$$

Reduction of the disulfide product also occurs via an ECE process that is facilitated by the presence of H_3O^+:

$$\overline{S(CH_2)_3S} + e^- \longrightarrow {}^-S(CH_2)_3S\cdot \xrightarrow{e^-} {}^-S(CH_2)_3S^- \qquad E_{p,c}, -1.77 \text{ V vs. SCE} \tag{11.20}$$

$$\overline{S(CH_2)_3S} + H_3O^+ + e^- \xrightarrow[H_2O]{} HS(CH_2)_3S\cdot \xrightarrow[H_2O]{H_3O^+,\, e^-} HS(CH_2)_3SH \qquad E_{p,c}, -0.25 \text{ V} \tag{11.21}$$

For the latter system, the primary electrochemistry is reduction of H_3O^+ to an H· atom, which is stabilized via formation of an S—H bond. Such a process is closely similar to the biological reduction of disulfides via H-atom transfer from NADH/flavoproteins.

11.3 NITROGEN

Nitric Oxide (•NO). Reduction of the oxides of nitrogen (·NO, $\cdot NO_2$, and N_2O) usually involves the addition of hydrogen atoms that are electrogenerated. Figure 11.6 illustrates the chronopotentiometric reduction of ·NO in aqueous media at pH 7.0 and pH 5.0.[9] The use of a mercury electrode inhibits the reduction of H_3O^+ to H_2 ($E°'$, −2.2 V vs. SCE at pH 5), but allows formation of H· when it couples with a substrate via covalent-bond formation:

$$\cdot N{=}O + H_3O^+ (\text{pH } 5.0) + e^- \longrightarrow [NO{-}H] \qquad E_{\tau/4}, -0.9 \text{ V vs. SCE}$$

$$\downarrow 2x$$

$$[HON{=}NOH] \longrightarrow \underset{(N_2O)}{N{\equiv}N{=}O} + H_2O \qquad (11.22)$$

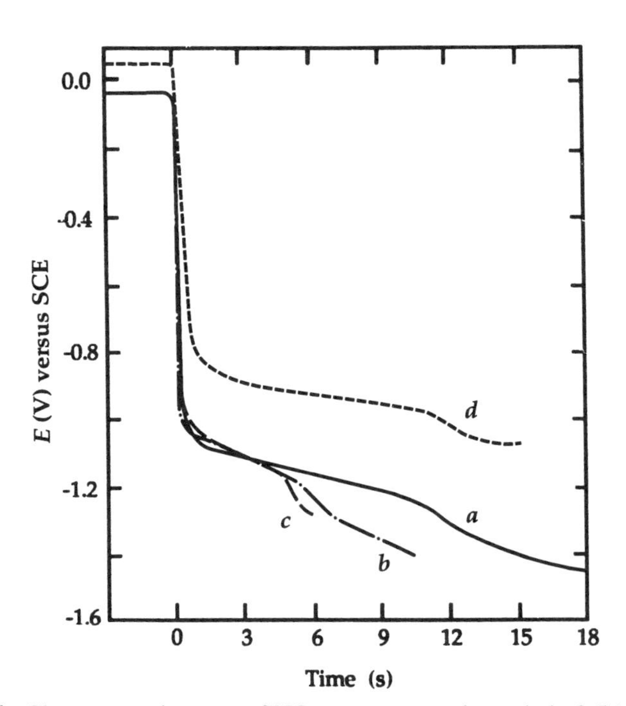

Figure 11.6 Chronopotentiograms of NO at a mercury electrode in 0.5 M Na_2HPO_4. (*a*), pH 7.0, 1.7 mM NO, 60 μA; (*b*), pH 7.0, 0.56 mM NO, 25 μA; (*c*), pH 7.0, 0.18 mM NO, 5 μA; (*d*), pH 5.0, 1.7 mM NO, 60 μA.

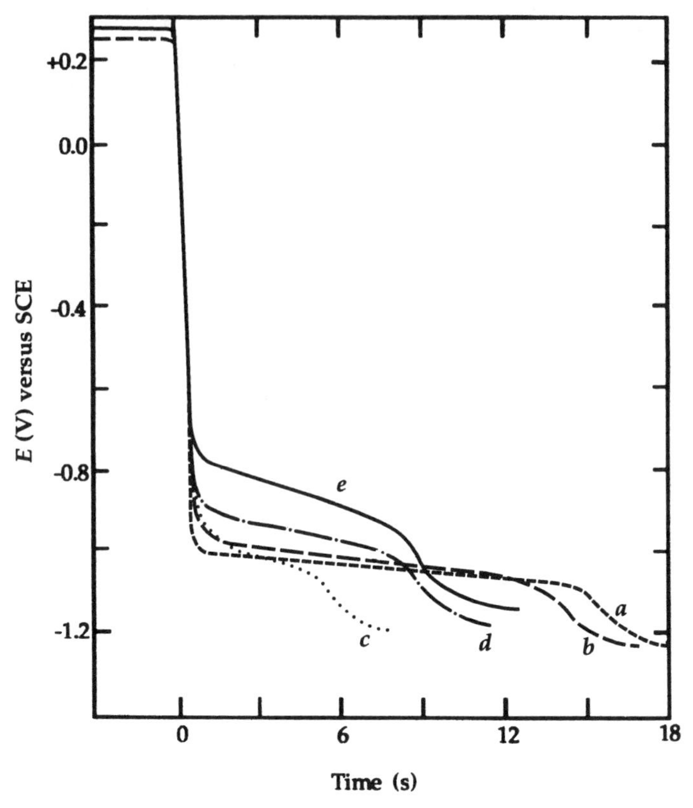

Figure 11.7 Chronopotentiograms of HONO at a mercury electrode in 0.5 M Na_2SO_4 and 0.5 M NaH_2PO_4: (*a*) pH 3.3, 5 mM HONO, 450 μA; (*b*) pH 3.3, 3 mM HONO, 300 μA; (*c*) pH 3.3, 1 mM HONO, 150 μA; (*d*) pH 3.0, 2.5 mM HONO, 300 μA; (*e*) pH 2.0, 2.5 mM HONO, 300 μA.

Nitrous Acid (HON=O). The initial step in the reductive electrochemistry for nitrous acid is closely similar to that for nitric oxide [Eq. (11.22)], but yields several products (Figure 11.7):[9]

$$\mathrm{HON{=}O} + \mathrm{H_3O^+}\,(\mathrm{pH}\ 3.0) + e^- \xrightarrow{(65\%)} [\mathrm{HO\dot{N}OH}] \qquad E_{\tau/4}, -0.9\ \mathrm{V\ vs.\ SCE}$$

$$\mathrm{HON{=}O} + \mathrm{H_3O^+} + e^- \xrightarrow{(35\%)} \cdot\mathrm{NO} + 2\,\mathrm{H_2O}$$

$$\mathrm{N_2O} + \mathrm{H_2O} \xleftarrow{2x} [\mathrm{NOH}] \xleftarrow[\mathrm{H_3O^+}]{e^-} \cdot\mathrm{NO} + 2\,\mathrm{H_2O}$$

$$[\mathrm{HO\dot{N}OH}] \xrightarrow{e^-,\ \mathrm{H_3O^+}} [\mathrm{HN{=}O}] + 2\,\mathrm{H_2O}$$

$$[\mathrm{HN{=}O}] \xrightarrow{2\,e^-,\ 2\,\mathrm{H_3O^+}} \mathrm{H_2NOH} + 2\,\mathrm{H_2O} \qquad (11.23)$$

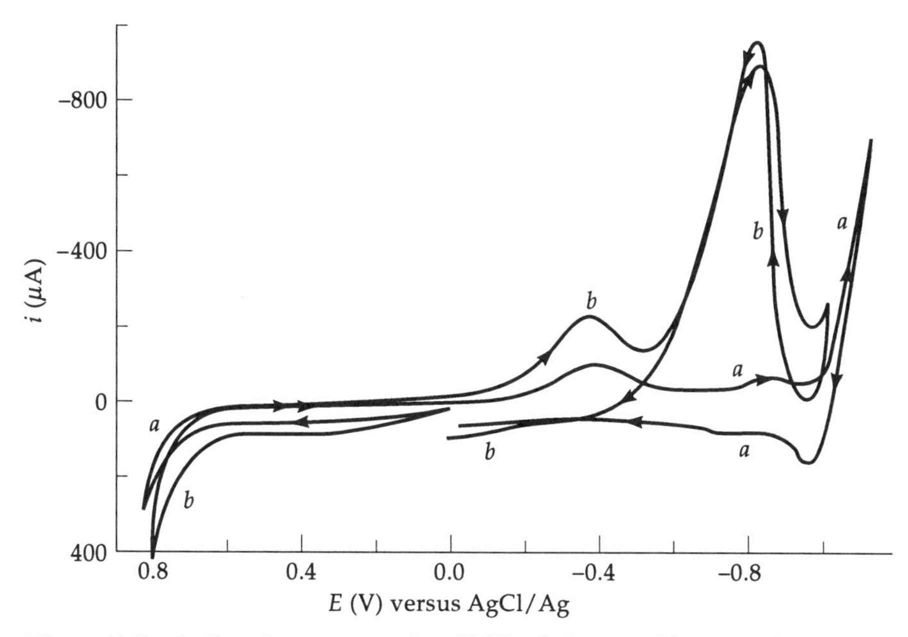

Figure 11.8 Cyclic voltammograms for a KOH solution (*a*) without and (*b*) with N_2O at 1 atm. Initial scan from 0.0 V in a positive direction, pH 14.0; 0.1 V s^{-1}.

Nitrous Oxide (N_2O). In contrast to nitric oxide, N_2O is not readily reduced by hydrogen atoms. However, it is a good O-atom transfer reagent and spontaneously reacts with metal surfaces at elevated temperatures. Under alkaline conditions with a freshly formed platinum surface (Pt*), N_2O induces a voltammetric reduction peak (Figure 11.8):[10]

$$N{\equiv}N{=}O + HO^- \rightleftharpoons N{\equiv}N(O^-)(OH) \xrightarrow[H_2O]{Pt^*(s)} Pt^{II}(OH)_2(s) + N_2 + HO^-$$

$$Pt^{II}(OH)_2(s) \xrightarrow{2e^-} Pt^* + 2\,HO^- \qquad E_{p,c}, -0.8 \text{ V vs. SCE} \qquad (11.24)$$

Because this is a surface reduction, the peaks are similar to those for adsorption.

Hydroxylamine (H_2NOH). The electrochemical oxidation of H_2NOH in dimethyl suloxide at a platinum electrode yields N_2O,[11] and as such is the reverse of the reduction of HON=O under acidic conditions [Eq. (11.23)]. Because the primary electron-transfer step is H_2O/HO^- oxidation ($2\,H_2O \xrightarrow{-e^-} H_3O^+$

+ HO•) and H_3O^+ deactivates the substrate, diffusion-controlled peak currents require at least a 2:1 HO^-/H_2NOH ratio:

$$H_2NOH + HO^- \xrightarrow{-e^-} H\dot{N}OH + H_2O \qquad E_{\tau/4}, +0.27 \text{ V vs. SCE}$$
(pH 10)

$$H\dot{N}OH \xrightarrow{2x} \begin{bmatrix} HNOH \\ | \\ HNOH \end{bmatrix} \longrightarrow H_2NOH + [NOH] \xrightarrow{2x} N_2O + H_2O \qquad (11.25)$$

The overall process is 80% efficient at a platinum electrode and 90% efficient at a gold electrode:

$$2\ H_2NOH + 4\ HO^- \xrightarrow{-4e^-} N_2O + 5\ H_2O \qquad (11.26)$$

Hydrazines and Amines. These substrates are directly oxidized in a base-free matrix (Me_2SO or MeCN) at platinum or glassy-carbon electrodes with the potential determined primarily by the RN—H bond energy, and secondarily by the basicity of the substrate (Figures 11.9 and 11.10);[15,16]

$$2\ \text{PhNHNHPh}\ (R\text{NH}) \xrightarrow[E_{p,a}, +0.6 \text{ V vs. SCE}]{-e^-} \text{Ph}\dot{\text{N}}\text{NHPh} + R\text{NH}_2^+$$

$$\text{Ph}\dot{\text{N}}\text{NHPh} \xrightarrow[R\text{NH}]{-e^-} \text{PhN{=}NPh} + R\text{NH}_2^+ \qquad (11.27)$$

$$2\ \text{PhNH}_2 \xrightarrow[E_{p,a}, +0.9 \text{ V vs. SCE}]{-e^-} \text{Ph}\dot{\text{N}}\text{H} + \text{PhNH}_3^+$$

$$\text{Ph}\dot{\text{N}}\text{H} \xrightarrow[-e^-]{2\ \text{PhNH}_2} \text{PhNHNHPh} + \text{PhNH}_3^+$$

$$\text{PhNHNHPh} \xrightarrow[-2\,e^-]{2\ \text{PhNH}_2} \text{PhN{=}NPh} + 2\ \text{PhNH}_3^+ \qquad (11.28)$$

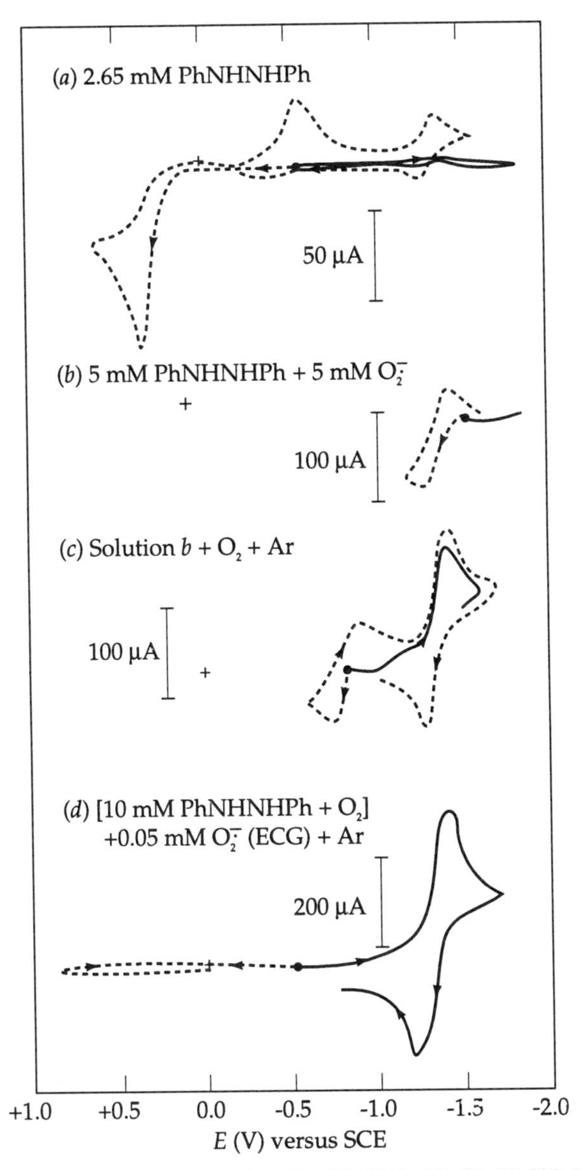

Figure 11.9 Cyclic voltammograms in Me_2SO [0.1 M $(Et_4N)ClO_4$] for (*a*) PhNHNHPh, (*b*) an equimolar mixture of PhNHNHPh and $O_2^{\cdot-}$ (electrogenerated) in an argon atmosphere, (*c*) solution *b* after exposure to O_2 (1 atm) for 3 min followed by a 5-min purge with argon, and (*d*) 10 mM PhNHNHPh in an O_2 (1 atm)-saturated solution after the in situ electrogeneration of 0.05 mM $O_2^{\cdot-}$ (1 min) followed by a 5-min purge with argon. Conditions: a platinum electrode (area 0.23 cm^2) at a scan rate of 0.1 V s^{-1}.

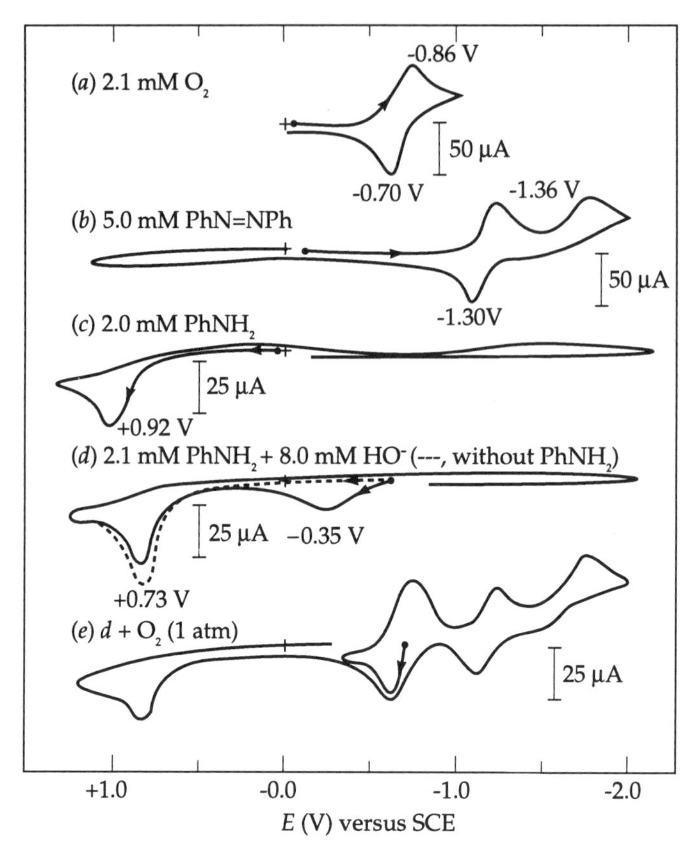

Figure 11.10 Cyclic voltammograms in Me_2SO (0.1 M TEAP) for (*a*) 2.1 mM O_2, (*b*) 5.0 mM PhN=NPh, (*c*) 2.0 mM $PhNH_2$, (*d*) 2.0 mM $PhNH_2$ plus 8.0 mM $(Bu_4N)OH$ (40%, in H_2O), and (*e*) solution *d* saturated with O_2 (1 atm, 2.1 mM) for 20 min and then purged with Ar for 20 min. Conditions: glassy-carbon electrode (area 0.06 cm^2); scan rate 0.1 V s^{-1}, 23°C.

The oxidation potentials in MeCN for several substituted hydrazines and amines are summarized in Table 11.1.

In contrast, when hydroxide ion HO^- is present, it is more easily oxidized than the amine substrates. In MeCN, in the absence of substrate, HO^- is oxidized at +0.7–0.9 V versus SCE. However, with hydrazines and amines present, as in the case of H_2NOH [Eq. (11.25)], the N—H bonds are homolytically cleaved by the $HO\cdot$ product of HO^- oxidation. The latter's oxidation potential is shifted by the difference in the HO—H and *R*N—H bond energies ($-\Delta G_{BF}$). Thus, the oxidation of PhNHNHPh is shifted by −1.7 V when HO^- becomes the electron-transfer mediator; with $PhNH_2$ the shift is −1.1 V:

$$\text{PhNHNHPh} + \text{HO}^- \xrightarrow[E_{p,a},\ -1.1\text{ V vs. SCE}]{-e^-} \text{Ph}\dot{\text{N}}\text{HNHPh} + \text{H}_2\text{O} \tag{11.29}$$

$$\xrightarrow[\text{HO}^-]{-e^-} \text{PhN{=}NPh} + \text{H}_2\text{O}$$

TABLE 11.1 Oxidation Potentials for 1–5 mM Hydrazines and Amines in MeCN [0.1 M $Et_4N(ClO_4)$] and in the Presence of an Equivalent of HO^-: Evaluation of *R*N—H Bond Energies ($-\Delta G_{BF}$)

	$E_{p,a}$ (V) versus SCE		$-\Delta G_{BF}$ (kJ mol^{-1})	
Substrate (*R*N—H)	*R*N—H	*R*N—H/HO^-	Redox Evaluation[a]	(Lit)[b]
Hydrazines				
PhNHNHPh	+0.6	−1.1	247	
MeNHNHMe	+0.1	−0.8	276	
$PhNHNH_2$	+0.4	−0.6	293	
Ph_2NNH_2	+0.5	−0.30	322	
$Ph(Me)NNH_2$		−0.26	326	
Me_2NNH_2	+0.3	+0.07	360	
H_2NNH_2		+0.07	360	335
Amines				
$PhNH_2$	+0.9	−0.18	335	335
Ph_2NH	+0.9	−0.30	322	
Ph(Me)NH		−0.25	331	331
indole[c]	+1.2	0.0	351	
pyrrole[d]	+1.3	0.0	351	
imidazole[e]	+1.0	+0.3	381	
benzimidazole[f]	+1.9	+0.7	423	
dihydrophenazine[g]	+0.1	−1.1	247	
dihydro-3-methyllumiflavin[h]	+0.1	−0.8	276	

[a] $-\Delta G_{BF} = [(E_{p,a})_{RN-H,HO^-/H_2O,RN\cdot} - (E_{p,a})_{H\cdot,HO^-/H_2O}]\ 96.5$; $(E_{p,a})_{H\cdot,HO^-/H_2O} = -3.65$ V versus SCE.

[b] $-\Delta G_{BF} = \Delta H_{DBE} - T\Delta S \approx (\Delta H_{DBE} - 33)$ kJ mol^{-1}; Ref. 12.

[c] (indole structure) [d] (pyrrole structure) [e] (imidazole structure) [f] (benzimidazole structure)

[g] (dihydrophenazine structure), Ref. 13. [h] (dihydro-3-methyllumiflavin structure), Ref. 14.

$$PhNH_2 + HO^- \xrightarrow[-0.2\text{ V vs. SCE}]{-e^-} Ph\dot{N}H + H_2O \quad (11.30)$$

$$Ph\dot{N}H \xrightarrow{HO^-,\ PhNH_2,\ -e^-} PhNHNHPh + H_2O$$

The oxidation potentials for other hydrazines and amines in combination with hydroxide (*R*NH/HO^-) are summarized in Table 11.1.

On the basis of the reduction potential for H_2O in MeCN (Chapter 9, $H_2O \xrightarrow{e^-} H\cdot + HO^-$), a reasonable estimate of the oxidation potential for HO^- in the presence of a free hydrogen atom ($H\cdot$) can be made:

$$HO^- + H\cdot \xrightarrow[GCE]{-e^-} HO{-}H \qquad E_{p,a}, -3.65 \text{ V vs. SCE (MeCN)}$$

$$-\Delta G_{BF}, 464 \text{ kJ mol}^{-1} \qquad -3.80 \text{ V } (Me_2SO) \quad (11.31)$$

Thus, the difference in the ease of oxidation of HO^- in presence of $H\cdot$ (−3.65 V vs. SCE) and PhNHNHPh (*R*NH, −1.1 V) is a measure of the *R*N—H bond energy ($-\Delta G_{BF}$):

$$-\Delta G_{BF} = [(E_{p,a})_{RNH} - (-3.65)]96.5$$
$$= (-1.1 + 3.65)96.5 = 247 \text{ kJ mol}^{-1} \quad (11.32)$$

Similar evaluations for hydrazines and amines are summarized in Table 11.1.

The weak N—H bonds of PhNHNHPh cause it to be an effective reductase (hydrogenase). For example, superoxide ion ($O_2^{\bar{\cdot}}$) transfers a hydrogen atom and a proton from PhNHNHPh:[15]

$$PhNHNHPh + O_2^{\bar{\cdot}} \longrightarrow HOOH + PhN\overset{\bar{\cdot}}{N}Ph \quad (11.33)$$

The product in turn reacts with O_2

$$PhN^{\bar{\cdot}}NPh + O_2 \longrightarrow PhN{=}NPh + O_2^{\bar{\cdot}} \quad (11.34)$$

Hence, bases (HO^-, $O_2^{\bar{\cdot}}$, e_{aq}^-) catalyze the autoxidation of hydrazines, dihydrophenazine (H_2Phen), and dihydroflavins (H_2Fl):

$$PhNHNHPh + O_2 \xrightarrow{HO^-} PhN{=}NPh + HOOH \quad (11.35)$$

This is a convenient system for the synthesis of HOOH and may be similar to the mechanistic path for HOOH synthesis in biology via dihydroflavin proteins. Aniline ($PhNH_2$) and phenylhydrazine ($PhNHNH_2$) in combination with HO^- and O_2 are effective reagents for the in situ synthesis of superoxide ion:[16]

$$2\,PhNH_2 + 2\,HO^- + 2\,O_2 \longrightarrow PhNHNHPh + 2\,O_2^{\bar{\cdot}} + 2\,H_2O \quad (11.36)$$

$$2\,PhNHNH_2 + 2\,HO^- + 3\,O_2 \longrightarrow 2\,O_2^{\bar{\cdot}} + HOOH + 2\,PhH$$
$$+ 2\,N_2 + 2\,H_2O \quad (11.37)$$

Phenazine (Phen). In aprotic solvents (DMF) this tricyclic aromatic heterocycle [phenazine structure] exhibits two one-electron reductions that are

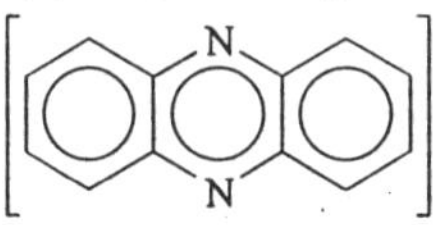

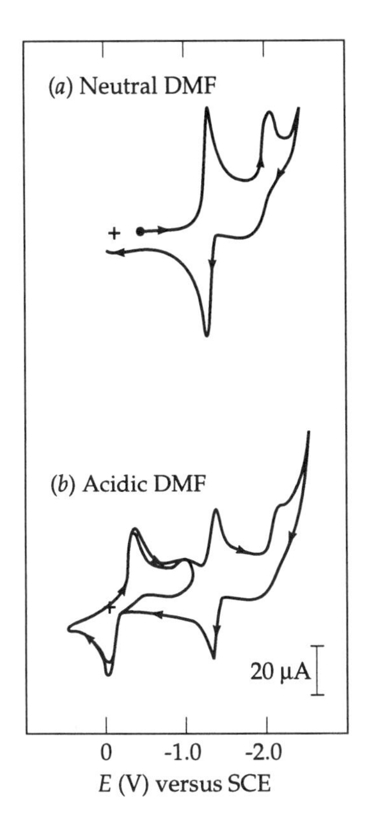

Figure 11.11 Cyclic voltammograms for 1 mM phenazine in neutral and acid dimethylformamide at a platinum electrode. Supporting electrolyte 0.1 M TEAP, scan rate 0.10 V s^{-1}.

similar to those for quinones (Figure 11.11):[17]

$$\text{Phen} + e^- \xrightarrow[E_{1/2},\ -1.14\ \text{V vs. SCE}]{} \text{Phen}\cdot^- \xrightarrow[-1.85\ \text{V}]{-e^-} \text{Phen}^{2-}$$

$$\text{Phen}\cdot^- \xrightarrow{\text{H}_2\text{O, Phen}\cdot^-} \text{Phen} + \text{HPhen}^- + \text{HO}^-$$

$$\text{Phen}^{2-} \xrightarrow{\text{H}_2\text{O}} \text{HPhen}^- + \text{HO}^- \xrightarrow[-1.11\ \text{V}]{-2e^-} \text{Phen} + \text{H}_2\text{O} \qquad (11.38)$$

When Brønsted acids (H_3O^+, HOAc, etc.) are present, they become the electron acceptor to product H· atoms that are stabilized by the aromatic nitrogens of Phen (Figure 11.11):[17]

$$\text{Phen} + \text{H}_3\text{O}^+(\text{DMF}) + e^- \xrightarrow[\text{H}_2\text{O}]{} \text{HPhen}\cdot \xrightarrow{\text{HPhen}\cdot} \text{H}_2\text{Phen} + \text{Phen} \xrightarrow{e^-, -1.14\text{ V}} \text{Phen}\cdot^-$$

$E_{p,c}$, −0.29 V vs. SCE

$$\text{H}_2\text{Phen} \xrightarrow[\text{H}_2\text{O},\ +0.03\text{ V}]{-2e^-} \text{Phen} + 2\,\text{H}_3\text{O}^+(\text{DMF}) \quad (11.39)$$

These H-atom transfer cycles closely resemble those of 3-methyl-lumiflavin and riboflavin, and may represent the redox mechanisms of flavin/hydroflavin/dihydroflavin proteins in biology.

11.4 CARBON

Although the electrochemistry of organic compounds is the focus of the next chapter, several carbon molecules are more appropriate to the present discussion of nonmetals. Carbon dioxide (CO_2, O=C=O) and carbon monoxide (CO, C≡O:) are the products of hydrocarbon combustion, and cynanide ion (CN^-; NC^-, $:N{\equiv}C:^-$) is unique with respect to a stable carbon-centered anion.

Carbon Dioxide (CO_2). Figures 11.12 and 11.13 illustrate the electron-transfer reduction of CO_2 in Me_2SO at gold, platinum, and mercury electrodes.[18, 19] Whereas the reduction at a gold electrode is a one-electron per CO_2 process on a voltammetric time scale, at mercury it is a sequential two-electron process. In both cases the overall reduction is two electrons per CO_2. The products for anhydrous conditions are CO_3^{2-} and CO, and with H_2O present $HOC(O)O^-$ and $HC(O)O^-$:

$$\text{CO}_2 + \text{Au(s)} + e^- \longrightarrow \text{Au}^{\text{I}}\text{—C}(\text{=O})\text{O}^-\,(\text{s}) \qquad E_{p,c}, -2.1\text{ V vs. SCE}$$

$$\text{Au}^{\text{I}}\text{—C}(\text{=O})\text{O}^-\,(\text{s}) \xrightarrow{\text{CO}_2} \text{Au}^{\text{I}}\text{—C(O)OC(O)O}^-(\text{s})$$

$$\text{Au}^{\text{I}}\text{—C(O)OC(O)O}^-(\text{s}) \xrightarrow[\text{H}_2\text{O}]{e^-} \text{Au(s)} + \text{HC(O)O}^- + \text{HOC(O)O}^-$$

$$\text{Au}^{\text{I}}\text{—C(O)OC(O)O}^-(\text{s}) \xrightarrow{e^-} \text{Au(s)} + \text{CO} + {}^-\text{OC(O)O}^- \quad (11.40)$$

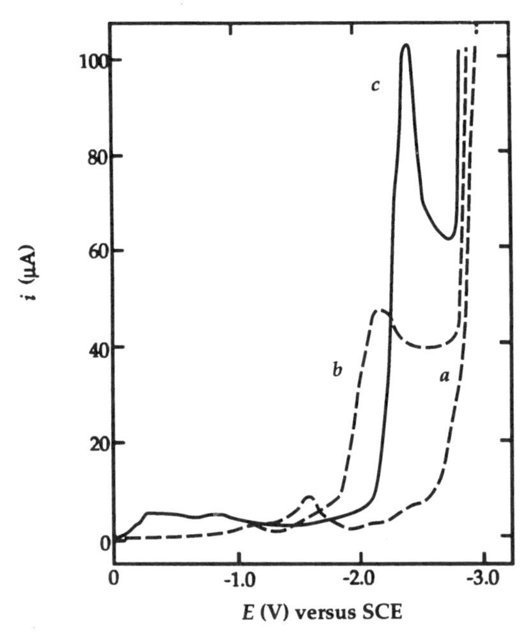

Figure 11.12 Voltammetric scans at gold amalgam electrodes for the reduction of CO_2 in dimethyl sulfoxide: (*a*) Au, N_2; (*b*) Au, 1 mM CO_2; (*c*) Hg(Au), 1 mM CO_2. Electrode area 2.43 cm^2, supporting electrolyte 0.03 M $(Et_4N)ClO_4$, scan rate 0.05 V s^{-1}.

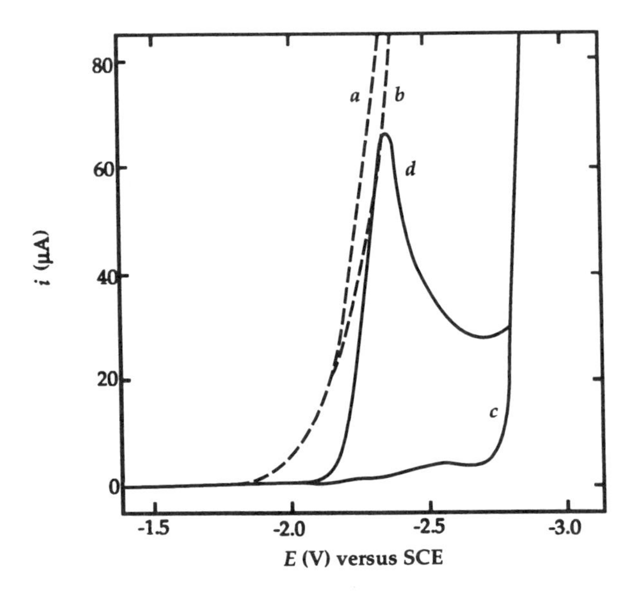

Figure 11.13 Voltammetric scans at gold amalgam electrodes for the reduction of CO_2 in dimethyl sulfoxide: (*a*) Pt, N_2; (*b*) Pt, 1 mM CO_2; (*c*) Hg(Pt), N_2; (*d*) Hg(Pt), 1 mM CO_2. Electrode area 0.215 cm^2, supporting electrolyte 0.03 M $(Et_4N)ClO_4$, scan rate 0.05 V s^{-1}.

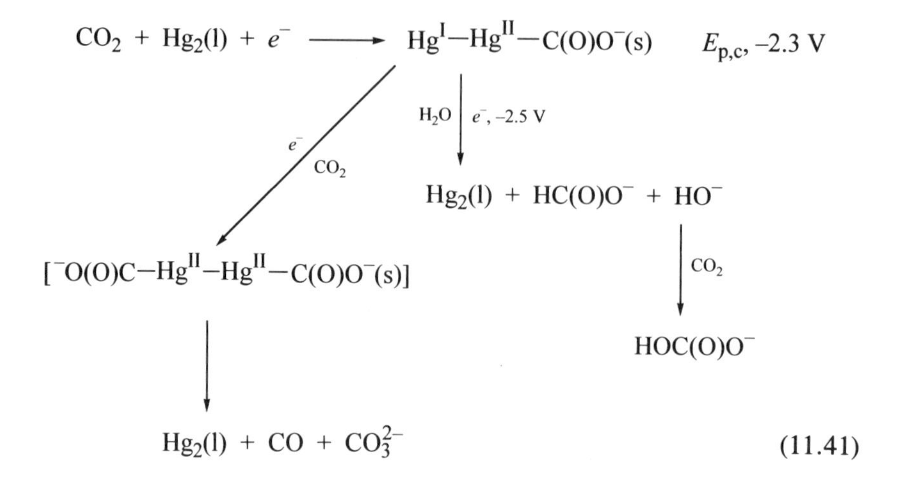

(11.41)

The apparent favored formation of $Au^{I}—C(O)O^-$ as a stabilized $\cdot C(O)O^-$ radical indicates that an Au-electrode/CO_2 combination should be an effective electrosynthesis system for the carboxylation of unsaturated carbon centers.

Carbon Monoxide (CO). At an activated gold electrode in alkaline (0.01 M NaOH) aqueous solution CO is oxidized to CO_3^{2-} via a HO^--centered ECE process (Figure 11.14):[20,21]

$$CO + HO^- \xrightarrow[E_{1/2},\ -0.55\ \text{V vs. SCE}]{-e^-,\ \text{Au}} [Au^{I}—C(O)OH] \xrightarrow[HO^-]{-e^-} (HO)_2C(O) \xrightarrow{2\ HO^-} CO_3^{2-} + 2\ H_2O \qquad (11.42)$$

Again, the activated gold surface stabilizes the carbon radical intermediate $[\cdot C(O)OH]$ and facilitates the second electron-transfer oxidation of HO^- to $HO\cdot$ via coupling to the carbon radical.

Cyanide Ion (NC^-). As is the case with HO^- ion, the NC^- anion is readily oxidized at metal electrodes. Figure 11.15 illustrates that NC^- gives a well-developed voltammetric oxidation at a platinum electrode to form a surface complex[22]

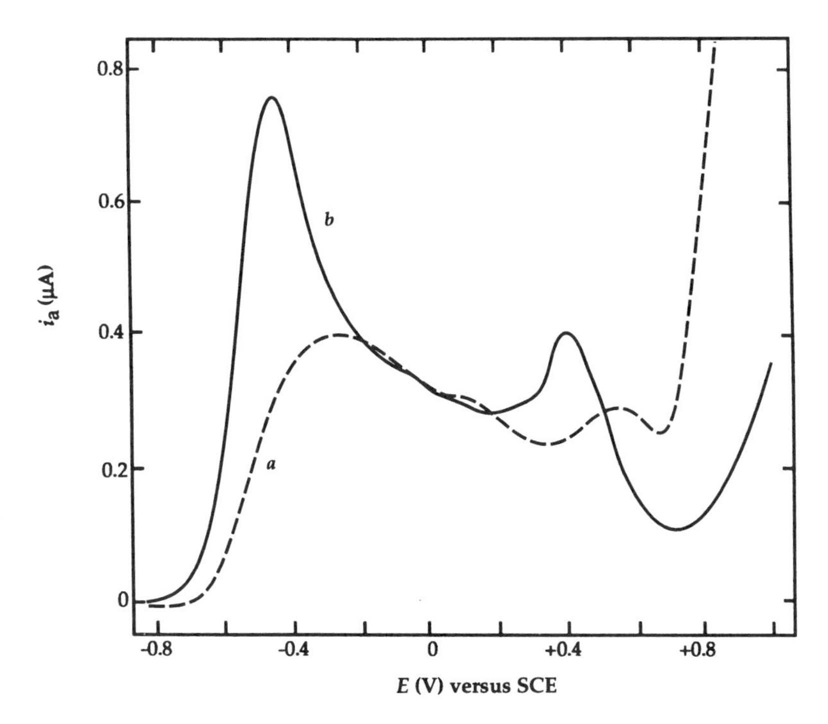

Figure 11.14 Voltammetric oxidation of carbon monoxide at a gold electrode: curve *a*, an inactive electrode; curve *b*, an activated electrode. Scan rate 0.05 V s^{-1}; supporting electrolyte, 0.01 M NaOH in 0.1 M K_2SO_4; solution saturated with 100% CO (~1 mM); electrode area 2.42 cm^2.

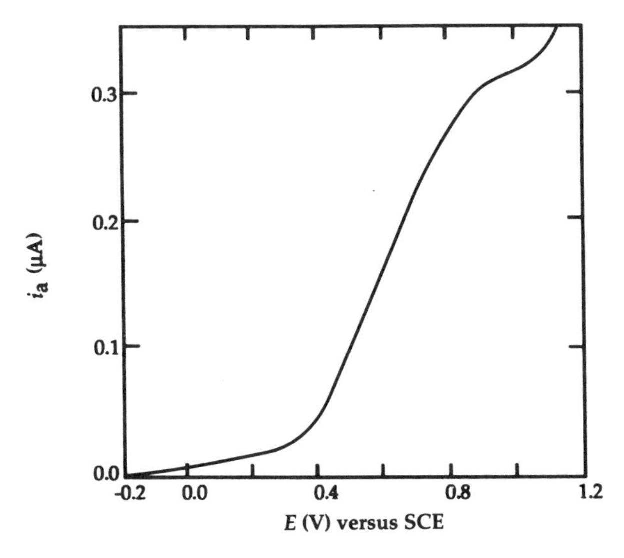

Figure 11.15 Voltammetric oxidation of cyanide ion at a platinum electrode. The solution contained 5 mM KCN and 0.1 M K_2SO_4, and was adjusted to pH 10.0. The scan rate was 0.3 V min^{-1}.

$$Pt^{II}O(s) + NC^- \rightleftharpoons (NC^-)Pt^{II}O(s) \xrightarrow[E_{1/2},\ +1.2\ \text{V vs. SCE}]{-e^-} NC{-}Pt^{III}O(s)$$

$$NC{-}Pt^{III}O(s) \xrightarrow{NC^-,\ -e^-} \left[(NC)_2 Pt^{IV}O \right] \longrightarrow \underset{(CN)_2}{Pt^{II}O(s) + NCCN}$$

(11.43)

which is oxidized in an ECE process to give cyanogen $[(CN)_2]$. Without the oxide film, a platinum electrode forms $Pt^{II}(CN)_4^{2-}$ during the anodic process and dissolves. This is the case for most metal electrodes.

The examples presented in this chapter illustrate that many molecules without metals undergo redox processes in which the voltammetric current is proportional to their concentration. Often these nonmetallic substrates give responses that are due to the facilitated electron-transfer reduction of H_3O^+/H_2O or oxidation of HO^-/H_2O. Hence, any substrate that forms a strong bond with $H\cdot$ or $HO\cdot$ (or has an $HO{-}R$ or an $R{-}H$ group with weak bonds to yield $H{-}OH$) will facilitate these electron-transfer processes at less extreme potentials to give peak currents that are proportional to the substrate concentration. The next two chapters (on organic compounds and organometallic compounds) include many more examples of matrix-centered electron-transfer redox processes.

REFERENCES

1. Martin, R. P.; Doub, W. H., Jr.; Roberts, J. L., Jr.; Sawyer, D. T., *Inorg. Chem.* **1973,** *12*, 1921.
2. Hojo, M.; Sawyer, D. T., *Inorg. Chem.* **1989,** *28*, 1201.
3. Cotton, F. A.; Wilkinson, G., *Advanced Inorganic Chemistry*, 4th ed., Wiley, New York, 1980, p. 513.
4. Martin, R. P.; Sawyer, D. T., *Inorg. Chem.* **1972,** *11*, 2644.
5. Jacobsen, I.; Sawyer, D. T., *J. Electroanal. Chem.* **1967,** *15*, 181.
6. Seo, E. T.; Sawyer, D. T., *Electrochimica Acta* **1965,** *10*, 239.
7. Sawyer, D. T.; George, R. S.; Rhodes, R. C., *Anal. Chem.* **1959,** *31*, 2.
8. Howie, J. K.; Houts, J. J.; Sawyer, D. T., *J. Am. Chem. Soc.* **1977,** *99*, 6323.
9. Ehman, D. L.; Sawyer, D. T., *J. Electroanal. Chem.* **1968,** *16*, 541.
10. Johnson, K. E.; Sawyer, D. T., *J. Electroanal. Chem.* **1974,** *49*, 95.
11. Goolsby, A. D.; Sawyer, D. T., *J. Electroanal. Chem.* **1968,** *19*, 405.
12. Lide, D. R., ed., *CRC Handbook of Chemistry and Physics*, 71st ed., CRC, Boca Raton, Fla., 1990, pp. **9**-86–98.

13. Sawyer, D. T.; Komai, R. Y., *Anal. Chem.* **1972,** *44*, 715.
14. Sawyer, D. T.; Komai, R. Y.; McCreery, R. L., *Experientia Suppl.* **1971,** *18*, 563.
15. Calderwood, T. S.; Johlman, C. L.; Roberts, J. L., Jr.; Wilkins, C. L.; Sawyer, D. T., *J. Am. Chem. Soc.* **1984,** *106*, 4683.
16. Jeon, S.; Sawyer, D. T., *Inorg. Chem.* **1990,** *29*, 4612.
17. Sawyer, D. T.; Komai, R. Y., *Anal. Chem.*, **1972,** *44*, 715.
18. Roberts, J. L., Jr.; Sawyer, D. T., *J. Electroanal. Chem.* **1965,** *9*, 1.
19. Haynes, L. V.; Sawyer, D. T., *Anal. Chem.* **1967,** *39*, 332.
20. Roberts, J. L., Jr.; Sawyer, D. T., *J. Electroanal. Chem.* **1964,** *7*, 315.
21. Roberts, J. L., Jr.; Sawyer, D. T., *Electrochimica Acta* **1965,** *10*, 989.
22. Sawyer, D. T.; Day, R. J., *J. Electroanal. Chem.* **1963,** *5*, 195.

CHAPTER 12

ORGANIC COMPOUNDS

12.1 INTRODUCTION

Most organic reactions are Lewis acid–base processes that involve the interaction of a nucleophilic center with an electrophilic center. Because electrochemistry provides the ultimate nucleophile via the electrons at the cathode surface and the ultimate electrophile via the electron–holes at the anode surface, it is the ideal methodology for the characterization of the electrophilicity and nucleophilicity of molecules. Thus, the carbon centers of saturated hydrocarbons (e.g., CH_4) are resistant to electrochemical reduction and oxidation because of their inert nature (all valence electrons are stabilized in sigma bonds; an absence of any Lewis acid–base character). However, organic molecules with electrophilic components [e.g., alkyl-, aryl-, and acyl-halides; carbonyl groups; unsaturated and aromatic hydrocarbons; nitro groups; Brønsted acids (HA)] react with the ultimate nucleophile (the solvated electron or its conjugate acid [$H\cdot$]) directly or via its reaction product hydronium ion (H_3O^+) or Brønsted acids (HA).

For those electrophiles (El) that undergo direct electron-transfer reduction at an inert electrode (glassy-carbon), the reduction potential (E_{red}) is a measure of their electron affinity and electrophilicity [relative to that for H_3O^+ (-2.10 V vs. NHE in aqueous media)] (the more positive the potential the more electrophilic the molecule; see Chapter 1]:

$$El + e^- \rightleftharpoons El^-\cdot \qquad E_{red} \tag{12.1}$$

Often the solution matrix contains Lewis acids (e.g, H_3O^+) that are more electrophilic than the substrate molecule, and in combination are even more electrophilic:

$$El + H_3O^+ + e^- \xrightarrow[H_2O]{} [HEl\cdot] \xrightarrow[H_2O]{e^-,\, H_3O^+} HElH \qquad (E_{red})_{EC} \tag{12.2}$$

The first-formed intermediate (with an unpaired electron) in combination with a second Lewis acid molecule has even greater electron affinity, and is reduced at a more positive potential to give a voltammogram that appears to be the result of an irreversible two-electron reduction process. In most cases it is an ECEC process in which each electron transfer (E part of the ECEC mechanism) to the Lewis acid (H_3O^+) is reversible to give a product ($H\cdot$) that forms a covalent bond with the substrate ($H{-}El\cdot$) (the C part of the mechanism).

Conversely, nucleophilic molecules (*Nu*) [Lewis bases; e.g., catechols, hydroquinones, phenols, alcohols, and thiols (and their anions); aromatic hydrocarbons and amines (benzene, toluene, pyridine, bipyridine)] can be oxidized by (1) direct electron-transfer oxidation [Eq. (12.3)] or (2) by coupling with the oxidation product of H_2O (or HO^-), hydroxyl radical ($HO\cdot$) [Eq. (12-4)]:

$$Nu{:} \overset{-e^-}{\rightleftarrows} Nu^{+\cdot} \qquad E_{ox} \tag{12.3}$$

The potential (E_{ox}) for those nucleophiles that undergo direct electron-transfer oxidation at an inert electrode is a quantitative measure of their nucleophilicity (the more negative the potential, the more nucleophilic the molecule; see Table 1.3 for representative values). In many cases water in the solvent matrix (or as the solvent) is more nucleophilic than the substrate molecule, and in combination is even easier to oxidize, which often results in an ECEC oxidation process:

$$Nu{:} + 2\,H_2O \xrightarrow[H_3O^+]{-e^-} [HO{-}Nu\cdot] \xrightarrow[H_3O^+]{-e^-,\, 2\,H_2O} Nu(OH)_2 \qquad (E_{ox})_{EC} \tag{12.4}$$

The electrochemical behavior of organic compounds has been of interest from the beginning of voltammetric methods. A summary of early polarographic investigations is available.[1] Two later compilations[2,3] review a great number of electroanalytical investigations.

The preparative aspects of organic electrochemistry are presented in the fundamental review by Lund and Baizer.[4] An earlier review[5] includes the technical aspects of electroorganic synthesis. Indirect methods in preparative organic electrochemistry also have been summarized;[6] these methods are used in the production of fine chemicals and pharmaceuticals. Electroenzymatic synthesis also has been reviewed.[7]

12.2 REDUCTION OF ELECTROPHILIC SUBSTRATES (LEWIS ACIDS)

Alkyl- and Aryl-Halides. Because the halo-groups of organic molecules have large electronegativities and electron affinities, all halo-carbon molecules are electrophilic. Their electrochemical reduction potential is a measure of their electrophilicity (and electron affinity), which is illustrated in Figure 12.1 for hexachlorobenzene (C_6Cl_6), 1,2,3,4-tetrachlorobenzene (1,2,3,4-$C_6H_2Cl_4$), and *n*-butyl iodide (*n*-BuI).[8,9] Table 12.1 summarizes the reduction potentials for several alkyl-halides and aryl-chlorides.[8–10]

In that absence of other Lewis acids alkyl-halides undergo direct electron addition at the electrode surface with subsequent stimulated electron transfer to (1) a second substrate and coupling (*R*—*R*) or (2) a Brønsted acid (H*A*; H_2O) to replace the C—*X* bond with a C—H bond. For example

$$n\text{-BuI} + e^- \longrightarrow [n\text{-Bu}\dot{\text{I}}^-] \quad E_{p,c}, -2.05 \text{ V vs. SCE} \tag{12.5}$$

$$(1)\quad [n\text{-Bu}\dot{\text{I}}^-] \xrightarrow{n\text{-BuI},\, e^-} n\text{-Bu—Bu-}n + 2\,\text{I}^-$$

$$(2)\quad [n\text{-Bu}\dot{\text{I}}^-] \xrightarrow{H_2O,\, e^-} n\text{-BuH} + \text{I}^- + \text{HO}^-$$

In both cases the overall process is an irreversible two-electron reduction via either (1) an EE path or (2) an ECEC path; the first electron transfer is the most difficult and depends on the substrates electrophilicity. In the presence of hydronium ion the primary electron transfer will be to the most electrophilic center, for instance

$$H_3O^+ + n\text{-BuI} + e^- \xrightarrow[-H_2O]{} [n\text{-Bu}\dot{\text{I}}\text{H}] \tag{12.6}$$

$$[n\text{-Bu}\dot{\text{I}}\text{H}] \xrightarrow[-H_2O]{H_3O^+,\, e^-} n\text{-BuH} + \text{HI}$$

Thus, the reduction of *n*-BuI is the equivalent of the addition of two hydrogen atoms [H·] (generated via the electrochemical reduction of the two hydronium ions).

With aryl chlorides (e.g., $PhCl_6$) a similar EEC process occurs as each chlorine atom is replaced with a hydrogen atom (Figure 12.1 and Table 12.1):

$$PhCl_6 + e^- \longrightarrow [PhCl_6^{\dot{}-}] \tag{12.7}$$

$$[PhCl_6^{\dot{}-}] \xrightarrow{e^-,\, H_2O} HPhCl_5 + Cl^- + HO^-$$

Hence, the $PhCl_6$ exhibits six irreversible two-electron reductions (each product species less electrophilic than its precursor) to yield at −2.8 V versus SCE benzene (PhH); an overall 12-electron process:

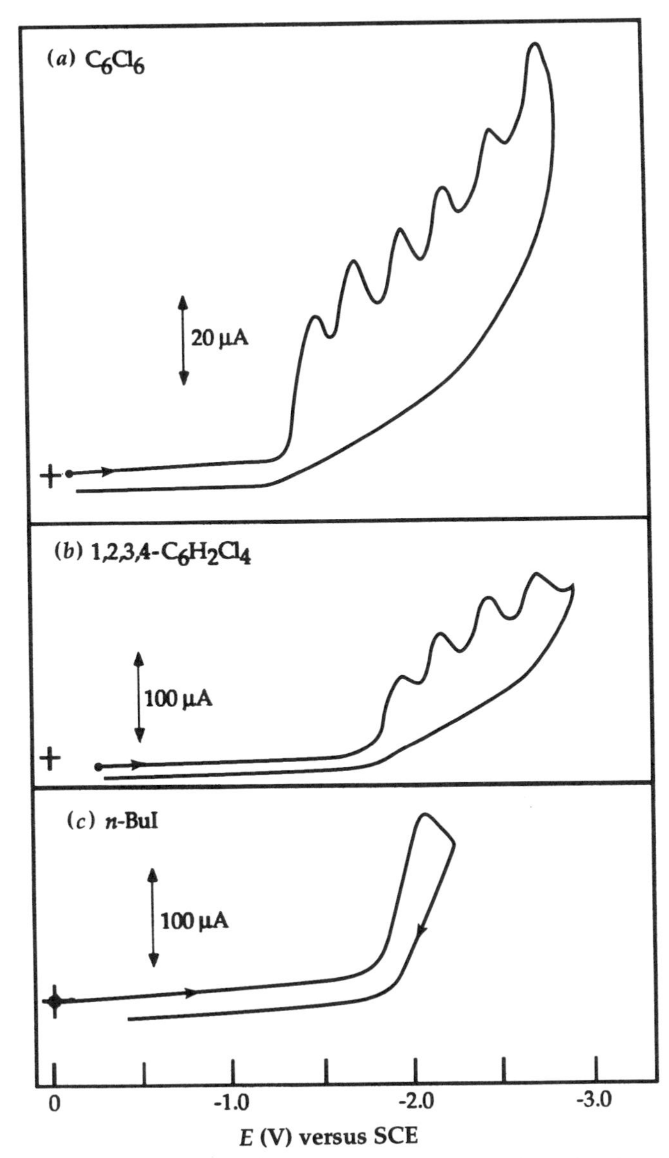

Figure 12.1 Cyclic voltammograms for chlorinated aromatic molecules and *n*-butyl iodide in dimethylformamide (0.1 M TEAP) at a glassy-carbon electrode (area 0.062 cm^2): (*a*) 1.1 mM C_6Cl_6; (*b*) 2.3 mM 1,2,3,4-$C_6H_2Cl_4$; (*c*) 20 mM *n*-BuI.

$$PhCl_6 + 6\,H_2O + 12\,e^- \rightarrow PhH + 6\,Cl^- + 6\,HO^- \qquad (12.8)$$

Although such electrolyses are done in aprotic solvents (e.g., DMF, DMSO, MeCN), even the most rigorously dried solvent contains 3–20 mM H_2O (50–350 ppm). If the solvent has a degree of Brønsted aciditity (e.g., alcohols and ketones), it can serve as a source of hydrogen atoms. An especially lucid

TABLE 12.1 Redox Potentials for Alkyl Halides (*RX*) and Aryl Chlorides (*Ar*Cl_x) in Dimethyl Formamide at a Glassy-Carbon Electrode[a]

A. Alkyl Halides

	$E_{p,c}$[b] (V) versus SCE[c]		
RX	I	Br	Cl
CH_3—	−2.10		
n-C_4H_9—	−2.05	−2.41	
sec-C_4H_9—	−1.92	−2.35	
t-C_4H_9—	−1.78	−2.25	
c-C_6H_{11}—	−2.03	−2.48	
$PhCH_2$—		−1.68	−1.90
ClH_2C—			−2.05
Cl_2HC—			−1.99
Cl_3C—			−1.13
FCl_2C—			−1.71
$(F_3C)Cl_2C$—			−1.31
$PhCl_2C$—			−1.47
[(*p*-ClPh)$_2$HC]Cl_2C—(DDT)			−1.50
[(*p*-ClPh)$_2$FC]Cl_2C—(F-DDT)			−1.44

B. Aryl Halides

ArCl_x	$E_{p,c}$ (V) versus SCE	*Ar*Cl_x	$E_{p,c}$ (V) versus SCE
PhCl (C_6H_5Cl)	−2.7	$PhCl_5$	−1.6
1,2-$PhCl_2$	−2.5	$PhCl_6$	−1.4
1,2,3-$PhCl_3$	−2.2	Cl_5Ph—$PhCl_5$	−1.5
1,2,3,4-$PhCl_4$	−1.9		

[a] 1 mM solutions in DMF [0.1 M $(Et_4N)ClO_4$]; Refs. 9 and 16.
[b] $E_{p,c}$, the reduction-peak potential.
[c] Saturated calomel electrode (SCE) versus NHE, +2.44 V.

discussion of this important mechanistic pathway for the reduction of unsaturated carbon centers has been presented.[11]

The electrolytic displacement of the chlorines of polychlorobiphenyl molecules (PCBs) has been demonstrated as effective technology for hazardous waste remediation,[8] for example

$$Cl_5Ph{-}PhCl_5 + 10\ H_2O + 20\ e^- \xrightarrow{-2.8\,V} \underset{\text{(biphenyl)}}{HPh{-}PhH} + 10\ Cl^- + 10\ HO^- \qquad (12.9)$$

Quinones, Semiquinones, and Catechols. All molecules with unsaturated bonds (olefins, acetylenes, aromatics, carbonyls, quinones, etc.) have a degree of electrophilicity and electron affinity. Within a class, the extent of conjugation increases the electron affinity (reduction of benzene occurs at a less negative potential than 1-butene) and the presence of unsaturated carbon–oxygen (carbonyl) functions within a conjugated system (e.g., quinones; O=⟨⟩=O also enhances the electron affinity of the molecule.

Figure 12.2 illustrates the cyclic voltammograms for (*a*) 3,5-di-*tert*-butyl-*o*-quinone (3,5-DBTQ), (*b*) its anion radical (3,5-DTBSQ$^{\cdot-}$), and (*c*) 3,5-DTBSQ$^{\cdot-}$) in the presence of an equivalent of HO^-.[12] The first reduction step is a reversible one-electron process that is followed by a second one-electron reduction, which can be reversible in rigorously anhydrous media to give catechol dianion (3,5-DTBC^{2-}):

$$\text{3,5-DTBQ} + e^- \rightleftharpoons \text{3,5-DTBSQ}^{\cdot-} \overset{e-}{\rightleftharpoons} \text{3,5-DTBC}^{2-} \quad (12.10)$$

Although DTBC^{2-} is a strong base that is hydrolyzed by residual H_2O

$$\text{DTBC}^{2-} + H_2O \rightleftharpoons \text{DTBCH}^- + OH^- \quad (12.11)$$

it is also a strong reductant that can reduce H_2O in DMF (Figure 12.2*b*):

$$\text{DTBC}^{2-} + H_2O \longrightarrow \text{DTBSQ}^{\cdot-} + \tfrac{1}{2}H_2 + HO^- \quad (12.12)$$

In many cases, the second reduction step of quinones [Eq. (12.10)] is irreversible and due to the facilitated reduction of residual H_2O:

$$\text{3,5-DTBSQ}^{\cdot-} + H_2O + e^- \longrightarrow \text{3,5-DTBCH}^- + HO^- \quad (12.13)$$

In the presence of hydronium ions (H_3O^+), the reduction of the quinones is an irreversible two-electron process (ECEC), in which the first step is the more difficult (requiring the more negative potential).

$$\text{3,5-DTBQ} + H_3O^+ + e^- \xrightarrow[-H_2O]{} [\text{3,5-DTBS}\dot{\text{Q}}\text{H}] \xrightarrow[-H_2O]{e^-,\ H_3O^+} \text{3,5-DTBCH}_2 \quad (12.14)$$

The redox potentials for several quionones (Q; 3,5-di-*tert*-butyl-*o*-quinone, *o*-benzoquinone, *p*-benzoquinone, and tetrafluoro-*o*-benzoquinone), their semiquinone anion radicals (SQ$^{\cdot-}$), and their fully reduced forms [catechols (H_2Cat) and catechol anions ($HCat^-$)] in four aprotic solvents [acetronitrile (AN), dimethylformamide (DMF), dimethylacetaminde (DMA), and dimethyl sulfoxide (Me_2SO)] are summarized in Table 12.2.[12]

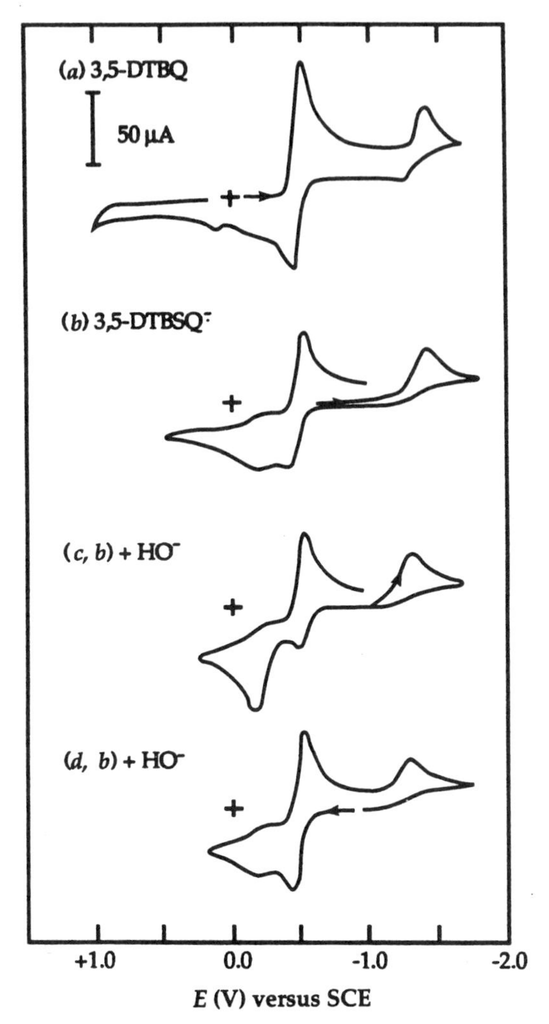

Figure 12.2 Cyclic voltammograms of (*a*) 2.0 mM 3,5-di-*tert*-butyl-*o*-benzoquinone; (*b*) 2.0 mM, 3,5-di-*tert*-butyl-*o*-semiquinone anion (formed by controlled potential electrolysis at −0.80 V; (*c*) solution *b* plus 2.0 mM tetraethylammonium hydroxide, initial cathodic scan; and (*d*) solution *c*, initial anodic scan. All solutions were in dimethylformamide that contained 0.1 M TEAP at a Pt electrode (surface area 0.23 cm^2). Scan rate was 0.1 V s^{-1}.

Although the dianion ($DTBC^{2-}$) in DMF is a strong base that is hydrolyzed by residual H_2O in DMF

$$Cat^{2-} + H_2O \rightleftharpoons HCat^- + HO^- \tag{12.15}$$

the experiments illustrated in Figure 12.2 indicate that it is also in an effective reductant of H_2O (addition of HO^- suppresses the acidity and reducibility of

TABLE 12.2 Summary of Redox Data for Quinones, Catechols, and Hydroquinone in Aprotic Media[a]

Reaction	Solvent[b]	Redox Potential (V) versus SCE[c]				
		DTBQ	*o*-Q	*p*-Q	TCQ	TFQ
(a) $Q + e^- \rightleftharpoons SQ^{\overline{\cdot}}$	AN	−0.51	−0.31	−0.44	+0.18	+0.17
	DMF	−0.49	−0.29	−0.42	+0.16	
	DMA	−0.44	−0.24	−0.37	+0.15	
	Me_2SO	−0.47	−0.26	−0.39	+0.14	
(b) $SQ^{\overline{\cdot}} + e^- \rightleftharpoons Cat^{2-} \xrightarrow{H_2O} SQ^{\overline{\cdot}} + \frac{1}{2} H_2 + HO^-$	AN	−1.30	−0.92	−1.12	−0.60	[d]
$SQ^{\overline{\cdot}} + H_2O + e^- \longrightarrow HCat^- + HO^-$	DMF	−1.34	−0.98	−1.31	−0.69	
	DMA	−1.18	−0.93	−1.37	−0.58	
	Me_2SO	−1.18	−1.03	−1.28	−0.61	
(c)[e] $H_2Cat + e^- \longrightarrow \frac{1}{2} H_2 + HCat^-$	AN	−1.22	−1.17	−1.17	−0.84	−0.51
	DMF	−1.47	−1.33	−1.92	−1.15	
	DMA	−1.25	−1.38	−1.00	−1.22	
	Me_2SO	−1.84	−1.80	[d]	−1.80	
(d)[e] $HCat^- + e^- \longrightarrow \frac{1}{2} H_2 + Cat^{2-}$	AN	[d]	−1.75	[d]	[d]	−0.80
	DMF	−2.25	−1.69	[d]	−1.85	
	DMA	−2.15	−1.79	−1.54	[d]	
	Me_2SO	[d]	[d]	[d]	[d]	
(e)[e] $Q + 2\ H_3O^+ + 2\ e^- \longrightarrow H_2Cat + 2\ H_2O$	AN	+0.40	+0.69	+0.42	+0.60	+0.47
	DMF	−0.32	−0.02	−0.19	+0.11	+0.10
	DMA	−0.26	−0.03	−0.24	+0.09	
	Me_2SO	−0.19	−0.11	−0.27	+0.08	

TABLE 12.2 (*Continued*)

Reaction	Solvent[b]	Redox Potential (V) versus SCE[c]				
		DTBQ	*o*-Q	*p*-Q	TCQ	TFQ
(f)[f] $HCat^- \xrightarrow{-e^-} HSQ\cdot \longrightarrow \frac{1}{2} H_2Cat + \frac{1}{2} Q$	AN	−0.12	+0.06	−0.37	[d]	[d]
	DMF	−0.14	−0.02	−0.39	+0.32	
	DMA	−0.14	+0.04	−0.42	+0.13	
	Me_2SO	−0.13	−0.07	−0.39	+0.14	
(g)[f] $H_2Cat \xrightarrow{H_2O} Q + 2\ H_3O^+ + 2\ e^-$	AN	+1.19	+1.32	+1.18	+1.36	+1.63
	DMF	+0.97	+1.12	+0.94	+1.27	
	DMA	+1.07	+1.02	+0.93	+1.22	
	Me_2SO	+0.96	+1.04	+0.92	+1.21	

[a] H_2Cat represents catechol or hydroquinone, Q quinone, $SQ^{\bar{\cdot}}$ semiquinone anion, $HCat^-$ catechol monoanion, and Cat^{2-} catechol dianion; DTBQ represents 3,5-di-*tert*-butyl-*o*-quinone, *o*-Q *o*-benzoquinone, *p*-Q *p*-benzoquinone, TCQ tetrachloro-*o*-benzoquinone, and TFQ tetrafluoro-*o*-benzoquinone.

[b] AN represents acetonitrile, DMF dimethylformamide, DMA dimethylacetamide, and Me_2SO dimethyl sulfoxide.

[c] For reversible processes, the redox potential is the mean of the cathodic and anodic peak potentials; for irreversible processes, it is the peak potential at a scan rate of 0.1 V s^{-1}.

[d] Not observed in this solvent.

[e] Irreversible; cathodic peak potential.

[f] Irreversible; anodic peak potential.

H_2O):

$$Cat^{2-} + H_2O \longrightarrow \tfrac{1}{2} H_2 + SQ^{\cdot -} + HO^- \qquad (12.16)$$

The acidic hydrogens of catechols (H_2Cat and $HCat^-$) can be reduced at a platinum electrode (Table 12.2), and the potential for reduction is a measure of their acidities (pK_a) in the various solvents (see Chapter 8). Thus, tetrachlorocatechol and tetrafluorocatechol in acetonitrile are the strongest Brønsted acids of the group, and 3,5-$DTBCH_2$ in dimethyl sulfoxide is the weakest acid. On the basis of the potentials in Table 12.2 and the evaluation procedure discussed in Chapter 8, the pK_a values in DMF for $TFCH_2$, $TCCH_2$, $CatH_2$, $DBTCH_2$, and HQ (hydroquinone) are 8.6, 11.0, 15.6, 19.9, and 24.9, respectively.[12]

Table 12.2 also summarizes the oxidation potentials for the fully reduced forms of the quinones (catechols; H_2Cat, $HCat^-$, and hydroquinones; H_2Q, HQ^-), with the first electron removal the most difficult [e.g, 3,5-$DTBCH^-$ in DMF (Figure 12.2*c*)] in an irreversible two-electron oxidation via an ECE mechanism:

$$HCat^- \xrightarrow{-e^-} HSQ\cdot \xrightarrow[HCat^-]{-e^-} Q + H_2Cat \qquad E_{p,a}, +0.14 \text{ V vs. SCE} \qquad (12.17)$$

Neutral catechols (H_2Cat) and hydroquinones (H_2Q) are much more resistant to electron removal because of their release of a proton via the solvent matrix; the more basic the solvent, the less positive the oxidation potential (Figure 12.3 and Table 12.2):

$$H_2Cat^- \xrightarrow[sol \;\searrow\; Hsol^+]{-e^-} [HSQ\cdot] \xrightarrow[sol \;\searrow\; Hsol^+]{-e^-} Q \qquad (E_{p,a})_{DMF}, +0.97 \text{ V}$$
$$(E_{p,a})_{AN}, +1.19 \text{ V} \qquad (12.18)$$

In acetonitrile, the residual H_2O is the stronger base, which results in the formation of H_3O^+.

The electrochemistry of quinones is surprisingly similar to that of dioxygen. It is as if a conjugated carbon link is inserted between two oxygen atoms:

$$Q \longrightarrow SQ^{\cdot -} \longrightarrow HCat^- \text{ vs. } O_2 \longrightarrow O_2^{\cdot -} \longrightarrow HOO^-;$$
$$Q \longrightarrow H_2Cat \text{ vs. } O_2 \longrightarrow HOOH.$$

Dehydroascorbic Acid and Ascorbic Acid. The oxidized and reduced forms of vitamin C (dehydroascorbic acid and ascorbic acid, respectively) have redox characteristics that are similar to those for *o*-quinone/catechol systems.[13] Al-

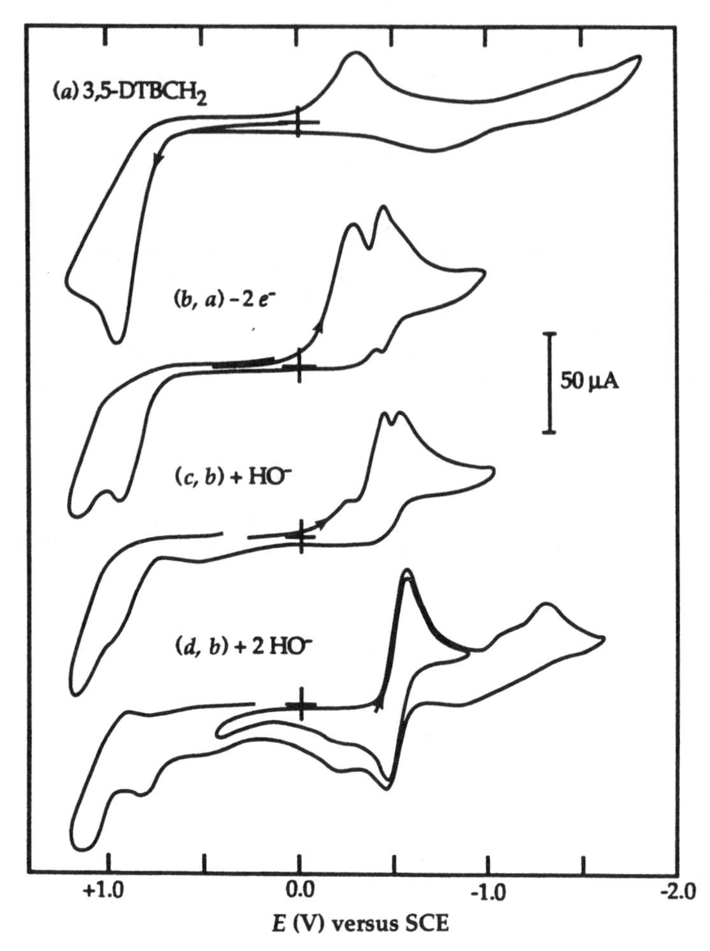

Figure 12.3 Cyclic voltammograms of (*a*) 1.45 mM 3,5-di-*tert*-butyl catechol, (*b*) solution *a* after exhaustive oxidation at +1.25 V, (*c*) solution *b* plus 1.45 mM tetraethylammonium hydroxide, and (*d*) solution *b* plus 2.90 mM tetraethylammonium hydroxide. All solutions were in dimethylformamide that contained 0.1 M tetraethylammonium perchlorate at a Pt electrode (surface area 0.23 cm^2). Scan rate was 0.1 V s^{-1}.

though dehydroascorbic acid (**A**) traditionally has been viewed as a tricarbonyl with a $C_1{-}C_4$ lactone ring

$$\begin{array}{c} \text{O} \\ \diagup \quad \diagdown \\ CH_2(OH)CH(OH)^4CH \qquad C^1{=}O \\ \diagdown \qquad \diagup \\ C{-\!-}C \\ \| \qquad \| \\ O \qquad O \end{array} \qquad (\mathbf{A})$$

NMR studies[14] have established that it is a symmetrical dimer (**A_2**) in dry DMF solutions. Addition of H_2O causes the dimer to dissociate to a hydrated, bicyclic monomer [**$A(OH_2)$**] and its dihydrate [**$A(OH_2)_2$**]:

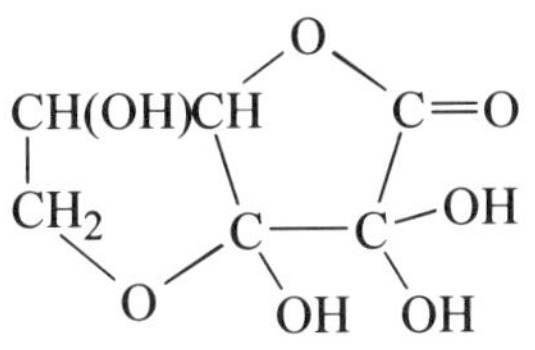

CH(OH)CH, O, C=O, CH_2, C—C—OH, O, OH OH **[$A(OH_2)$]**

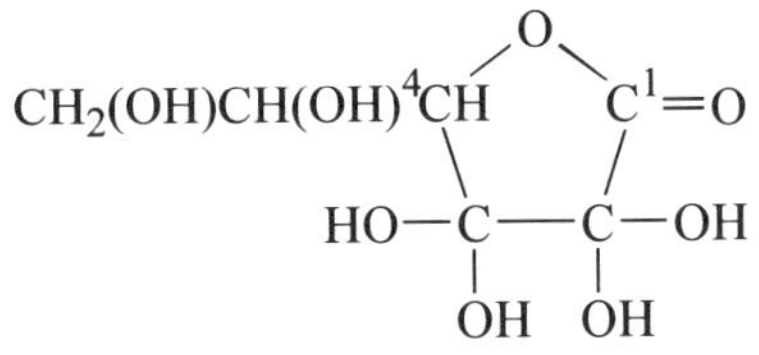

$CH_2(OH)CH(OH)^4CH$, O, $C^1=O$, HO—C—C—OH, OH OH **[$A(OH_2)_2$]**

Chemical or electrochemical reduction of **$A(OH_2)$** yields ascorbic acid (**H_2A**) in a two-hydrogen-atom ($2\ H_3O^+ + 2\ e^-$) process:

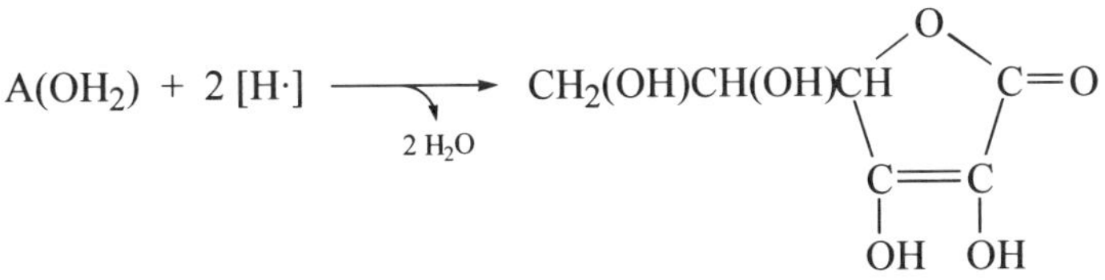

$$A(OH_2) + 2\,[H\cdot] \xrightarrow[-2\,H_2O]{} CH_2(OH)CH(OH)CH\text{—}O\text{—}C{=}O,\ C{=}C(OH)(OH) \quad (\mathbf{H_2A}) \tag{12.19}$$

Figure 12.4 illustrates the electrochemical characteristics of ascorbic acid and its anion (**H_2A** and **HA^-**) and of dehydroascorbic acid (**A**) and its anion radical (**$A^{\cdot-}$**) in DMF at a platinum electrode. The hydrogens of **H_2A** are moderately acidic (in H_2O; $pK_1 = 4.1$, $pK_2 = 11.8$), which accounts for the cathodic voltammograms (Figure 12.4*a*):

$$\mathbf{H_2A} + e^- \longrightarrow \mathbf{HA^-} + \tfrac{1}{2}\,H_2 \qquad E_{p,c},\ -0.98\ \text{V vs. SCE} \tag{12.20}$$

Ascorbic acid also exhibits an irreversible two-electron oxidation to give dehydroascorbic acid:

$$\mathbf{H_2A} + 2\,H_2O \xrightarrow{-2e^-} \mathbf{A} + 2\,H_3O^+ \qquad E_{p,a},\ +1.13\ \text{V vs. SCE} \tag{12.21}$$

With its anion (**HA^-**) the oxidation process occurs at a potential that is almost 1 V less positive (in H_2O at pH 7 **HA^-** is the dominant form of ascorbic acid):

$$\mathbf{HA^-} \xrightarrow{-e^-} [\mathbf{HA}\cdot] \xrightarrow[\text{HA}^-]{-e^-} \mathbf{A} + \mathbf{H_2A} \qquad E_{p,a},\ +0.17\ \text{V vs. SCE} \tag{12.22}$$

In contrast to **H_2A,** electrochemical reduction of **A** yields the anion radical of dehydroascorbic (**$A^{\cdot-}$**) in an irreversible, complex electron-transfer process (Figure 12.4*c*, *d*):

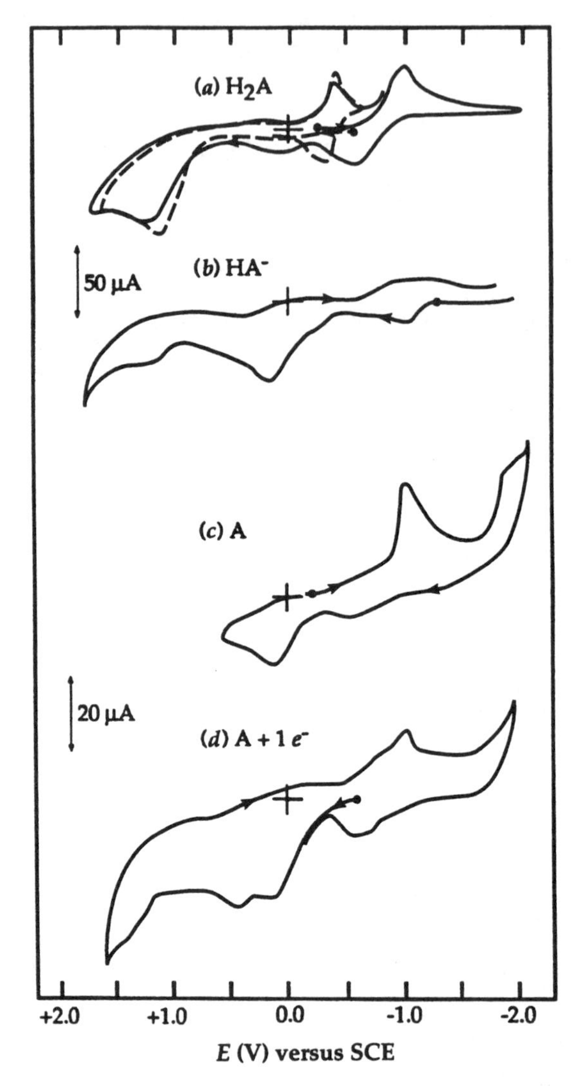

Figure 12.4 Cyclic voltammograms in dimethylformamide (0.1 M tetraethylammonium perchlorate) of 2 mM ascorbic acid (H_2A), 2 mM HA^- (from 2 mM H_2A and 2 mM TEAOH), 2 mM dehydroascorbic acid (A), and the product from the one-electron reduction of 2 mM A at -1.2 V versus SCE. Measurements were made with a platinum electrode (area 0.23 cm^2) at a scan rate of 0.1 V s^{-1}, temperature 25°C; $E_{NHE} = E_{SCE} + 0.244$ V.

$$[1/2\,(\mathbf{A_2}) \xrightarrow{H_2O} \mathbf{A(OH_2)}] + e^- \xrightarrow[H_2O]{} \mathbf{A}^{\overline{\cdot}} \qquad E_{p,c}, -1.01 \text{ V vs. SCE} \tag{12.23}$$

Both the electrochemistry and spectroscopic measurements confirm that $\mathbf{A}^{\overline{\cdot}}$ is reasonably stable in dry DMF.[13] However, water appears to promote the formation of an $\mathbf{A(HA^-)}$ adduct via a disproportionation process:

$$2\,\mathbf{A}^{\overline{\cdot}} + H_2O \rightleftharpoons \mathbf{A(HA^-)} + HO^- \tag{12.24}$$

The adduct of this process is transformed by an irreversible two-electron oxidation to two **A** molecules:

$$\mathbf{A(HA^-)} + HO^- \xrightarrow{-2e^-} 2\,\mathbf{A} + H_2O \qquad E_{p,a}, +0.44 \text{ V vs. SCE} \tag{12.25}$$

The product solution from the electrochemical oxidation of $\mathbf{H_2A}$ [Eq. (12.21)] exhibits a reversible one-electron reduction:

$$\mathbf{A(OH_2)_2} + H_3O^+ + e^- \rightleftharpoons \mathbf{HA\cdot} + 3\,H_2O \qquad E_{1/2}, -0.4 \text{ V vs. SCE} \tag{12.26}$$

Carbonyl Groups, Olefins, and Aromatic Hydrocarbons. Unsaturated carbon centers possess a limited degree of electrophilicity and will accept an electron at potentials significantly more negative than their chloro derivatives (usually at least −2.5 V vs. SCE in rigorously anhydrous solvents). When water is present, its reduction is synergistically facilitated via unsaturated carbon. For example, in acetronitrile at a glassy-carbon electrode:

$$H_2O + e^- \longrightarrow [H\cdot] + HO^- \qquad E_{p,c}, -3.9 \text{ V vs. SCE} \tag{12.27}$$

$$\text{(naphthalene)} + e^- \longrightarrow NP^{\overline{\cdot}} \qquad E_{p,c}, -2.6 \text{ V vs. SCE} \tag{12.28}$$

$$NP + H_2O + e^- \xrightarrow[HO^-]{} [HNP\cdot] \xrightarrow[H_2O \; \searrow HO^-]{e^-} H_2NP \qquad E_{p,c}, \sim -2.0 \text{ V vs. SCE} \tag{12.29}$$

Similar synergism is observed for other Brønsted acids (H*A*) in the presence of unsaturated carbon centers:

$$\mathrm{PhCH(O)} + \mathrm{H}A + e^- \xrightarrow[A^-]{} [\mathrm{Ph\dot{C}HOH}] \xrightarrow[\mathrm{H}A,\ A^-]{e^-} \mathrm{PhCH_2OH} \qquad (12.30)$$

For aromatic hydrocarbons, the more extensive the conjugation, the less negative the potential for direct electron transfer:

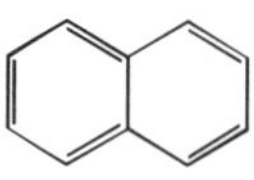

[naphthalene], –2.6 V vs. SCE; [anthracene], –2.1 V;

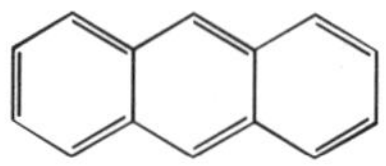

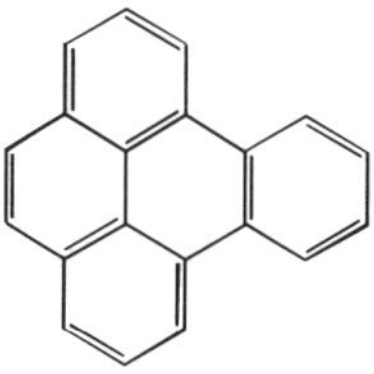

[triphenylene], –1.9 V; [perylene], –1.8 V

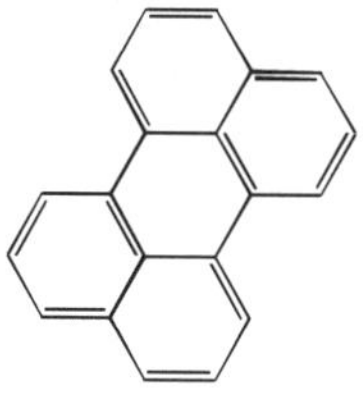

The hydrolysis of carbonyl centers yields acidic hydrogens that can be reduced at platinum electrodes, for instance

$$\mathrm{PhCH(O)} + \mathrm{H_2O} \rightleftharpoons \mathrm{PhCH(OH)_2} \underset{\mathrm{Pt}}{\overset{e^-}{\rightleftharpoons}} \tfrac{1}{2}\,\mathrm{H_2} + \mathrm{PhCH(OH)O^-} \qquad (12.31)$$

Likewise, the enolized form of ketones is a Brønsted acid that can be reduced at a platinum electrode:

$$\mathrm{CH_3C(O)CH_2CH_3} \rightleftharpoons \mathrm{CH_3C(OH){=}CHCH_3} \xrightarrow[\mathrm{Pt}]{e^-}$$

$$\tfrac{1}{2}\,\mathrm{H_2} + \mathrm{CH_3C(O^-){=}CHCH_3} \qquad (12.32)$$

In both cases, the reduction potential is directly related to the Brønsted acidity, which depends on the basicity of the solvent [e.g, phenol (PhOH) has a pK_a of 16 in MeCN and 21 in Me_2SO; see Chapter 8].

The azo function [e.g., azobenzene (PhN=NPh)] is reduced in a manner that is similar to that for quinones (discussed above). The electrochemistry for azo groups is a part of the discussion of the nitrogen compounds in Chapter 11 (Figure 11.10 illustrates the cyclic voltammogram for azobenzene).

Brønsted Acids. Carboxylic acids, phenols, and alcohols are electrochemically reduced by means of their Brønsted acidity at a reduction potential that is a direct measure of their acidity (pK_a) in a given solvent (see Chapter 8):

$$HA + e^- \longrightarrow \tfrac{1}{2}H_2 + A^- \qquad E_{1/2} \approx -0.059(pK_a) \text{ V vs. NHE} \qquad (12.33)$$

Viologens. The viologen class of electrophiles $R{-}\overset{+}{N}C_5H_4{-}C_5H_4\overset{+}{N}{-}R$; R = Me, $PhCH_2$, $CH_2{=}CH$, $HOCH_2CH_2$, $CH_3CH{=}CH$] are unique in their ability to undergo two reversible one-electron reductions in almost any medium (Table 12.3).[15] This property makes the viologens especially useful mediators between one-electron and two-electron processes. This property also is the basis for the herbicidal activity of methyl viologen (paraquat), which disrupts the electron-transport chain in green-plant photosynthesis.

12.3 OXIDATION OF NUCLEOPHILIC SUBSTRATES AND LEWIS BASES

All molecules with nonbonding electron pairs (e.g., H_2O, ROH, ROR, $R\mathrm{NH}_2$, RSH, RSR) are, by definition, Lewis bases with a degree of nucleophilicity. Their electrochemical oxidation potential is a measure of (1) the ease of removal for one of the electron pair of electrons and (2) relative nucleophilicity (the less positive the potential, the more nucleophilic). Aromatic molecules with Lewis base substituents are easier to oxidize than the aliphatic forms of the substituents (e.g., PhOMe, +1.75 V vs. SCE; MeOH, +2.5 V vs. SCE) because the aromatic ring provides a means for delocalizing the positive charge and electron spin that would result from electron removal (in the case of PhOMe, there are five additional hydrogen atoms to share the positive charge and six

TABLE 12.3 Reduction Potentials for Viologens ($R{-}\overset{+}{N}C_5H_4{-}C_5H_4\overset{+}{N}{-}R$) **in Aqueous Solutions at pH 7.0**

Viologen (R)	$(E^{\circ\prime})_1$ (V) versus NHE	$(E^{\circ\prime})_2$ (V) versus NHE
Me	−0.45	−0.77
$PhCH_2$	−0.36	−0.54
$CH_2{=}CH$	−0.36	−0.69
$CH_3CH{=}CH$	−0.56	−0.82
$HOCH_2CH_2$	−0.41	−0.75

Source: Ref. 15.

unsaturated carbon centers to share the spin density). Within this context, the water molecule in a nonbasic solvent matrix is the most resistant to electron removal:

$$2\ H_2O \xrightarrow[\text{MeCN}]{-e^-} [(H_2O)^+H_2O\cdot] \longrightarrow HO\cdot + H_3O^+$$

$$E_{p,a},\ +2.8\ \text{V vs. SCE} \qquad (12.34)$$

Because aliphatic alcohols can be viewed as "organic water" (but with a greater basicity and a weaker O—H bond), they are almost as difficult to oxidize:

$$\text{MeOH} \xrightarrow[\text{MeCN}(H_2O)]{-e^-} [(H_2O)^+H\dot{O}Me] \longrightarrow MeO\cdot + H_3O^+$$

$$E_{p,a},\ +2.5\ \text{V vs. SCE} \qquad (12.35)$$

Other aliphatic bases (amines and thiols) are oxidized by similar pathways; their electrochemistry is discussed to a limited extent in Chapter 11.

Catechols and Hydroquinones. Just as quinones are ideal examples of electrophilic substrates, their fully reduced form (catechols and hydroquinones) illustrates the electrochemical oxidation of aromatic nucleophilic substrates (Lewis bases). Figure 12.3*a*, *b* illustrates the oxidation of 3,5-di-*tert*-butylcatechol ($DTBCH_2$) via an irreversible two-electron process (ECEC) to give the *o*-quinone (DTBQ):[12]

$$DTBCH_2 \xrightarrow[BH^+]{-e^-,\,B} [DTBS\dot{Q}H] \xrightarrow[BH^+]{-e^-,\,B} DTBQ \qquad E_{p,a},\ +0.97\ \text{V vs. SCE}$$

$$B = H_2O \text{ or solvent} \qquad (12.36)$$

When 2 equiv of HO^- are available, the anodic process occurs at +0.14 V versus SCE with $DTBCH^-$ the electroactive species. Table 12.2f, g summarizes the effect of solvent basicity on the oxidation potentials for several catechols and benzohydroquinone. The basicity of the solvent as well as of the substituents on the aromatic ring of the catechols causes their oxidation potentials to shift to less positive values (Table 12.2).

Whereas hydroquinone (*p*-HOPhOH) in acetonitrile is oxidized via an irreversible two-electron process at +1.18 V versus SCE (Eq. 12.34 and Table 12.2), its dimethyl ether (*p*-MeOPhOMe) is significantly more resistant with a reversible one-electron oxidation at +1.30 V versus SCE (Figure 12.5).[16] The initial oxidation of the latter is followed by a second irreversible one-electron oxidation (+1.81 V vs. SCE) that yields a product that is reduced at +0.59 V versus SCE [consistent with the reduction of benzoquinone in the presence of hydronium ions (Table 12.2)]:

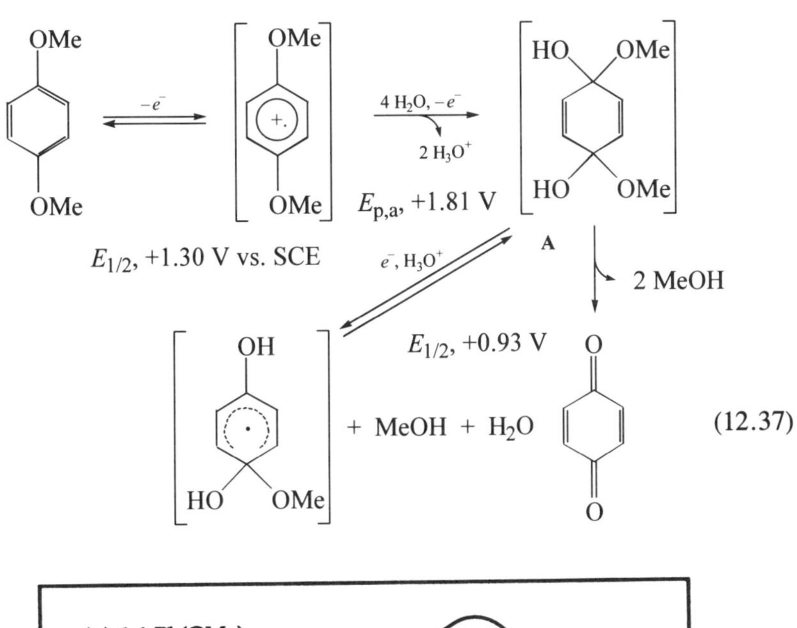

(12.37)

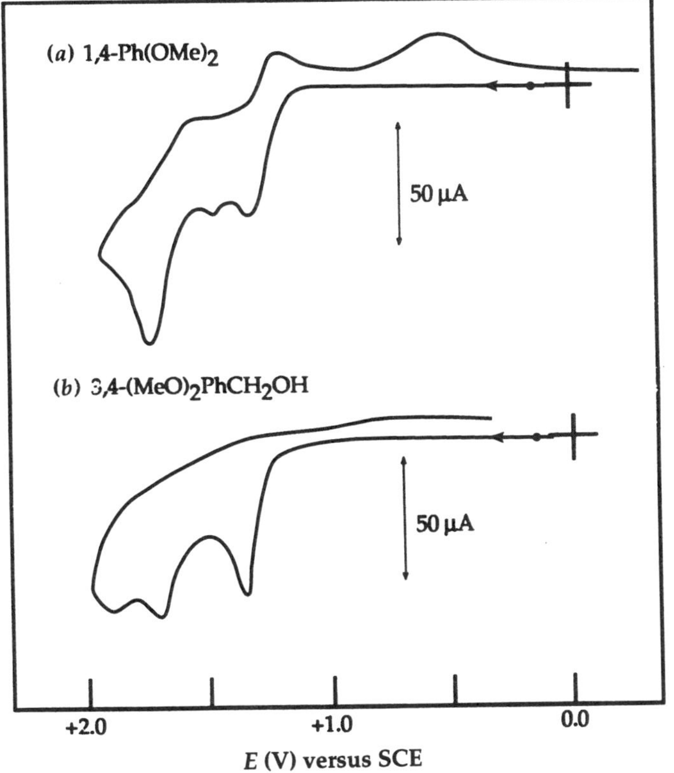

Figure 12.5 Cyclic voltammograms of (*a*) 1.5 mM 1,4-Ph$(OMe)_2$ and (*b*) 3,4-$(MeO)_2PhCH_2OH$ in MeCN [0.1 M $(Et_4N)ClO_4$]. Scan rate 0.1 V s^{-1}; GCE (0.09 cm^2); SCE versus NHE, + 0.244 V.

TABLE 12.4 Voltammetric Oxidation Potentials and Peak Currents (μA/mM) of Alkoxy-Substituted Benzenes, Phenols, and Benzyl Alcohols, and Their Derivatives in MeCN (0.1 M TEAP) at a Glassy-Carbon Electrode

Substrate	$E_{p,a}$ (V) versus SCE	i_p (μA/mM)
$PhOCH_3$	1.75	80
1,4-$(MeO)_2Ph$	1.30 (rev[a]), 1.81 (0.90 ↔ 0.96)	37
1,2,4-$(MeO)_3Ph$	1.10 (rev), 175 (0.87 ↔ 0.94)	55, 30, 30
2-Cl-1,4-$(MeO)_2Ph$	1.45 (rev)	100
1,2-$(MeO)_2Ph$	1.44 (1.00 ↔ 1.05)	80 (30)
1,3-$(MeO)_2Ph$	1.45	65
1,2,3-$(MeO)_3Ph$	1.41	45
1,3,5-$(MeO)_3Ph$	1.35	36
PhOH	1.55	75
4-CH_3PhOH	1.35	77
2,6-$(CH_3)_2PhOH$	1.30	80
2,6-$(MeO)_2PhOH$	1.05	85
$PhCH_2OH$	2.00	60
3-$MeOPhCH_2OH$	1.75	100
2-$MeOPhCH_2OH$	1.63	100
4-$MeOPHCH_2OH$	1.55	50
2,3,4-$(MeO)_3PhCH_2OH$	1.43	60
3,4-$(MeO)_2PhCH_2OH$	1.35, 1.73	60, 30
3,4,5-$(MeO)_3PhCH_2OH$	1.35	40
3,4-$(MeO)_2PhCH(OH)CH(CH_2OH)OPh$-2,4-$(MeO)_2$	1.36, 1.75	
3,4-$(EtO)_2PhCH(OH)CH(CH_3)Ph$-4-$(MeO)$	1.20, 1.68	
$PhCH_2OCH_3$	2.10	62
$PhCH_3$	2.15	60
PhH	2.45	
H_2O (5 mM)	2.8	

[a]Reversible.

Table 12.4 summarizes the voltammetric oxidation potentials and peak currents for 1,4-$(MeO)_2Ph$ and other alkoxy-substituted benzenes, phenols, and benzyl alcohols. Only the 1,4-$(MeO)_2PhX$ members of the series exhibit an initial irreversible anodic cyclic voltammogram via the sequence of Eq. (12.37). These plus the 1,2-$(MeO)_2Ph$ isomer yield a metastable product from the second oxidation [species **A**, Eq. (12.37)] that undergoes a reversible reduction. Thus, the two-electron oxidation of dimethoxy benzenes yields the corresponding quinone.

Phenols. In contrast to aliphatic alcohols, phenols are much easier to oxidize (even though they are much stronger Brønsted acids). This must be the result of their ability to distribute the positive charge over the six hydrogen atoms and the unpaired-electron spin density over six carbons and oxygen:

$$\text{PhOH} \xrightarrow{-e^-} \left[\text{PhOH}^{\cdot +} \xrightarrow[H_3O^+]{H_2O} \text{PhO}^{\cdot} \right] \xrightarrow[H_3O^+]{2\,H_2O,\,-e^-} [\ldots] \rightarrow \text{catechol (also hydroquinone)}$$

$E_{p,a}$, +1.55 V vs. SCE

(12.38)

However, because catechol and hydroquinone are oxidized at less positive potentials (+1.32 V and +1.18 V, respectively), the only detectable products are the *o*- and *p*-quinones via an overall four-electron (ECECECEC) oxidation. The data for substituted phenols (Table 12.4) indicate that electron-donating substituents cause phenol to become more nucleophilic (a stronger Lewis base) and easier to oxidize. Anisole (PhOMe) is more resistant to electron removal (+1.75 V vs. SCE) than phenol (+1.55 V vs. SCE), but undergoes electron removal via a similar path [Eq. (12.38)] to yield the same quinone products.

Benzyl Alcohols. The ease of oxidation for this class of alcohols ($PhCH_2OH$, +2.00 V vs. SCE) lies between aliphatic alcohols (MeOH, +2.50 V) and phenols (PhOH, +1.55 V), with the adjacent aromatic ring able to stabilize positive charge and spin density. Again, electron-donating substituents lower the potential for electron removal. The 3,4-dimethoxy derivative [3,4-$(MeO)_2PhCH_2OH$, veratryl alcohol; $E_{p,a}$, +1.36 V vs. SCE, Table 12.4] is a model-substrate monomer for lignin, which is the Earth's second most abundant plant product (after cellulose).[17] Figure 12.5*b* illustrates its anodic voltammogram in acetonitrile. The ultimate product is the aldehyde [3,4-$(MeO)_2$-PhCH(O)]. Scheme 12.1 outlines a reasonable pathway via an ECEC process with the net result that two of the hydrogen atoms are removed from the alcohol to yield the aldehyde, which is equivalent to the biological process with hydrogen peroxide and *lignin peroxidase* the catalyst (Fenton reagents mimic the enzyme).[17]

Aromatic Hydrocarbons. The π-electron cloud of the aromatic ring is much more susceptible to electron removal than a saturated hydrocarbon. Thus, benzene is oxidized at +2.45 V versus SCE; electron-donating substituents reduce

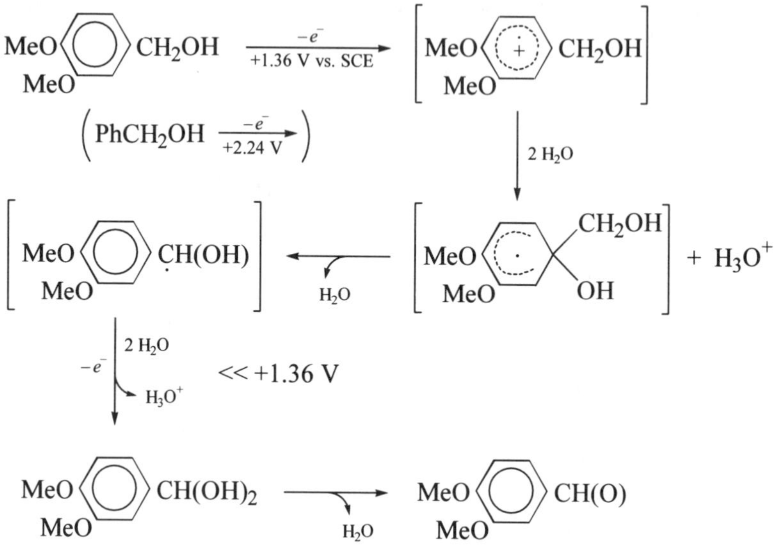

$$\text{MeO, MeO-C}_6\text{H}_3\text{-CH}_2\text{OH} \xrightarrow[+1.36 \text{ V vs. SCE}]{-e^-} [\text{MeO, MeO-C}_6\text{H}_3^{+\cdot}\text{-CH}_2\text{OH}]$$

$$\left(\text{PhCH}_2\text{OH} \xrightarrow[+2.24 \text{ V}]{-e^-}\right)$$

$$[\text{MeO, MeO-C}_6\text{H}_3^{+\cdot}\text{-CH}_2\text{OH}] \xrightarrow{2\,\text{H}_2\text{O}} [\text{MeO, MeO-C}_6\text{H}_3^{\cdot}\text{(CH}_2\text{OH)(OH)}] + \text{H}_3\text{O}^+$$

$$[\text{MeO, MeO-C}_6\text{H}_3^{\cdot}\text{(CH}_2\text{OH)(OH)}] \xrightarrow{-\text{H}_2\text{O}} [\text{MeO, MeO-C}_6\text{H}_3\text{-}\dot{\text{C}}\text{H(OH)}]$$

$$[\text{MeO, MeO-C}_6\text{H}_3\text{-}\dot{\text{C}}\text{H(OH)}] \xrightarrow[-e^-,\ -\text{H}_3\text{O}^+]{2\,\text{H}_2\text{O}} \quad \ll +1.36 \text{ V}$$

$$\text{MeO, MeO-C}_6\text{H}_3\text{-CH(OH)}_2 \xrightarrow{-\text{H}_2\text{O}} \text{MeO, MeO-C}_6\text{H}_3\text{-CH(O)}$$

Scheme 12.1 Electron-transfer oxidation of 3,4-$(MeO)_2PhCH_2OH$ in acetonitrile (with trace H_2O).

the potential by as much as 0.7 V (Table 12.4). This effect is illustrated in Figure 12.6, which correlates the oxidation potential for substituted benzenes with the substituent constants (σ).[16,18] Although a rigorously base-free solvent matrix should result in a reversible oxidation of benzene, the resulting cation radical ($PhH^{+\cdot}$) is a strong electrophile and Lewis acid that will react with most solvents and water (complete separation of water from benzene is difficult; 18 ppm H_2O is a 1 mM concentration):

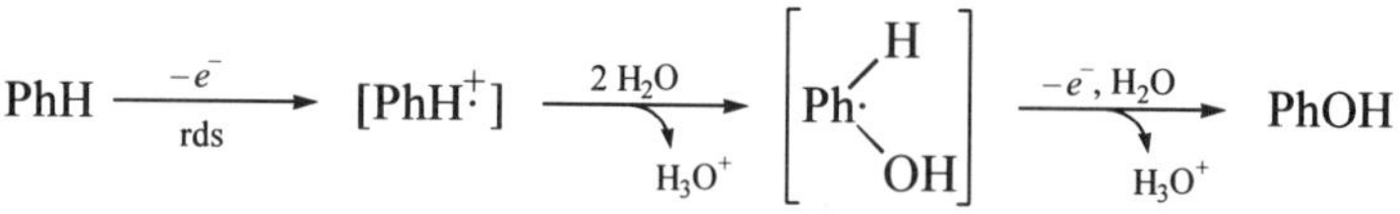

$$\text{PhH} \xrightarrow[\text{rds}]{-e^-} [\text{PhH}^{+\cdot}] \xrightarrow[-\text{H}_3\text{O}^+]{2\,\text{H}_2\text{O}} \left[\text{Ph}^{\cdot}\begin{smallmatrix}\text{H}\\ \text{OH}\end{smallmatrix}\right] \xrightarrow[-\text{H}_3\text{O}^+]{-e^-,\ \text{H}_2\text{O}} \text{PhOH} \tag{12.39}$$

In acetonitrile 5 mM H_2O is oxidized at +2.80 V versus SCE; the presence of benzene would facilitate electron removal by stabilizing the resulting hydroxyl radical to give an ECEC process:

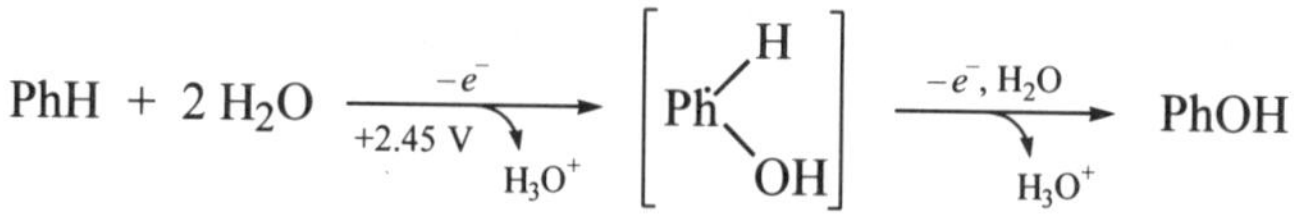

$$\text{PhH} + 2\,\text{H}_2\text{O} \xrightarrow[+2.45 \text{ V},\ -\text{H}_3\text{O}^+]{-e^-} \left[\text{Ph}^{\cdot}\begin{smallmatrix}\text{H}\\ \text{OH}\end{smallmatrix}\right] \xrightarrow[-\text{H}_3\text{O}^+]{-e^-,\ \text{H}_2\text{O}} \text{PhOH} \tag{12.40}$$

In this case the potential is determined by the acidity of H_3O^+ in the solvent matrix (pK_a, −8.8 in MeCN; pK_a, 0.00 in H_2O).

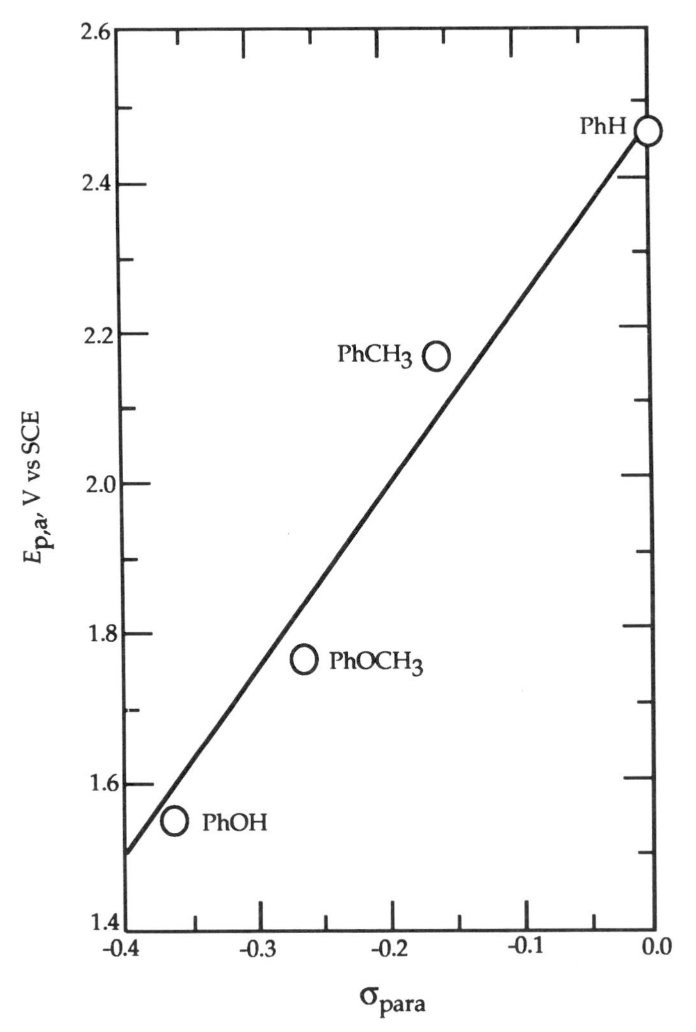

Figure 12.6 Voltammetric oxidation potentials ($E_{p,a}$) of substituted benzenes versus their substituent constants (σ_{para}, Ref. 16); slope, 2.43 V σ^{-1}. Conditions: 1 mM substrate in MeCN [0.1 M $(Et_4N)ClO_4$]; scan rate 0.1 V s^{-1}; GCE (0.09 cm^2) (Ref. 18).

Dithiols. Thiols and mercaptans (RSH) are similar to alcohols, as oxidizable reductants, but are much stronger nucleophiles and usually couple to form disulfides:

$$2\,RSH \xrightarrow[H_2O]{-2e^-} RSSR + 2\,H_3O^+ \tag{12.41}$$

When the substrate is a dithiol (e.g., propane dithiol, $HSCH_2CH_2CH_2SH$), it undergoes an apparent reversible oxidation at a gold electrode:[19]

$$\mathrm{HSCH_2CH_2CH_2SH} \underset{E_{1/2},\ +0.089\ \mathrm{V\ vs.\ SCE}}{\xrightleftharpoons{-e^-,\ H_2O}} [\mathrm{HSCH_2CH_2CH_2S\cdot}] + \mathrm{H_3O^+}$$

$$\longrightarrow \tfrac{1}{2}\mathrm{HSCH_2CH_2CH_2SH} + \tfrac{1}{2}\ \text{(1,2-dithiolane, S–S ring)} \qquad (12.42)$$

If 2 equiv of HO^- are added, a broad irreversible two-electron oxidation occurs; the disulfide product is reduced by two electrons:

$$\text{(CH}_2)_3(\text{S}^-)_2 \xrightarrow[-0.73\ \mathrm{V\ vs.\ SCE}]{-2e^-} \text{(CH}_2)_3\text{S–S (ring)} \xrightarrow[-1.77\ \mathrm{V}]{2e^-} \text{(CH}_2)_3(\text{S}^-)_2 \qquad (12.43)$$

In biology the oxidation of two cysteine amino acid residues (cyst-SH) follows a path similar to that of Eq. (12.42):

$$2\ \text{cyst—SH} \xrightarrow[2B^-]{-2e^-} \text{cyst—SS—cyst} + 2\ \text{BH}$$

$$+\ 0.1\ \text{V vs. SCE (aqueous, pH 7)} \qquad (12.44)$$

where B^- usually is bicarbonate ion [HOC(O)O^-].

In summary, the electrochemical oxidation–reduction of organic molecules usually is mediated by the water ($H_2O/H_3O^+/HO^-$) in the solvent matrix, especially if the substrate is more nucleophilic or electrophilic than water. Many reductions are catalyzed hydrogenations, and many oxidations are facilitated dehydrogenations. For these reasons, meaningful electrochemical measurements depend on the use of an inert electrode surface (*not* Pt, Pd, Au, or any other metal; highly polished glassy carbon is the preferred choice in most cases).

REFERENCES

1. Schwabe, K., *Polarographic und Chemische Konstitution Organischer Verbindungen*, Akademic Verlag, Berlin, 1957.
2. Meites, L.; Zuman, P.; et al., eds., *Electrochemical Data*, Wiley, New York, 1974.
3. Meites, L.; Zuman, P.; et al., eds., *CRC Handbook Series in Organic Electrochemistry*, Vols. 1–4, CRC Press, Boca Raton, Fla., 1980–1986.
4. Lund, H.; Baizer, M. M., eds., *Organic Electrochemistry*, 3rd ed., Marcel Dekker, New York, 1991.

5. Weinberg, N. L., ed., *Techniques of Electroorganic Synthesis*, Wiley, New York, 1974.
6. Steckhan, E., *Topics Curr. Chem.* **1987,** *142*, 1.
7. Steckhan, E., *Topics Curr. Chem.* **1994,** *170*, 83.
8. Sugimoto, H.,; Matsumoto, S.; Sawyer, D. T., *Environ. Sci. Technol.* **1988,** *22*, 1182.
9. Qui, A., Ph.D. dissertation, Texas A&M University, College Station, Tex., 1992, pp. 20–23.
10. Roberts, J. L., Jr.; Calderwood, T. S.; Sawyer, D. T., *J. Am. Chem. Soc.* **1983,** *105*, 7691.
11. M'Halla, F.; Pinson, J.; Savéant, J.-M., *J. Am. Chem. Soc.* **1980,** *102*, 4120.
12. Stallings, M. D.; Morrison, M. M.; Sawyer, D. T., *Inorg. Chem.* **1981,** *20*, 2655.
13. Sawyer, D. T.; Chiercato, G., Jr.; Tsuchiya, T., *J. Am. Chem. Soc.* **1982,** *104*, 6273.
14. Hvoslef, J.; Pedersen, B., *Acta Chem. Scand., Ser. B* **1979,** *B33*, 503.
15. Szentrimay, R.; Yey, P.; Kuwana, T., in *Electrochemical Studies of Biological Systems*, Sawyer, D. T., ed., ACS Symposium Series No. 38, American Chemical Society, Washington, D.C., 1977, pp. 143–169.
16. Qui, A., Texas A&M University, unpublished results, 1992.
17. Tung, H.-C.; Sawyer, D. T., *FEBS Lett.* **1992,** *311*, 165.
18. Lowrey, T. H.; Richardson, K. S., *Mechanism and Theory in Organic Chemistry*, 2nd ed., Harper & Row, New York, 1981.
19. Howie, J. K.; Houts, J. J.; Sawyer, D. T., *J. Am. Chem. Soc.* **1977,** *99*, 6323.

CHAPTER 13

ORGANOMETALLIC COMPOUNDS AND METALLOPORPHYRINS

13.1 INTRODUCTION

The defining characteristic of organometallic molecules is the presence of one or more metal–carbon bonds. In contrast to the acid–base character of coordination complexes of metal ions (with their ligand-centered redox chemistry; see Chapter 10), the metal–carbon center is highly covalent with limited polarity (similar to carbon–carbon, carbon–nitrogen, or carbon–oxygen centers). As a result, the electrochemistry of organometallic molecules is more closedly related to that of organic molecules than inorganic coordination complexes. The use of electrochemical methods for the characterization of organometallic molecules has been reviewed.[1,2]

Although metalloporphyrins are traditionally viewed as coordination complexes, their uncharged nature and covalent metal–nitrogen bonding are closely similar to organometallics, for instance

$Fe(NH_3)_4^{2+}$ versus

Tetrahedral; $S = \frac{4}{2}(d^6sp)$; two Fe—N sigma bonds in four resonance hybrids

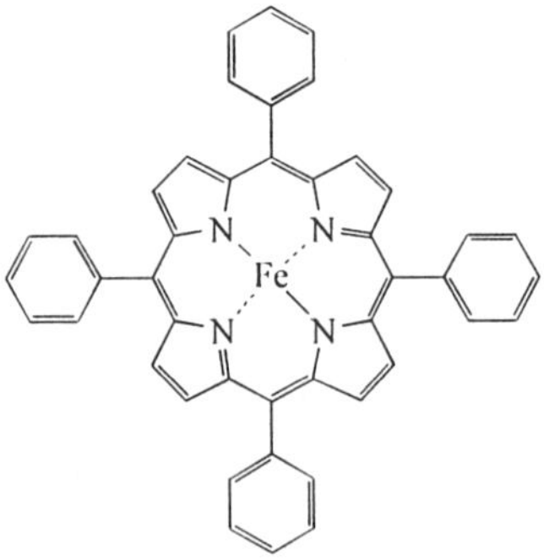

Planar; $S = \frac{4}{2}(d^6sp)$; two Fe—N σ bonds in two resonance hybrids

As such, they are much more resistant to acid- or base-induced dissociation or hydrolysis than are coordination complexes [e.g., $Fe^{II}(TPP)$ is stable in the presence of 1 M H_3O^+ or 1 M HO^-, whereas $Fe^{II}(NH_3)_4^{2+}$ is not]. An especially useful review of the electrochemistry of metalloporphyrins in nonaqueous media is available,[3] which is complemented by a review of the electrochemistry of metalloproteins.[4]

13.2 ORGANOMETALLIC MOLECULES

The "foundation stone" of organometallic chemistry is bis(cyclopendienyl)iron(II) [ferrocene, $(Cp)Fe^{II}(Cp)$], an iron atom sandwiched between two five-membered carbon rings [Cp, $C_5H_5\cdot$; each carbon with a *p* electron to give (1) two π-bonds delocalized around the carbon ring and (2) an unpaired electron to give a covalent bond that is shared by the five carbons of the ring.] Thus, the $Fe^{II}(Cp)_2$ molecule has the iron on a line that connects the centers of two parallel planar $Cp\cdot$ groups to give an "iron sandwich."

Figure 13.1 illustrates the electrochemical redox chemistry in acetronitrile for several coordination complexes of iron [$Fe^{II}(MeCN))_4^{2+}$, $Fe^{III}Cl_3$, and $Fe^{III}(acac)_3$ (acac = acetylacetonate; see Chapter 10)] in relation to that for two iron organometallics [$Fe^{II}(Cp)_2$ and $Fe^{III}(CO)_5$ (iron–pentacarbonyl); both are stable 18-electron systems].[5] In MeCN $Fe^{II}(MeCN)_4^{2+}$ is the only charged species of the group. It is reversibly oxidized (II/III couple; $E_{1/2}$, +1.6 V vs. SCE). The uncharged $Fe^{III}Cl_3$ molecule is reversibly reduced (III/II couple; $E_{1/2}$, +0.2 V vs. SCE) to give $Fe^{II}Cl_3^-$, which is reduced by an irreversible two-electron process to iron metal ($E_{p,c}$, −1.5 V vs. SCE). The more basic $Fe^{III}(acac)_3$ molecule is reversibly reduced (III/II couple; $E_{1/2}$, −0.7 V vs. SCE), but does not exhibit a second reduction peak. The III/II reduction potentials for these three coordination complexes are a measure of their relative electrophilicity (Lewis acidity).

Ferrocene. The $Fe^{II}(Cp)_2$ molecule is resistant to reduction, but exhibits a highly reversible one-electron oxidation

$$Fe^{II}(Cp)_2 \xrightarrow[MeCN]{-e^-} Fe^{III}(Cp)^+ \qquad E_{1/2},\ +0.45\ \text{V vs. SCE} \qquad (13.1)$$

$$S = 0 \qquad\qquad S = \tfrac{1}{2}$$

with the single positive charge delocalized over the 10 equivalent $(Cp)_2$ hydrogens (+0.1 each). For a time there was a belief that the $Fe^{II}(Cp)_2/Fe^{III}(Cp)_2^+$ couple's potential was independent of solvent, and thus an ideal reference electrode with which to measure solvent effects for other redox couples. However, subsequent measurements have shown that the $Fe^{III}(Cp)_2^+$ ion possesses considerable acidity, which causes some solvent effects. The more serious problem is the limited solubility of $Fe^{II}(Cp)_2$ in H_2O. The respective $E°$ values for the $Fe^{III}(Cp)_2^+/Fe^{II}(Cp)_2$ couple are MeCN, +0.69 V versus NHE

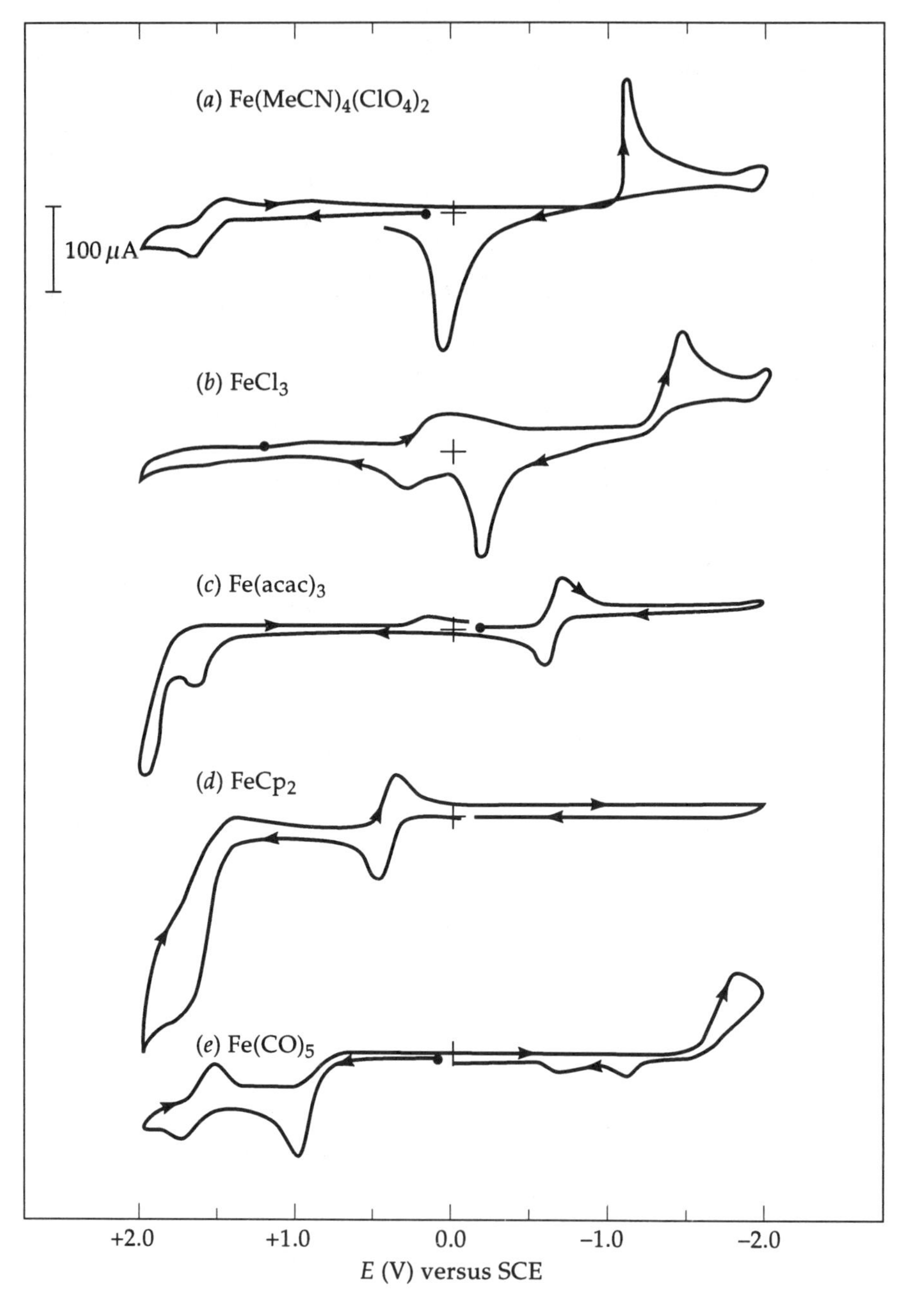

Figure 13.1 Cyclic voltammograms: (a) 3 mM $[Fe(MeCN)_4](ClO_4)_2$; (b) 3 mM $FeCl_3$; (c) 3 mM $Fe(acac)_3$; (d) 3 mM $Fe(Cp)_2$; (e) 3 mM $Fe(CO)_5$ in MeCN (0.1 M tetraethylammonium perchlorate). Conditions: scan rate 0.1 V s^{-1}; ambient temperature; glassy-carbon working electrode (0.09 cm^2); saturated calomel electrode (SCE) versus NHE, +0.244 V.

(+0.45 V vs. SCE); DMF, +0.72 V; py, +0.76 V; Me_2SO, +0.68 V; H_2O, +0.40 V.[6]

The large multielectron oxidation of $Fe^{III}(Cp)_2^+$ at +1.6 V versus SCE (Figure 13.1*d*) apparently is due to an electrocatalytic oxidation of residual water:

$$2\,H_2O \xrightarrow[Fe^{III}(Cp)_2^+]{-e^-} H_3O^+ + [HO\cdot] \qquad E_{p,a},\ +1.6\text{ V vs. SCE} \tag{13.2}$$

$$[HO\cdot] \xrightarrow{\text{solvent}} \text{products}$$

In the absence of $Fe^{III}(Cp)_2^+$, residual H_2O in MeCN is oxidized at +2.8 V.

Iron–Pentacarbonyl. The $Fe^{VIII}(CO)_5$ molecule is equally fundamental to organometallic chemistry and electrochemistry, and, like $Fe^{II}(Cp)_2$, is a diamagnetic 18-electron system. It exhibits (a) an irreversible two-electron oxidation and (b) an irreversible two-electron reduction (Figure 13.1*e*). In each case $Fe(CO)_5$ has a synergistic effect on (1) the reduction of residual H_2O and (2) the oxidation of solvent molecules:

$$Fe(CO)_5 + e^- \xrightarrow[HO^-]{H_2O} \left[(CO)_4Fe{-}C{\lesssim}^{H}_{O}\right] \xrightarrow[HO^-]{e^-,\ H_2O} Fe(s) + H_2C(O) + 4\,CO$$

$$E_{p,c},\ -1.8\text{ V vs. SCE} \tag{13.3}$$

By an analogous process the CO adduct of an iron(II) porphyrin $[(Cl_8TPP)Fe^{IV}(CO)]$ is reduced to $H_2C(O)$ at −0.87 V (discussed in the next section of this chapter). The oxidation of $Fe(CO)_5$ in MeCN yields $Fe^{II}(MeCN)_4^{2+}$ in a two-electron process (Figure 13.1*e*):

$$Fe(CO)_5 + 4\,MeCN \xrightarrow{-2e^-} Fe^{II}(MeCN)_4^{2+} + 5\,CO$$

$$E_{p,a},\ +0.97\text{ V vs. SCE} \tag{13.4}$$

On the basis that Fe(s) is oxidized to $Fe^{II}(MeCN)^{4+}$ at ~0.0 V (Figure 13.1*a*), the carbonyls of $Fe(CO)_5$ stabilize the iron against oxidation by about 96 kJ mol^{-1} [$\Delta E \times 96.5$ kJ mol^{-1} $(eV)^{-1}$; 0.97×96.5]. The $(Cl_8TPP)Fe^{IV}(CO)$ molecule is oxidized at +0.75 V versus +0.32 V for $(Cl_8TPP)Fe^{II}$; a stabilization by the CO of about 42 kJ mol^{-1}.

Gold-Catalyzed Oxidation of Carbon Monoxide. The oxidation of gold electrodes in alkaline media has been discussed in Chapter 10; and the electrochemical oxidation of CO, in Chapter 11. Because the intermediate involves an Au—C bond, the relevant anodic processes are reiterated here.[7] At an inert glassy-carbon electrode CO is totally resistant to electrochemical oxidation:

$$HO^- \xrightarrow[\text{MeCN, GCE}]{-e^-} HO\cdot \qquad E_{p,a}, +0.68 \text{ V vs. SCE} \tag{13.5}$$

$$Au(s) + HO^- \xrightarrow[\text{MeCN}]{-e^-} Au^I{-}OH \qquad E_{p,a}, -0.35 \text{ V} \tag{13.6}$$

$$Au(s) + HO^- + CO \xrightarrow[\text{MeCN}]{-e^-} \left[Au{-}C\begin{matrix} /OH \\ \backslash\backslash O \end{matrix} \right] \xrightarrow[HO^-]{-e^-}$$

$$Au(s) + (HO)_2C{=}O \qquad E_{p,a}, -0.45 \text{ V}$$

$$\downarrow\uparrow$$

$$CO_2/H_2O \tag{13.7}$$

Similar metal-facilitated oxidations of CO are observed at palladium and platinum electrodes, but are much less efficient and less reproducible.

These examples of the electrochemical character of organometallics are limited, but illustrate that their oxidation–reduction is closely similar to that for organic molecules (Chapter 12). Thus, the electron transfer is never carbon-centered, and often involves residual water [H-atom addition via reduction and (HO·) addition or H-atom abstraction via oxidation] or solvent components:

$$H_2O + e^- \longrightarrow [H\cdot] + HO^- \tag{13.8}$$

$$2\,H_2O \xrightarrow{-e^-} [HO\cdot] + H_3O^+ \tag{13.9}$$

13.3 METALLOPORPHYRINS

These organometallic-like molecules have been the object of electrochemical characterization ever since adequate instrumentation and solvents became available in the 1960s. The iron porphyrins are important cofactors in electron-transfer enzymes and the critical component in the oxygen-binding proteins (hemoglobin and myoglobin), and the cobalt prophyrin in vitamin B_{12} is essential for its carbon-transfer function. Several authoritative summaries of the electron-transfer characteristics and thermodynamics (with and without axial groups) have appeared during the past 15 years,[3,8–13] but systematic intercomparisons have been hindered by variations in the porphyrin, solvent–solution conditions, and electrode material.

A recent study[14] has overcome this limitation via the use of a rugged, sterically hindered porphine [*meso*-tetrakis(2,6-dichlorophenyl)porphine, $(Cl_8TPP)H_2$][15] in noncoordinating solvents (H_2CCl_2 or MeCN) at a glassy-carbon electrode. As a result, the electron-transfer thermodynamics can be quantitatively compared for $(Cl_8TPP)H_2$ and its metalloporphyrins [$(Cl_8TPP)M$; M = Mn, Fe, Co, Ni, Cu, Ag, Zn] as well as for the oxene, O_2, and CO adducts of the iron, cobalt, and manganese porphyrins. This provides a convenient means for evaluation of the molecular activation (catalytic) properties of these important metal-centered systems. The cyclic voltammograms of $(Cl_8TPP)H_2$ and of the $(Cl_8TPP)M$ (M = Zn, Co, Ni, Cu) complexes in H_2CCl_2 are illustrated in Figure 13.2. Similar data for $(Cl_8TPP)H_2$, $(Cl_8TPP)Zn$, $(Cl_8TPP)Cu$, and $(Cl_8TPP)Ag$ in dimethylformamide (DMF) are presented in Figure 13.3. The $(Cl_8TPP)Zn$, $(Cl_8TPP)Cu$, and $(Cl_8TPP)Ag$ complexes also have one reversible oxidation couple each. In H_2CCl_2, $(Cl_8TPP)H_2$, $(Cl_8TPP)Zn$, and $(Cl_8TPP)Cu$ exhibit four reversible redox couples, and $(Cl_8TPP)Ag$ has the same electrochemical behavior as in DMF. The $E_{1/2}$ values for the oxidation and reduction couples of these porphyrin complexes in H_2CCl_2 (and in MeCN and DMF) are summarized in Table 13.1.

The redox potentials for the $(Cl_8TPP)M^{III}(OH_2)_2^+$, $(Cl_8TPP)M^{III}(py)_2^+$, $(Cl_8TPP)M^{III}OH$, and $(Cl_8TPP)M^{III}Cl$ complexes (M = Mn, Fe, Co, Ni, Cu, Ag) in MeCN and in H_2CCl_2 are summarized in Tables 13.2 and 13.3. In addition, the effect of excess Cl^- or H_3O^+ on the redox chemistry of $(Cl_8TPP)M$ complexes is included. Figure 13.4 illustrates the cyclic voltammograms for $(Cl_8TPP)Fe^{III}OH$ and $(Cl_8TPP)Fe^{III}SCH_2Ph$ in DMF.[16] The effect of CO (1 atm) and of O_2 (1 atm) on the electrochemistry of $(Cl_8TPP)M$ (M = Fe, Co, Mn) in DMF is illustrated in Figure 13.5 and summarized in Table 13.4.

Figure 13.6 illustrates the cyclic voltammograms in acetonitrile at −35°C for $(Cl_8TPP)Mn^{III}(ClO_4)$, $(Cl_8TPP)Fe^{III}(ClO_4)$, and $(Cl_8TPP)Co^{III}(ClO_4)$ and for the oxene adduct of the iron(III) porphyrin [$(\cdot Cl_8TPP^+)Fe^{IV}(O)$].[17] The potentials of the redox couples for these porphyrins and for the oxene adduct of $(Cl_8TPP)Mn(ClO_4)$ are summarized in Table 13.1.

The data of Table 13.1 (and Figures 13.2 and 13.3) indicate that the oxidation and reduction couples for $(Cl_8TPP)H_2$, $(Cl_8TPP)Zn$, and $(Cl_8TPP)Ni$ occur at similar potentials and are consistent with porphyrin centered electron-transfer processes. In our view[18] neutral porphyrin complexes [M(por)] consist of uncharged metal centers [Zn($d^{10}sp$), Mn(d^5sp), Fe(d^6sp), Co(d^7sp), and Ni(d^8sp)] bonded via two metal–nitrogen covalent bonds (sp–p) with uncharged porphyrin (all four nitrogens equivalent via resonance). Thus, the removal and addition of electrons for $(Cl_8TPP)H_2$, $(Cl_8TPP)Zn$, and $(Cl_8TPP)Ni$ (Figure 13.2 and Table 13.1) occur within the π-electron manifold of the porphyrin ring. The first reduction potential is directly porportional to the relative electron affinity of the metalloporphyrin centers, with $(Cl_8TPP)H_2$ having the greatest electron affinity for this group (Table 13.1).

The $(Cl_8TPP)^{II}Co\cdot$ (d^7sp) complex is unique with two metal–nitrogen (sp–p) covalent bonds and a metal-centered unpaired electron, which facilitates

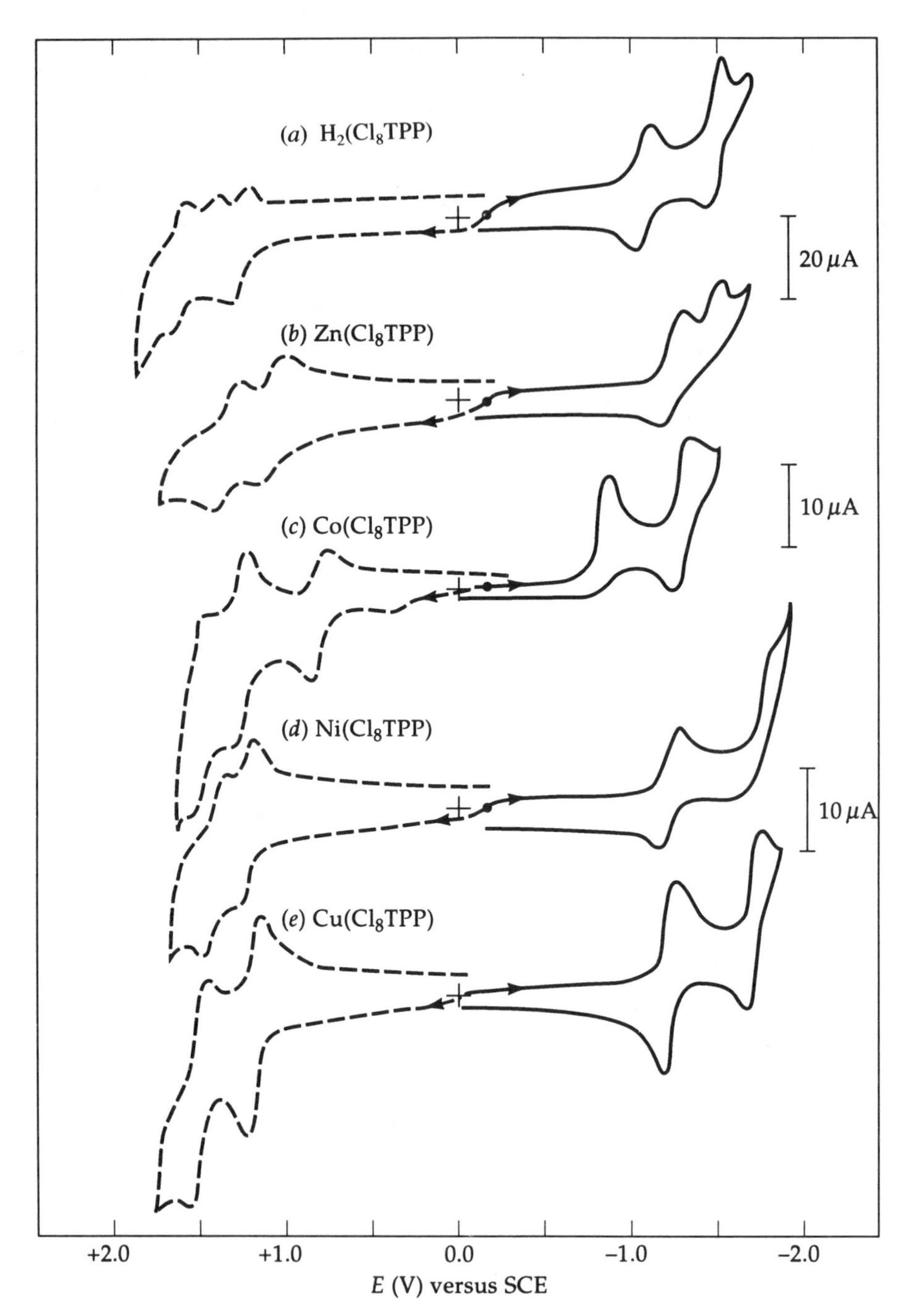

Figure 13.2 Cyclic voltammograms [at a glassy carbon electrode in methylene chloride (0.1 M tetrabutylammonium perchlorate) at room temperature] for (a) 0.6 mM $(Cl_8TPP)H_2$, (b) 0.6 mM $(Cl_8TPP)Zn$; (c) 0.6 mM $(Cl_8TPP)Co$, (d) 0.4 mM $(Cl_8TPP)Ni$, and (e) 0.6 mM $(Cl_8TPP)Cu$. Scan rate was 0.1 V s^{-1}.

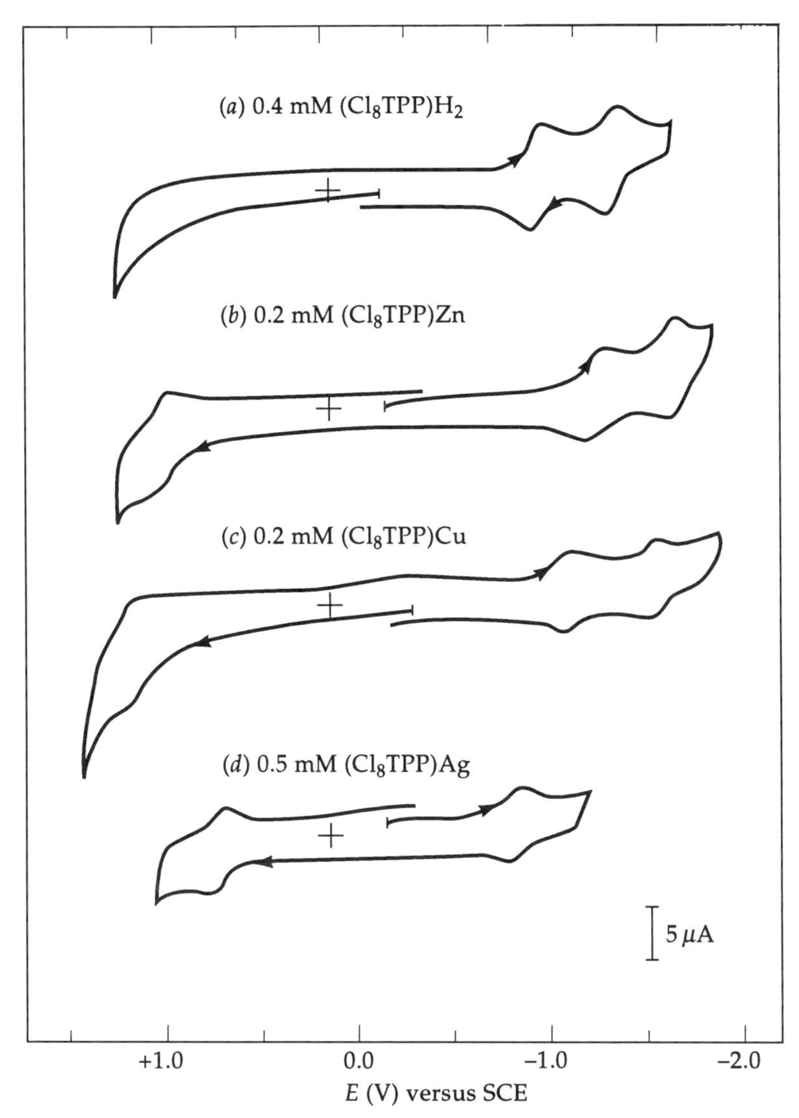

Figure 13.3 Cyclic voltammograms in DMF (0.1 M TEAP) of $(Cl_8TPP)H_2$, $(Cl_8TPP)Zn$, $(Cl_8TPP)Cu$, and $(Cl_8TPP)Ag$ at a glassy carbon electrode (area 0.07 cm^2). Scan rate was 0.1 V s^{-1}.

the addition of an electron to give $(Cl_8TPP)^{II}Co^-(d^8sp)$. [The Roman numeral superscript represents the covalence (number of covalent bonds) rather than the oxidation number or state of the metal.] Removal of an electron occurs within the porphyrin and is facilitated [first oxidation at +0.82 V vs. SCE in contrast to +1.23 V for $(Cl_8TPP)Ni$; see Figure 13.2 and Table 13.1] via stabilization of the porphyrin cation radical by formation of third cobalt–nitrogen covalent

TABLE 13.1 Redox Potentials for $H_2(Cl_8TPP)$, for Its Metal Complexes [*MP* and *MP*(ClO_4)], and for the Oxene Adducts of [(Cl_8TPP)Fe](ClO_4) and [(Cl_8TPP)Mn](ClO_4) in H_2CCl_2 (and MeCN) [and DMF]

	$E_{1/2}$ (V) versus SCE						
MP^a	$^+(sol)MP^{2+} \leftarrow$	$M^{II}P^{2+} \leftarrow M^{II}P^{+\cdot}$	$M^{II}P^{+\cdot} \leftarrow M^{II}P$	$^+(sol)M^{III}P \leftarrow M^{II}P$	$M^{II}P \rightarrow {}^-M^{II}P$	$M^{II}P \rightarrow M^{II}P^{\cdot-}$	$M^{II}P^{\cdot-} \rightarrow M^{II}P^{2-}$
H_2P		+1.63	+1.24			−1.10 [−0.93]	−1.54 [−1.30]
$Zn^{II}P$ ($d^{10}sp$)		+1.33	+1.03 [+1.01]			−1.26 [−1.22]	−1.72 [−1.64]
$NI^{II}P$ (d^8sp)		+1.41	+1.23			−1.20	−1.72
$\cdot Co^{II}P$ (d^7sp)	+1.52	+1.28 ($Co^{II}P^{\cdot 2+}$)	+0.82 ($Co^{III}P^+$) (d^6sp^2)		−0.88	−1.30 ($^-Co^{II}P^{\cdot-}$)	
$(Co^{III}P)^+$ (d^6sp^2)	+1.54 (+1.68)	+1.26 ($Co^{III}P^{\cdot 2+}$) (+1.40)	+0.93 ($Co^{III}P^+ \rightarrow$)	(+0.16)	−0.86 (−0.68)	−1.34 ($^-Co^{II}P^{\cdot-}$)	
$[(sol)^+Fe^{III}P]^+$ (d^5sp^2)	+1.52 (+1.72)	+1.37 (sol)$^+Fe^{III}P^{+\cdot}$) (+1.46)		+0.34 (→) (+0.32)	−0.93 (−1.00)	−1.17 ($^-Fe^{II}P^{\cdot-}$)	−1.61 ($^-Fe^{II}P^{2-}$)
$^{+\cdot}PFe^{IV}O$ (d^6sp)	(+1.72)	(+1.59) ($^{+\cdot}PFe^{IV}O$ ←)	(+1.28)(→$PFe^{IV}O$) (d^6sp)		(−0.54 ($PFe^{IV}O \rightarrow PFe^{III}O^-$) ($d^5sp^2$)		
$(Mn^{III}P)^+$ (d^5sp)	(+1.85)	+1.46 (+1.57) (d^5sp, coval 4)	−0.03 ($Mn^{III}P^+ \rightarrow$)	(+0.04)		−1.27 (−1.17)	(−1.69)
$^+PMn^VO$ (d^5sp)			(+0.52) (→ $PMn^{IV}O$) (d^5sp)		(−0.80) ($PMn^{IV}O \rightarrow PMn^{III}O^-$) ($d^5sp$)		
$Cu^{III}P\cdot$ (d^8sp^2)		+1.40 ($Cu^{III}P^{\cdot 2+}$)	+1.20 ($Cu^{III}P^+$) [+1.17] (d^8sp^2)			−1.09 ($Cu^{III}P^-$) [−1.08] (d^8sp^2)	−1.70 ($Cu^{III}P^{2\cdot-}$) [−1.55]
$Ag^{III}P\cdot$ (d^8sp^2)			+0.77 ($Ag^{III}P^+$) [+0.75] (d^8sp^2)			−1.01 ($Ag^{III}P^-$) [+0.81] (d^8sp^2)	

[a]The superscript Roman numerals indicate the covalence (number of covalent bonds) for the metal; *not* the charge state.

TABLE 13.2 Redox Potentials for $(Cl_8TPP)M^{III}(OH_2)_2^+$, $(Cl_8TPP)M^{III}(py)_2^+$, and $(Cl_8TPP)M^{III}OH$ Complexes in MeCN (0.1 M TEAP); Catalyzed Reduction of H_3O^+

	$E_{1/2}$ (V) versus SCE[a]				
Complex	(←)	(←)	(→)	(→)	(→)
$PMn^{III}(OH_2)_2^+$		+1.57	−0.03	−1.16	−1.68
+ H_3O^+			−0.02	[−0.86]	[−1.18]
$PMn^{III}(py)_2^+$		[+1.56]	−0.06	−1.17	−1.69
$PMn^{III}OH$			−0.4		
$PFe^{III}(OH_2)_2^+$		+1.45	+0.31	[−0.95] (→ PFe^-)	[−1.60]
+ H_3O^+			+0.25	[−0.87]	
$PFe^{III}(py)_2^+$		+1.34	+0.23	−1.18	[−1.62]
$PFe^{III}OH$	+1.64	+1.35	−0.80	−1.31	−1.63
$PCo^{III}OH$ (DMF)	[+1.08]	[+0.80]	−0.45	−0.75 (→ PCo^-)	[−1.31]
PCo^{II} (H_2CCl_2)				[−0.85] (→ PCo^-)	−1.26
+ H_2O^+				[−0.75]	−1.29
PNi^{II} (H_2CCl_2)				−1.19	[−1.78]
+ H_3O^+			[−0.12]	[−0.96]	
$\cdot PCu^{III}$ (H_2CCl_2)				−1.20	−1.70
+ H_3O^+				[−0.91]	

[a]Brackets [ ] indicate irreversible process; peak potential.

bond (cobalt radical–porphyrin radical coupling) to give a diamagnetic species, $(Cl_8TPP^+)Co^{III}$ (d^6sp^2). This difference in oxidation potentials for *P*Ni and $PCo\cdot$, $\Delta E_{1/2}$, is an approximate measure of the free energy of formation for the third cobalt–nitrogen covalent bond ($-\Delta G_{BF}$) of $+PCo^{III}$:[17,18]

$$-\Delta G_{BF} = [(E_{1/2})_{PNi} - (E_{1/2})_{PCo\cdot}] \times 96.5 \text{ kJ mol}^{-1}$$

$$= (1.23 - 0.82) \times 96.5 = 40 \text{ kJ mol}^{-1} \qquad (13.10)$$

The $(Cl_8TPP)Fe^{III}(sol)^+$ complexes (metal valence-electron hybridization, d^5sp^2) also is unique with two metal–nitrogen (*sp–p*) and one metal-$(sol)^+$ (*sp–p*) covalent bonds. The first reduction is (sol^+) centered to give $(Cl_8TPP)Fe^{II}(d^6sp)$, followed by a metal-centered reduction to give $(Cl_8TPP)^{II}Fe^-$ (d^7sp). In contrast, oxidation of $(Cl_8TPP)Fe^{III}(sol)^+$ (d^5sp^2) is ligand-centered to give the porphyrin cation radical, $(Cl_8TPP\cdot^+)Fe^{III}(sol)^+$ (d^5sp^2). Oxidation of $(Cl_8TPP)Fe^{II}$ in the presence of ligands (L = H_2O, py, HO^-, Cl^-, or DMF) yields $(Cl_8TPP)Fe^{III}(OH_2)_2^+$ (d^5sp^2), $(Cl_8TPP)Fe^{III}(py)_2^+$ (d^5sp^2), $(Cl_8TPP)Fe^{III}OH$, $(Cl_8TPP)Fe^{III}Cl$, and $(Cl_8TPP)Fe^{III}(DMF)^+$; see Tables 13.2 and 13.3.[18]

Reference to the data of Table 13.1 and the results from an earlier study[18] indicate that the $(Cl_8TPP^+)Mn^{III}$ complex has unique bonding with three manganese–nitrogen ($d^5sp–p$) covalent bonds with the positive charge at the porphyrin ring ($S = \frac{4}{2}$). Reference to Tables 13.2 and 13.3 confirms that the presence of py (or Cl^-) has no effect on the reduction potential of

TABLE 13.3 Redox Potentials for $H_2(Cl_8TPP)$ and for Its Metal/Chloro Complexes in H_2CCl_2

	$E_{1/2}$ (V) versus SCE[a]				
MP	$^{\dot{+}}PM^{III}\text{-Cl} \leftarrow PM^{III}Cl$	$PM^{III}\text{-Cl} \leftarrow PM^{II}(Cl^-)$	$PM^{III}Cl \rightarrow PM^{II}(Cl^-)$	$PM^{II}(Cl^-) \rightarrow {}^-M^{II}P$	${}^-M^{II}P \rightarrow {}^-M^{II}P^{\dot{-}}$
$H_2P + Cl^-$	[+1.64] ($HP^+ \leftarrow$)	+1.27 ($HP\cdot + HCl \leftarrow$)		−1.05 ($\rightarrow HP^{\dot{-}}$)	−1.44 ($\rightarrow H_2P^{2-}$)
$PZn^{II}(Cl^-)$	[+1.19] ($^+PZn^{II}Cl \leftarrow$)	+0.81 ($\cdot PZn^{II}Cl \leftarrow$)		−1.22 ($\rightarrow Zn^{II}P^{\dot{-}}$)	−1.64 ($\rightarrow Zn^{II}P^{2-}$)
$[^+PMn^{III}]Cl$	+1.49 ($^+PMn^{IV}Cl \leftarrow$)		−0.06	−1.34 ($\rightarrow Mn^{II}P^{\dot{-}}$)	
$PFe^{III}Cl$	+1.64 ($^{2+}PFe^{III}Cl$)	+1.35 ($^{\dot{+}}PFe^{III}Cl \leftarrow$)	−0.29	−0.97	−1.63
$PCo^{III}Cl$			[−0.14]	−0.81	−1.29
$PNi^{II} + Cl^-$	+1.45	+1.21		−1.22 ($\rightarrow Ni^{II}P^{\dot{-}}$)	
$\cdot PCu^{III} + Cl^-$	+1.54	+1.22		−1.20 ($\rightarrow Cu^{III}P^-$)	−1.64 ($\rightarrow Cu^{III}P^{2\dot{-}}$)
$\cdot PAg^{III} + Cl^-$	.	+0.77		−1.01 ($\rightarrow Ag^{III}P^-$)	

[a]Brackets [ ] indicate irreversible redox process; peak potential.

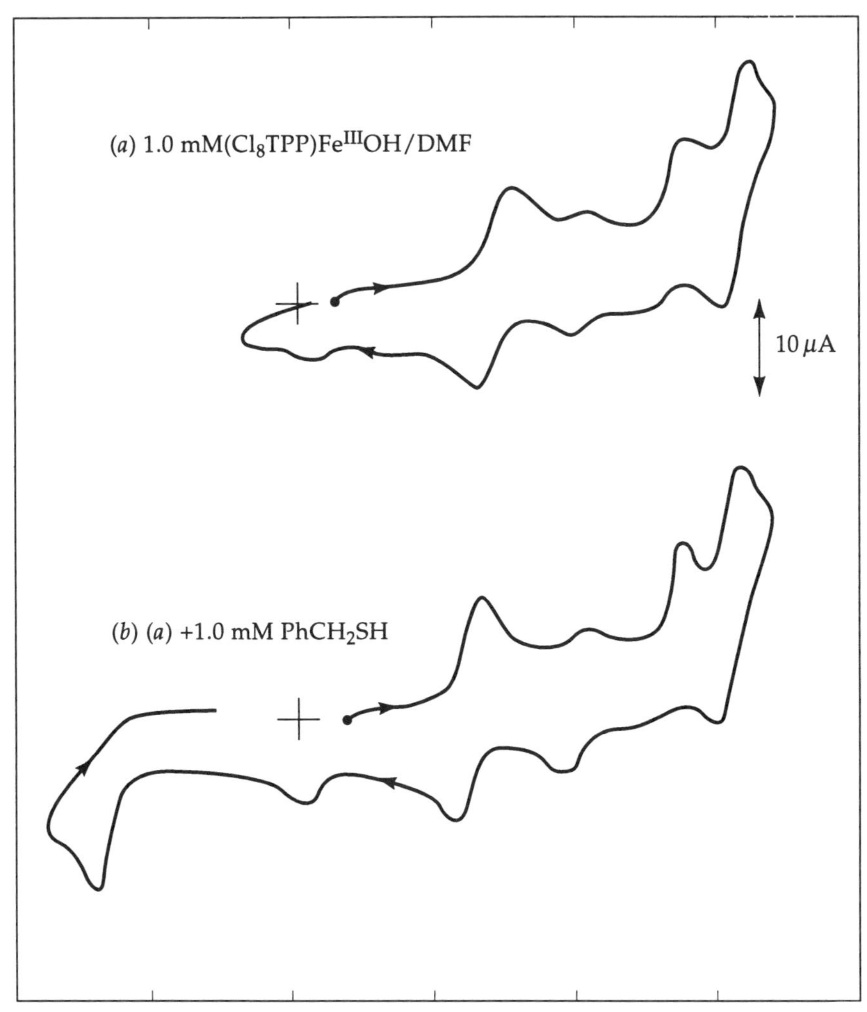

Figure 13.4 Cyclic voltammograms for $(Cl_8TPP)Fe^{III}OH$ and $(Cl_8TPP)Fe^{III}SCH_2Ph$ in DMF. Conditions: 0.1 M TEAP, scan rate 0.1 V s^{-1}.

$(Cl_8TPP^+)Mn^{III}$. Addition of an electron occurs at the porphyrin ring (−0.04 V vs. SCE, Figure 13.6) to give $(Cl_8TPP)Mn^{II}$ (d^5sp, $S = \frac{5}{2}$) with two metal–nitrogen *sp–p* covalent bonds [analogous to $(Cl_8TPP)Zn^{II}$, $(Cl_8TPP)Ni^{II}$]. Oxidation of $(Cl_8TPP^+)Mn^{III}$ occurs at the porphyrin to give $(Cl_8TPP^{2+})Mn^{IV}$ (d^5sp) with four d^2sp–p manganese–nitrogen covalent bonds (two quarternized pyrrole nitrogens). As with $(Cl_8TPP)Co^{II}\cdot$, oxidation of $(Cl_8TPP)Mn^{II}$ occurs at the porphyrin ring to give a cation radical, which is stabilized via coupling with one of the five unpaired *d* electrons of manganese. This facilitated removal of an electron (−0.03 V vs. SCE) relative to that for $(Cl_8TPP)Ni^{II}$ (+1.23 V)

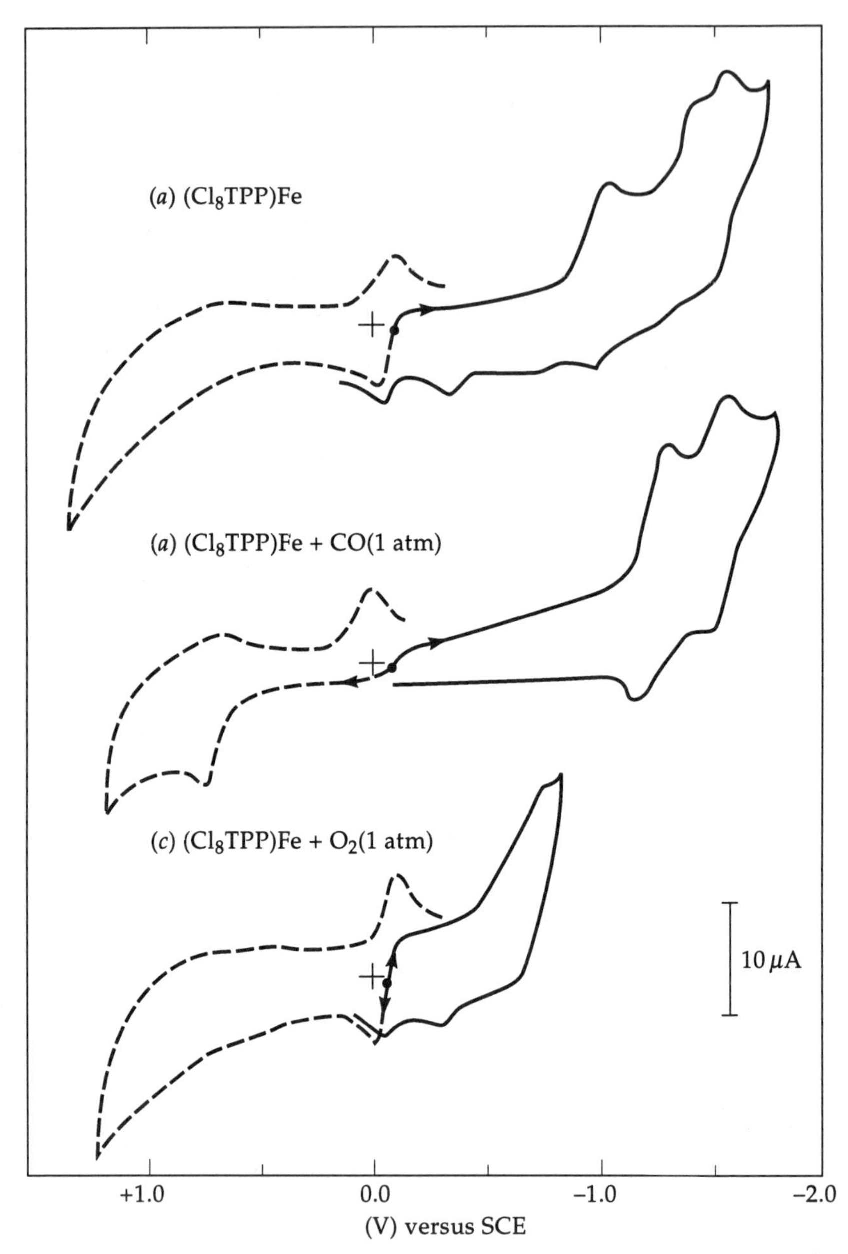

Figure 13.5 Cyclic voltammograms at a glassy carbon electrode (area 0.09 cm^2) in DMF (0.1 M TEAP) of (a) 0.5 mM $(Cl_8TPP)Fe$, (b) 0.5 mM $(Cl_8TPP)Fe$ in the presence of CO (1 atm), and (c) 0.5 mM $(Cl_8TPP)Fe$ in the presence of O_2 (1 atm). Scan rate was 0.1 V s^{-1}.

TABLE 13.4 Redox Potentials in DMF (0.1 M TEAP) for $(Cl_8TPP)M(py)$ Complexes in the Presence of Ar, O_2, and CO.

	$E_{1/2}$ (V) versus SCE[a]					
	$PMn^{II}(py)$		$PFe^{II}(py)$		$PCo^{II}(py)$	
Gas	(←)	(→)	(←)	(→)	(←)	(→)
Ar	−0.15	−1.18	+0.04	−1.04	[+0.10]	−0.75
O_2	−0.15	−1.18	+0.05	[−0.60]	[+0.13]	[−0.62]
CO	−0.15	−1.18	[+0.75]	−0.87	[+0.10]	−0.75

[a]Brackets [ ] indicate irreversible redox process; peak potential.

is an approximate measure of the bond energy for the third manganese–nitrogen bond [$-\Delta G_{BF}$, 121 kJ mol^{-1}; Eq. (13.10)].

The uncharged $(Cl_8TPP)Cu$ (d^8sp^2) and $(Cl_8TPP)Ag$ (d^8sp^2) complexes contain an odd number of electrons, with magnetic moments that are consistent with one unpaired electron, and ESR spectra that indicate that the spin density is concentrated within the porphyrin ring.[14] The one-electron reductions of $(\cdot Cl_8TPP)Cu^{III}$ and $(\cdot Cl_8TPP)Ag^{III}$ (Figure 13.3 and Table 13.1) are facilitated [relative to those for $(Cl_8TPP)Zn^{II}$] by the presence of an unpaired electron in the porphyrin. Hence, the results are in accord with d^8sp^2 hybridization for the Cu and Ag atoms within the porphyrin ring and three metal–nitrogen (sp^2–p) covalent bonds. Because the spin density is delocalized, the $(\cdot Cl_8TPP)Cu^{III}$ and $(\cdot Cl_8TPP)Ag^{III}$ molecules do not interact with $\cdot O_2\cdot$ or electrogenerated $HO\cdot$, which is in sharp contrast with transition-metal complexes with metal-centered spin density.[18]

Oxidation of $(\cdot Cl_8TPP)Cu^{III}$ and $(\cdot TPP)Cu^{III}$ occurs at potentials more positive than those for $(Cl_8TPP)Zn^{II}$ and $(TPP)Zn^{II}$ (Table 13.1), but electron removal from $(\cdot Cl_8TPP)Ag^{III}$ and $(\cdot TPP)Ag^{III}$ is easier than from their zinc analogs. Because the potentials for the direct removal of an electron from Cu (−0.05 V vs. SCE) and Ag (+0.30 V) in MeCN are much less positive (and in the reverse order), oxidation of $(\cdot Cl_8TPP)Cu^{III}$, $(\cdot Cl_8TPP)Ag^{III}$, $(\cdot TPP)Cu^{III}$, and $(\cdot TPP)Ag^{III}$ must involve the removal of the unpaired electron of the porphyrin ring, for example

$$\underset{(d^8sp^2,\ S = \frac{1}{2})}{(\cdot Cl_8TPP)Ag^{III}} \longrightarrow \underset{(d^8sp^2,\ S = 0)}{(^+Cl_8TPP)Ag^{III}} + e^- \tag{13.11}$$

to give a porphyrin-centered diamagnetic cation with three metal–nitrogen (sp–p) covalent bonds. Reduction of $(\cdot Cl_8TPP)Cu^{III}$ and $(\cdot Cl_8TPP)Ag^{III}$ is prophyrin-centered to give diamagnetic $(Cl_8TPP^-)Cu^{III}$ (d^8sp^2) and $(Cl_8TPP^-)Ag^{III}$ (d^8sp^2), each with three sp^2–p covalent bonds.

The redox data of Tables 13.1–13.4 can be used to estimate the metal–ligand bond energies for $(Cl_8TPP)M^{III}L$ complexes, as, for instance, for $(Cl_8TPP)Fe^{III}—OH$:[18]

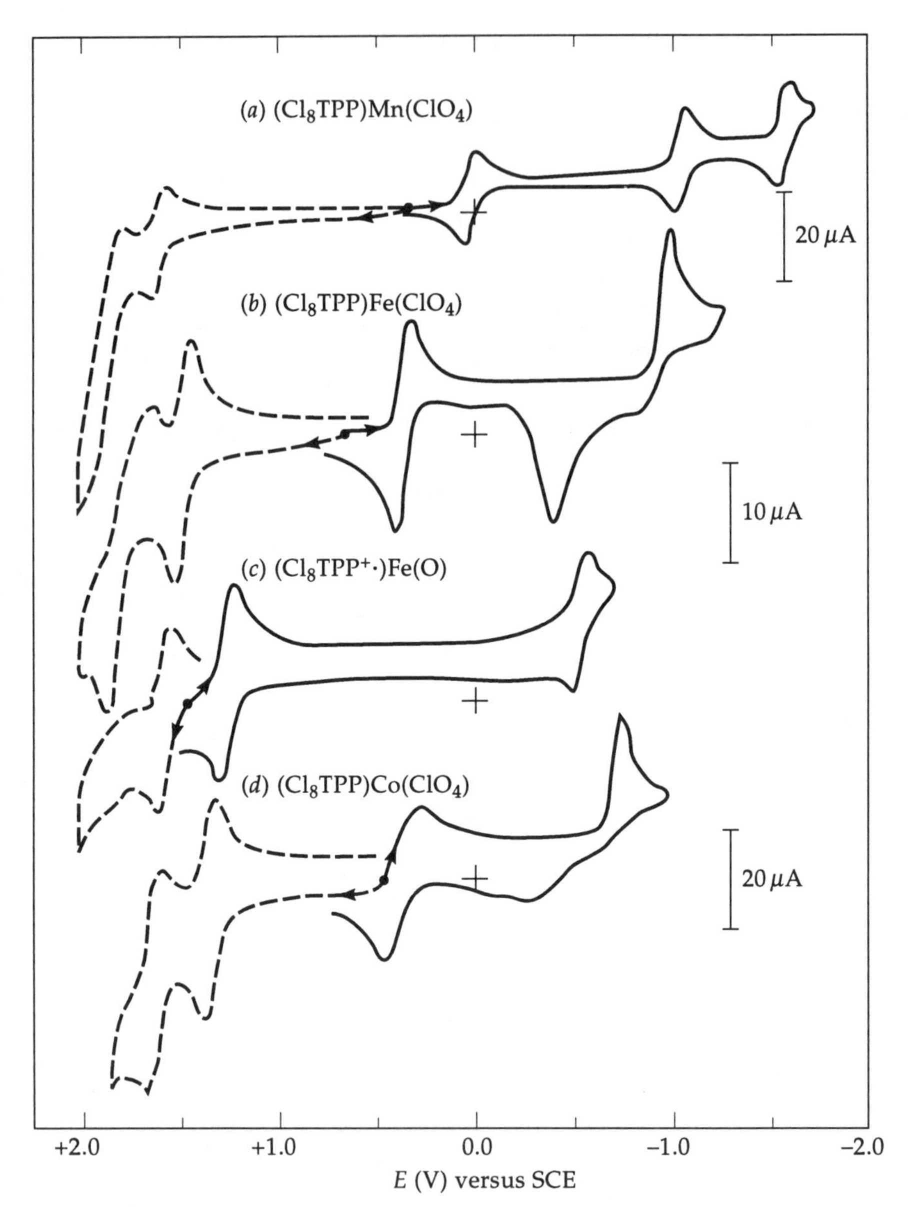

Figure 13.6 Cyclic voltammograms [at a glassy carbon electrode in acetonitrile (0.1 M tetrabutylammonium perchlorate) at −35°C] for (a) 1 mM (Cl_8TTP)Mn(ClO_4), (b) 1 mM (Cl_8TPP)Fe(ClO_4), (c) solution b plus 0.03 atm O_3(g) (1 atm) for 10 s (solution purged with argon), and (d) 1 mM (Cl_8TPP)Co(ClO_4). Scan rate was 0.1 V s^{-1}.

$$(-\Delta G_{BF}) = [E_{p,a,(HO^-/HO\cdot)} - E_{p,a,(PFe,HO^-/PFe^{III}OH)}] \times 96.5$$
$$= [+0.68 \text{ V} - (-0.70 \text{ V})]\, 96.5 = 130 \text{ kJ mol}^{-1} \quad (13.12)$$

Table 13.5 summarizes the approximate $(-\Delta G_{BF})$ values for a variety of $(Cl_8TPP)M{-}L$ bonds on the basis of analogous calculations.

Addition of an oxygen atom from O_3 to $(Cl_8TPP)Fe^{III}(sol)^+$ (d^5sp^2) and $(Cl_8TPP^+)Mn^{III}$ (d^5sp) in acetonitrile at −35°C (Figure 13.6 and Table 13.1) yields $(\cdot Cl_8TPP^+)Fe^{IV}{=}O$ (d^6sp, $S = \frac{3}{2}$, ferryl porphyrin cation radical, and model for compound I of horseradish peroxidase)[17] and $(Cl_8TPP^+)Mn^V{=}O$ (d^5sp, $S = \frac{2}{2}$). Addition of an electron to each is ligand-centered to give $(Cl_8TPP)Fe^{IV}{=}O$ (d^6sp, $S = \frac{2}{2}$, ferryl, compound II of HRP)[17] and $(Cl_8TPP)Mn^{IV} = O$ (d^5sp, $S = \frac{3}{2}$). Formation of the latter via reduction of $(Cl_8TPP^+)Mn^V{=}O$ at +0.52 V versus SCE (Table 13.1) breaks the third manganese–nitrogen bond. Because reduction of $(\cdot Cl_8TPP^+)Fe^{IV}{=}O$ occurs at +1.28 V versus SCE, the difference is an approximate measure of the Mn—N bond energy in $(Cl_8TPP^+)Mn^V{=}O$; $-\Delta G_{BF} = 75$ kJ mol^{-1}. Addition of a second electron to each is oxygen-centered to give $(Cl_8TPP)Fe^{III}{-}O^-$ (d^5sp^2)[17,18] and $(Cl_8TPP)Mn^{III}{-}O^-$ (d^5sp).

The $M{=}O$ bond energies $(-\Delta G_{BF})$ for these four oxene adducts have been estimated on the basis of their reduction potentials (Table 13.1)[17,18] relative to

TABLE 13.5 Bond Energies $(-\Delta G_{BF})$ for $(Cl_8TPP)M$-L Compounds

Bond	$-\Delta G_{BF}$ (kJ mol^{-1})
$(Cl_8TPP)Mn^{III}{-}OH$[a]	109
$(Cl_8TPP)Fe^{III}{-}OH$	130
$(Cl_8TPP)Co^{III}{-}OH$	96
$(Cl_8TPP)Fe^{III}{-}Cl$	209
$(Cl_8TPP)Co^{III}{-}Cl$	192
$(Cl_8TPP)Mn^{III}{-}H$[b]	113
$(Cl_8TPP)Fe^{III}{-}H$	113
$(Cl_8TPP)Co^{III}{-}H$	126
$(Cl_8TPP^+)Mn^V{=}O$	155
$(Cl_8TPP^+)Fe^{IV}{=}O$	192
$(Cl_8TPP)Mn^{IV}{=}O$	364
$(Cl_8TPP)Fe^{IV}{=}O$[c]	280
$(Cl_8TPP)Fe^{IV}{=}CO$[d]	59
$(Cl_8TPP)Fe^{IV}{=}(O_2)$	4
$(Cl_8TPP)Co^{III}{-}OO\cdot$	4

[a] $(-\Delta G_{BF}) = [E_{p,a(HO^-/HO\cdot)} - E_{p,a(PM,HO^-/PMOH]} \times 96.5.$
[b] $(-\Delta G_{BF}) = [E_{p,c(PM,H_3O^+/H_2O)}] \times 96.5$
[c] $(-\Delta G_{BF}) = 3 \times (-\Delta G_{BF})_{\pi\,bond} = 3 \times 93.3$ kJ mol^{-1} $[E_{p,c(O/O^{\cdot-})} - E_{p,c(PFe=O)/PFe-O^-)}] \times 96.5.$
[d] $(-\Delta G_{BF}) = [E_{p,a(PFe=CO/PFe^+,CO)} - E_{p,a(PFe/PFe^+)}] \times 96.5.$

the reduction potential for $O_{(g)}$ (+0.43 V vs SCE).[19] Thus, the free energies of bond formation ($-\Delta G_{BF}$) for these oxene adducts are estimated to be as follows: ($\cdot Cl_8TPP^+$)Fe^{IV}=O, 193 kJ mol^{-1}; (Cl_8TPP)Fe^{IV}=O, 280 kJ mol^{-1}; (Cl_8TPP^+)Mn^V=O, 155 kJ mol^{-1}; (Cl_8TPP)Mn^{IV}=O, 364 kJ mol^{-1}. Their reactivities for the epoxidation of olefins[17,20] (a 351-kJ-mol^{-1} exothemic process)[21] are inversely correlated with their M=O bond energies. Hence, (Cl_8TPP^+)Mn=O is the most reactive, and its reduced form, (Cl_8-TPP)Mn^{IV}=O, is inert.

Nucleophilic Character of Iron- and Cobalt-Porphyrin Anions; Evaluation of Their Metal-Carbon Bond Energies. Reference to Table 13.1 indicates that reduction of iron (II)- and cobalt(II)-porphyrins yields their metal-centered anions [EPR (electron paramagnetic resonance) data for the $(TPP)^{II}Fe^-$ anion confirm the anionic character of the iron (d^7sp, $S = \frac{1}{2}$; electronically equivalent to Co^{II}(TPP)].[22] Figure 13.7 illustrates the reductive electrochemistry for (Cl_8TPP)Fe^{III}(ClO_4), *n*-BuI, their combination (curve *c*), and the combination of (Cl_8TPP)Fe^{III}(ClO_4) with excess *n*-BuBr in dimethylformamide.

The (Cl_8TPP)Fe^{III}(ClO_4) porphyrin exhibits three reversible reduction peaks at −0.01 V, −0.97 V, and −1.54 V versus SCE for the (Cl_8TPP)$Fe^{III}(DMF)^+$/(Cl_8TPP)Fe^{II}, (Cl_8TPP)Fe^{II}/(Cl_8TPP)Fe^-, and (Cl_8TPP)Fe^-/($Cl_8TPP^{\cdot-}$)Fe^- couples, respectively (Figure 13.7*a*).[23] The reduction of *n*-BuI is an irreversible two-electron ECE process ($E_{p,c,}$ −2.05 V, Figure 13.7*b*). Table 13.6 summarizes the redox potentials ($E_{1/2}$) for several iron and cobalt porphyrins in DMF. The reduction potential for (OEP)Fe^{II} is the most negative, and that for ($F_{20}TPP$)Fe^{II} is the least negative of the iron porphyrins. Although the $E_{1/2}$ value for the (por)Fe^{II}/(por)Fe^- couple is more negative than that for (por)Co^{II}/(por)Co^-, the $E_{1/2}$ value for the (por)Co^-/($por^{\cdot-}$)Co^- couple is more negative than that for (por)Fe^-/($por^{\cdot-}$)Fe^-. Table 13.7 summarizes the reduction peak potentials for several alkyl halides (*RX*) in DMF. With the same alkyl group (*R*) the cathodic potentials are in the order *R*Cl > *R*Br > *R*I (*R*Cl, the most negative).

With (Cl_8TPP)Fe^{III}(ClO_4) in the presence of *n*-BuBr, a new reversible anodic peak ($E_{1/2}$, −0.75 V) appears after an initial negative scan (Figure 13.7*c*, *d*), as well as an irreversible oxidation peak ($E_{p,a,}$ +0.41 V). The results indicate that (Cl_8TPP)Fe^- reacts with *n*-BuI (but not with *n*-BuBr) to form (Cl_8-TPP)$Fe^{III}R$, which is further reduced to form ($Cl_8TPP^{\cdot-}$)$Fe^{III}R$ anion. The dianion [($Cl_8TPP^{\cdot-}$)Fe^-] reacts with *n*-BuBr to form ($Cl_8TPP^{\cdot-}$)$Fe^{III}R$, which is oxidized at −0.72 V to give (Cl_8TPP)Fe^{III}Bu-*n*. The latter is oxidized at +0.41 V. Table 13.8 summarizes the redox potentials for several (por)$M^{III}R$ complexes. Although ($por^{\cdot-}$)Fe^{III}Bu-*t* is formed (oxidation peak, Table 13.8), the absence of an oxidation peak for (por)Fe^{III}Bu-*t* indicates that it is unstable. The $E_{1/2}$ values for the ($por^{\cdot-}$)$Fe^{III}R$ redox couples are in the order (OEP)$Fe^{III}R$ (most negative) > [$(MeO)_4TPP$]$Fe^{III}R$ > (TPP)$Fe^{III}R$ > (Cl_8TPP)$Fe^{III}R$ > ($F_{20}TPP$)$Fe^{III}R$. The electrochemical characteristics of ($F_{20}TPP$)$Fe^{III}R$ are different from other porphyrin complexes because the ($F_{20}TPP^{\cdot-}$)$Fe^{III}R$ anion has

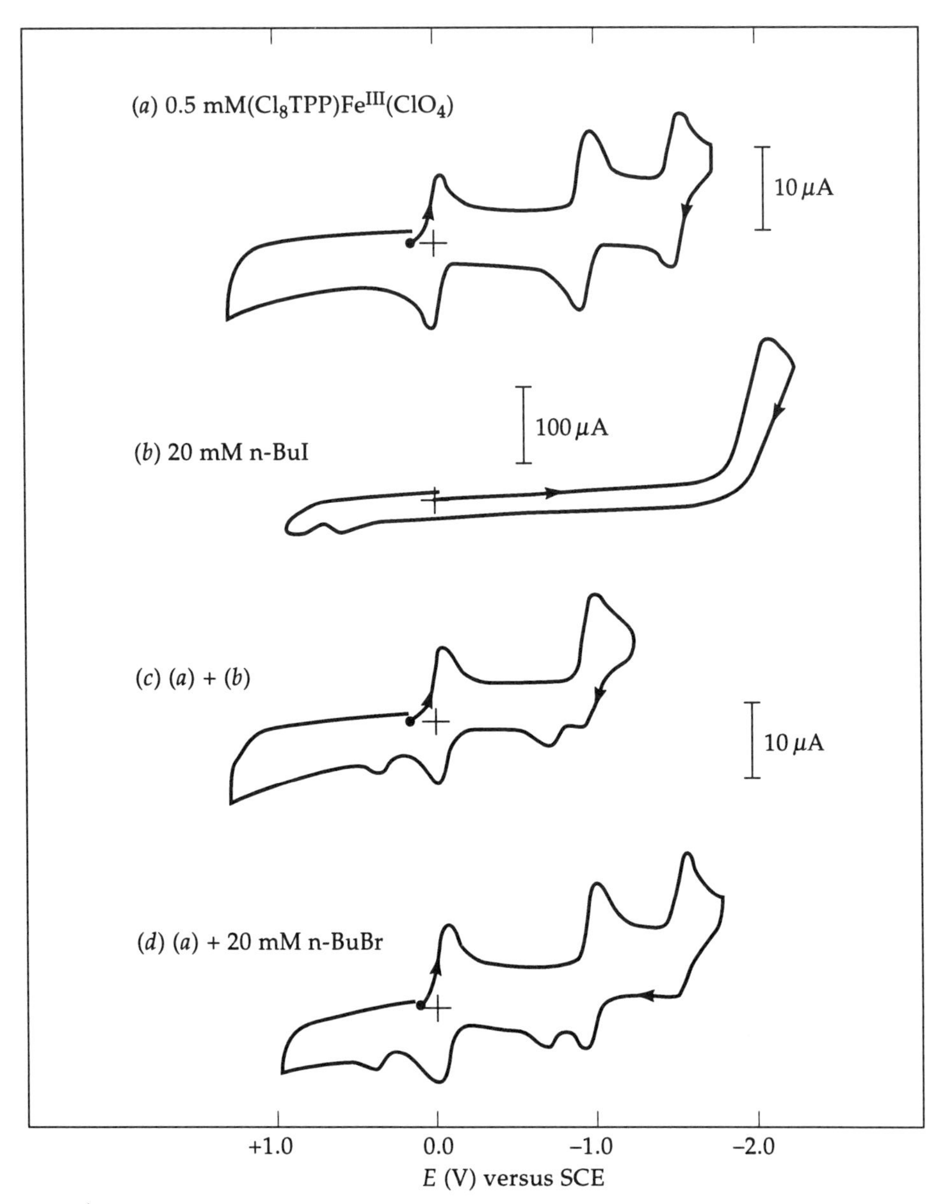

Figure 13.7 Cyclic voltammograms for (a) 0.5 mM $(Cl_8TPP)Fe^{III}(ClO_4)$; (b) 20 mM *n*-BuI; (c) 0.5 mM $(Cl_8TPP)Fe^{III}(ClO_4)$ plus 20 mM *n*-BuI; and (d) 0.5 mM $(Cl_8TPP)Fe^{III}(ClO_4)$ plus 20 mM *n*-BuBr in DMF [0.1 M $(Et_4N)ClO_4$]. Scan rate 0.1 V s^{-1}; glassy-carbon electrode (GCE) (0.09 cm^2); SCE versus NHE, +0.244 V.

TABLE 13.6 Redox Potentials for (por)Fe^{II} and (por)Co^{II} in Dimethylformamide[a]

	$E_{1/2}$[b] (V) versus SCE[c]	
(por)M	(por)M/(por)M^-	(por)M^-/(por$^{\cdot-}$)M^-
(OEP)Fe^{II}	−1.27	−1.95
[$(MeO)_4$TPP]Fe^{II}	−1.07	−1.67
(TPP)Fe^{II}	−1.04	−1.65
(Cl_8TPP)Fe^{II}	−0.97	−1.54
(F_{20}TPP)Fe^{II}	−0.84	−1.32
[$(MeO)_4$TPP]Co^{II}	−0.79	−1.93
(TPP)Co^{II}	−0.77	−1.87
(Cl_8TPP)Co^{II}	−0.73	−1.77

[a]0.5 mM solutions in DMF [0.1 M $(Et_4N)ClO_4$].
[b]$E_{1/2}$ taken as $(E_{p,a} + E_{p,c})/2$ for reversible redox processes (Ref. 36).
[c]Saturated calomel electrode (SCE) versus NHE, +0.244 V.

the least negative oxidation potential of the alkylated metalloporphyrins [(por$^{\cdot-}$)M—R], and exhibits an additional redox couple ($E_{1/2}$, −1.63 V) in the presence of n-BuI.

The redox potential for the (por)$Co^{III}R$/(por$\cdot$)$Co^{III}R$ couple is more negative than the corresponding (por)$Fe^{III}R$/(por$^{\cdot-}$)$Fe^{III}R$ couple (Table 13.8). In contrast to alkyl-iron porphyrins, (por)$Co^{III}R$ is not reduced by the (por)Co^- anion. Instead, it is reduced to (por$^{\cdot-}$)$Co^{III}R$ anion by a more negative scan ($E_{1/2}$, −1.17 to −1.36 V).

Controlled-potential electrolysis of (por)M^{II}/RX combinations yields dialkyl products (R—R). Electrolysis at −1.25 V of the (TPP)Fe^{II}/$PhCH_2Br$ combination yields $PhCH_2CH_2Ph$ (current efficiency, 92 ± 5%; two electrons per product).

TABLE 13.7 Reduction Potentials for Alkyl Halides (RX) in Dimethylformamide[a]

	$E_{p,c}$[b] (V) versus SCE[c]		
RX	I	Br	Cl
CH_3	−2.10		
n-C_4H_9	−2.05	−2.41	
sec-C_4H_9	−1.92	−2.35	
t-C_4H_9	−1.78	−2.25	
c-C_6H_{11}	−2.03	−2.48	
CH_2Cl			−2.0
$PhCH_2$		−1.68	−1.90

[a]1 mM solutions in DMF [0.1 M $(Et_4N)ClO_4$].
[b]$E_{p,c}$, the reduction-peak potential (scan rate, 0.1 V s^{-1}.
[c]Saturated calomel electrode (SCE) versus NHE, +0.244 V.

TABLE 13.8 Redox Potentials for Alkylated Porphyrins [(por)$M^{III}-R$] in Dimethyl Formamide

	$E_{1/2}$[a] (V) versus SCE	$E_{p,a}$[a] (V) versus SCE[b]
(por)$Fe^{III}-R$	(por)Fe$-R$/(por$^{\bar{\cdot}}$)Fe$-R$	(por)Fe$-R$ Oxidation
(OEP)$Fe^{III}-C_4H_9$-*n*	−1.05	0.12
(OEP)$Fe^{III}-C_4H_9$-*sec*	−1.08	0.10
(OEP)$Fe^{III}-C_6H_{11}$-*c*	−1.08	0.05
(OEP)$Fe^{III}-CH_2Ph$	−1.07	0.10
[$(MeO)_4$TPP]$Fe^{III}-C_4H_9$-*n*	−0.82	0.30
[$(MeO)_4$TPP]$Fe^{III}-C_4H_9$-*sec*	−0.84	0.20
[$(MeO)_4$TPP]$Fe^{III}-C_4H_9$-*t*	−0.85	*c*
[$(MeO)_4$TPP]$Fe^{III}-C_6H_{11}$-*c*	−0.83	0.20
[$(MeO)_4$TPP]$Fe^{III}-CH_2Ph$	−0.78	0.21
(TPP)$Fe^{III}-CH_3$	−0.77	0.28
(TPP)$Fe^{III}-C_4H_9$-*n*	−0.78	0.29
(TPP)$Fe^{III}-C_4H_9$-*sec*	−0.80	0.26
(TPP)$Fe^{III}-C_4H_9$-*t*	−0.78	*c*
(TPP)$Fe^{III}-C_6H_{11}$-*c*	−0.74	0.30
(TPP)$Fe^{III}-CH_2Ph$	−0.76	0.25
(Cl_8TPP)$Fe^{III}-C_4H_9$-*n*	−0.75	0.41
(Cl_8TPP)$Fe^{III}-C_4H_9$-*sec*	−0.72	0.34
(Cl_8TPP)$Fe^{III}-C_4H_9$-*t*	−0.70	*c*
(Cl_8TPP)$Fe^{III}-C_6H_{11}$-*c*	−0.70	0.41
(F_{20}TPP)$Fe^{III}-C_4H_9$-*n*	−0.56	0.54
(F_{20}TPP)$Fe^{III}-C_4H_9$-*sec*	−0.53	0.55
(F_{20}TPP)$Fe^{III}-C_4H_9$-*t*	−0.54	*c*
(F_{20}TPP)$Fe^{III}-C_6H_{11}$-*c*	−0.55	0.47
(F_{20}TPP)$Fe^{III}-CH_2Ph$	−0.54	*c*
[$(MeO)_4$TPP]$Co^{III}-C_4H_9$-*n*	−1.36	0.85
[$(MeO)_4$TPP]$Co^{III}-CH_2Ph$	−1.35	0.80
(TPP)$Co^{III}-C_6H_{13}$-*n*	−1.30	0.91
(TPP)$Co^{III}-C_4H_9$-*n*	−1.35	0.91
(TPP)$Co^{III}-C_4H_9$-*sec*	−1.25	0.74
(TPP)$Co^{III}-C_4H_9$-*t*	−1.30	0.76
(TPP)$Co^{III}-C_6H_{11}$-*c*	−1.32	0.74
(TPP)$Co^{III}-CH_2Ph$	−1.17	0.80
(Cl_8TPP)$Co^{III}-C_4H_9$-*n*	−1.20	0.97

[a] $E_{1/2}$ taken as $(E_{p,a} + E_{p,c})/2$ for reversible couples; $E_{p,a}$ is the anodic peak potential for the irreversible oxidation (scan rate, 0.1 V s^{-1}).
[b] Saturated calomel electrode (SCE) versus NHE, +0.244 V.
[c] No observed oxidation peak; apparently (por)Fe^{III}—Bu-*t* is unstable.

Figure 13.8 illustrates the reductive electrochemistry for (TPP)Co^{II} in CH_2Cl_2, DMF, and THF, and provides clear evidence that (TPP)Co^- reacts with CH_2Cl_2. Electrogenerated (por$^{\bar{\cdot}}$)M^- reacts with CO_2, which is illustrated by the cyclic voltammograms for (TPP)Co^{II}, carbon dioxide, and their combination in DMF (Figure 13.9). The reduction of carbon dioxide is an irrevers-

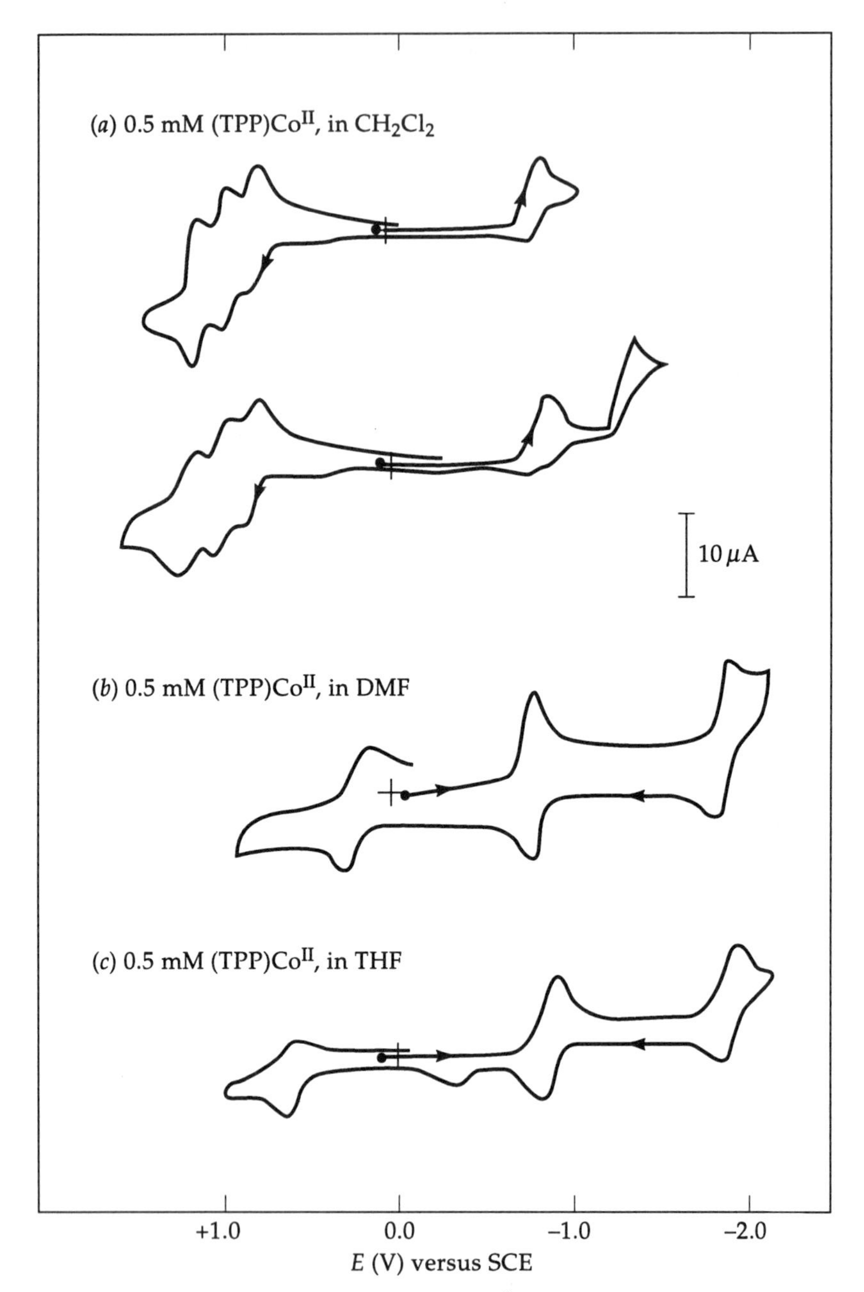

Figure 13.8 Cyclic voltammograms for $(TPP)Co^{II}$: (a) in CH_2Cl_2 [0.1 M $(Bu_4N)ClO_4$]; (b) in DMF [0.1 M$(Et_4N)ClO_4$]; and (c) in THF [0.1 M $(Bu_4N)ClO_4$]. Scan rate, 0.1 V s^{-1}; glassy-carbon electrode (GCE) (0.09 cm^2); SCE versus NHE, +0.244 V.

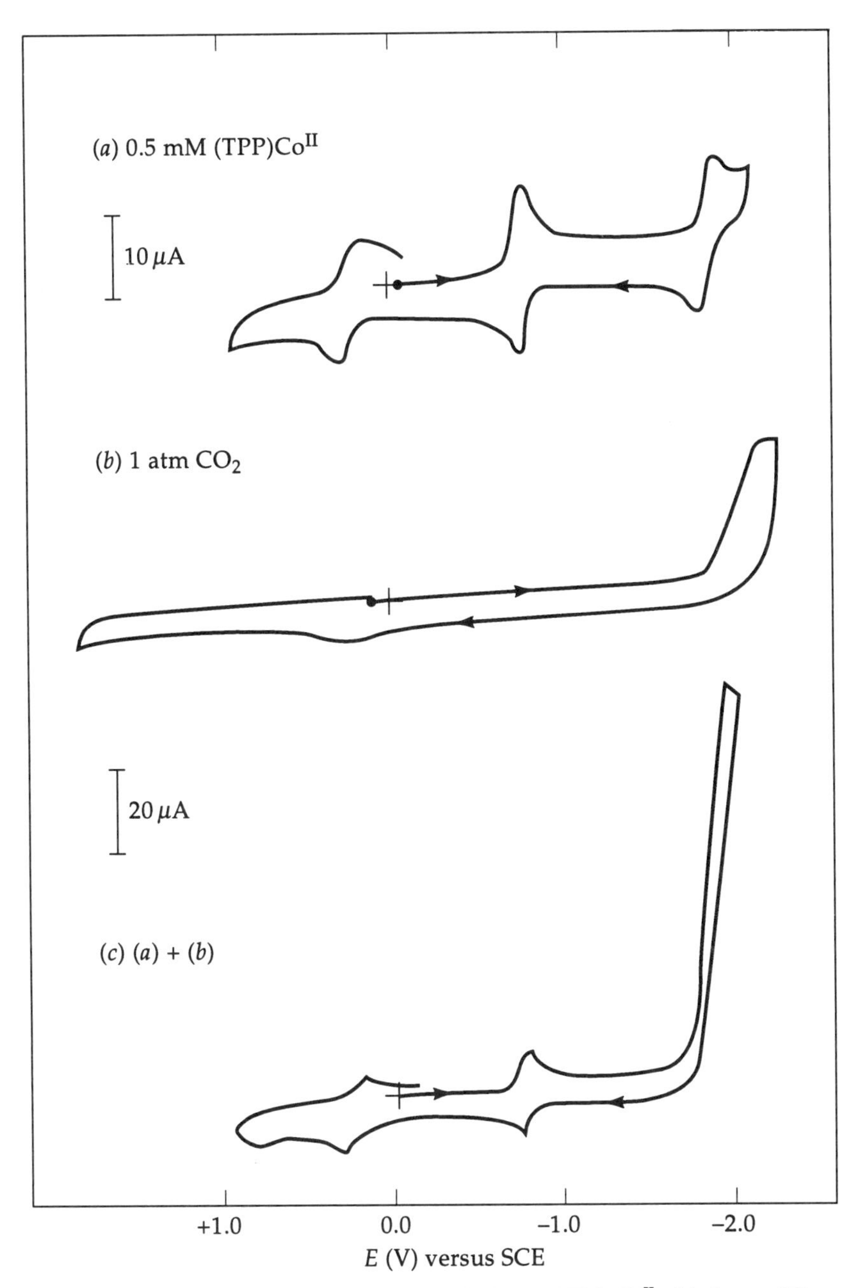

Figure 13.9 Cyclic voltammograms for (a) 0.5 mM $(TPP)Co^{II}$; (b) 1 atm CO_2; and (c) 0.5 mM $(TPP)Co^{II}$ plus 1 atm CO_2 in DMF [0.1 M $(Et_4N)ClO_4$]. Scan rate 0.1 V s^{-1}; glassy-carbon electrode (GCE) (0.09 cm^2); SCE versus NHE, +0.244 V.

ible process at −2.23 V (curve *b*). The reduction potential and peak current for the $(por)M^{II}/(por)M^-$ couple are unaffected by the presence of CO_2. However, the reduction peak current for the $(por)M^-/(por^{\overline{\cdot}})M^-$ couple is enhanced by the presence of CO_2. The current for the second reduction peak of $(TPP)Co^{II}$ ($E_{p,c,}$ −1.96 V) is increased by a factor of 8. A new irreversible oxidation peak appears at +0.85 V on reversal of an initial negative scan. With $(TPP)Fe^{III}Cl$ in the presence of CO_2, the peak current for the reduction of $(TPP)Fe^-$ ($E_{p,c,}$ −1.70 V) is doubled.

The electrochemically generated anions $[(por)^{II}M^-$ or $(por^{\overline{\cdot}})^{II}M^-]$ react with alkyl halides to form σ-alkyl-metal porphyrin derivatives.[24] The tendency of the nucleophilic substitution reaction for these anions with RX depends on the $R{-}X$ bond energy and the nucleophilicity of the metalloporphyrin anions (compare voltammograms *c* and *d* of Figure 13.7). The nucleophilicity of the $(por)^{II}M^-$ and $(por^{\overline{\cdot}})M^-$ anions is directly related to the reduction potential for their formation $[(OEP)^{II}Fe^-$ is the strongest anionic nucleophile because its formation potential is the most negative (Table 13.6)]. The porphyrin dianion $[(por^{\overline{\cdot}})Fe^-]$ is a stronger nucleophile than the $[(por)Fe^-]$ anion. Thus, $(por)Fe^-$ reacts with *n*-BuI but not with *n*-BuBr. In contrast, $(por^{\overline{\cdot}})Fe^-$ reacts with both *n*-BuI and *n*-BuBr to give $(por^{\overline{\cdot}})Fe^{III}R$, which is reversibly oxidized at −0.72 V [for $(Cl_8TPP)Fe^{II}$].

The present results are closely similar to those of the Savéant group, and are in accord and consistent with their detailed mechanistic analysis.[25,26] The reaction sequence begins with the electrochemical formation of the metalloporphyrin anion (−1.27 to −0.73 V, Table 13.6):

$$(por)Fe^{II} + e^- \rightleftharpoons (por)^{II}Fe^- \qquad E_{13} \tag{13.13}$$

In turn, the anion acts as a nucleophile toward electrophilic alkyl-halides to produce $(por)Fe^{III}R$:

$$(por)^{II}Fe^- + RX \longrightarrow (por)Fe^{III}R + X^- \tag{13.14}$$

The $(por)Fe^{III}R$ [produced by an EC (electrochemical–chemical) process] is reduced directly or by electrogenerated $(por)^{II}Fe^-$ to yield $(por^{\overline{\cdot}})Fe^{III}R$:

$$(por)Fe^{III}R + e^- \rightleftharpoons (por^{\overline{\cdot}})Fe^{III}R \tag{13.15}$$

$$(por)Fe^{III}R + (por)^{II}Fe^- \longrightarrow (por^{\overline{\cdot}})Fe^{III}R + (por)Fe^{II} \tag{13.16}$$

Alternatively, $(por)^{II}Fe^-$ is reduced to produce the dianion, which is a substantially stronger nucleophile (−1.95 to −1.32 V, Table 13.6) that reacts with RX to produce $(por^{\overline{\cdot}})Fe^{III}R$:

$$(por)^{II}Fe^- + e^- \rightleftharpoons (por^{\overline{\cdot}})Fe^- \tag{13.17}$$

$$(por^{\overline{\cdot}})^{II}Fe^- + RX \longrightarrow (por^{\overline{\cdot}})Fe^{III}R + X^- \tag{13.18}$$

Only the $(F_{20}TPP^{\bar{\cdot}})Fe^{III}R$ alkyl-metal porphyrin anion exhibits another reversible couple ($E_{1/2}$, -1.63 V) to yield a dianion [the four electron-withdrawing pentafluorophenyl groups on the porphyrin ring make possible the addition of an electron to give the $(F_{20}TPP^{2-})Fe^{III}R$ dianion within the voltage limit of the solvent].

The alkyl-metal porphyrin anion $[(por^{\bar{\cdot}})Fe^{III}R]$ is oxidized reversibly on reversal of a cathodic scan (-1.36 to -0.54 V, Table 13.8). The further oxidation of $(por)Fe^{III}R$ does not occur when R is a tertiary-butyl group [apparently steric effects weaken the $(por)Fe^{III}{-}R$ bond]. The oxidation potentials for the iron porphyrins (with the same porphyrin ring) are in the order $(por)Fe^{III}R > (por)Fe^{II} > (por^{\bar{\cdot}})Fe^{III}R > (por)^{II}Fe^{-} > (por^{\bar{\cdot}})^{II}Fe^{-}$ (most negative).

The $(por)^{II}Fe^{-}$ anion is a stronger nucleophile than $(por)^{II}Co^{-}$, but the $(por^{\bar{\cdot}})^{II}Co^{-}$ dianion is stronger than $(por^{\bar{\cdot}})^{II}Fe^{-}$ (with the same porphyrin ring; compare the $E_{1/2}$ values for iron and cobalt porphyrins in Table 13.6). The $(por)^{II}Co^{-}$ anion reacts with RX to give $(por)Co^{III}R$, which is reduced to $(por^{\bar{\cdot}})^{II}CoR$ at a more negative potential. The oxidation potentials for the cobalt porphyrins (with the same porphyrin ring) are in the order $(por)Co^{III}R > (por)Co^{II} > (por)^{II}Co^{-} > (por^{\bar{\cdot}})Co^{III}R > (por^{\bar{\cdot}})^{II}Co^{-}$ (most negative). The $(por^{\bar{\cdot}})Co^{III}R$ anion is a stronger nucleophile than $(por)Co^{-}$, but $(por^{\bar{\cdot}})Fe^{III}R$ is a weaker nucleophile than $(por)^{II}Fe^{-}$. The $(por)^{II}Co^{-}$ anion reacts with CH_2Cl_2 to form $(por)Co^{III}CH_2Cl$, which is reduced at -1.41 V.

Nucleophilic substitution reactions are driven by (1) the difference between the electron affinity of the electrophile (E_E, one-electron reduction potential) and the electron-donating propensity of the nucleophile (E_N, one-electron oxidation potential) ($E_E - E_N$) and (2) the bond energy of the newly formed bond.[27] For example, the nucleophilic attack by HO^- in MeCN:

$$\begin{array}{llll} HO^- + BuI & \longrightarrow & HO{-}Bu + I^- & -\Delta G_{reac} \\ (N^-) \quad (E) & & (-\Delta G_{BF}) & \end{array} \tag{13.19}$$

where $E_E = -2.05$ V vs SCE, $E_{N^-} = +0.68$ V, and $-\Delta G_{BF} = 343$ kJ mol^{-1}. The reaction tendency ($-\Delta G_{reac}$) is given by the relation

$$\begin{aligned} -\Delta G_{reac} &= 96.5\,(E_E - E_{N^-}) + (-\Delta G_{BF})_{Bu-OH} \\ &= 96.5\,(-2.05 - 0.68) + 343 \\ &= 79 \text{ kJ mol}^{-1} \end{aligned} \tag{13.20}$$

If a given nucleophile does not react with an electrophile, it is because the bond energy and/or the electron affinity are too small ($-\Delta G_{reac} < 0$).

The metal-centered reduction of iron and cobalt porphyrins $[(por)M^{II}]$ yields metalloporphyrin anions [Eq. (13.13)]. The reduction potential for this reaction is E_{13}, and is equivalent to the E_{N^-} value for the oxidation of the metal-centered nucleophile $[(por)^{II}M^{-}]$. The one-electron reduction of alkyl halides yields the alkyl radical and the halide anion, and has a reduction potential, E_{21}, that is equivalent to E_E:

$$RX + e^- \longrightarrow R\cdot + X^- \qquad E_{21} \tag{13.21}$$

The reaction between $(\text{por})^{\text{II}}\text{Fe}^-$ or $(\text{por}^{\overline{\cdot}})^{\text{II}}\text{Fe}^-$ and RX [Eq. (13.14) or (13.18)] obeys the relation of Eq. (13.20), and is more complete as the value of $(-\Delta G_{\text{reac}})$ is more positive. Therefore, the lower and upper limits of the free energy for bond formation $(-\Delta G_{\text{BF}})$ in σ-alkyl-metal (Fe or Co) porphyrins are given by the relations

$$-\Delta G_{(\text{por})M-R} \geq 96.5\ (E_{13} - E_{21})$$

$$[\text{if (por)}M^- \text{ reacts with } RX] \tag{13.22}$$

$$-\Delta G_{(\text{por})M-R} \leq 96.5\ (E_{13} - E_{21})$$

$$[\text{if (por)}M^- \text{ doesn't react with } RX] \tag{13.23}$$

For example, the $(\text{Cl}_8\text{TPP})\text{Fe}^-$ anion reacts with n-BuI, which provides a measure of the lower limit for the iron–carbon bond energy:

$$E_{13} = E_{1/2[(\text{Cl}_8\text{TPP})\text{Fe}]} = -0.97\ \text{V} \qquad (\text{Table 13.6}) \tag{13.24}$$

$$E_{21} = E_{\text{p,c/2}(n\text{-BuI})} + 0.03$$

$$= -1.98 + 0.03 = -1.95\ \text{V} \qquad (\text{Table 13.7}) \tag{13.25}$$

$$-\Delta G_{(\text{por})M-\text{Bu-}n} \geq 96.5\ (-0.97 + 1.95) = 95\ \text{kJ mol}^{-1} \tag{13.26}$$

Because the $(\text{F}_{20}\text{TPP})\text{Fe}^-$ anion doesn't react with n-BuBr, a similar calculation yields the upper the limit for the bond energy:

$$E_{13} = E_{1/2[(\text{F}_{20}\text{TPP})\text{Fe}]} = -0.84\ \text{V} \qquad (\text{Table 13.6}) \tag{13.27}$$

$$E_{21} = E_{\text{p,c/2}(n\text{-BuBr})} + 0.03$$

$$= -2.33 + 0.03 = -2.3\ \text{V} \qquad (\text{Table 13.7}) \tag{13.28}$$

$$-\Delta G_{(\text{por})M-\text{Bu-}n} \leq 96.5\ (-0.84 + 2.3) = 142\ \text{kJ mol}^{-1} \tag{13.29}$$

Hence, the value of $-\Delta G_{(\text{por})\text{Fe}-\text{Bu-}n}$ is between 95 and 142 kJ mol^{-1} (117 ± 25 kJ mol^{-1}, Table 13.9), and depends to some degree on the electron density of the porphyrin ring. The data of Tables 13.6 and 13.7 have been used to estimate the metal–carbon bond energies for various alkyl groups via similar calculations [Eqs. (13.24)–(13.29)], and are summarized in Table 13.9.

The values of $-\Delta G_{\text{BF}}$ for $(\text{por})\text{Co}^{\text{III}}-R$ are larger than those for $(\text{por})\text{Fe}^{\text{III}}-R$, which is in accord with earlier estimates and gas-phase metal–

TABLE 13.9 The Bond-Formation Free Energies ($-\Delta G_{BF}$) for (por)$Fe^{III}-R$ and (por)$Co^{III}-R$ Bonds

R	(por)$Fe^{III}-R$ (kJ mol^{-1})	(por)$Co^{III}-R$ (kJ mol^{-1})
n-C_4H_9	117 ± 25	151
sec-C_4H_9	105 ± 25	121 ± 21
t-C_4H_9	89 ± 33	109 ± 25
c-C_6H_{11}	121 ± 25	138 ± 21
$PhCH_2$	84 ± 13	105

carbon bond energies.[28] The value of $-\Delta G_{BF}$ for (por)$Fe^{III}-R$ increases as the electron density increases in the porphyrin ring [(OEP)$M^{III}-R$ has the strongest bonds because the eight ethyl substituents are electron-donor groups]. In contrast, (F_{20}TPP)$M^{III}-R$ has the weakest bond of the listed porphyrins because the fluoro substituents are electron-withdrawing groups (relative to hydrogen). The strength of the σ-alkyl-metal bond also is affected by the steric character of the alkyl group ($1° > 2° > 3°$).

The results of reductive electrolysis of a (TPP)$Fe^{III}Cl/PhCH_2Br$ combination confirm that (TPP)$Fe^{III}CH_2Ph$ is an intermediate product [Eqs. (13.13) and (13.14)]. However, it undergoes further reduction to give a nucleophile [($TPP^{\bar{\cdot}}$)$Fe^{III}CH_2Ph$; Eqs. (13.14) and (13.16)] that reacts with a second $PhCH_2Br$ to produce a carbon–carbon coupling:

$$(TPP^{\bar{\cdot}})Fe^{III}CH_2Ph + PhCH_2Br \longrightarrow (TPP)Fe^{II} + PhCH_2CH_2Ph + Br^- \quad (13.30)$$

In aprotic solvents, direct electrochemical reduction of CO_2 (−2.23 V vs. SCE) yields carbon monoxide and carbonate ion.[29] The ($por^{\bar{\cdot}}$)Fe^- dianion also reduces CO_2 to CO, but at a less negative potential (−1.70 V).[30] Hence, the estimated iron–carbon bond energy ($-\Delta G_{BF}$) for the ($por^{\bar{\cdot}}$)$Fe^{III}-C(O)O^-$ dianion is at least 50 kJ mol^{-1} [$-\Delta G_{BF} \geq 96.5(-1.70 + 2.23)$].

The second reduction couple of (TPP)Co^{II}(−1.95 V) produces ($TPP^{\bar{\cdot}}$)Co^-, which also facilitates the transformation of CO_2 to CO and CO_3^{2-} (Figure 13.9). Scheme 13.1 outlines a reasonable mechanistic path that has the first-formed adduct, (TPP^-)$Co^{III}C(O)O^-$, acting as a nucleophile toward a second CO_2 molecule. This adduct dissociates heterolytically to give CO_3^{2-} and bound CO.

In summary, the electrochemical results indicate that the alkyl-metal bond-formation free energies range from 54 to 146 kJ mol^{-1} for iron porphyrins and from 84 to 159 kJ mol^{-1} for cobalt porphyrins. The maximum bond energies are for primary alkyl groups bonded to [$(MeO)_4$TPP]Co^{II} and (OEP)Fe^{II} porphyrins. The porphyrin dianions [$(por^{\bar{\cdot}})^{II}Fe^-$ and $(por^{\bar{\cdot}})^{II}Co^-$] facilitate the reduction of CO_2 to CO via the transient formation of a metal–carbon bond [($por^{\bar{\cdot}}$)$M-C(O)O^-$; $-\Delta G_{BF} \geq 50$ kJ mol^{-1} for iron porphyrins]. Thus, iron and cobalt porphyrins are especially effective electrocatalysts for the reduction of CO_2:

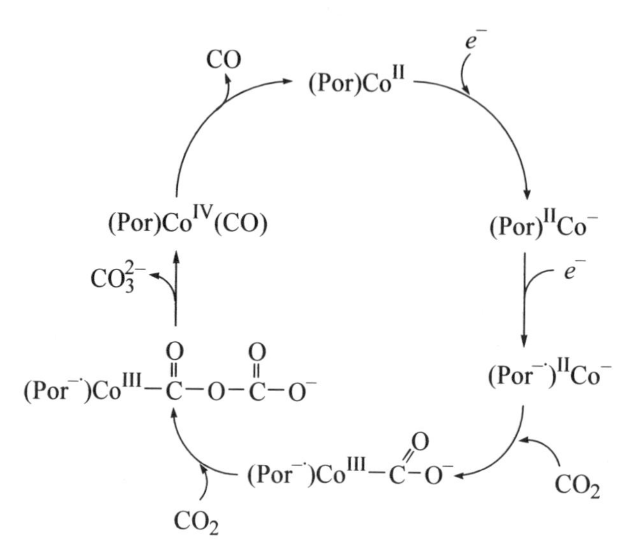

Scheme 13.1 Reduction of CO_2 by the $(por^{-}\cdot)^{II}Co^{-}$ dianion in DMF (Figure 13.9).

$$2\ CO_2 + 2\ e^- \xrightarrow{(por)M^{II}} CO + CO_3^{2-} \tag{13.31}$$

In summary, the electrochemistry of organometallic and metalloporphyrins is dominated by synergistic electron transfer of extramolecular solution components (H_2O, O_2, electrophiles, and nucleophiles). This provides a convenient means for evaluation of the molecular activation (catalytic) properties of these important metal-centered systems. Only in the case of iron (II)- and cobalt(II)-porphyrins (previous section) is metal-centered electron transfer observed $[(por)^{II}M \xrightarrow{e} (por)^{II}M^{-}]$.

REFERENCES

1. Magdesieva, T. V.; Butin, K. P.; Reutov, O. A., *Russ. Chem. Rev.* **1988,** *57,* 864.
2. Ocella, D., *Materials Chem. Phys.* **1991,** *29,* 117.
3. Kadish, K. M., in *Progress in Inorganic Chemistry*; Lippard, S. J., ed., Wiley-Interscience, New York, 1986, Vol. 34, pp. 435–605.
4. Hill, H. O. A., *Pure Appl. Chem.* **1990,** *62,* 1047.
5. Richert, S. A., Ph.D. dissertation, Texas A&M University, College Station, Tex., 1989, pp. 49–53.
6. Barrette, W. C., Jr.; Johnson, H. W., Jr.; Sawyer, D. T., *Anal. Chem.* **1984,** *56,* 1890.
7. Roberts, J. L., Jr.; Sawyer, D. T., *Electrochim. Acta* **1965,** *10,* 989.

8. Guilard, R.; Lecomte, C.; Kadish, K. M., *Structure and Bonding* **1987,** *64,* 205–268.
9. Bottomley, L. A.; Olson, L.; Kadish, K. M., *Adv. Chem. Ser.* **1982,** No. 201, 279–311.
10. Kadish, K. M.; Lin, X. Q.; Ding, J. Q.; Wu, Y. T.; Araullo, C., *Inorg. Chem.* **1986,** *25,* 3236.
11. Castro, C. E., in *The Porphyrins*; Dolphin, D., ed., Academic Press, New York, 1979, Vol. V, pp. 16–27, and references cited therein.
12. Hartzell, C. R.; Schroedl, N. A., in *Biochemical and Clinical Aspects of Oxygen*, Caughey, W. S., ed., Academic Press, New York, 1979; pp. 337–354.
13. Ferguson-Miller, S.; Brautigan, D. L.; Margoliash, E., in *The Porphyrins*, Dolphin, D., ed., Academic Press, New York, 1979, Vol. VII, pp. 149–240, and references cited therein.
14. Tung, H.-C.; Chooto, P.; Sawyer, D. T., *Langmuir* **1991,** *7,* 1635.
15. Traylor, P. S.; Dolphin, D.; Traylor, T. G., *J. Chem. Soc., Chem. Commun.* **1984,** *5,* 279.
16. Redman, C. E., Ph.D. dissertation, Texas A&M University, 1994, pp. 40–77.
17. Sugimoto, H.; Tung, H.-C.; Sawyer, D. T., *J. Am. Chem. Soc.* **1988,** *110,* 2465.
18. Richert, S. A.; Tsang, P. K. S.; Sawyer, D. T., *Inorg. Chem.* **1989,** *28,* 2471.
19. Sawyer, D. T.; Chooto, P.; Tsang, P. K. S., *Langmuir* **1989,** *5,* 84.
20. Meunier, B.; Guilmet, E.; DeCarvalho, M.-E.; Poiblanc, R., *J. Am. Chem. Soc.* **1984,** *106* 6668.
21. Tolman, C. A., Central Research and Development Department, du Pont de Nemours & Co., Experimental Station, Wilmington, Del., 1989; private communication, March 1987.
22. Srivasta, G. S.; Sawyer, D. T.; Boldt, N. J.; Bocian, D. F., *Inorg. Chem.* **1985,** *24,* 2123.
23. Qiu, A.; Sawyer, D. T. *Inorg. Chem.*, submitted, February, 1995.
24. Guilard, R.; Lecomete, C.; Kadish, K. M., *Structure and Bonding* **1987,** *64,* 205.
25. Battioni, J. P.; Lexa, D.; Mansuy, D.; Savéant, J.-M., *J. Am. Chem. Soc.* **1983,** *105,* 207.
26. Lexa, D.; Savéant, J.-M.; Wang, D. L., *Organometallics* **1986,** *5,* 1428.
27. Sawyer, D. T., *J. Phys. Chem.* **1989,** *93,* 7977.
28. Armentrout, P. B., in *C-H Activation*, Liebman, J.; Greenberg, A., eds., VCH Publishers, New York, 1990.
29. Haynes, L. V.; Sawyer, D. T., *Anal. Chem.* **1967,** *39,* 332.
30. Hammouche, M.; Lexa, D.; Savéant, J. M., *J. Electroanal. Chem.* **1988,** *249,* 347.

INDEX